Unmanned Aerial Vehicles

Unmanned Aerial Vehicles

Editors

Sylvain Bertrand
Hyo-Sang Shin

MDPI

Basel • Beijing • Wuhan • Barcelona • Belgrade • Novi Sad • Cluj • Manchester

Editors
Sylvain Bertrand
ONERA – Paris-Saclay University
France

Hyo-Sang Shin
Cranfield University
UK

Editorial Office
MDPI
St. Alban-Anlage 66
4052 Basel, Switzerland

This is a reprint of articles from the Special Issue published online in the open access journal *Applied Sciences* (ISSN 2076-3417) (available at: https://www.mdpi.com/journal/applsci/special_issues/Unmanned_Aerial_Vehicles).

For citation purposes, cite each article independently as indicated on the article page online and as indicated below:

Lastname, A.A.; Lastname, B.B. Article Title. *Journal Name* **Year**, *Volume Number*, Page Range.

ISBN 978-3-0365-7824-8 (Hbk)
ISBN 978-3-0365-7825-5 (PDF)
doi.org/10.3390/books978-3-0365-7825-5

Contents

About the Editors vii

Sylvain Bertrand and Hyo-Sang Shin
Special Issue on Unmanned Aerial Vehicles
Reprinted from: *Appl. Sci.* **2023**, *13*, 4134, doi:10.3390/app13074134 1

Yao Lei, Yiqiang Ye and Zhiyong Chen
Horizontal Wind Effect on the Aerodynamic Performance of Coaxial Tri-Rotor MAV
Reprinted from: *Appl. Sci.* **2020**, *10*, 8612, doi:10.3390/app10238612 5

Mirosław Nowakowski, Krzysztof Sibilski, Anna Sibilska-Mroziewicz and Andrzej Żyluk
Bifurcation Flight Dynamic Analysis of a Strake-Wing Micro Aerial Vehicle
Reprinted from: *Appl. Sci.* **2021**, *11*, 1524, doi:10.3390/app11041524 21

Muwnaika Jdiobe, Kurt Rouser, Ryan Paul and Austin Rouser
Validation of a Wind Tunnel Propeller Dynamometer for Group 2 Unmanned Aircraft
Reprinted from: *Appl. Sci.* **2022**, *12*, 8908, doi:10.3390/app12178908 43

Yisak Debele, Ha-Young Shi, Assefinew Wondosen, Jin-Hee Kim and Beom-Soo Kang
Multirotor Unmanned Aerial Vehicle Configuration Optimization Approach for Development of Actuator Fault-Tolerant Structure
Reprinted from: *Appl. Sci.* **2022**, *12*, 6781, doi:10.3390/app12136781 83

Michał Okulski and Maciej Ławryńczuk
A Small UAV Optimized for Efficient Long-Range and VTOL Missions: An Experimental Tandem-Wing Quadplane Drone
Reprinted from: *Appl. Sci.* **2022**, *12*, 7059, doi:10.3390/app12147059 107

Bedada Endale, Abera Tullu, Hayoung Shi and Beom-Soo Kang
Robust Approach to Supervised Deep Neural Network Training for Real-Time Object Classification in Cluttered Indoor Environment
Reprinted from: *Appl. Sci.* **2021**, *11*, 7148, doi:10.3390/app11157148 131

Saud S. Alotaibi, Hanan Abdullah Mengash, Noha Negm, Radwa Marzouk, Anwer Mustafa Hilal, Mohamed A. Shamseldin, et al.
Swarm Intelligence with Deep Transfer Learning Driven Aerial Image Classification Model on UAV Networks
Reprinted from: *Appl. Sci.* **2022**, *12*, 6488, doi:10.3390/app12136488 151

Yalew Zelalem Jembre, Yuniarto Wimbo Nugroho, Muhammad Toaha Raza Khan, Muhammad Attique, Rajib Paul, Syed Hassan Ahmed Shah and Beomjoon Kim
Evaluation of Reinforcement and Deep Learning Algorithms in Controlling Unmanned Aerial Vehicles
Reprinted from: *Appl. Sci.* **2021**, *11*, 7240, doi:10.3390/app11167240 171

Chen-Huan Pi, Wei-Yuan Ye and Stone Cheng
Robust Quadrotor Control through Reinforcement Learning with Disturbance Compensation
Reprinted from: *Appl. Sci.* **2021**, *11*, 3257, doi:10.3390/app11073257 185

Zhanyuan Jiang, Jianquan Ge, Qiangqiang Xu and Tao Yang
Terminal Distributed Cooperative Guidance Law for Multiple UAVs Based on Consistency Theory
Reprinted from: *Appl. Sci.* **2021**, *11*, 8326, doi:10.3390/app11188326 199

Rui Qiao, Guili Xu, Yuehua Cheng, Zhengyu Ye and Jinlong Huang
Simulation and Analysis of Grid Formation Method for UAV Clusters Based on the 3 × 3Magic Square and the Chain Rules of Visual Reference
Reprinted from: *Appl. Sci.* **2021**, *11*, 11560, doi:10.3390/app112311560 . . . . **221**

Abera Tullu, Bedada Endale, Assefinew Wondosen and Ho-Yon Hwang
Machine Learning Approach to Real-Time 3D Path Planning for Autonomous Navigation of Unmanned Aerial Vehicle
Reprinted from: *Appl. Sci.* **2021**, *11*, 4706, doi:10.3390/app11104706 . . . . **245**

Abdul Majeed and Seong Oun Hwang
Path Planning Method for UAVs Based on Constrained Polygonal Space and an Extremely Sparse Waypoint Graph
Reprinted from: *Appl. Sci.* **2021**, *11*, 5340, doi:10.3390/app11125340 . . . . **265**

Krzysztof Andrzej Gromada and Wojciech Marcin Stecz
Designing a Reliable UAV Architecture Operating in a Real Environment
Reprinted from: *Appl. Sci.* **2022**, *12*, 294, doi:10.3390/app12010294 . . . . **291**

Carlos Capitán, Héctor Pérez-León, Jesús Capitán, Ángel Castaño and Aníbal Ollero
Unmanned Aerial Traffic Management System Architecture for U-Space In-Flight Services
Reprinted from: *Appl. Sci.* **2021**, *11*, 3995, doi:10.3390/app11093995 . . . . **311**

Joseph Kim and Ella Atkins
Airspace Geofencing and Flight Planning for Low-Altitude, Urban, Small Unmanned Aircraft Systems
Reprinted from: *Appl. Sci.* **2022**, *12*, 576, doi:10.3390/app12020576 . . . . **335**

Jérôme Morio, Baptiste Levasseur and Sylvain Bertrand
Drone Ground Impact Footprints with Importance Sampling: Estimation and Sensitivity Analysis
Reprinted from: *Appl. Sci.* **2021**, *11*, 3871, doi:10.3390/app11093871 . . . . **359**

Antonio Guillen-Perez, Ana-Maria Montoya, Juan-Carlos Sanchez-Aarnoutse and Maria-Dolores Cano
A Comparative Performance Evaluation of Routing Protocols for Flying Ad-Hoc Networks in Real Conditions
Reprinted from: *Appl. Sci.* **2021**, *11*, 4363, doi:10.3390/app11104363 . . . . **373**

Laura Pierucci
Hybrid Direction of Arrival Precoding for Multiple Unmanned Aerial Vehicles Aided Non-Orthogonal Multiple Access in 6G Networks
Reprinted from: *Appl. Sci.* **2022**, *12*, 895, doi:10.3390/app12020895 . . . . **389**

Yali Zhang, Xinrong Huang, Yubin Lan, Linlin Wang, Xiaoyang Lu, Kangting Yan, et al.
Development and Prospect of UAV-Based Aerial Electrostatic Spray Technology in China
Reprinted from: *Appl. Sci.* **2021**, *11*, 4071, doi:10.3390/app11094071 . . . . **401**

About the Editors

Sylvain Bertrand

Dr. Sylvain Bertrand is a Senior Research Scientist at ONERA-The French AerospaceLab, University of Paris-Saclay. He graduated from Ecole Centrale de Lille and University of Lille, France, in 2004 and received his MSc from University of Lille, France, in the same year. He obtained his PhD in 2007 from the University of Nice Sophia Antipolis. He is the co-author of more than 80 publications in journals, conference proceedings or book chapters. He is a regular lecturer in several graduate schools of engineering, with a particular interest in control education and teaching activities in automatic control, multi-agent systems and robotics. His research activities concern optimization-based control and estimation for constrained dynamical systems, distributed control and event-triggered control of multi-agent systems, and distributed state estimation for sensor networks. He is also interested in applications to autonomous, aerospace and robotic systems and risk analysis for unmanned aerial vehicles.

Hyo-Sang Shin

Professor Hyo-Sang Shin is a Professor of Guidance, Control and Navigation Systems at Cranfield University. He gained an MSc on flight dynamics, guidance and control in Aerospace Engineering from KAIST and a PhD on cooperative missile guidance from Cranfield University in 2005 and 2011, respectively. He has published over 200 journal and conference papers and has been invited for many lectures, invited talks and keynotes. He has significant experience and a track record in the fields of flight control, guidance, and sensor fusion-related research and development projects. He is a member of various technical committees and program and editorial boards, and also an Associate Editor of various international journals. Professor Shin is Head of Autonomous and Intelligent Groups and leads relevant research areas. His current research activities include data-centric guidance and control, decision making, information-driven sensing and fusion, target tracking, and multiple agent cooperation.

applied sciences

MDPI

Editorial

Special Issue on Unmanned Aerial Vehicles

Sylvain Bertrand [1,*] and Hyo-Sang Shin [2,*]

[1] Traitement de l'Information et Systèmes, ONERA, Université Paris-Saclay, 91123 Palaiseau, France
[2] Centre for Autonomous and Cyber-Physical Systems, Cranfield University, Cranfield MK43 0AL, UK
* Correspondence: sylvain.bertrand@onera.fr (S.B.); h.shin@cranfield.ac.uk (H.-S.S.)

1. Introduction

Unmanned Aerial Vehicles (UAVs) are recognized as very useful tools to replace, help, or assist humans in various missions, such as inspection and monitoring, surveillance, search and rescue, exploration, logistics and transportation, etc. Practical uses for such missions in both civilian and defense contexts have experienced a significant growth thanks to recent technological progresses. Nevertheless, some challenges and open issues remain to ensure a full operational use of UAVs.

This Special Issue aims to present recent advances in technologies and algorithms to improve the levels of autonomy, reliability, and safety of UAVs. Different topics are addressed in this Special Issue, covering vehicle design and characterization (aerodynamics, flight dynamics, design optimization, communications), algorithms for autonomy (guidance and control, path planning, machine learning, computer vision, perception), traffic and risk management (Unmanned Traffic Management, reliability, risk assessment). Open issues related to new missions such as precision agriculture or telecommunication relays are also considered.

A total of twenty papers (nineteen research papers and one review paper) are presented in this Special Issue.

2. Vehicle Design and Characterization

Fight mechanics and aerodynamics studies can be done to derive accurate dynamical models of UAVs. When dealing with specific configurations of UAVs, these model can be useful for performance evaluation and vehicle design, control algorithm synthesis, etc.

In [1], aerodynamic characterization of a coaxial tri-rotor Micro Air vehicle (MAV) is performed, with particular attention to the influence of wind effect. Another type of MAV configuration, namely a strake-wing MAV, is considered in [2], where bifurcation theory is applied to study the open loop flight dynamics of the vehicle.

Actuators design and characterization also plays an important role in performance analysis and design of the vehicle and control algorithms. Regarding characterization of propellers (efficiency, thrust coefficient), an approach is proposed in [3] to validate a wind tunnel propeller dynamometer. The choice of the number and types of actuators, and how they are used in the vehicle design may improve its robustness wrt to faults. In [4], an optimization framework is presented to design a novel actuator fault-tolerant multrirotor MAV.

Optimization of the vehicle configuration can also be considered to account for specific requirements of the mission. Optimized for long hover and long-range missions, a new tandem-wing quadplane UAV configuration is proposed in [5].

3. Algorithms for Autonomy

Navigation, guidance and control of Unmanned Aerial Vehicles rely on different types of algorithms that must realize automatic/autonomous functions of perception, decision making, path planing, motion control, etc.

Citation: Bertrand, S.; Shin, H.-S. Special Issue on Unmanned Aerial Vehicles. *Appl. Sci.* **2023**, *13*, 4134. https://doi.org/10.3390/app13074134

Received: 6 March 2023
Accepted: 10 March 2023
Published: 24 March 2023

Machine Learning algorithms are now widely used in perception and computer vision, especially for classification and decision making. In [6], a light-weight deep neural network architecture is proposed for real-time object classification, considering mission specific input data augmentation techniques. In [7], a classifier is designed for aerial images via deep transfer learning for UAV networks.

Reinforcement learning algorithms are proposed in [8] for solving the position control problem of a quadrotor. In case of wind, a robust controller is developed in [9] through Reinforcement Learning and disturbance compensation.

In the case of multiple vehicles, distributed cooperative control laws are proposed in [10] for the problem of interception of static and maneuvering targets by several UAVs. In [11], a distributed formation controller is presented using specific index patterns and chain rules of visual references among the vehicles of the fleet, resulting in a good robustness wrt losses of vehicle(s).

When moving in cluttered environments, collision-free reference trajectories are to be sought, to be followed by the vehicles. As path planing can be computationally demanding, a new light-weight planner is developed in [12] based on relative position of detected obstacles that can be used in real-time in a perception and control loop. Another approach is proposed in [13] that exploits obstacle geometry information to give priority to search in sub-spaces where a solution can be found quickly.

4. Traffic and Risk Management

When operating in real world environments, reliability and safety requirements have to be satisfied for the UAV and the operation to ensure mitigation of risks wrt third parties: other manned or unmanned platforms in the airspace, people at ground, etc.

The work in [14] proposes a method to design reliable UAV architectures accounting for modeling of emergency situations such as collision risk avoidance behaviors.

Regarding Unmanned aerial system Traffic Management (UTM), an open source software architecture is presented in [15] to track aerial operations an monitor the airspace during in real time. Furthermore, the system is capable of in-flight emergency management and tactical deconfliction. For low-altitude UTM, a 3D flight volumization algorithm along with path planning solutions is presented in [16] for definition and management of geofenced airspaces that would contain compatible UAV trajectories and ensure avoidance of no-fly zones.

To deal with risk wrt third parties at ground, a method based on Importance Sampling is proposed in [17] to generate reliable ground impact footprints that contains a high percentile of the drone impact points.

5. New Missions for UAVs

Unmanned Aerial Vehicles offer new capabilities that can be employed for innovative usages. Acting as communication relays is one type of new missions that can be envisaged for UAVs. Classification of routing protocols for Flying Ad-Hoc networks is proposed in [18], along with a comparison of several protocols for WiFi technology. Regarding UAVs as relays to be integrated in the future 6G cellular network, the work in [19] proposes a method to detect the directions of arrival of each UAV relay in a network supporting an uplink non-orthogonal multiple access cellular system.

Precision agriculture is another type of mission for which UAVs are at the center of the attention. In this context, aerial electrostatic spray is a technology of interest for reducing environmental pollution from application of pesticides. A review on the development of such a technology in China is presented in [20].

Acknowledgments: We would like to thank all the authors for their valuable contributions to this Special Issue 'Unmanned Aerial Vehicles' as well as peer reviewers for their hard work and valuable comments to the authors. Thanks to the dedicated editorial team of *Applied Sciences*, all the staff and people involved in this Special Issue.

Conflicts of Interest: The authors declare no conflict of interest.

References

1. Lei, Y.; Ye, Y.; Chen, Z. Horizontal Wind Effect on the Aerodynamic Performance of Coaxial Tri-Rotor MAV. *Appl. Sci.* **2020**, *10*, 8612. [CrossRef]
2. Nowakowski, M.; Sibilski, K.; Sibilska-Mroziewicz, A.; Żyluk, A. Bifurcation Flight Dynamic Analysis of a Strake-Wing Micro Aerial Vehicle. *Appl. Sci.* **2021**, *11*, 1524. [CrossRef]
3. Jdiobe, M.; Rouser, K.; Paul, R.; Rouser, A. Validation of a Wind Tunnel Propeller Dynamometer for Group 2 Unmanned Aircraft. *Appl. Sci.* **2022**, *12*, 8908. [CrossRef]
4. Debele, Y.; Shi, H.; Wondosen, A.; Kim, J.; Kang, B. Multirotor Unmanned Aerial Vehicle Configuration Optimization Approach for Development of Actuator Fault-Tolerant Structure. *Appl. Sci.* **2022**, *12*, 6781. [CrossRef]
5. Okulski, M.; Ławryńczuk, M. A Small UAV Optimized for Efficient Long-Range and VTOL Missions: An Experimental Tandem-Wing Quadplane Drone. *Appl. Sci.* **2022**, *12*, 7059. [CrossRef]
6. Endale, B.; Tullu, A.; Shi, H.; Kang, B. Robust Approach to Supervised Deep Neural Network Training for Real-Time Object Classification in Cluttered Indoor Environment. *Appl. Sci.* **2021**, *11*, 7148. [CrossRef]
7. Alotaibi, S.S.; Abdullah Mengash, H.; Negm, N.; Marzouk, R.; Hilal, A.; Shamseldin, M.; Motwakel, A.; Yaseen, I.; Rizwanullah, M.; Zamani, A. Swarm Intelligence with Deep Transfer Learning Driven Aerial Image Classification Model on UAV Networks. *Appl. Sci.* **2022**, *12*, 6488. [CrossRef]
8. Jembre, Y.; Nugroho, Y.; Khan, M.; Attique, M.; Paul, R.; Shah, S.; Kim, B. Evaluation of Reinforcement and Deep Learning Algorithms in Controlling Unmanned Aerial Vehicles. *Appl. Sci.* **2021**, *11*, 7240. [CrossRef]
9. Pi, C.; Ye, W.; Cheng, S. Robust Quadrotor Control through Reinforcement Learning with Disturbance Compensation. *Appl. Sci.* **2021**, *11*, 3257. [CrossRef]
10. Jiang, Z.; Ge, J.; Xu, Q.; Yang, T. Terminal Distributed Cooperative Guidance Law for Multiple UAVs Based on Consistency Theory. *Appl. Sci.* **2021**, *11*, 8326. [CrossRef]
11. Qiao, R.; Xu, G.; Cheng, Y.; Ye, Z.; Huang, J. Simulation and Analysis of Grid Formation Method for UAV Clusters Based on the 3×3 Magic Square and the Chain Rules of Visual Reference. *Appl. Sci.* **2021**, *11*, 11560. [CrossRef]
12. Tullu, A.; Endale, B.; Wondosen, A.; Hwang, H. Machine Learning Approach to Real-Time 3D Path Planning for Autonomous Navigation of Unmanned Aerial Vehicle. *Appl. Sci.* **2021**, *11*, 4706. [CrossRef]
13. Majeed, A.; Hwang, S. Path Planning Method for UAVs Based on Constrained Polygonal Space and an Extremely Sparse Waypoint Graph. *Appl. Sci.* **2021**, *11*, 5340. [CrossRef]
14. Gromada, K.; Stecz, W. Designing a Reliable UAV Architecture Operating in a Real Environment. *Appl. Sci.* **2022**, *12*, 294. [CrossRef]
15. Capitán, C.; Pérez-León, H.; Capitán, J.; Castaño, Á.; Ollero, A. Unmanned Aerial Traffic Management System Architecture for U-Space In-Flight Services. *Appl. Sci.* **2021**, *11*, 3995. [CrossRef]
16. Kim, J.; Atkins, E. Airspace Geofencing and Flight Planning for Low-Altitude, Urban, Small Unmanned Aircraft Systems. *Appl. Sci.* **2022**, *12*, 576. [CrossRef]
17. Morio, J.; Levasseur, B.; Bertrand, S. Drone Ground Impact Footprints with Importance Sampling: Estimation and Sensitivity Analysis. *Appl. Sci.* **2021**, *11*, 3871. [CrossRef]
18. Guillen-Perez, A.; Montoya, A.; Sanchez-Aarnoutse, J.; Cano, M. A Comparative Performance Evaluation of Routing Protocols for Flying Ad-Hoc Networks in Real Conditions. *Appl. Sci.* **2021**, *11*, 4363. [CrossRef]
19. Pierucci, L. Hybrid Direction of Arrival Precoding for Multiple Unmanned Aerial Vehicles Aided Non-Orthogonal Multiple Access in 6G Networks. *Appl. Sci.* **2022**, *12*, 895. [CrossRef]
20. Zhang, Y.; Huang, X.; Lan, Y.; Wang, L.; Lu, X.; Yan, K.; Deng, J.; Zeng, W. Development and Prospect of UAV-Based Aerial Electrostatic Spray Technology in China. *Appl. Sci.* **2021**, *11*, 4071. [CrossRef]

applied sciences

MDPI

Article

Horizontal Wind Effect on the Aerodynamic Performance of Coaxial Tri-Rotor MAV

Yao Lei [1,2,*], Yiqiang Ye [1] and Zhiyong Chen [1]

1 School of Mechanical Engineering and Automation, Fuzhou University, Fuzhou 350116, China; yiqiang_ye@outlook.com (Y.Y.); zychen@fzu.edu.cn (Z.C.)
2 Hydrodynamic and electrohydraulic Intelligent Control Key Laboratory of Fujian University, Fuzhou University, Fuzhou 350116, China
* Correspondence: yaolei@fzu.edu.cn; Tel.: +86-0591-2286-6316

Received: 11 November 2020; Accepted: 27 November 2020; Published: 1 December 2020

Abstract: The coaxial Tri-rotor micro air vehicle (MAV) is composed of three coaxial rotors where the aerodynamic characteristics of is complicated in flight especially when the wind effect is introduced. In this paper, the hovering performance of a full-scale coaxial Tri-rotor MAV is analyzed with both the simulations and wind tunnel experiments. Firstly, the wind effect on the aerodynamic performance of coaxial Tri-rotor MAV is established with different rotor speed (1500–2300 rpm) and horizontal wind (0–4 m/s). Secondly, the thrust and power consumption of coaxial Tri-rotor (L/D = 1.6) were obtained with low-speed wind tunnel experiments. Furthermore, the streamline distribution, pressure distribution, velocity contour and vortex distribution with different horizontal wind conditions are obtained by numerical simulations. Finally, combining the experiment results and simulation results, it is noted that the horizontal wind may accelerate the aerodynamic coupling, which resulting in the greater thrust variation up to 9% of the coaxial Tri-rotor MAV at a lower rotor speed. Moreover, the aerodynamic performance is decreased with more power consumption at higher rotor speed where the wind and the downwash flow are interacted with each other. Compared with no wind flow, the shape of the downwash flow and the deformation of the vortex affect the power loading and figure of metric accordingly.

Keywords: coaxial Tri-rotor MAV; horizontal wind; low-speed wind tunnel; numerical simulation

1. Introduction

Compared with traditional Quad-rotor or Hex-rotor, the coaxial Tri-rotor the coaxial Tri-rotor has a much wider class including a compact structure without redundancy device since the vehicle mass is related to the rotor arm where the Quad-rotor or Hex-rotor is limited with more rotors to avoid rotor conflict. Also, it provides the unique capability of being able to resist any applied wrench or wind gust or failure tolerance with coaxial rotors (If one rotor, even three rotors, fails the system, it still has the freedom of movement). For a coaxial Tri-rotor MAV, the three coaxial rotors are evenly distributed along with the vehicle center. The aerodynamic interference is mainly including two parts: the rotor interference between the upper rotor and lower rotor and adjacent coaxial rotors [1–6]. When the horizontal wind is introduced, the whole aerodynamic environment will be affected by the horizontal wind during flight. Therefore, the objective of the present work is to explore wind effect on the aerodynamic performance on a full-scale coaxial Tri-rotor MAV.

Currently, the research on the multi-rotor MAV is mainly focused on the control strategies. Pflimlin, Zhang and Kirsch [7–9] designed the adaptive backstepping sliding mode controllers and realized attitude, velocity and position control of the Hex-rotor MAV. Salazar and Arellano-Muro [10,11] adopted the dynamic model of a multi rotor MAV to analyze the attitude and translation and estimate the aerodynamic forces and moments acting of a hexarotor MAV in flight. Shi [12] presented an indoor path planning algorithm and overcome the drawbacks of Global Positioning System (GPS). So the shortest trajectory for the Hex-rotor MAV in the complex terrains is obtained. Ma [13] designed a 4 channels PID controller to achieve the attitude control of a miniature Hex-rotor MAV. Chen [14] proposed a controller with cascaded structure, which has the ability to maintain the flight state of MAV. Zhao [15] presented a novel Hex-Rotor MAV based on the unique configuration of its six driving rotors and overcame the effect of the under-actuation and strong coupling characteristics on the flight performance. Salazar-Cruz [16] proposed the dynamical model of an original coaxial Tri-rotor MAV and applied a nested saturations control law to control the roll and the forward displacement, resulting in the better behavior of controller. Mohamed [17] and Chiou [18] proposed the design and control of the single tilt tri-rotor and shown the effectiveness of the controllers design scheme through nonlinear simulation model. Brossard, Mystkowski and Tunik [19–21] discussed a nonlinear robust control design procedure to micro air vehicle, resulting in the stable flight of MAV in the presence of perturbations.

Therefore, only a few studies lay emphasis on the aerodynamic characteristics of a Multi-rotor MAV or even to consider the wind effect. Lei et al. studied the hover performance of a Multi-rotor MAV by means of the combination of experiment and simulation [22,23]. Zhao promoted a method to analyze the effects of airflow disturbance and rotor interference on the control scheme, which is based on the dynamic experiment of the Hex-rotor MAV [24]. Hrishikeshavan reviewed the hover capability of MAV with varying solidity, collective, operating RPM and planform [25]. The results of the above studies are all conducive to analyze the aerodynamic characteristics of the coaxial Tri-rotor MAV. However, there are no aerodynamic studies of the coaxial Tri-rotor MAV with the Horizontal wind so far. Hence, this paper presents the aerodynamic characteristics of a coaxial Tri-rotor MAV with the effect of the horizontal wind.

2. Theoretical Analysis

2.1. Structure

Structure of the coaxial Tri-rotor MAV is shown in Figure 1.

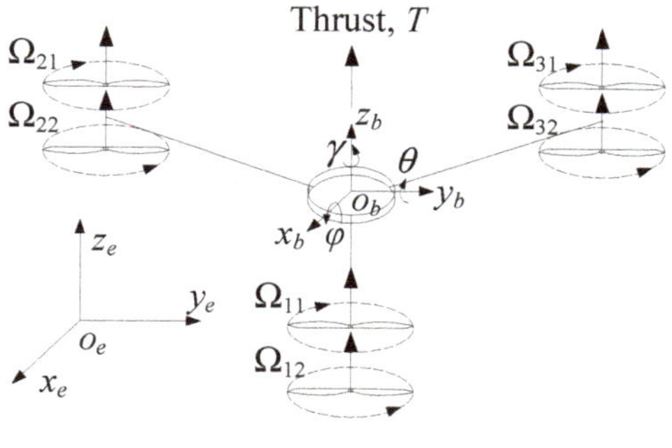

Figure 1. Structure of coaxial Tri-rotor micro air vehicle (MAV).

Figure 1 shows the structural model of the coaxial Tri-rotor MAV. An inertial reference frame o_e (x_e, y_e, z_e), an Euler angle in inertial frame $(\varphi, \theta, \gamma)$ and a body reference frame o_b (x_b, y_b, z_b) that indicates a set of coordinate fixed to the MAV, are defined in Figure 1. The coaxial Tri-rotor MAV is composed of three coaxial rotors units. The connecting line between the centers of three coaxial rotors forms an equilateral triangle, which is center-symmetric. In addition, compared with traditional rotor arrangement, the Tri-rotor MAV is more compact with less rotors in a same plane.

2.2. Flow Field Model

Flow model of the coaxial Tri-rotor MAV considering the horizontal wind is shown in Figure 2.

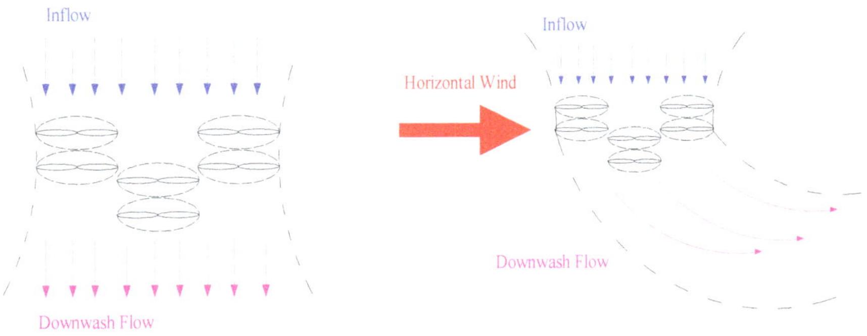

(a) (b)

Figure 2. Flow model of coaxial Tri-rotor MAV in the horizontal wind: (**a**) No wind; (**b**) Horizontal wind.

In the Figure 2, it can be observed that the flow field of the coaxial Tri-rotor system will shift along the incoming flow direction in a horizontal wind. Compared with no wind effect, it can be found that the downwash flows are coupled with each other besides the interference between upper and lower rotor of coaxial rotors. In this case, the wake vortices of the front rotors directly affect the flow of the rear rotors. In this case, the wind may aggravate the aerodynamic interference among rotors with varied power consumption accordingly.

In the natural environment, the wind speed is usually less than 5.0 m/ s. Furthermore, the light breeze (1.6–3.3 m/ s) and the gentle breeze (3.4–5.4 m/s) frequently appear in the natural environment. The average values of the wind, 2.5 m/s and 4 m/s are selected as the horizontal incoming wind speed. Therefore, it is conducive to study the influence of the horizontal wind on the aerodynamic performance of the rotor system. In the meanwhile, the situation of the horizontal wind at 0 m/s is also taken as the comparison to analyze the effect of the horizontal wind.

2.3. Force Analysis

Taking a rotor as an example, the airflow model with the presence of the horizontal airflow is shown in Figure 3.

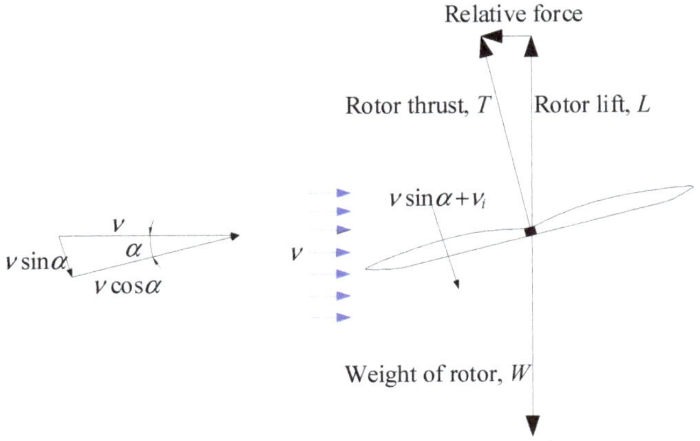

Figure 3. The airflow model of the rotor disk in the horizontal wind.

In Figure 3, v is the horizontal wind velocity, W is the weight of rotor, T is the thrust of rotor, L is the relative lift of rotor and α is the angle of attack. With the horizontal airflow, the rotor disk will tilt at a certain angle. Because the rotor is required to generate relative force (to balance the horizontal force) and lift (to overcome the rotor gravity), resulting in stable hovering state. Clearly, the induced power changes with the angle of attack. To derive the effect of the horizontal wind velocity on induced power, the induced velocity v_i for a rotor can be obtained as follow [26]:

$$v_i = \frac{{v_h}^2}{\sqrt{(v \cos \alpha)^2 + (v \sin \alpha + v_i)^2}}, \tag{1}$$

where α is the angle of attack, v is the horizontal wind velocity, v_h is the induced velocity in hover [26].

$$v_h = \sqrt{\frac{T_1}{2\rho A}}, \tag{2}$$

where ρ is the air density, kg/m3; A is rotor disk area, m^2. By applying the energy conservation, the power required is obtained as follow:

$$P = T(v_i + v \sin \alpha). \tag{3}$$

Therefore, the horizontal wind may cause more power consumption resulted by the induced velocity and affect the flight efficiency eventually.

2.4. The Parameters of Aerodynamic Performance

2.4.1. Power Loading

The total hover efficiency of a MAV can be quantified by means of effective power loading (*PL*). The *PL* is defined as the ratio of the thrust to power required [27]:

$$PL = \frac{T}{P}. \tag{4}$$

The thrust coefficient C_T and power coefficient C_P are defined as [28]:

$$C_T = \frac{T}{\rho A \Omega^2 R^2}, \tag{5}$$

$$C_P = \frac{P}{\rho A \Omega^3 R^3} = \frac{Q\Omega}{\rho A \Omega^3 R^3} = \frac{Q}{\rho A \Omega^2 R^3}. \tag{6}$$

Therefore, the power loading (*PL*) can be written as:

$$PL = \frac{C_T}{\Omega R C_P} = \frac{T}{Q\Omega}, \tag{7}$$

where T is the thrust, N; A is the area of the rotor, m^2; P is the power, W; Ω is rotational speed of the rotor, r/min; R is the rotor radius, m; Q is the torque, Nm; ρ is the fluid density, kg/m3; C_T is the thrust coefficient; C_P is the power coefficient.

To maximize the *PL*, that is, for a given thrust, the power demand is minimum. When designing a vehicle it is wanted to maximize the power loading such that energy requirements are minimized. This will give the vehicle the best endurance or payload capabilities possible.

2.4.2 Hover Efficiency

A figure of merit (*FM*) is adopted to characterize the hover efficiency. It is regarded as the ratio of the ideal power demand to the actual power demand. In addition, by means of the measured quantities, the figure of merit equation is defined as [29]:

$$FM = \frac{{C_T}^{3/2}}{\sqrt{2}C_P} = \frac{T^{3/2}}{Q\Omega\sqrt{2\rho A}}, \tag{8}$$

where T is the thrust, N; Ω is rotational speed of the rotor, r/min; Q is the torque, Nm; C_T is the thrust coefficient; C_P is the power coefficient.

3. Experiment

3.1. Experiment Setup

In order to obtain the performance of the MAV in the horizontal wind, wind tunnel tests were carried out to simulate the environment of MAV at 0 m/s, 2.5 m/s and 4 m/s. Experiment process of the coaxial Tri-rotor MAV considering the horizontal wind is shown in Figure 4.

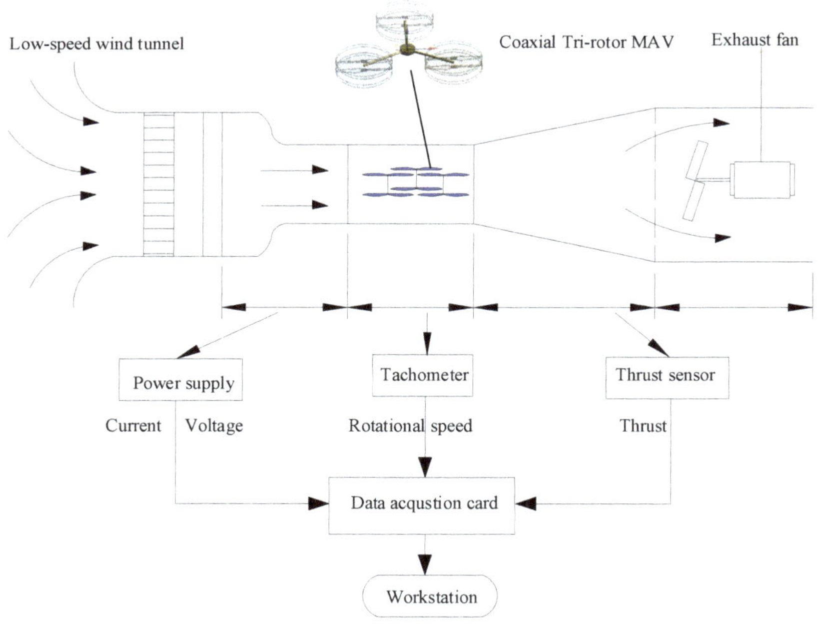

Figure 4. Experimental process.

The dimensions of rectangular test section of wind tunnel are 3 m (length) × 3 m (width) × 2.5 m (height), ensuring sufficient space for maneuvering the multirotor platform. A settling chamber is attached before the test section to characterize the output wind. Wind is generated with two 3 m diameter, 45 kW fans. For the current low Reynolds number experiments the maximum testing velocity is 12.5 m/s. According to the theoretical analysis, the power, rotor speed and thrust of the coaxial Tri-rotor MAV with the horizontal wind are obtained accordingly to convert into the power loading and *FM*. In the test, propeller is specially made with unidirectional carbon fiber fabrics as stiffener based on the airfoil of C5.5/4.5, with 15.7 cm of pitch and 2.8 cm of chord at 75% position. The rotational speed range of rotor is 1500–2300 r/min. The motor is brushless DC motor(model: MSYSLRK 195.03).

The main measuring equipment is as follows: (1) speed controller (model: BL-6); (2) tachometer (model: DT-2234C, accuracy: 6 ± (0.05% + 1D)); (3) thrust sensor (model: CZL605, accuracy: 0.02% F.S.).

3.2. Experimental Results

Figure 5 shows the thrust and power variation.

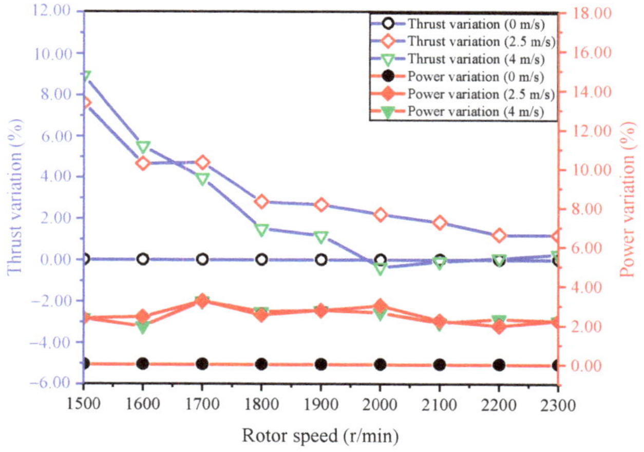

Figure 5. Thrust and power variation.

The thrust and power consumption at 0 m/s is set as the reference value to obtain the increment the variation with the wind effect. According to the Figure 5, it can be observed that the thrust increased with the wind effect, especially at 4 m/s. At the same time, it can be noted that the thrust increment approached to 0 m/s at a higher rotor speed for 2.5 m/s. This is because the rotor interference is much stronger at 2.5 m/s for a rotor speed ranging from 2000 to 2300 r/min. With an increased wind speed, this interference is not domain the aerodynamic environment. In addition, it also can be observed that the required power of coaxial Tri-rotor is also increased 2–4% with the wind speed. This extra power consumption may be generated by the introduced rotor interference when the horizontal wind is considered in this case. However, the thrust increment is higher than the power increment, especially at a lower rotor speed. Clearly, it is advantageous to promote the power loading and the coupling interference is offset to the minimum in the horizontal wind.

Figure 6 shows the variation of the power loading.

In Figure 6, it can be noted that the variation of power loading gradually decreases with the rotor speed especially for 4 m/s. Also, it can be observed that the power loading with incoming flow is greater than that of no wind effect between 1500 and 1800 r/min. In this case, the horizontal airflow will improve the aerodynamic performance of the coaxial Tri-rotors to a certain extent. However, the decreased power loading at a higher rotor speed indicated that the rotor interference is coupled with each other. At this point, the external airflow aggravates the aerodynamic interference between the rotors. For a light breeze, the coaxial Tri-rotor presented a good wind resistance. When the working speed is 2200 r/min, the power loading variation at 2.5 m/s and 4 m/s is about −0.5% and −2%, respectively.

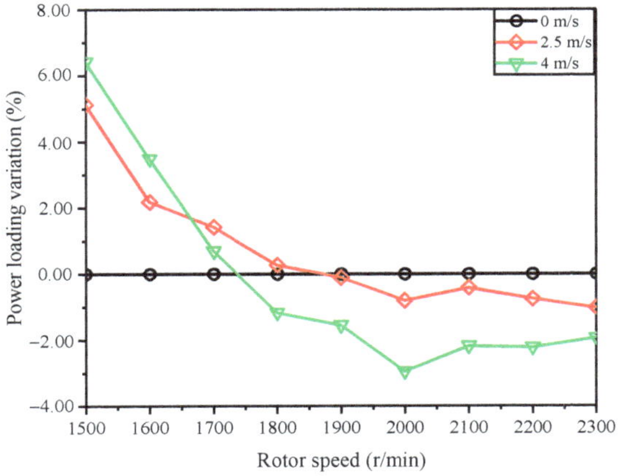

Figure 6. Power loading variation.

Figure 7 shows the *FM* increment with the wind effect.

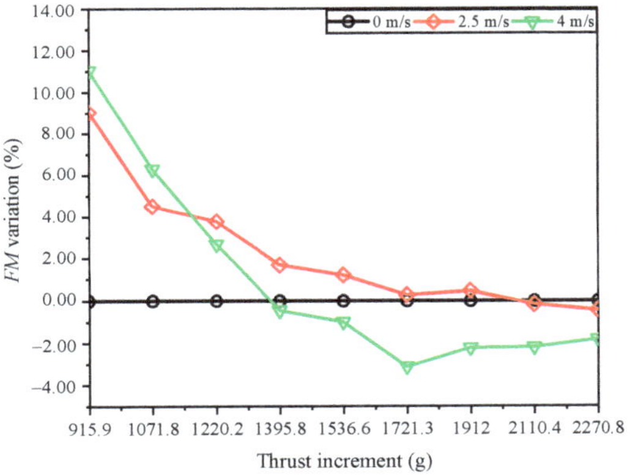

Figure 7. Figure of merit (*FM*) variation.

In Figure 7, it can be noted that the hover efficiency in the horizontal wind is higher at lower thrust where the power increment is relatively low. It can be seen that the intervention of the horizontal airflow can promote the aerodynamic coupling between rotors and improve the hover efficiency of the coaxial Tri-rotor. In addition, it can be observed that the hover efficiency is slightly higher when the horizontal airflow velocity at 2.5 m/s, which indicates that the hover efficiency of the coaxial Tri-rotor can be improved by the intervention of the horizontal wind.

4. Simulation Analysis

4.1. Computational Fluid Dynamics (CFD) Setup

Sliding-mesh is applied to solve for the motion of the rotors due to the highly unsteady nature of flow involved in the study and the time-step size is 10e-5. The meshing distribution of the entire computing domain is shown in Figure 8.

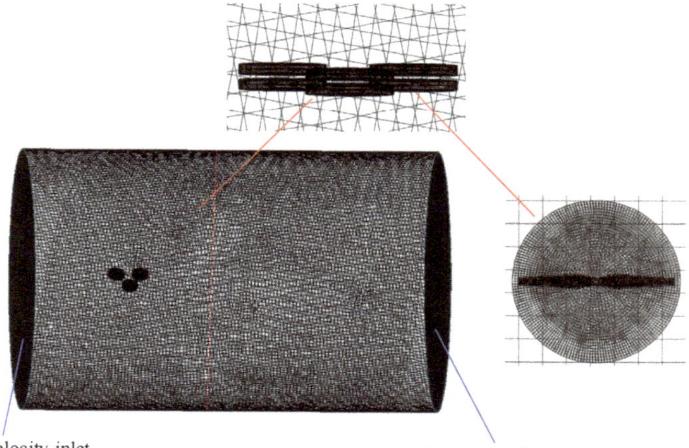

Figure 8. Mesh distribution.

The whole computational domain is divided into 7 regions including 1 cylinder stationary region and 6 cylinder rotating regions to capture the flow detail of rotors with refined mesh. Also, the MAV is located at the left region of the domain to obtain the detail of the downwash flow along with the wind direction. Mesh parameters are showed in the Table 1.

Table 1. Mesh parameters.

Nodes	Elements	Average Skewness	Turbulence Model	Pressure Interpolation	Spatial Discretization
1164307	6472340	0.21734	Spalart-Allmaras	Standard	Second-order upwind

To validate the effectiveness of the CFD method, the comparison of CFD and experimental results is showed in Table 2. Both the C_T and C_P in experiment and simulation showed that they are generally in good agreement.

Table 2. Comparison of Computational Fluid Dynamics (CFD) and experiments.

Cases	C_T Experiment	C_T Simulation	Relative Error (%)	C_P Experiment	C_P Simulation	Relative Error (%)
1	0.404	0.385	4.94	0.198	0.181	9.39
2	0.578	0.562	2.85	0.215	0.196	9.69
3	0.783	0.742	5.53	0.218	0.197	10.66
4	0.854	0.878	−2.73	0.274	0.248	10.48
5	0.924	0.952	−2.94	0.408	0.375	8.80
6	0.968	0.988	−2.02	0.524	0.572	8.39
7	0.925	0.945	−2.12	0.798	0.764	4.45
8	0.965	0.925	4.32	0.834	0.848	−1.65

4.2. Simulation Results

4.2.1. Velocity Contour

The velocity contour of the coaxial Tri-rotor considering the horizontal wind is shown in Figure 9.

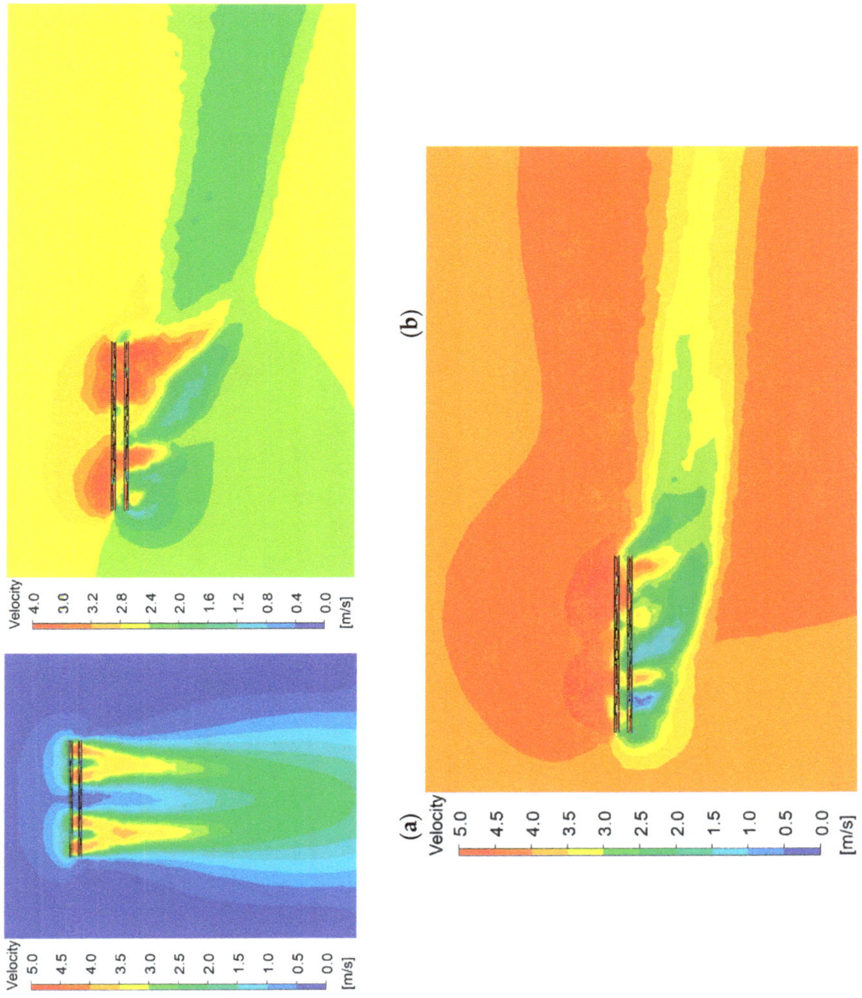

Figure 9. The velocity contour of coaxial Tri-rotor (2200 r/min): (**a**) 0 m/s; (**b**) 2.5 m/s; (**c**) 4 m/s.

In Figure 9, it can be clearly noted that the downwash flow is moved along with the wind direction. Compared with the case with no wind effect, the velocity variation becomes even more complex, which will affect the aerodynamic performance of the coaxial Tri-rotor. Moreover, with the increase of the horizontal wind velocity, the downwash velocity of the coaxial Tri-rotor gradually decreases and the velocity gradient arrangement of downwash becomes closer, leading to the strong rotor interference. It can be seen that the aerodynamic coupling will be affected by the external airflow and aggravate the aerodynamic interference between the rotors. Moreover, the enhancement of aerodynamic interference will bring the increase of required power. This also verifies that in Figures 5 and 7, the overall power consumption of the coaxial Tri-rotor system with the influence of external airflow is significantly greater than that without airflow.

4.2.2. Streamline Distribution

The streamline distribution of the coaxial Tri-rotor with the horizontal airflow is shown in Figure 10.

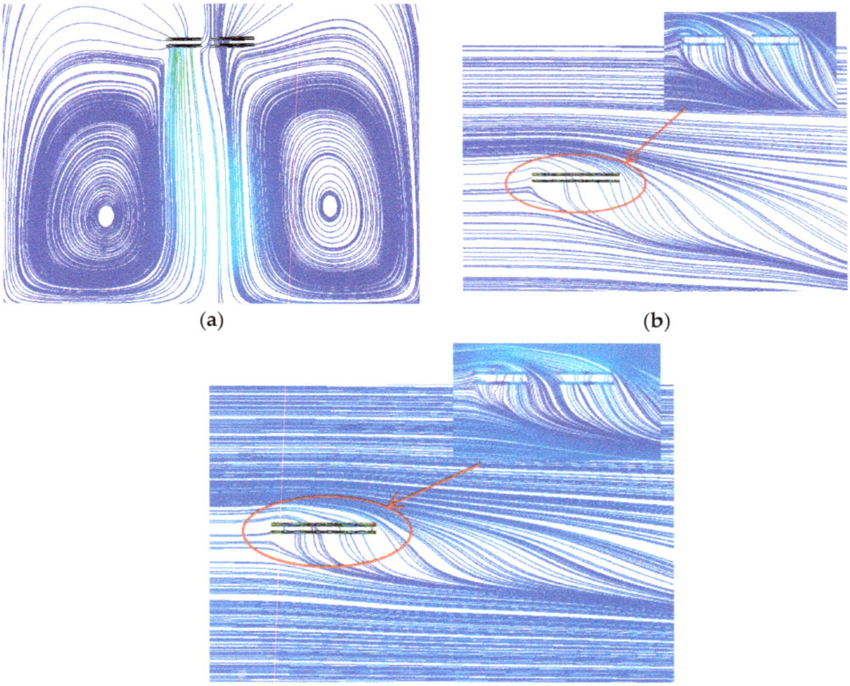

(c)

Figure 10. The streamline distribution of the coaxial Tri-rotor (2200 r/min): (**a**) 0 m/s; (**b**) 2.5 m/s; (**c**) 4 m/s.

In Figure 10, it can be observed that compared with no-flow environment, the streamline deformed with more vortices around the rotor tip in the horizontal wind environment. The streamline is squeezed and deformed, resulting in the vortices under the rotor being deformed and the streamline inclined distribution. At the same time, with the increase of the horizontal wind speed, the streamline arrangement is more compact and the aerodynamic interference between rotors is more intense in this case. Therefore, it can be seen that the horizontal airflow will move the coupling interference between rotors and affect the overall aerodynamic performance. In addition, it can be also noted that compared with the horizontal wind at 4 m/s, the streamline distribution with the horizontal wind at 2.5 m/s is more uniform. The more uniform the streamline arrangement, the better the aerodynamic performance, which also verifies that the power loading of the horizontal wind at 2.5 m/s is greater than that of the horizontal wind at 4 m/s in Figure 6.

4.2.3. Vortex Distribution

The vortex distribution of the coaxial Tri-rotor considering the horizontal wind is shown in Figure 11.

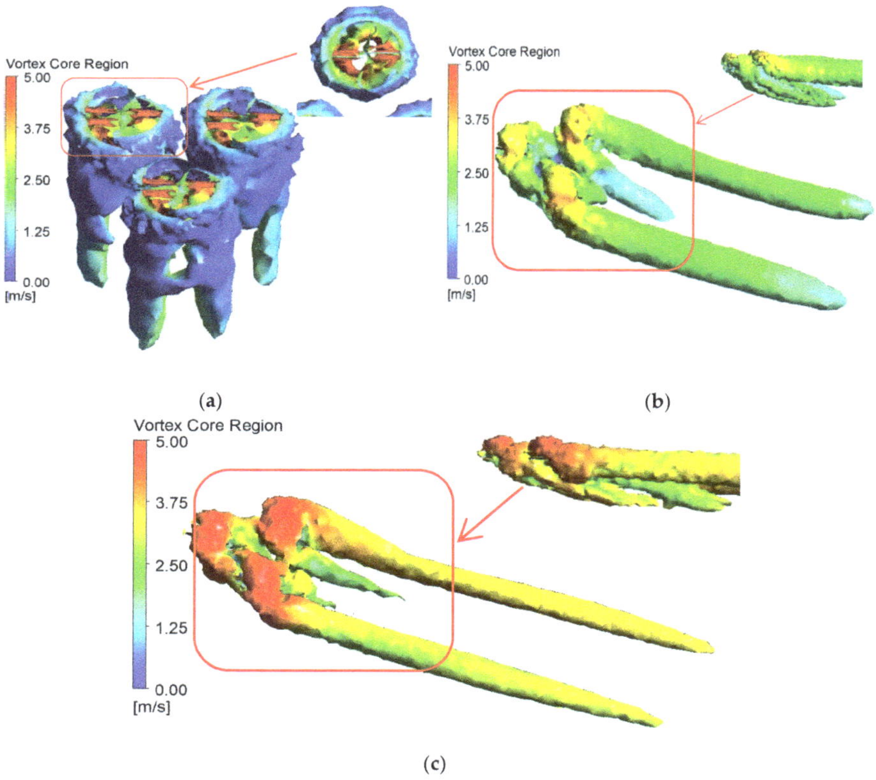

Figure 11. The vortex distribution of the coaxial Tri-rotor (2200 r/min): (**a**) 0 m/s; (**b**) 2.5 m/s; (**c**) 4 m/s.

As shown in the Figure 11, it is observed that the vortex will shift to the rear when it is affected by the horizontal airflow. With the increase of the horizontal wind velocity, the vortex shape inclines to the rotor plane. In the meanwhile, it can be noted that with the increase of horizontal wind velocity, the vortex shape becomes slenderer and the vortex overlap area of rotors is larger. Hence, it can be seen that the coupling interference between rotors will be moved with the horizontal airflow, so as to aggravate the aerodynamic interference, which will affect the overall aerodynamic performance of the coaxial Tri-rotor. Above also verifies that when working speed at 2200 r/min, the power loading of the horizontal airflow at 2.5 m/s is greater than that of the horizontal wind at 4 m/s in Figure 6.

4.2.4. Pressure Contour

The pressure contour of rotor tip in coaxial rotors is shown in Figure 12.

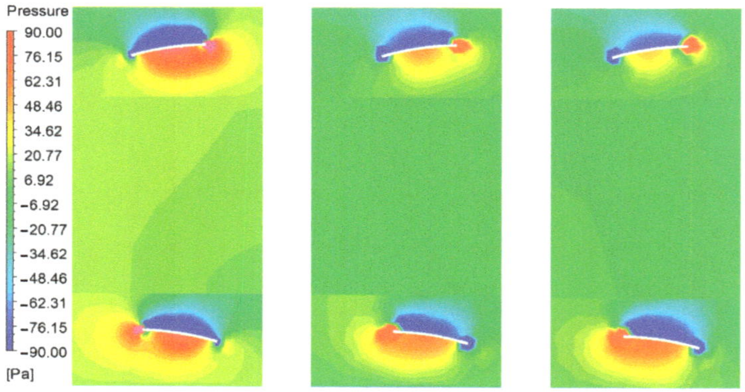

(a) (b) (c)

Figure 12. The vertical pressure contour of rotor tip in a coaxial rotors unit (2200 r/min): (**a**) 0 m/s; (**b**) 2.5 m/s; (**c**) 4 m/s.

In Figure 12, it can be observed that the pressure difference between the upper and lower surfaces of the rotor tip is higher with the horizontal wind which indicated a higher thrust. Also, it can be seen that a part of the thrust produced by the coaxial Tri-rotor MAV with the horizontal airflow needs to be adopted to balance the external force, resulting in the weakened thrust of the rotor. This also verifies that in Figures 6 and 7. With the influence of the horizontal wind, the thrust growth rate of the coaxial Tri-rotor is low, resulting in the poor hover efficiency. In addition, with the increase of the horizontal wind speed, the thrust growth rate of the coaxial Tri-rotor becomes lower and lower.

4.2.5. Velocity Variation Figures

The velocity variation of the lower rotor plane is shown in the Figure 13.

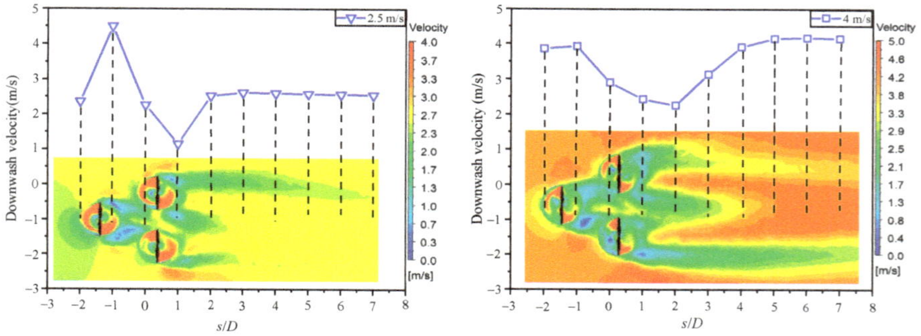

(a) (b)

Figure 13. The velocity variation of the lower rotor plane: (**a**) 2.5 m/s; (**b**) 4 m/s.

In Figure 13, the lower rotor plane of the coaxial Tri-rotor system is applied to obtain the velocity distribution contour of the horizontal wind at 2.5 m/s and 4 m/s and extract the relevant velocity values for analysis. The distance between the reference point and the coordinate origin is expressed by s. It is interesting to note that the minimum of the downwash velocity will move with the horizontal airflow. When the horizontal airflow is larger, the airflow around the rotor will flow faster and the rotor needs to increase the angle of attack to maintain its overall stability. This also verifies that in Figure 5, with the influence of the horizontal airflow, the thrust growth rate of the coaxial Tri-rotor is relatively low.

5. Conclusions

In this paper, low-speed wind tunnel tests and numerical simulations are performed to obtain the aerodynamic performance of the coaxial Tri-rotor MAV with the horizontal wind ranged from 0 to 5 m/s. Conclusions are as follows:

(1) For a lower rotor speed ranging from 1500 to 1800 r/min, the power required is constant, while the thrust increased up to 9%, which indicated that the coaxial Tri-rotor system with lower speed has larger power loading and better aerodynamic performance. In fact, part of the rotor interference is offset by the horizontal inflow.

(2) The velocity and streamline distribution proved that the required power increment is the result of the downwash deformation with the horizontal wind effect. At the same time, the greater the deformation of downwash comes along with a larger the horizontal wind, which will decrease the whole flight efficiency. Combined with the pressure distribution, it also can be seen that the aerodynamic performance is related to the instantaneous thrust variation.

(3) Compared with the case of no wind, the horizontal wind can promote the aerodynamic coupling between the rotors and improve the aerodynamic performance of the coaxial Tri-rotor system at a lower speed. Conversely, the interaction between rotor tip vortices is stronger with a higher rotor speed, thus the interaction between rotors is transferred by the horizontal wind, resulting in reduced thrust. In this case, the stronger coupling interference directly affects the rear rotor with the action of the horizontal wind, which may lead to the rotor vibration with extra power consumption.

(4) For the rotor speed ranging from 1900 to 2300 r/min, part of the horizontal flow is interacted with the downwash flow, resulting in stronger interference to form an unstable flight. Hence, further study will focus on the compensation of the control strategy, considering a lager wind speed.

Author Contributions: Y.L. carried out experiments; Y.Y. wrote the manuscript with assistant of Y.L.; Z.C. is the funding raiser. All authors have read and agreed to the published version of the manuscript.

Funding: This research was supported by the National Natural Science Foundation of China (Grant No. 51505087 and 11502052), Fuzhou University Jinjiang Science and Education Park (No. 2019-JJFDKY-59) and Fujian Provincial Industrial Robot Basic Components Technology Research and Development Center (2014H21010011).

Acknowledgments: The authors thank the Key Laboratory of Fluid Power and Intelligent Electro-Hydraulic Control (Fuzhou University), Fujian Province University for applying the experimental field.

Conflicts of Interest: The authors declare no conflict of interest.

References

1. Ali, Z.A.; Wang, D.; Masroor, S.; Loya, M.S. Attitude and altitude control of trirotor UAV by using adaptive hybrid controller. *J. Control Sci. Eng.* **2016**, *2016*, 6459891. [CrossRef]
2. Kerrow, P.M. Modeling the draganflyer four-rotor helicopter. In Proceedings of the IEEE International Conference on Robotics & Automation, New Orleans, LA, USA, 6 July 2004; pp. 3596–3601.
3. Yoo, D.W.; Oh, H.D.; Won, D.Y.; Tahk, M. Dynamic modeling and control system design for Tri-rotor UAV. In Proceedings of the 2010 3rd International Symposium System and Control in Aeronautics and Astronautics, Harbin, China, 8–10 June 2012; pp. 762–767.
4. Mai, Y.; Zhao, H.; Guo, S. The analysis of image stabilization technology based on small-UAV airborne video. In Proceedings of the 2012 International Conference on Computer Science and Electronics Engineering, Hangzhou, China, 23 March 2012.
5. Hoffmann, G.; Huang, H.; Waslande, S.; Tomlin, C. Quadrotor helicopter flight dynamics and control: Theory and experiment. In Proceedings of the AIAA Guidance, Navigation, & Control Conference and Exhibit, Hilton Head, SC, USA, 20 August 2007.

6. Pounds, P.; Mahony, R.; Corke, P. Modelling and control of a large quadrotor robot. *Control Eng. Pract.* **2010**, *18*, 691–699. [CrossRef]
7. Pflimlin, J.M.; Soueres, P.; Hamel, T. Position control of a ducted fan VTOL UAV in crosswind. *Int. J. Control* **2007**, *80*, 666–683. [CrossRef]
8. Zhang, Z.; Liu, Z.; Wen, N. Research on adaptive backstepping sliding mode control method for a hex-rotor unmanned aerial vehicle. In Proceedings of the 2016 IEEE Chinese Guidance, Navigation and Control Conference (IEEE CGNCC2016), Nanjing, China, 12–14 August 2016.
9. Kirsch, B.; Alexopoulos, A.; Badreddin, E. Non-linear model based control and parameter identification of a hex-rotor UAV. In Proceedings of the IEEE International Conference on Systems, Man, and Cybernetics (SMC2016), Budapest, Hungary, 9–12 October 2016.
10. Salazar, S.; Romero, H.; Lozano, R.; Castillo, P. Modeling and real-time stabilization of an aircraft having eight rotors. *J. Intell. Robot. Syst.* **2009**, *54*, 455–470. [CrossRef]
11. Arellano-Muro, C.A.; Luque-Vega, L.F.; Castillo-Toledo, B.; Loukianov, A.G. Backstepping control with sliding mode estimation for a hexacopter. In Proceedings of the 2013 10th International Conference on Electrical Engineering, Computing Science and Automatic Control (CCE), Mexico City, Mexico, 30 September–4 October 2013.
12. Shi, T.; Wang, H.; Cui, W.; Ren, L. Indoor path planning for hex-rotor aircraft with landmark-based visual navigation. In Proceedings of the 12th International Conference on Natural Computation, Fuzzy Systems and Knowledge Discovery (ICNC-FSKD), Changsha, China, 13–15 August 2016.
13. Ma, Q.; Sun, Z.; Wu, J.; Zhang, W. Dynamic modeling for a miniature six-rotor unmanned aerial vehicle. *Appl. Mech. Mater.* **2013**, *321–324*, 819–823. [CrossRef]
14. Chen, X.; Wang, L. Cascaded model predictive control of a quadrotor UAV. In Proceedings of the Control Conference, Fremantle, WA, Australia, 4–5 November 2013.
15. Zhao, C.; Bai, Y.; Gong, X.; Xu, D.; Xu, Z. Control system design of a hex-rotor aircraft based on the neural network sliding mode method. *Adv. Mater. Res.* **2004**, *971*, 418–421. [CrossRef]
16. Salazar-Cruz, S.; Lozano, R.; Escareno, J. Stabilization and nonlinear control for a novel trirotor mini-aircraft. *Control Eng. Pract.* **2009**, *17*, 886–894. [CrossRef]
17. Mohamed, K.M.; Lanzon, A. Design and control of novel tri-rotor UAV. In Proceedings of the UKACC International Conference on Control 2012, Cardiff, UK, 3 September 2012. [CrossRef]
18. Chiou, J.S.; Tran, H.K.; Peng, S.T. Attitude control of a single tilt Tri-rotor UAV system: Dynamic modeling and each channel's nonlinear controllers design. *Math. Probl. Eng.* **2013**, *2013*, 275905. [CrossRef] [PubMed]
19. Brossard, J.; Bensoussan, D.; Landry, R. Robustness studies on quadrotor control. In Proceedings of the 2019 International Conference on Unmanned Aircraft Systems Association, ICUAS 2019, Atlanta, GA, USA, 11–14 June 2019; pp. 344–352.
20. Mystkowski, A. An application of mu-synthesis for control of a small air vehicle and simulation results. *J. Vibroengineering* **2012**, *14*, 79–86.
21. Tunik, A.A.; Nadsadnaya, O.I. A flight control system for small unmanned aerial vehicle. *Int. Appl. Mech.* **2018**, *54*, 239–247. [CrossRef]
22. Lei, Y.; Lin, R. Effect of wind disturbance on the aerodynamic performance of coaxial rotors during hovering. *Meas. Control.* **2019**, *52*, 665–674. [CrossRef]
23. Lei, Y.; Wang, J. Aerodynamic Performance of Quadrotor UAV with Non-Planar Rotors. *Appl. Sci.* **2019**, *9*, 2779. [CrossRef]
24. Zhao, C.; Yue, B.; Hun, G.; Cheng, P. Hex-rotor unmanned aerial vehicle controller and its flight experiment under aerodynamic disturbance. *Opt. Precis. Eng.* **2015**, *23*, 1088–1095. [CrossRef]
25. Hrishikeshavan, V. Experimental Investigation of a Shrouded Rotor Micro Air Vehicle in Hover and in Edgewise Gusts. Ph.D. Thesis, University of Maryland, College Park, MD, USA, 2011.
26. Leishman, J.G. *Principles of Helicopter Aerodynamics*; Cambridge University Press: New York, NY, USA, 2000.
27. Bohorquez, F.; Samuel, P.; Sirohi, J.; Pines, D.; Rudd, L.; Perel, R. Design, analysis and hover performance of a rotary wing micro air vehicle. *J. Am. Helicopter Soc.* **2003**, *48*, 80–81. [CrossRef]
28. Lei, Y.; Bai, Y.; Xu, Z.; Gao, Q.; Zhao, C. An experimental investigation on aerodynamic performance of a coaxial rotor system with different rotor spacing and wind speed. *Exp. Therm. Fluid Sci.* **2013**, *44*, 779–785. [CrossRef]

29. Bohorquez, F. Rotor Hover Performance and System Design of an Efficient Coaxial Rotary Wing Micro Air Vehicle. Ph.D. Thesis, University of Maryland, College Park, MD, USA, 2007.

applied sciences

MDPI

Article

Bifurcation Flight Dynamic Analysis of a Strake-Wing Micro Aerial Vehicle

Mirosław Nowakowski [1,*], Krzysztof Sibilski [2,*], Anna Sibilska-Mroziewicz [3] and Andrzej Żyluk [1]

[1] Air Force Institute of Technology, 01-494 Warsaw, Poland; andrzej.zyluk@itwl.pl
[2] Faculty of Power and Aviation Engineering, Institute of Aviation Engineering and Applied Mechanics, Warsaw University of Technology, 00-665 Warsaw, Poland
[3] Faculty of Mechatronics, Warsaw University of Technology, 02-525 Warsaw, Poland; anna.mroziewicz@pw.edu.pl
* Correspondence: mirosław.nowakowski@itwl.pl (M.N.); krzysztof.sibilski@pw.edu.pl (K.S.); Tel.: +48-604-417540 (M.N. & K.S.)

Abstract: Non-linear phenomena are particularly important in -flight dynamics of micro-class unmanned aerial vehicles. Susceptibility to atmospheric turbulence and high manoeuvrability of such aircraft under critical flight conditions cover non-linear aerodynamics and inertia coupling. The theory of dynamical systems provides methodology for studying systems of non-linear ordinary differential equations. The bifurcation theory forms part of this theory and deals with stability changes leading to qualitatively different system responses. These changes are called bifurcations. There is a number of papers, the authors of which applied the bifurcation theory for analysing aircraft flight dynamics. This article analyses the dynamics of critical micro aerial vehicle flight regimes. The flight dynamics under such conditions is highly non-linear, therefore the bifurcation theory can be applied in the course of the analysis. The application of the theory of dynamical systems enabled predicting the nature of micro aerial vehicle motion instability caused by bifurcations and analysing the post-bifurcation microdrone motion. This article presents the application of bifurcation analysis, complemented with time-domain simulations, to understand the open-loop dynamics of strake-wing micro aerial vehicle model by identifying the attractors of the dynamic system that manages upset behaviour. A number of factors have been identified to cause potential critical states, including non-oscillating spirals and oscillatory spins. The analysis shows that these spirals and spins are connected in a one-parameter space and that due to improper operation of the autopilot on the spiral, it is possible to enter the oscillatory spin.

Keywords: nonlinear dynamics of flight; bifurcation theory; micro aerial vehicles; strake-wing micro drones

Citation: Nowakowski, M.; Sibilski, K.; Sibilska-Mroziewicz, A.; Żyluk, A. Bifurcation Flight Dynamic Analysis of a Strake-Wing Micro Aerial Vehicle. *Appl. Sci.* **2021**, *11*, 1524. https://doi.org/10.3390/app11041524

Academic Editor: Sylvain Bertrand
Received: 10 January 2021
Accepted: 2 February 2021
Published: 8 February 2021

1. Introduction

Classic methods of testing dynamic aircraft stability enable analysing their dynamic properties in the framework of minor disturbances of a steady straight flight. However, Micro Aerial Vehicles (MAV) operating in open space, where instantaneous gusts of wind can exceed 10 m/s (which amounts for 25% of their cruising velocity) are exposed to sudden flight parameter changes, the angles of attack and slip, in particular. At strong gusts of wind, the disturbance velocity reaches values comparable to the flight velocity. This issue has been reviewed in, among others, the publications [1–3]. In average weather conditions, the changes of attack angle are sudden, and their amplitude amounts from +60° to −30°. The velocity changes from 20 to 130 km/h, and the altitude from 250 to 25 m [3]. This is why there was a need to analyse the dynamic properties of micro aerial vehicles over the entire operational range, including the range of near-critical and super-critical angles of attack. Due to the high non-linearity of the differential equations describing flight dynamics, well-developed and described methods of modal analysis cannot be applied in this case. There was a need to develop a new method for testing aircraft motion stability, which would enable studying its dynamic properties throughout the entire range of usable angles of attack and slip. In the second half of the 20th century, researchers suggested a continuation method based on the bifurcation theory that enabled analysing dynamic properties of an aircraft, over a wide range of flight states. Articles on this matter, published in the 1970s include: [4,5]. The cited research work concerned the analysis of a fast inertial barrel roll and spin. Numerous studies devoted to the bifurcation analysis of aircraft flight dynamics and aerodynamics appeared in the 1970s, 80s and 90s. These include the AGARD report from 1985 [6], and many papers on the issues of non-stationary aerodynamics and non-linear dynamics of flight in the perspective of the dynamical system theory and the bifurcation theory: [7–14].

The dynamical system theory enabled a global analysis of a system of strongly non-linear ordinary differential equations describing the state of a dynamical system, and in this aspect it is a generalization of the aforementioned dynamic stability analysis involving a steady straight flight of an aircraft. The first step in this method is the assessment of quasi-steady state stability. The quasi-steady state is determined by equating the derivatives of state vector to zero and solving a system of non-linear algebraic equations. The local stability of a dynamical system can be assessed based on the Hartman-Grobman [15] theorem. This stability is determined by the eigenvalues of a locally linearised equations of MAV flight dynamics. In our case, this system is linearized around the equilibrium position, [7–14]. If even one of eigenvalues has a positive real part, then the equilibrium position is unstable. It can be proved that if a linearized equation system is non-singular, then the steady state of a dynamical system is a continuous function of state parameters [15]. It can also be proved that steady states of differential equations describing an aircraft motion are continuous functions of rudder surface deflections [16]. Stability changes occur when at least one eigenvalue of a locally linearized differential equation system of aircraft motion changes its sign. The changes in the steady state stability lead to a qualitatively various system response and are called bifurcations. Stability limits can be determined through searching for eigenvalues with zero real zero part [17,18]. In the bifurcation theory, the stability of quasi-steady states is tested as a function of the so-called bifurcation parameters [15,19,20]. The usual assumption in the course of a flight dynamics bifurcation analysis is that such parameters are the rudder surface deflections within the range of change from the minimum to the maximum value of these angles. It enables obtaining an image of all steady states (or quasi-steady) of aircraft motion. The source literature often refers to this computation process as "*global analysis*" (cf. work [10,12,21–29]). This is why, this type of aircraft flight analysis will be called a "global analysis of equilibrium position stability". Of course, classic states defining the stability of a steady and straight micro aerial vehicle flight are one of the points determined in the course of a global analysis of equilibrium position stability. In this perspective, this analysis is a generalization of the classical aircraft stability analysis.

This article presents the application of bifurcation analysis, complemented with time-domain simulations, to understand the open-loop dynamics of strake-wing micro aerial vehicle model by identifying the attractors of the dynamic system that manages upset behaviour. A number of factors have been identified to cause potential critical states, including non-oscillating spirals and oscillatory spins. The analysis shows that these spirals and spins are connected in a one-parameter space and that due to improper operation of the autopilot on the spiral, it is possible to enter the oscillatory spin.

The Cobra maneuver was also studied from the point of view of the bifurcation theory and the results of the time histories of the flight parameters during this maneuver were presented.

2. Bifurcation and Continuation Analysis

A bifurcation can be defined as a qualitative change in the system dynamics as a parameter is varied. In flight dynamics, a qualitative change is usually understood as a change in the stability of the aircraft. Bifurcation analysis and continuation methods are based on the principles of the dynamical system theory. The basis of the dynamical systems theory are described in many books (it can be mention here: [15–17,20,30,31]). These methods consist in finding and tracking solutions numerically, in a selected range of parameters, in order to generate bifurcation diagrams. Bifurcation diagrams allow to identify qualitative changes in the dynamic response of the system.

2.1. Bifurcation Theory and Bifurcation Types

In our case bifurcation analysis is applied to autonomous dynamical system of general form:

$$\dot{\mathbf{x}} = \mathbf{f}(\mathbf{x}, \boldsymbol{\mu}) \tag{1}$$

where $\mathbf{x} \in \Re^n$ is vector of n state variables, $\boldsymbol{\mu} \in \Re^m$ is vector of m control parameters, $\mathbf{f}$ is nonlinear vector field governing system dynamics. The bifurcation for Equation (1) is every qualitative change of the phase portrait occurring upon the passage of parameter $\boldsymbol{\mu}$ through a certain point $\boldsymbol{\mu}_0$. Point $\boldsymbol{\mu}_0$ is called the bifurcation point for Equation (1). By establishing the set of parameters $\boldsymbol{\mu} = \boldsymbol{\mu}_0$ and selecting the initial conditions $\mathbf{x}(t = 0) = \mathbf{x}_0$, the system of differential Equation (1) can be integrated for the selected initial condition. In this way, one can study the evolution of vector field $\mathbf{x}$ over time. In order to thoroughly study the dynamics of the system, this process can be repeated an infinite number of times starting from different initial conditions and for an innumerable fixed combination of parameter values $\boldsymbol{\mu}$. This task is exhausting and tedious. The major problem in this exercise is the selection of initial conditions, which is a non-trivial task when dealing with systems that are nonlinear. An alternative and more efficient approach to the analysis of nonlinear dynamical systems described by Equation (1) is based on the asymptotic bifurcation and continuation method. The bifurcation and continuation method begins with the calculation of steady states of the equilibrium type of the Equation (1), which comes down to solving a system of nonlinear algebraic equations:

$$\dot{\mathbf{x}} = 0 \rightarrow \mathbf{f}(\mathbf{x}, \boldsymbol{\mu}) = 0 \tag{2}$$

and computing the eigenvalues of the Jacobi matrix:

$$\mathbf{J} = \frac{\partial \mathbf{f}}{\partial \mathbf{x}} \tag{3}$$

in each equilibrium state. The numerical scheme for solving both problems is called the continuation algorithm. It is an algorithm of the predictor-corrector type. A continuation algorithm is used to solve the system of Equation (2) as a function of a single parameter of the system $\mu \in \boldsymbol{\mu}$. In other words, the task comes down to determining the zeros of family $\mathbf{f} : \Re^n \rightarrow \Re^n$, namely, to determine the solutions to stationary, non-linear algebraic Equation (2). The dimension of the $\boldsymbol{\mu}$ parameter vector is called the bifurcation dimension. For a one dimensional case $\boldsymbol{\mu} \equiv \mu \in \Re^1$.

The bifurcation theory concerning non-linear ordinary differential equations deals with a system of first-order differential Equation (1), which is a mathematical model of a dynamical system in an n-dimensional Euclidean space $\Re^n$. If the system (1) has asymptotically stable stationary solutions $\mathbf{x}$=0, then for all solutions $\mathbf{x}(0)$ belonging to this neighbourhood:

- trajectory $\mathbf{x}(t)$ fulfils the condition: $|\mathbf{x}(t)| < \varepsilon$ for $t > 0$;
- $|\mathbf{x}(t)| \rightarrow 0$ for $t \rightarrow \infty$.

Solving the problem involves finding answers to the question of how a parameter change $\mu \in \boldsymbol{\mu}$ locally affects the neighbourhood of point $\mathbf{x} = \mathbf{x}_0$. Due to the fact that for all $\boldsymbol{\mu}$ the Equation (2) is satisfied, and Equation (1) can be expressed as:

$$\dot{\mathbf{x}} = \mathbf{R}\mathbf{x} + \sigma(\mathbf{x}, \boldsymbol{\mu}) \tag{4}$$

where in **R** is a square characteristic matrix (Jacobian matrix) with elements provided by the equation:

$$[\mathbf{R}]_{i,j} = \frac{\partial f_i(\mathbf{x}_0, \boldsymbol{\mu})}{\partial x_j}, \tag{5}$$

and non-linear vector function $\boldsymbol{\sigma}$ fulfils the conditions:

$$\boldsymbol{\sigma}(\mathbf{x}_0, \boldsymbol{\mu}) = 0, \frac{\partial \sigma_i(\mathbf{x}_0, \boldsymbol{\mu})}{\partial x_j} = 0. \tag{6}$$

and finally:

$$\dot{\mathbf{x}} = \mathbf{R}\mathbf{x} \tag{7}$$

The Hartman-Grobman theorem applies within the process of studying the stability of a stationary solutions to equation [15,16,19,20]. It states that if all eigenvalues of a Jacobian matrix **R** of a linearized system (1) lie in the left complex half-plane, i.e.,

$$\mathrm{Re}(\lambda_j) < 0, i = 1, 2, \ldots, n \tag{8}$$

then there is a certain continuous, homomorphic transformation of variables reducing a locally non-linear system of Equation (1) to a linear system. This means that if a stationary solution to a linearized system of equations is asymptotically stable, then also the solution to a non-linear system is stable. The Hartman-Grobman theorem also indicates that every qualitative change in the nature of solutions to a system of non-linear equations describing a dynamical system is indicated by the appearance of zero real parts of the eigenvalues of a linearized system characteristic matrix **R**.

In mathematical terms, bifurcation of equilibrium positions takes place, when a eigenvalue of Jacobi matrix (3), (5) of system (1), estimated in the state of equilibrium, intersects the imaginary axis. Similarly, in the case of an oscillating solution, bifurcation occurs when the Floquet multiplier intersects the unit circle. The results shown in this article discuss five bifurcation types; they all have a codification of one, which means that they are encountered when a single continuation parameter changes.

Basic bifurcation types are [32]:

A. A saddle-node bifurcation, also called a saddle limit point, occurs when the real eigenvalue of a Jacobian matrix (5) estimated in a state of equilibrium, intersects the imaginary axis. There is no equilibrium on one side of the bifurcation point (locally), whereas there are two equilibrium branches on the other (e.g., one stable and one unstable) (Figure 1).

B. Hopf bifurcation occurs, when a complex pair of Jacobian (5) eigenvalues, assessed at equilibrium, intersect the imaginary axis. In this case, the equilibrium changes stability and a periodic orbit is formed, which can be stable or unstable (Figure 1).

C. The limit point or periodic orbit fold bifurcation occur when the real Floquet multiplier intersects the unit circle at 1; as for the states of equilibrium, then there are no periodic orbits on one side of the bifurcation (locally), whereas there are two periodic orbits on the other (Figure 1).

D. A period-doubling bifurcation occurs when the real Floquet multiplier intersects the unit circle at −1. The periodic orbit loses stability when a new period orbit appears with a period (approximately) twice as long (Figure 1).

E. The Neimark-Sacker bifurcation or torus bifurcation appears, when the periodic orbit becomes unstable, namely, when a pair of complex Floquet multipliers intersects the unit circle and an additional oscillation frequency is introduced. The outcome is a torus dynamic, which can be periodic (blocked) or quasi-periodic (Figure 1).

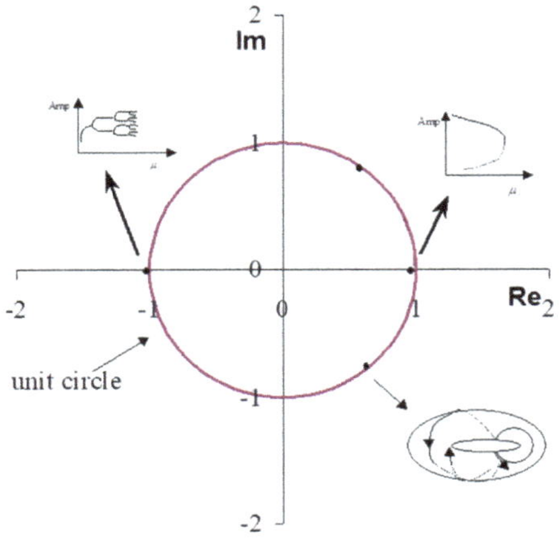

Figure 1. Unit circle—diagram showing basic periodic orbit bifurcations—periodic orbit bifurcations. 1. The actual eigenvalue exceeds +1. Periodic limit points appear in this case. 2. The actual eigenvalue exceeds −1. Period doubling bifurcation occurs in this case. In the proximity of this point, a stable periodic orbit with a period of T becomes unstable and a new stable periodic orbit, with a period of $2T$ appears. This type of stability loss leads to chaotic motions. 3. Two conjugated eigenvalues leave the unit circle. Stable or unstable trajectories surround an unstable bifurcation orbit [32].

To sum up, it can be concluded that:

The bifurcation theory is a set of mathematical results aimed at analysing and explaining unexpected modifications in the asymptotic behaviour of non-linear systems of differential equations, when their parameters slowly change.

Starting from the asymptotic state approximation, for a set value of bifurcation parameters, the computer code, in the course of a continuation process, determines curves for the solution of $\mathbf{x}(\mu)$, which constitute a set of solutions to non-linear algebraic equations (2). In the case of each of the points of the solutions to system (2) and (7), the stability of solutions is determined:

$$\begin{cases} \text{Equilibrium points}: & \mathbf{f}(\mathbf{x},\boldsymbol{\mu}) = 0 \\ \text{Limit points}: & \mathbf{f}(\mathbf{x},\boldsymbol{\mu}) = 0 \\ \text{and at least one of eigenvalues}: & \lambda = 0 \\ \text{Hopf points}: & \mathbf{f}(\mathbf{x},\boldsymbol{\mu}) = 0 \\ \text{and at least } \mathrm{Re}(\lambda_{ij}) = 0 \text{ of pair of complex eigenvluwes}: & \lambda_{ij} = \pm 2i\pi/T \\ \text{Periodic orbits}: & \mathbf{x}(T) = \mathbf{x}(0) + \int_0^T \mathbf{f}(\mathbf{x},\boldsymbol{\mu})dt \end{cases} \quad (9)$$

The continuation process assumes that all functions of the system of Equation (2) are continuous and differentiable.

Summarising, the numerical continuation methods are a set of tools that enables obtaining information required to conduct bifurcation analyses. These methods utilize the predictor-corrector technique for finding, tracking and constructing equilibrium curves or periodic orbits of a differential Equation (1), as solutions of a properly defined system of algebraic Equation (2). Information on the stability is calculated based on analysing the J eigenvalues of Jacobian matrix (5) for equilibrium states, or based on Floquet multipliers (in the case of periodic orbits). Bifurcations can be detected, as well as tracked at various bifurcation parameter values.

2.2. Continuation Software

Numerical methods for solving bifurcation problems appeared relatively recently and complement the analytical achievements in this field. The following fundamental difficulties encountered when applying these methods can be distinguished:

- numerical instabilities associated with calculations in close proximity to bifurcation points,
- issues related to parametrization in close proximity to bifurcation points and limit points,
- structures of bifurcation branches,
- determination whether bifurcation actually takes place,
- problems associated with the convergence of the Newton-Raphson method at singular points.

The system of Equation (2) can have numerous solutions. These can be isolated solutions and they may not exist at all. There is no theory, which could be used as a base to determine which case you are dealing with. Therefore, this issue is quite challenging, because when using numerical methods, the question whether all solutions have already been found still stands. Among the many methods, the ones applied the most when the **f** functions are smooth is the Newton-Raphson method.

At the time, there are several software packages available, which are designed to analyse non-linear dynamical systems, e.g., MatCont [33,34] or KRIT [35]. The most recognizable computer program designed for bifurcation analysis of non-linear dynamical systems is the one developed at the Canadian Concordia University, by a team led by Prof. Eusebius Doedel, the AOTO97 package—program developed in the FORTRAN language and AUTO2000—developed in the C language. The latest (as of 2021) version of these packages is the AUTO07P [36]. The description of subsequent versions of the AUTO package can be found in: [37–39]. AUTO07P was developed in the FORTRAN language, for the UNIX operating system environment. A team at the University of Pittsburgh in America, led by prof. Bard Ermentrout, developed XPPAUT [40], which is a version of the AUTO package compatible with the WINDOWS system. A comprehensive description of this package, together with a manual, can be found in the textbook by prof. Bard Ermentrout [41]. A MATLAB system toolbox—*Dynamical System Toolbox* [42] was also created based on the AUTO [43].

The best-known program designed for the bifurcation analysis of homogeneous ordinary differential equations is that developed by a team lead by Prof. Doedel from the Canadian Concordia University called AUTO. Differential equations describing MAV flight dynamics were analysed using an AUTO version implemented in MATLAB (Dynamical System Toolbox [42]).

3. Micro Aerial Vehicle Mathematical Model

3.1. Reference Frames

Coordinate systems rigidly associated with the object (MAV for example) or associated with the inflow of air streams, and their combinations are usually selected to represent motion equations in aviation. Figure 2 shows coordinate systems, used to derive aircraft motion equations. Details of deriving aircraft motion equations can be found in numerous textbooks and monographs (e.g., [44–49]), which is why will not discuss them in this work. Below you can find micro aerial vehicle motion equations expressed in a so-called-semi constrained coordinate system. This means that centre of mass motion equations are written within a velocity system of coordinates (Figure 2b), while the equations of MAV spherical motion relative to the centre of mass are written within a system related to MAV axes of inertia.

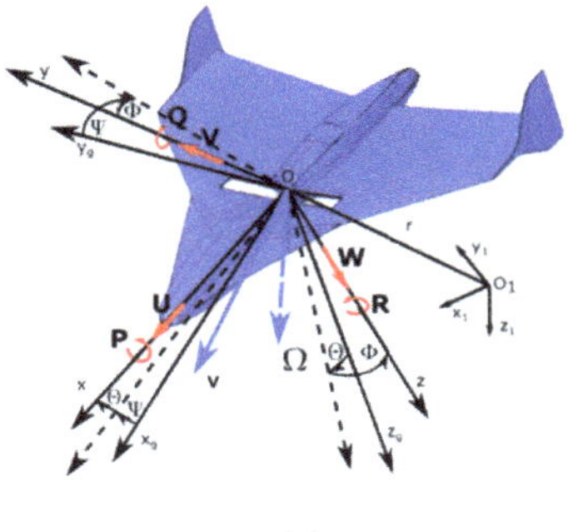

(a)

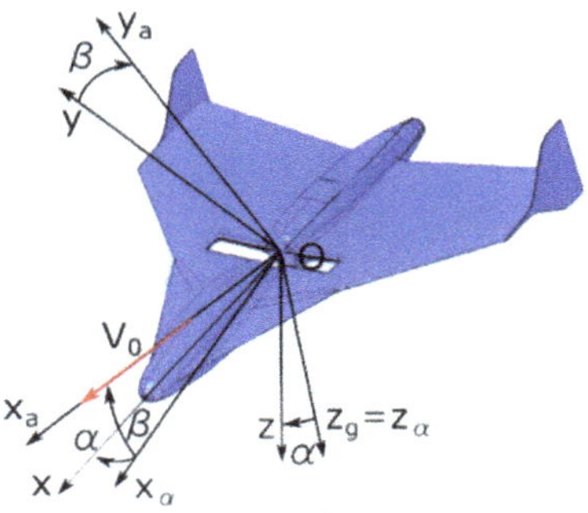

(b)

Figure 2. Coordinate systems with written micro aerial vehicle motion equations; (**a**) systems related to axes of inertia (so-called "aircraft system"); (**b**) coordinate system related to flow (so-called "velocity system" or "wind axis system of co-ordinates").

3.2. Equations of Motion

In the course of implementing a mathematical dynamics model of an aircraft into MATLAB Dynamical System Toolbox, it is more convenient to present aircraft centre of mass motion equations in a velocity coordinate system $Ox_ay_az_a$ (Figure 2) [49]. Furthermore, the equations omitted the aircraft yaw (heading) angle Ψ and the aircraft centre of mass position relative to the system related to the Earth, since these values do not influence the dynamic properties of an aircraft [25]. This enabled the state vector dimension to be reduced to eight components:

$$\mathbf{x} = [V, \alpha, \beta, P, Q, R, \Theta, \Phi]^T \tag{10}$$

where:

V flight velocity,
α angle of attack
β slip angle,
P banking angular velocity,
Q pitching angular velocity,
R yawing angular velocity,
Θ pitch angle,
ϕ bank angle.

In the general case of bifurcation analysis of aircraft flight dynamics, the components of the vector of bifurcation parameters m may be the control surface deflection angles (aileron, elevator, rudder, thrust) and other parameters such as the position of the center of gravity. In this particular case of bifurcation analysis of strake-wing microdrone flight dynamics, the bifurcation parameters vector has following form:

$$\boldsymbol{\mu} = [T, \delta_e, \delta_{elv}]^T \tag{11}$$

where:

T propeller thrust
δ_e angle of symmetrical elevon deflection
δ_{elv} angle of asymmetrical elevon deflection,

The general form of the micro-airplane motion equations takes the form of autonomous differential Equation (1). Components of vector **f** have the form [44–48]:

$$\begin{aligned}
f_1 &= \tfrac{1}{m}[T\cos\alpha - P_{X_a}] - g[\cos\Theta\sin\Phi\sin\beta - (\sin\Theta\cos\alpha - \cos\Theta\cos\Phi\sin\alpha)\cos\beta] \\
f_2 &= Q - (P\cos\alpha + R\sin\alpha)\tan\beta + \\
&\quad - \tfrac{1}{mV_0\cos\beta}[T\sin\alpha + P_{Z_a} - mg(\sin\Theta\sin\alpha + \cos\Theta\cos\Phi\cos\alpha)] \\
f_3 &= P\sin\alpha - R\cos\alpha - \tfrac{1}{mV}[T\cos\alpha - mg(\sin\Theta\cos\alpha - \cos\Theta\cos\Phi\sin\alpha)]\sin\beta + \\
&\quad - mg\cos\Theta\cos\Phi\cos\beta - P_{Y_a}\}
\end{aligned} \tag{12}$$

$$\begin{aligned}
f_4 &= \left(\tfrac{J_X - J_Z}{J_X} - \tfrac{J^2{}_{XZ}}{J_X J_Z}\right)\tfrac{QR}{D} + \left(1 - \tfrac{J_Y - J_X}{J_Z}\right)\tfrac{J_{XZ}PQ}{J_X D} + \\
&+ \tfrac{qSb}{J_X D}\left\{C_{l_0}(\alpha,\beta) + \tfrac{\partial C_l}{\partial \overline{R}}\tfrac{\overline{R}b}{2V_0} + \tfrac{\partial C_l}{\partial \delta_{elv}}\delta_{elv} + \left[\tfrac{\partial C_l}{\partial \overline{P}} + \left(\tfrac{\partial^2 C_l}{\partial \overline{P}\partial\beta}\beta + \tfrac{\partial^2 C_l}{\partial \overline{P}^2}\tfrac{\overline{P}b}{2V_0}\right)\right]\tfrac{\overline{P}b}{2V_0}\right\} + \tfrac{L_T}{J_x D} + \\
&+ \tfrac{J_{XZ}qSb}{J_X J_Z D}\left\{C_{n_0}(\alpha,\beta) + \tfrac{\partial C_n}{\partial \overline{R}}\tfrac{\overline{R}b}{2V_0} + \tfrac{\partial C_n}{\partial \delta_{elv}}\delta_{elv} + \left[\tfrac{\partial C_n}{\partial \overline{P}} + \left(\tfrac{\partial^2 C_n}{\partial \overline{P}\beta}\beta + \tfrac{\partial^2 C_n}{\partial \overline{P}\beta}\tfrac{\overline{P}b}{2V_0}\right)\right]\tfrac{\overline{P}b}{2V}\right\} + \tfrac{J_{xz}N_T}{J_x J_z D} \\
f_5 &= \tfrac{qSc_A}{J_Y}\left\{C_{m_0}(\alpha,\beta) + \tfrac{\partial C_m}{\partial \dot{\alpha}}\tfrac{1}{V_0} + \tfrac{\partial C_m}{\partial \overline{Q}}\tfrac{\overline{Q}c_A}{2V_0} + \tfrac{\partial C_m}{\partial \delta_e}\delta_e\right\} + \tfrac{M_T}{J_Y} + \tfrac{J_Z - J_X}{J_Y}QP + \tfrac{J_{XZ}}{J_Y}(R^2 - P^2) \\
f_6 &= \left(\tfrac{J^2{}_{XZ}}{J_X J_Z} - \tfrac{J_Y - J_X}{J_Z}\right)\tfrac{PQ}{D} + \left(\tfrac{J_Y - J_Z}{J_X} - 1\right)\tfrac{J_{XZ}QR}{J_Z D} + \\
&+ \tfrac{J_{XZ}}{J_X J_Z}\left\{C_{l_0}(\alpha,\beta) + \tfrac{\partial C_l}{\partial \overline{R}}\tfrac{\overline{R}b}{2V_0} + \tfrac{\partial C_l}{\partial \delta_{elv}}\delta_{elv} + \left[\tfrac{\partial C_l}{\partial \overline{P}} + \left(\tfrac{\partial^2 C_l}{\partial \overline{P}\beta}\beta + \tfrac{\partial^2 C_l}{\partial \overline{P}^2}\tfrac{\overline{P}b}{2V_0}\right)\right]\tfrac{\overline{P}b}{2V_0}\right\} + \tfrac{J_{XZ}L_T}{J_X J_Z} + \\
&+ \tfrac{qSb}{J_Z D}\left\{C_{n_0}(\alpha,\beta) + \tfrac{\partial C_n}{\partial \overline{R}}\tfrac{\overline{R}b}{2V_0} + \tfrac{\partial C_n}{\partial \delta_{elv}}\delta_{elv} + \left[\tfrac{\partial C_n}{\partial \overline{P}} + \left(\tfrac{\partial^2 C_n}{\partial \overline{P}\beta}\beta + \tfrac{\partial^2 C_n}{\partial \overline{P}\beta}\tfrac{\overline{P}b}{2V_0}\right)\right]\tfrac{\overline{P}b}{2V}\right\}
\end{aligned} \tag{13}$$

$$\begin{aligned}
f_7 &= P + Q\sin\Phi\tan\Theta + R\cos\Phi\tan\Theta \\
f_8 &= Q\cos\Phi - R\sin\Phi
\end{aligned} \tag{14}$$

where:

$D = 1 - \frac{J^2{}_{XZ}}{J_X J_Z}$,

$q = \frac{1}{2}\rho V_0^2$ is the dynamic pressure,

$\overline{P} = \frac{bP}{2V_0}$, $\overline{Q} = \frac{c_A Q}{2V_0}$, $\overline{R} = \frac{bP}{2V_0}$ are dimensionless angular velocities.

The aerodynamic characteristics of the "Bee" MAV shown in Figure 3 were identified based on aerodynamic water tunnel testing. The static and dynamic aerodynamic loads were measured using a five-component aerodynamic balance. The testing was conducted over a wide range of angles of attack and slip. A wide description of these tests can be found in the works [49,50]. The identified aerodynamic characteristics presented in the work [Sibilski et al.] were shown in the form of graphs and were available in tabular form. Examples of aerodynamic derivative waveforms are shown in Figure 4.

Figure 3. Strake–wing MAV "*Pszczoła*" (Bee) [49,50].

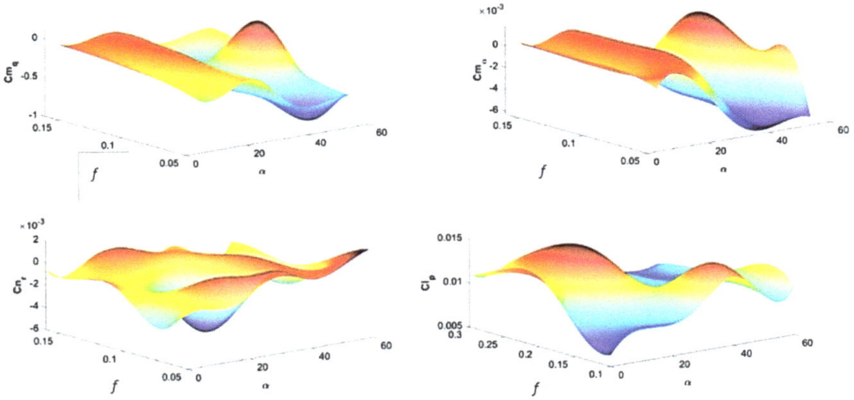

Figure 4. Examples of aerodynamic derivative waveforms as function of angle of attack α, and reduced frequency f [50].

The linear interpolation of aerodynamic data tables commonly used in simulation studies, due to the lack of derivative continuity, cannot be used in the case of continuation tests, since the continuation software of the AUTO type can misidentify bifurcation points. This is why a different method for interpolating aerodynamic data had to be used. Smooth, differentiable state parameter function were created as a result of interpolating aerodynamic characteristics. The *pchip* function interpolation in the MATLAB software was used for this purpose, while maintaining linear interpolation and a multi-variative orthogonal function. Block interpolation was also implemented using a 3rd-order spline function. Owing to the structure of the "Bee", which only has elevons, the bifurcation parameters were the elevon deflection angles: δ_e and δ_{elv}.

4. Methodology of Bifurcation Tests in Aircraft Flight Dynamics

MAV motion is described through a system of highly non-linear ordinary differential equations. For a classical model of a non-deformable micro aerial vehicle with movable control surfaces, motion equations are presented by relationships (1), (12), (13), and (14).

The dynamical system theory enables analysing solutions to a system of highly non-linear ordinary differential equations describing aircraft motion, depending on slow parameter changes (so-called bifurcation parameters). When analysing aircraft flight dynamics, it is assumed that the bifurcation parameters are the control vector components (i.e., deflection angles of the elevator, rudder, ailerons, thrust vector, etc.). The first stage in the analysis of a non-linear system of differential equations in the dynamical system theory is assessing the stability of steady states of a system of differential equations describing aircraft flight dynamics (1), (12), (13), and (14). The steady state is determined by equating the derivatives to zero and solving a system of algebraic Equation (2).

Given the experience from numerous research (based on the bifurcation analysis and continuation technique), the following, three-stage test diagram for a non-linear aircraft motion was formulated [8–14,18,21–28,34,48,49,51–56]

1. The first stage involves determining all parameters of a dynamical system. The fundamental task is to study all possible equilibrium states and periodic orbits, and the analysis of their local stability. This test should be very thorough. The outcome of the attempted test should be a determined global structure of the state space (e.g., phase portraits) of all discovered attractors (steady states and closed orbits). Approximated graphic representations of the calculations are crucial in this case, since they enable diagnosing the obtained results.
2. The second stage involves, based on information on the evolution of phase portraits together with parameter changes, predicting dynamical system behaviour. Next,

based on the knowledge on the type of present bifurcations and the current position of system parameters relative to stable areas, further aircraft behaviour is predicted. Information on the range of parameter changes is also important for these analyses and predictions. Rapid parameter changes and higher differences between steady and transient states are also observed.

3. The third step involves a numerical simulation, which enables verification of the expected aircraft behaviour. Waveforms of transient system characteristics for significant state parameter changes upon a dynamical system parameter change are obtained.

5. Bifurcation Flight Dynamic Analysis of a Micro Aerial Vehicle

Figures 5 and 6 show single-parameter bifurcation diagrams for equilibrium positions: angles of attack α and angular velocity of banking P for different values of the bifurcation parameter, namely, elevator deflection angles δ_e from a range of $\delta_e \in (-35°, 30°)$ for angles of attack equilibrium position, $\delta_e \in (-35°, 40°)$ for banking angular velocity equilibrium positions. Figure 5 shows enlarged bifurcation diagrams for elevator deflections in the range of $\delta_e \in (3°, 10°)$, corresponding to steady states at low and moderate angles of attack for $\alpha \in (0°, 54°)$, the second area corresponds to a range of high angles of attack $\alpha \in (65°, 83°)$. Various types of micro aerial vehicle dynamics, corresponding to seven flight regimes, can be classified in these areas. These regimes are marked with the letters A to G. Regime A means steady symmetrical flight equilibrium states, for angles of attack in the range of $\alpha \in (2°, 23°)$. Area B, for angles of attack in the range of $\alpha \in (15°, 50°)$, corresponds to various motion states with stable and unstable orbits. Area C, for angles of attack in the range of $\alpha \in (35°, 55°)$, corresponds to deep steady spirals. Area D corresponds to inverted spirals. Area E corresponds to steady spins. Areas F and G correspond to transient spins.

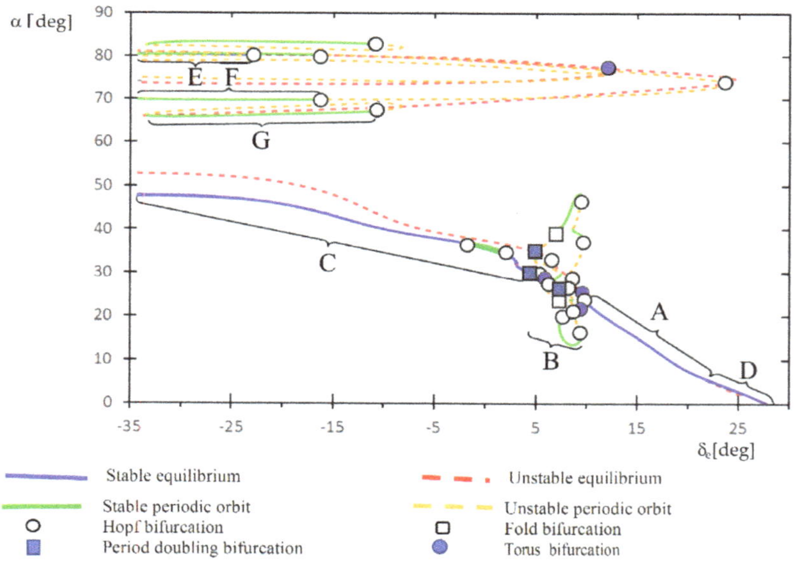

Figure 5. Bifurcation diagram. Quasi-steady states for various elevator deflection angles; $\alpha(\delta_e)$.

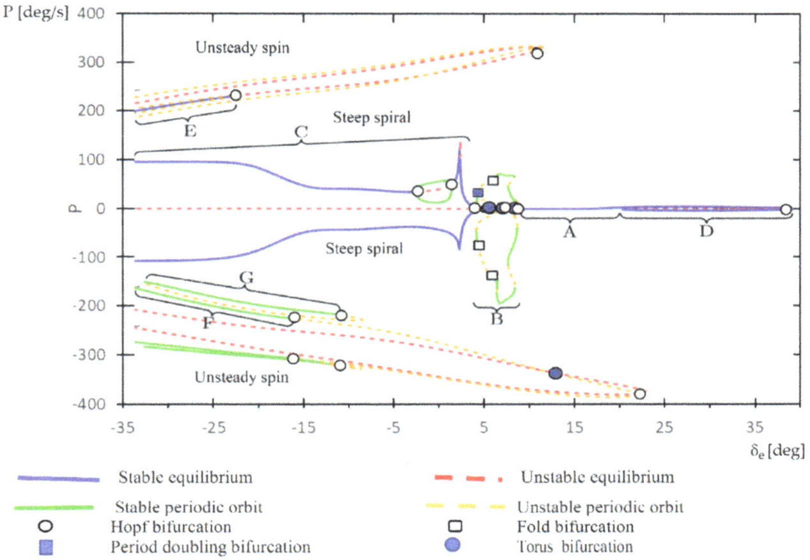

Figure 6. Bifurcation diagram. Quasi-steady states for various elevator deflection angles; $P(\delta_e)$.

The aircraft equilibrium positions in each of these two areas were analysed depending on the bifurcation parameter, namely elevator deflection angles (symmetrical elevon deflections) δ_e and asymmetrical elevon deflections δ_{elv}. During continuation tests, following the methodology described in Section 4, it was assumed that the propeller thrust does not change, and elevon deflections are only symmetrical (or $\delta_{elv} = 0$). Continuation analyses were conducted for positive and negative changes in the bifurcation parameter (symmetrical elevon deflections δ_e), starting with $\delta_e = 20°$. The calculations were conducted for the elevator deflection angle range of change of $-35° \leq \delta_e \leq 40°$, so that it was possible to detect almost all solutions corresponding to the quasi-steady flight states.

At the starting point of the continuation analysis ($\delta_e = 20°$, $\alpha = 3°$), the aircraft was conducting a steady symmetrical flight. As the elevator deflection angle changed, the elevator angle of attack initially increased, and the deflection angular velocity remained at zero. This dynamic regime was denoted with the letter A. Its corresponding range of angles of attack is $2° < \alpha < 25°$ (Figures 5 and 7). The range of steady states denoted with the letter A corresponds to a symmetrical flight of the MAV (angular velocity $P = 0$). When the elevator deflection angle decreases below $\delta_e \approx 10$ and the angle of attack $\alpha \approx 28$ degrees, the micro aerial vehicle enters the dynamics regime denoted with the letter B, with both anti-symmetrical, as well as symmetrical motions. They can be associated with unstable Dutch roll, wing-rock oscillations, spiral motion instabilities, as well as unstable phugoids. Ant-symmetrical oscillations result from Hopf bifurcations in B. In this area, when the elevator deflection angle continues to decrease, reaching a value of $\delta_e \approx 4°$ and $\alpha \approx 34°$, the spiral mode loses stability, which leads to the bifurcation of the stable and almost symmetrical solution. Two asymmetrical equilibrium position branches appear. At higher angles of attack, the spiral model becomes unstable along the almost symmetrical solution branch.

Development of equilibrium position asymmetry can be observed at continued reduction in the elevator deflection angle, and at $\delta_e \approx 4°$, the micro aerial vehicle enters the range of equilibrium states marked with C, which corresponds to spiral motions with a high amplitude of bank angles (Figure 7). Stable position branches corresponding to spiral motions exist on both sides of the straight line defining unstable horizontal flight equilibrium positions. Due to the aerodynamic load asymmetry, the micro aerial vehicle banks to the left wing (with a negative banking angular velocity P), since the equilibrium branch representing the breaking away is connected with A (Figure 6). The equilibrium branch representing rightward MAV banking is practically detached from branch A. The bifurcations therein lead to the appearance of unstable equilibrium positions, where $P \approx 0$ [°/s]. However, it should be noted that the spiral motion direction is determined in practice by the nature of the disturbance or the transient dynamics state of a micro aerial vehicle. As shown in Figures 5 and 6, equilibrium branches of dynamics regime C (within the physical range of elevator deflections), do not converge on the straight line corresponding to the horizontal flight conditions. The MAV will remain at one of two spiral branches only up to a value of $\delta_e = -35°$. Branch C, shown on the bifurcation diagrams (Figures 5 and 6), also represents undesirable steep spirals. Steep spirals can be deemed undesirable flight conditions, therefore, it is important to determine a recovery strategy. Based on a bifurcation diagram (Figure 5), a micro aerial vehicle can be easily recovered from such a flight state by reducing the angle of attack to a level below which no spiral branches are formed. This can be achieved by increasing the elevator deflection angle. The ability to find and determine branches, such as steep spirals, stresses the advantage of continuation and bifurcation analyses over classical linear methods. Although classical linear methods for analysing flight stability can identify stability changes, they are unable to find stable and unstable branches of quasi-steady flight states (Figure 8).

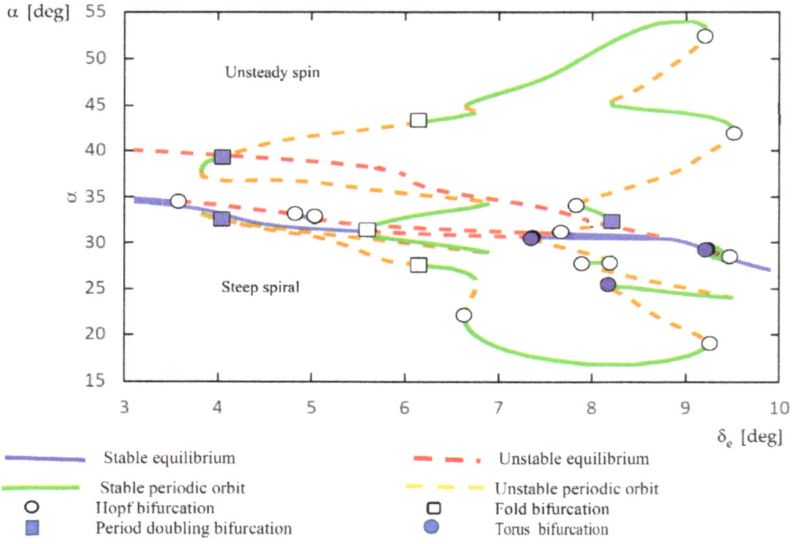

Figure 7. Bifurcation diagram. Quasi-steady states for various elevator deflection angles; $\alpha(\delta_e)$.

The slight lateral instability is also present at low angles of attack ($\alpha \leq 2°$), (branch D in Figures 5 and 6). In the case of the bifurcation analysis and continuation test pattern in question, which assumes constant thrust, flight velocity reaches the highest and not always realistic values at low, negative angles of attack, which entails an increase in the negative angle of MAV pitch, until inverted flight conditions. Thus, although regime D can be deemed an inverted spiral, it is not representative for realistic flight conditions and will not be discussed in more detail (Figure 8).

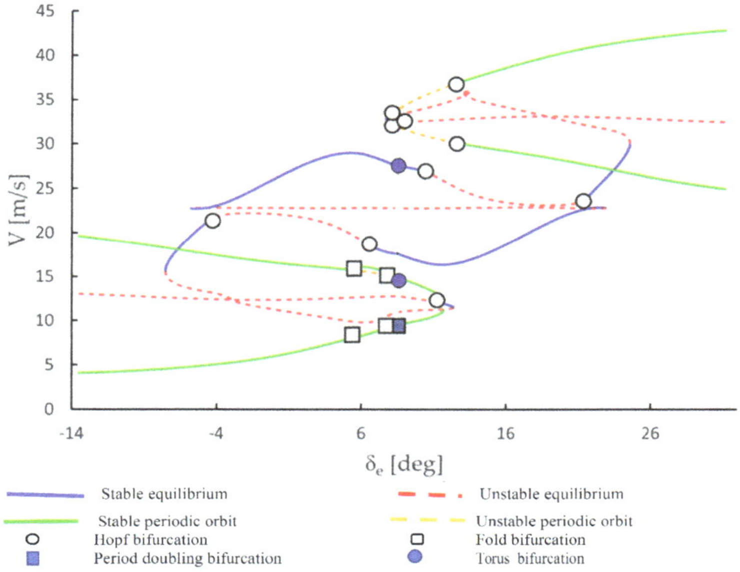

Figure 8. Bifurcation diagram. Quasi-steady states for various elevator deflection angles; V(δ_e).

Equilibrium positions of quasi-steady flight states also appear at high angles of attack 65° < α < 85° (bifurcation diagrams shown in Figures 5 and 6). There are unstable equilibrium position branches for negative deflection angular velocities and angles of attack within a range of 80° < α < 85°, and elevator deflections angles in the range of −35° < δ_e < 10°. There is a certain area of stable equilibrium positions for positive values of deflection angular velocity. The differences arise from the asymmetry of MAV loads. Left- and right-handed spins differ in terms of the state of equilibrium of inertial and aerodynamic forces, which undoubtedly impacts the local stability of spin branches (just as in the case of lower angles of attack α). Stable equilibrium positions, corresponding to spins executed at a constant angular velocity, are represented by branch E of the bifurcation diagrams. It can be seen that there is also a slight area of stable spins near the elevator deflection angle of $\delta_e \approx 12^o$. The waveform of flight parameters indicates that attractors are present within branch E, hence, this equilibrium branch does not play a significant role in the dynamics of disturbed spins [26,27].

For an angle of attack range of 65° < α < 85°, most equilibrium positions are unstable. There can be no steady spins on unstable branches, while one can expect solutions of deterministic chaos nature on these branches [13]. The spin dynamics is significantly impacted by two Hopf bifurcations present at δ_e = 23.4° and δ_e = −13.2° and a period-doubling bifurcation present at δ_e = 12.5°. All three bifurcations appear for angles of attack of α > 65°. In the case of an elevator deflection angle of δ_e = −13.2°, the periodic orbits created through the Hopf bifurcation are unstable throughout the entire range of bifurcation parameters. The period-doubling bifurcation appearing at δ_e = 12.5° can lead to a spin of chaotic nature [13]. The periodic orbit created through Hopf bifurcation at δ_e = 23.4° is more significant. It is initially unstable and then branches into a stable periodic orbit through the torus bifurcation at δ_e = −15.9°. The stable periodic orbit of branch F (Figure 5) is significant, since in this case, the average angle of attack is $\alpha \approx 70°$, with high values of deflection angular velocities. Therefore, it can be concluded that branch F corresponds to steep oscillation spin states. It can also be inferred that MAV oscillation spins occur at angles of attack of approximately 70°. Also, a second stable periodic orbit can be identified on branch G (Figure 6) within the range of high angles of attack. This solution is true for a closed curve that is not connected with other curves. It is the outcome of torus bifurcation present on branch F (for $\delta_e = -15.9°$). Due to torus bifurcations, the periodic orbit becomes unstable, and MAV flight dynamics equation solutions are attracted by branch G periodic orbit. Solutions to micro aerial vehicle motion equations over branch G represent more rapid, oscillation spins, relative to branch F solutions [26,27].

6. Numerical Verification of Predicted MAV Behaviour

According to the bifurcation test methodology, the third step in continuation analysis involves numerical simulations, which enable verifying the predicted aircraft behaviour. Waveforms of transient system characteristics for significant state parameter changes upon a dynamical system parameter change are obtained. Examples of "wing rock" oscillation, unsteady spin and the "Cobra" manoeuvre simulations are shown below.

6.1. "Wing Rock" Oscillation Simulation

A typical phenomenon encountered when flying at large angles of attack is the so-called "wing rock". It is a self-excited phenomenon, which occurs at subcritical and supercritical angles of attack. Despite *"wing-rock"* generally not being dangerous, it is advisable to examine it more thoroughly. Studying the dynamics of this motion was attempted in numerous research work (for example: [10,55–60]. All of the quoted work adopted quasi-static aerodynamic aircraft characteristics or presented different identification methods. However, due to the oscillatory changes in the angle of attack occurring at high, near-critical and supercritical angles of attack, the occurrence of phenomena associated with flow non-stationarity, including hysteresis, is obvious. *"Wing rock"* oscillation simulations were conducted based on aerodynamic derivatives obtained during water tunnel testing and identified through a pulse function.

Examples of simulation results are shown in Figure 9. The motion disturbance involved elevator displacement from 8° to 7.8°. This caused MAV displacement into an area of unstable steady states (Hopf bifurcation). The outcomes were gradually increasing oscillations, turning into a limit cycle, with a period of about 0.2 s.

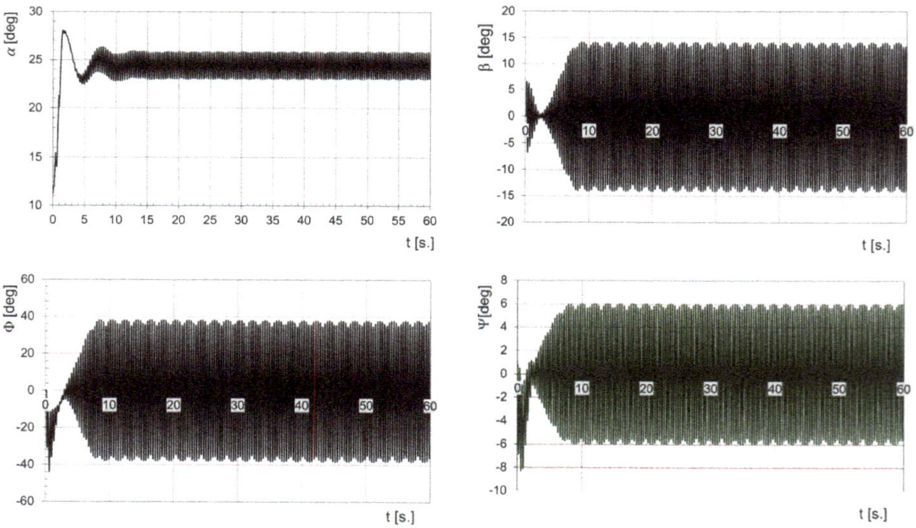

Figure 9. Simulation of *Wing-rock* oscillations; waveforms of selected flight parameters.

Characteristic oscillations appeared as a result of the micro aerial vehicle moving into the area of unstable steady states. They primarily involve rocking of the MAV from wing to wing. That rocking has an amplitude of approximately 39° (for the banking angle), and 6° for the angle of attack and slip. The flight velocity is practically constant. Whereas the period of oscillations corresponding to MAV pitch angle change is equal to the period of phugoid motions. These oscillations are of limit cycle vibration character.

6.2. MAV Spin Simulation

The G and F branches of quasi-steady states (Figures 6 and 7) correspond to states of steady spins. The bifurcation analysis enables "control matching", which allows to recover from critical flight states. A sketch of a "bifurcation control matching" is shown in Figure 10. The equilibrium surface of banking angular velocities P and the bifurcation graph on the deflection plane of the elevator δ_e and elevons δ_{elv} show critical autorotation areas (excluding the spin and a rapid inertia barrel). Recovery trajectories encounter an "apex catastrophe". Skipping through this singularity allows to "smoothly" achieve the desired point of zero banking rate.

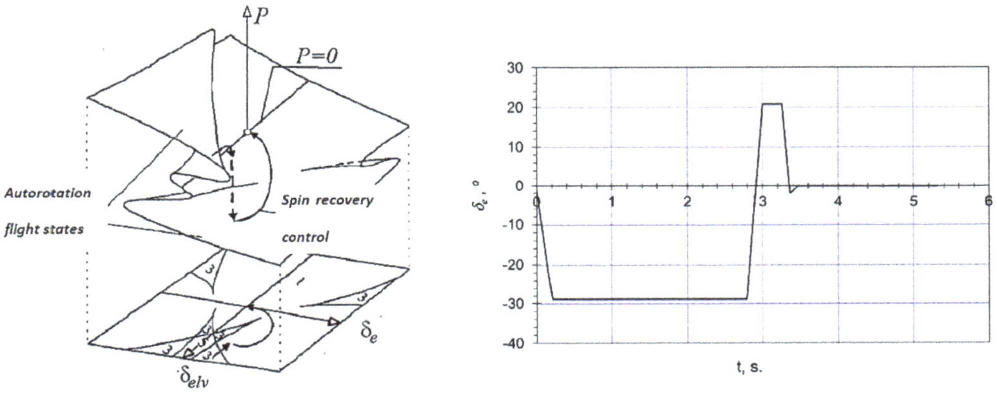

Figure 10. Diagram of bifurcation control matching, which enables recovery from a spin (skipping through the so-called "cusp catastrophe") [25] with the waveform of elevon deflection angle changes.

Figure 11 show the results of a spin recovery simulation. The method of spin recovery was matched using the method shown in Figure 10 and involved displacement of control surface positions beyond the area of unstable equilibrium states and limit cycles. The analysis of simulation results shows the effectiveness of the "bifurcation control matching" method.

6.3. "Cobra" Manoeuvre Simulation

"Cobra" is one of aerobatic figures. The manoeuvre was first executed at the turn of the 1950s and 60s, by the pilots of the Swedish Air Force (*Svenska Flygvapnet*) flying J-35 Draken fighters. It has a Swedish name of "kort parad"—"short parade" and was part of standard short-range manoeuvring combat training for Swedish fighter pilots [48,49]. In the 1980s, the manoeuvre was executed by OKB Sukhoi test pilot, Igor Volk, during Su-27 spin tests. The Cobra was first demonstrated in public by Viktor Pugachev at the Le Bourget air show in 1989, on a MiG-29 fighter. This manoeuvre can be executed with an aircraft exhibiting excellent manoeuvrability and low thrust load or equipped with thrust vector control engines. Besides the spectacular impression it makes at air shows, enabling to demonstrate the manoeuvrability and turning abilities of modern combat aircraft, this manoeuvre, used in direct air combat, is primarily aimed at forcing the chasing foe to overtake through rapid deceleration, thus providing the chased aircraft with a convenient position to open fire. The effect of sudden deceleration is achieved owing to rapidly increasing aircraft drag resulting from vertical positioning of the aircraft body in the upward direction (perpendicular to the previous flight direction).

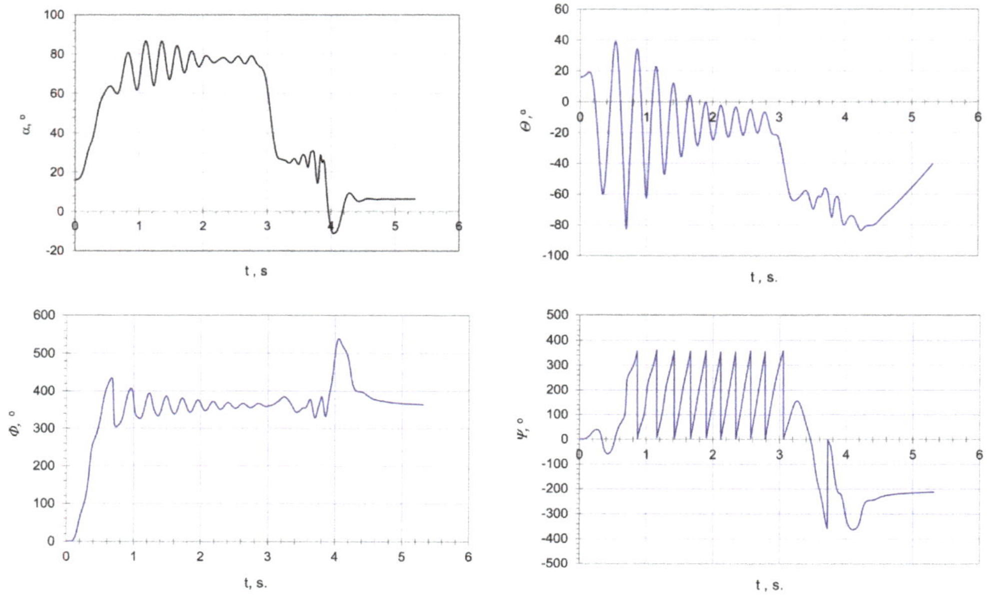

Figure 11. Spin simulation. Angle waveforms for selected flight parameters.

The "Cobra" is conducted at supercritical angles of attack, significantly exceeding the value under normal operating conditions. Figure 12 shows a diagram of the "Cobra". This manoeuvre is divided into three basic stages:

1. Transition from horizontal flight into the phase of increasing the aircraft pitch angle, due to very rapid increase in the elevator deflection angle (sudden pulling of the control stick to a maximum), while simultaneously throttling the engine or engines,
2. manoeuvre phase in which, as a result of such action by the pilot, the aircraft nose rapidly rises up, until it reaches a very high angle of attack (even up to 120°),
3. the exit phase, which involves increasing the thrust and releasing the control stick, leading to the aircraft rapidly increasing its pitch angle, while simultaneously accelerating and returning to horizontal flight, with a minor altitude loss.

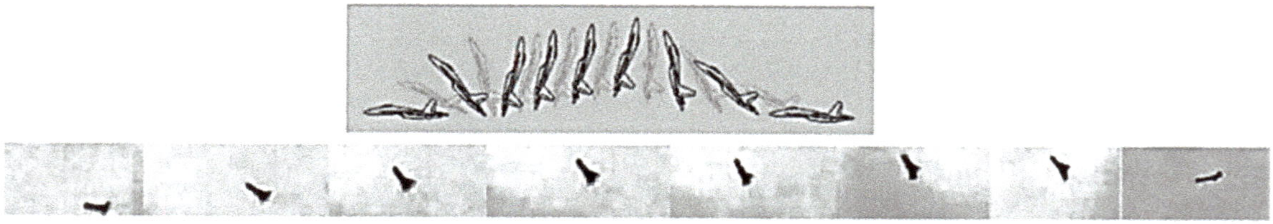

Figure 12. "Cobra" manoeuvre phases (**top**), "Bee" MAV during a "Cobra" manoeuvre, time-lapse photos (**bottom**) [61].

The "Bee" MAV has a stake-wing and its dynamics is similar to the dynamics of modern high-manoeuvring combat aircraft. Based on in-flight tests, it was concluded that it was able to execute a "Cobra" manoeuvre (Figure 12) [61]. The "Cobra" manoeuvre simulations were conducted based on aerodynamic and mass data of the "Bee" micro aerial vehicle. Simple analyses show that executing the "Cobra" manoeuvre will be possible without changing the flight altitude, when the vertical projection of the sum of external forces acting on the MAV in the course of the manoeuvre should be equal to zero. However, due to the fact that the computational thrust force value was initially negative, it was assumed that the MAV starts the manoeuvre with the engine off (zero thrust). It was also assumed that, in the course of executing a manoeuvre, the thrust depends on the difference in the aircraft's angles of pitch Θ and attack α, and the flight velocity V [62,63].

Due to the fact that aircraft equipped with strake wings are characterized by wing-rock instability and are not spirally stable, it was impossible to obtain a fully symmetrical flight parameter waveform. This is associated with the occurrence of Hopf bifurcation and torus creation bifurcation on G, E and F branches (Figures 5 and 6). The "Cobra" manoeuvre is initiated with a sudden downward displacement of elevons ($\delta_e = -35°$). This causes a rapid transition of the MAV through C, G, E and F branches of steady flight states (Figures 5 and 6).

Figures 13 and 14 show the results of a digital simulation of the "Cobra" manoeuvre, taking into account the occurrence of limit cycles. Based on the analysis of these graphs, it can be concluded that all flight parameters are significantly changed during a Cobra figure, increasing their values. Using the terminology of the Dynamical System Theory, it can be said that the "Cobra" is unstable due to the presence of Hopf bifurcation (at an elevator displacement angle of $\delta_e = -23.1°$) and torus bifurcation (for $\delta_e = -16.5°$). Figure 14 shows Poincarè maps for selected state parameters. It can be concluded when the non-stationarity (hysteresis) of aerodynamic coefficients were taken into account, MAV motion equation solution irregularities of quasi-harmonic nature were obtained. The digital simulation took into account the fact that at high angles of attack, the aircraft is spirally unstable (branch C, Figures 5 and 6) and has a tendency to wing-rock oscillations (branch B, Figures 6 and 7). In the course of a manoeuvre, the aircraft attack and pitch angles rapidly increase, reaching maximum values of approximately 84° (for an attack angle of α) and 100° (for a pitch angle of Θ). The MAV bank and yaw angle waveforms (Figure 13) indicate that these angles increase over time. This is associated with the present area of spiral instability (branch C in bifurcation diagrams—Figures 5 and 6). The development of wing-rock oscillations is also visible. The period of these oscillations varies at approx. 0.2 s. It should be noted that the amplitude and frequency of these oscillations are irregular (quasi-harmonic).

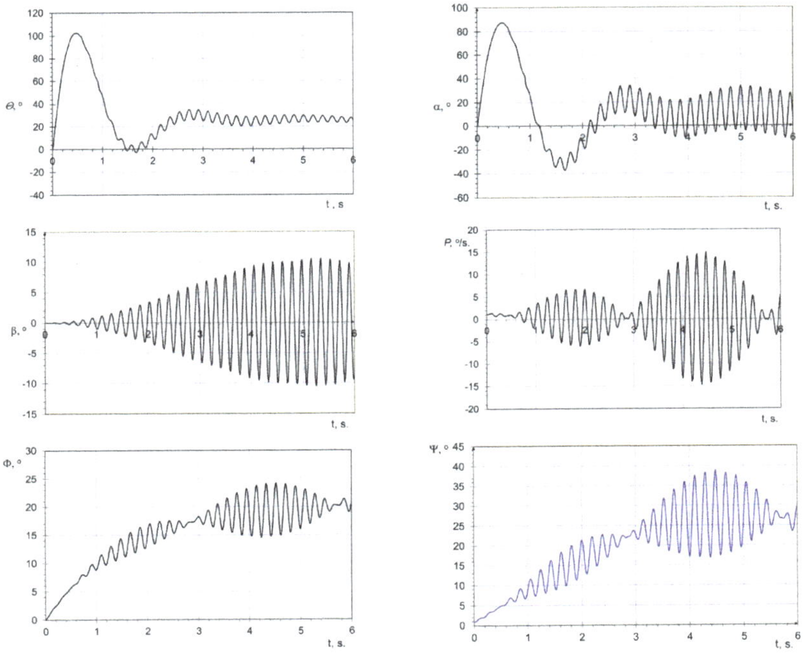

Figure 13. "Cobra" manoeuvre simulation. Time waveforms of selected flight parameters.

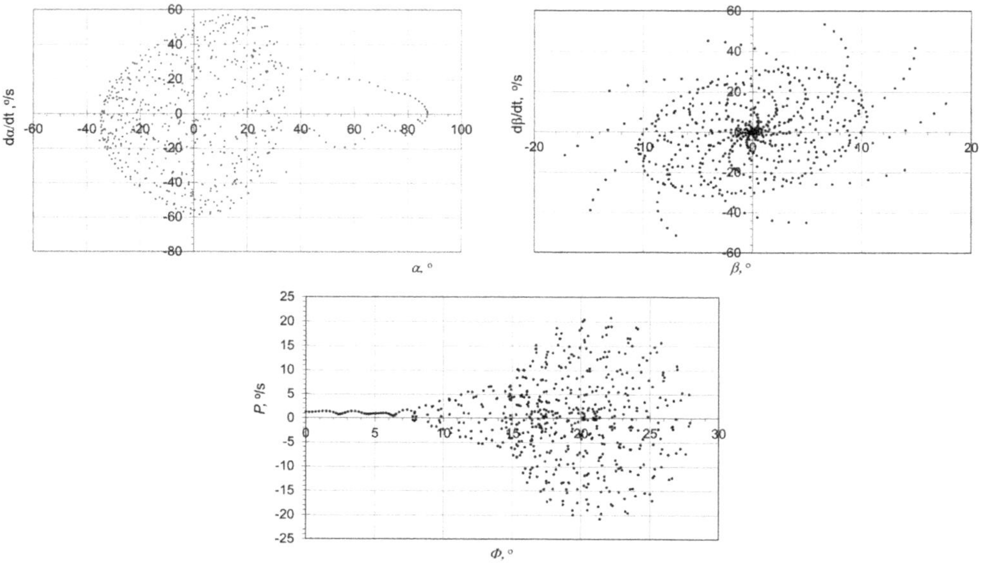

Figure 14. "Cobra" manoeuvre. Poincare maps.

7. Conclusions

The article presents application examples of the theory of dynamical systems relative to studying the flight dynamic specifics of a micro aerial vehicle, constructed as a fixed-wing aircraft with a strake-wing. Such microdrons are characterized by high manoeuvrability and can fly with high, supercritical angles of attack. The bifurcation analysis shown in the article enabled identifying some of the numerous factors impacting the behaviour of a strake-wing MAV. More specifically, this analysis allowed to discover a number of stable attractors, associated with disturbances to the states of equilibrium. Compared with purely simulational studies, which require long-term computations leading to the disappearance of motion history transitions on weak attractors, creating bifurcation diagrams is much less computation-demanding. Furthermore, bifurcation diagrams show the fundamental structure of a dynamical system, hence they suggest, where and when the time domain simulation should be conducted in order to better explain the behaviour of a dynamical system (in this case, a micro aerial vehicle). Time domain simulations were presented as supplementary to bifurcation diagrams, in order to gain more clarity in terms of the nature of various attractor dynamics regimes. It was shown that an MAV was susceptible to steep spiral motion disturbances, induced by loss of stability in this flight range. Bifurcation analysis was also able to identify that elevator displacement aimed at recovery from a steep spiral glide can lead to an oscillating spin. Bifurcation diagram analysis indicates that the correct reaction is to restore elevator position to the value corresponding to balancing conditions in straight flight.

The approach adopted in this article can be extended in several ways. Firstly, MAV behaviour in an open control loop can be further tested through expanding the parameter space in question; this can cover additional control signals, as well as combining the centre of gravity and various damage scenarios, which can be added to the model. Another potentially beneficial extension is the use of bifurcation methods to analyse micro aerial vehicle models expanded with closed control loops, including bifurcation matching that enables recovery from critical flight states.

Author Contributions: Conceptualization, K.S. and A.Ż.; methodology, M.N.; software, A.S.-M.; validation, K.S. and M.N.; formal analysis, K.S., and A.Ż.; writing—original draft preparation, M.N. and K.S.; writing—review and editing, K.S.; visualization, A.S-M.; supervision, M.N., and A.Ż.; project administration, K.S.; funding acquisition, M.N., and A.Ż. All authors have read and agreed to the published version of the manuscript.

Funding: This research was partially supported by Air Force Institute of Technology by subsidy from financial resources for the maintenance and development of research potential, and the statutory funds of the Faculty Mechatronics of the Warsaw University of Technology granted for 2020/2021.

Institutional Review Board Statement: Not applicable.

Informed Consent Statement: Not applicable.

Data Availability Statement: Not applicable.

Conflicts of Interest: The authors declare no conflict of interest.

Abbreviations

C_l	body axis rolling moment coefficient
C_{lp}	rolling moment coefficient derivative with respect to to rolling rate
C_m	body axis pitching moment coefficient
$C_{m\alpha}$	pitching moment coefficient derivative with respect to angle of attack
C_{mq}	pitching moment coefficient derivative with respect to pitching angular rate
C_n	body axis yawing moment coefficient
C_{nr}	yawing moment coefficient derivative with respect to to yawing rate
f	reduced frequency of model oscillation in water tunnel
$\mathbf{f}$	vector of generic nonlinear function
$f_{i,\, i=1,\dots 8}$	components of **f** vector describing microdrone flight dynamics
g	acceleration of gravity
$\mathbf{J}$	Jacobi matrix
J_X, J_Y, J_Z, J_{XZ}	moments of inertia of microdrone
L_T	body axis banking moment due to propulsion
m	mass of microdrone
M_T	body axis pitching moment due to propulsion
N_T	body axis yawing moment due to propulsion
V, V_0	flight velocity
P	body axia roll (banking) rate
P_{Xa}	x wind axis aerodynamic force component
P_{Ya}	y wind axis aerodynamic force component
P_{Za}	z wind axis aerodynamic force component
q	dynamic pressure
Q	body axis pitching rate
R	body axis yawing rate
$\mathbf{R}$	state matrix (Jacobian of linearised aircraft equations of motion)
S	wing area
T	propeller thrust
α	angle of attack
β	slip angle
δ_e	angle of symmetrical elevon deflection
δ_{elv}	angle of asymmetrical elevon deflection
Θ	pitch angle
λ	eigenvalue
$\boldsymbol{\mu}$	vector of bifurcation parameters (in this case microdrone control vector)
μ	single bifurcation parameter
ρ	air density
ϕ	bank (roll) angle
Ψ	yaw (heading) angle
$\dot{()} = \frac{d}{dt}$	time derivative

$\bar{()}$	dimensionless quantity
MAV	micro aerial vehicle, microdrone, micro aircraft

References

1. Abdulrahim, M.; Watkins, S.; Segal, R.; Sheridan, J. Dynamic sensitivity to atmospheric turbulence of Fixed-Wing UAV with varying configuration. *J. Aircr.* **2010**, *47*, 1873–1883. [CrossRef]
2. Wróblewski, W.; Sibilski, K.; Garbowski, M.; Żyluk, A. The gust resistant MAV—Aerodynamic measurements, performance analysis, and flight tests (AIAA2015—1684 CP). In Proceedings of the AIAA SciTech Forum and AIAA Atmospheric Flight Mechanics Conference, Kissimmee, FL, USA, 5–9 January 2015. [CrossRef]
3. Kowalski, M.; Sibilski, K.; Żyluk, A. Studies and tests of micro aerial vehicle during flight. *J. KONES* **2015**, *22*, 155–162. [CrossRef]
4. Adams, W.W. *SPINEQ: A Program for Determining Aircraft Equilibrium Spin Characteristics Including Stability, 1979, NASA TM 78759*; The National Aeronautics and Space Administration: Washington, DC, USA, 1978. Available online: https://ntrs.nasa.gov/api/citations/19790002903/downloads/19790002903.pdf (accessed on 5 January 2021).
5. Schy, A.A.; Hannaht, M.F. Prediction of Jump Phenomena in Roll-Coupled Maneuvers of Airplanes. *J. Aircr.* **1977**, *14*, 375–382. [CrossRef]
6. Roberts, L.; Hamel, P.; Orlik-Ruckeman, K.J. (Eds.) *AGARD CP-386 Unsteady Aerodynamics-Fundamental and Applications to Aircraft Dynamics*; North Atlantic Treaty Organization Advisory Group for Aerospace Research and Development: Neuilly-sur-Seine, France, 1985; ISBN 92-835-0382-1. Available online: https://apps.dtic.mil/dtic/tr/fulltext/u2/a165045.pdf (accessed on 3 January 2021).
7. Tobak, M.; Schiff, L.B. *On the Formulation of the Aerodynamic Characteristics in Aircraft Dynamics, NASA TR-R-456*; The National Aeronautics and Space Administration: Washington, DC, USA, 1976. Available online: https://ntrs.nasa.gov/api/citations/19760007994/downloads/19760007994.pdf (accessed on 5 January 2021).
8. Tobak, M.; Schiff, L.B. *The Role of Time-History Effects in the Formulation of the Aerodynamics of Aircraft Dynamics, NASA TM 78471*; The National Aeronautics and Space Administration: Washington, DC, USA, 1978. Available online: https://ntrs.nasa.gov/citations/19780011113 (accessed on 5 January 2021).
9. Hui, W.; Tobak, M. Bifurcation analysis of nonlinear stability of aircraft at high angles of attack, (AIAA 82-244 CP). In Proceedings of the 20th AIAA Aerospace Sciences Meeting, Orlando, FL, USA, 11–14 January 1982. [CrossRef]
10. Carroll, J.V.; Mehra, R.K. Bifurcation analysis of non-linear aircraft dynamics. *J. Guid. Contro Dyn.* **1982**, *5*, 529–536. [CrossRef]
11. Tobak, M.; Ünal, A. *Bifurcation in Unsteady Aerodynamics, NASA TM 8316*; The National Aeronautics and Space Administration: Washington, DC, USA, 1986. Available online: https://ntrs.nasa.gov/api/citations/19870002264/downloads/19870002264.pdf (accessed on 5 January 2021).
12. Guicheteau, P. Bifurcation theory in flight dynamics and application to a real combat aircraft (ICAS-90-5.10.4 CP). In Proceedings of the 17th ICAS Congress, Stockholm, Sweden, 9–17 September 1990. Available online: https://www.icas.org/ICAS_ARCHIVE/ICAS1990/ICAS-90-5.10.4.pdf (accessed on 5 January 2021).
13. Guicheteau, P. Bifurcation theory: A tool for nonlinear flight dynamics. *Phil. Trans. R. Soc. Lond. A* **1998**, *356*, 2181–2201. [CrossRef]
14. Mehra, R.; Prasanth, R. Bifurcation and limit cycle analysis of nonlinear pilot induced oscillations (AIAA 98-4249 CP). In Proceedings of the 23rd AIAA Atmospheric Flight Mechanics Conference, Boston, MA, USA, 10–12 August 1998. [CrossRef]
15. Wiggins, S. *Introduction to Applied Nonlinear Dynamical Systems and Chaos*, 2nd ed.; Springer: New York, NY, USA, 2003; ISBN 0-387-00177-8. Available online: https://www.springer.com/gp/book/9780387001777 (accessed on 5 January 2021).
16. Tobak, M. *On the Use of the Indicial Function Concept in the Analysis of Unsteady Motions of Wings and Wing-Tail Combinations*; NACA Report 1188; The National Aeronautics and Space Administration: Washington, DC, USA, 1954. Available online: https://digital.library.unt.edu/ark:/67531/metadc65696/ (accessed on 5 January 2021).
17. Kuznetsov, Y.A. *Elements of Applied Bifurcation Theory*; Springer: New York, NY, USA, 1998; ISBN 0-387-98382-1. Available online: https://www.springer.com/gp/book/9780387219066 (accessed on 5 January 2021).
18. Ioos, G.; Joseph, D. *Elementary Stability and Bifurcation Theory*, 2nd ed.; Springer: New York, NY, USA, 2002; ISBN 978-1-4612-7020-2. Available online: https://link.springer.com/book/10.1007/978-1-4612-1140-2 (accessed on 5 January 2021).
19. Keller, H.B.; Langford, W.F. Iterations, perturbations and multiplicities for nonlinear bifurcation problems. *Arch. Ration. Mech. Anal.* **1972**, *48*, 83–108. [CrossRef]
20. Keller, H.B. *Lecture Notes on Numerical Methods in Bifurcation Problems*; Springer: New York, NY, USA, 1987; ISBN 3-540-20228-5.
21. Abed, E.H.; Lee, H. Nonlinear Stabilization of High Angle-of-Attack Flight Dynamics using Bifurcation Control. In Proceedings of the 1990 American Control Conference, San Diego, CA, USA, 23–25 May 1990; pp. 2235–2238. [CrossRef]
22. Avanzini, G.; de Matteis, G. Bifurcation analysis of a highly augmented aircraft model. *J. Guid. Control Dyn.* **1997**, *20*, 754–759. [CrossRef]
23. Charles, G.; Lowenberg, M.; Stoten, D.; Wang, X.; di Bernardo, M. Aircraft Flight Dynamics Analysis and Controller Design Using Bifurcation Tailoring (AIAA-2002-4751 CP). In Proceedings of the AIAA Guidance, Navigation, and Control Conference and Exhibit, Monterey, CA, USA, 5–8 August 2002. [CrossRef]
24. Goman, M.G.; Khramtsovsky, A.V. Application of continuation and bifurcation methods to the design of control systems. *Philos. Trans. R. Soc. Lond. A* **1998**, *356*, 2277–2294. [CrossRef]

25. Goman, M.G.; Zagainov, G.I.; Khramtsovsky, A.V. Application of bifurcation methods to nonlinear flight dynamics problems. *Prog. Aerosp. Sci.* **1997**, *33*, 9–10, 539–586. [CrossRef]
26. Gill, S.J.; Lowenberg, M.H.; Neild, S.A.; Krauskopf, B.; Puyou, G.; Coetzee, E. Upset Dynamics of an Airliner Model: A Nonlinear Bifurcation Analysis. *J. Aircr.* **2013**, *50*, 1832–1842. [CrossRef]
27. Gill, S.J.; Lowenberg, M.H.; Neild, S.A.; Crespo, L.G.; Krauskopf, B.; Puyou, G. Nonlinear Dynamics of Aircraft Controller Characteristics Outside the Standard Flight Envelope, *J. Guid. Control Dyn.* **2015**, *38*, 2301–2308. [CrossRef]
28. Eaton, A.J.; Howcroft, C.; Coetzee, E.B.; Neild, S.A.; Lowenberg, M.H.; Cooper, J.E. Numerical Continuation of Limit Cycle Oscillations and Bifurcations in High-Aspect-Ratio Wings. *Aerospace* **2018**, *5*, 78. [CrossRef]
29. Angiulli, G.; Calcagno, S.; De Carlo, D.; Laganá, F.; Versaci, M. Second-Order Parabolic Equation to Model, Analyze, and Forecast Thermal-Stress Distribution in Aircraft Plate Attack Wing–Fuselage. *Mathematics* **2020**, *8*, 6. [CrossRef]
30. Awrejcewicz, J.; Pyryev, J.; Kudra, G.; Olejnik, P. *Mathematical and Numerical Methods of Bifurcation and Chaotic Dynamics Analysis of Mechanical Systems with Friction and Impact*; Publishing House of the Lodz University of Technology: Łódź, Poland, 2006; ISBN 83-7283-173-4. (In Polish)
31. Guckenheimer, J.; Holmes, P. *Non-Linear Oscillators, Dynamical Systems, and Bifurcations of Vector Fields*; Springer: New York, NY, USA, 2002; ISBN 978-1-4612-7020-1. Available online: https://link.springer.com/book/10.1007/978-1-4612-1140-2 (accessed on 5 January 2021).
32. Magnitskii, N.A.; Sidorov, S.V. *New Methods for Chaotic Dynamics*; World Scientific Series on Nonlinear Science Series A: Singapore, 2006; Volume 58, ISBN 978-981-256-817-5. [CrossRef]
33. Available online: https://sourceforge.net/projects/matcont/files/matcont/ (accessed on 5 January 2021).
34. Dhooge, A.W.; Govaerts, W.; Kuznetsov Yu, A. Matconta: Matlab Package for Numerical Bifurcation Analysis of ODEs. *ACM Trans. Math. Softw.* **2003**, *29*, 141–164. [CrossRef]
35. Goman, M.G.; Khramtsovsky, A.V. Computational framework for investigation of aircraft nonlinear dynamics. *Adv. Eng. Softw.* **2008**, *39*, 167–177. [CrossRef]
36. Available online: http://indy.cs.concordia.ca/auto/ (accessed on 5 January 2021).
37. Doedel, E.J.; Fairgrieve, T.F.; Champneys, A.R.; Sandstede, B.; Kuznetsov, Y.A.; Wang, X. *Auto97: Continuation and Bifurcation Software for Ordinary Differential Equations (with HomCont)*; Technical Report for Concordia University: Montreal, QC, Canada, 1998. Available online: http://citeseerx.ist.psu.edu/viewdoc/summary?doi=10.1.1.44.9955 (accessed on 5 January 2021).
38. Doedel, E.; Keller, H.B.; Kernevez, J.P. Numerical analysis and control of bifurcation problems. *Int. J. Bifurc. Chaos* **1991**, *1*, 493–520. [CrossRef]
39. Doedel, E.J.; Oldeman, B.E. *AUTO-07P: Continuation and Bifurcation Software for Ordinary Differential Equations*; Technical Report for Concordia University: Montreal, QC, Canada, 2009. Available online: https://www.macs.hw.ac.uk/~{}gabriel/auto07/auto.html (accessed on 5 January 2021).
40. Available online: http://www.math.pitt.edu/~{}bard/xpp/xpp.html (accessed on 5 January 2021).
41. Ermentrout, B. *Simulating, Analyzing, and Animating Dynamical Systems. A Guide to XPPAUT for Researchers and Students*; SIAM: Philadephia, PA, USA, 2002; ISBN 978-0-89871-819-5. Available online: https://epubs.siam.org/doi/abs/10.1137/1.9780898718195 (accessed on 5 January 2021).
42. Available online: https://www.mathworks.com/matlabcentral/fileexchange/32210-dynamical-systems-toolbox (accessed on 5 January 2021).
43. Coetzee, E.B.; Krauskopf, B.; Lowenberg, M.H. The Dynamical Systems Toolbox: Integrating AUTO into MATLAB. In Proceedings of the 16th U.S. National Congress of Theoretical and Applied Mechanics, State College, PA, USA, 27 June–2 July 2010.
44. Etkin, B.; Reid, L.D. *Dynamics of Atmospheric Flight*, 3rd ed.; John Willey & Sons Inc.: Hoboken, NJ, USA, 1996; ISBN 0-471-03418-5.
45. Pamadi, B.N. *Performance, Stability, Dynamics, and Control of Airplanes*, 2nd ed.; AIAA: Reston, VA, USA, 2003; ISBN 978-1-62410-274-5. [CrossRef]
46. Schmidt, L. *Introduction to Flight Dynamics*; AIAA, Ed.; Series AIAA: Reston, VA, USA, 1998; ISBN 1-56347-226-0. [CrossRef]
47. Zipfel, P.H. *Modeling and Simulation of Aerospace Vehicle Dynamics*; AIAA, Ed.; Series AIAA: Reston, VA, USA, 2003; ISBN 978-1-62410-250-9. [CrossRef]
48. Sibilski, K. *Modeling and Simulation of Flying Vehicles Dynamics*; MH Publishing House: Warsaw, Poland, 2004; ISBN 83-906620-8-6.
49. Sibilski, K.; Lasek, M.; Sibilska-Mroziewicz, A.; Garbowski, M. *Dynamics of Flight of Fixed Wings Micro Aerial Vehicle*; Warsaw University of Technology Publishing House: Warsaw, Poland, 2020; ISBN 978-83-8156-124-2.
50. Sibilski, K.; Nowakowski, M.; Rykaczewski, D.; Szczepaniak, P.; Żyluk, A.; Sibilska-Mroziewicz, A.; Garbowski, M.; Wróblewski, W. Identification of Fixed-Wing Micro Aerial Vehicle Aerodynamic Derivatives from Dynamic Water Tunnel Tests. *Aerospace* **2020**, *7*, 116. [CrossRef]
51. Abramov, N.; Goman, M.; Khrabrov, A. Aircraft dynamics at high incidence flight with account of unsteady aerodynamic effects. In Proceedings of the AIAA Meeting Papers, AIAA Aymospheric Flight Mrchanics Conference and Exhibit, AIAA 2004-5274 CP, Providence, RI, USA, 16–19 August 2004. [CrossRef]
52. Abramov, A.N.; Goman, M.; Khrabrov, A.; Kolesnikov, E.; Fucke, L.; Soemarwoto, B.; Smaili, H. Pushing Ahead—SUPRA Airplane Model for Upset Recovery. In Proceedings of the AIAA Modeling and Simulation Technologies Conference, (AIAA 2012-4631 CP), Minneapolis, MN, USA, 13–16 August 2012. [CrossRef]

53. Pauck, S.; Jacobus Engelbrecht, J. Bifurcation Analysis of the Generic Transport Model with a view to Upset Recovery. In Proceedings of the AIAA Meeting Papers, AIAA Atmospheric Flight Mechanics Conference, AIAA 2012-4646, Minneapolis, MN, USA, 13–16 August 2012. [CrossRef]
54. Cunis, T.; Condomines, J.-P.; Burlion, L.; la Cour-Harbo, A. Dynamic Stability Analysis of Aircraft Flight in Deep Stall. *J. Aircr.* **2020**, *57*, 143–155. [CrossRef]
55. Jahnke, C.C.; Culick, F.E.C. Application of Bifurcation Theory to the High-Angle-of-Attack Dynamics of the F-14. *J. Aircr.* **1993**, *31*, 26–34. [CrossRef]
56. Jahnke, C.C. On the Roll-Coupling Instabilities of High-Performance Aircraft. *Phil. Trans. R. Soc. Lond. A* **1998**, *356*, 2223–2239. [CrossRef]
57. Dul, F.; Lichota, P.; Rusowicz, A. Generalized Linear Quadratic Control for Full Tracking Problem in Aviation. *Sensors* **2020**, *20*, 2955. [CrossRef] [PubMed]
58. Lichota, P. Multi-Axis Inputs for Identification of a Reconfigurable Fixed-Wing UAV. *Aerospace* **2020**, *7*, 113. [CrossRef]
59. Liebst, B.S. The dynamics, prediction, and control of wing rock in high-performance aircraft. *Phil. Trans. R. Soc. Lond. A* **1998**, *356*, 2257–2276. [CrossRef]
60. Pietrucha, J. Modern Techniques for Active Modification of the Aircraft Dynamic Behaviour. *J. Theor. Appl. Mech.* **2000**, *38*, 132–156. Available online: http://www.ptmts.org.pl/jtam/index.php/jtam/article/view/v38n1p131 (accessed on 5 January 2021).
61. Galiński, C.; Mieloszyk, J. Results of the Gust resistant MAV Programme. In Proceedings of the 28th International Congress of the International Council of the Aeronautical Sciences, Brisbane, Australia, 23–28 September 2012; Paper ICAS 2012-3.1.1. Available online: http://www.icas.org/ICAS_ARCHIVE/ICAS2012/PAPERS/186.PDF (accessed on 5 January 2021).
62. Dżygadło, Z.; Kowaleczko, G.; Sibliski, K. Method of Control of a Straked Wing Aircraft for Cobra Manoeuvres. In Proceedings of the 20th Congress of International Council of Aeronautical Sciences, ICAS'96, Sorrento, Italy, 8–13 September 1996; Paper ICAS-96-3. 7.4. pp. 1566–1573, ISBN 1 56347-219-8. Available online: https://www.icas.org/ICAS_ARCHIVE/ICAS1996/ICAS-96-3.7.4.pdf (accessed on 5 January 2021).
63. Sibilski, K. An Agile Aircraft Non-Linear Dynamics by Continuation Methods and Bifurcation Theory. In Proceedings of the 22nd Congress of International Council of Aeronautical Sciences, ICAS'2000, Harrogate, UK, 27 August–1 September 2000; Paper ICAS-2000-3. 11.2. ISBN 0 9533991 2 5. Available online: https://www.icas.org/ICAS_ARCHIVE/ICAS2000/ABSTRACTS/ICA3112.HTM (accessed on 5 January 2021).

applied sciences

MDPI

Article

Validation of a Wind Tunnel Propeller Dynamometer for Group 2 Unmanned Aircraft

Muwanika Jdiobe *, Kurt Rouser, Ryan Paul and Austin Rouser

School of Mechanical and Aerospace Engineering, Oklahoma State University, Stillwater, OK 74078, USA
* Correspondence: muwanika.jdiobe@okstate.edu

Abstract: This paper presents an approach to validate a wind tunnel propeller dynamometer applicable to Group 2 unmanned aircraft. The intended use of such a dynamometer is to characterize propellers over a relevant range of sizes and operating conditions, under which such propellers are susceptible to low-Reynolds-number effects that can be challenging to experimentally detect in a wind tunnel. Even though uncertainty analysis may inspire confidence in dynamometer data, it is possible that a dynamometer design or experimental arrangement (e.g., configuration and instrumentation) is not able to detect significant propeller characteristics and may even impart artifacts in the results. The validation method proposed here compares analytical results from Blade Element Momentum Theory (BEMT) to experimental data to verify that a dynamometer captures basic propeller physics, as well as self-similar experimental results to verify that a dynamometer is able to resolve differences in propeller diameter and pitch. Two studies were conducted to verify that dynamometer experimental data match the performance predicted by BEMT. The first study considered three propellers with the same 18-inch (0.457 m) diameter and varied pitch from 10 to 14 inches (0.254 to 0.356 m). The second study held pitch constant and varied diameter from 14 to 18 inches (0.356 to 0.457 m). During testing, wind tunnel speeds ranged from 25 ft/s to 50 ft/s (7.62 to 15.24 m/s), and propeller rotational speeds varied from 1500 to 5500 revolutions per minute (RPM). Analytical results from a BEMT code were compared to available experimental data from previous work to show proper application of the code to predict performance. Dynamometer experimental results for thrust coefficient and propeller efficiency were then compared to BEMT results. Experimental results were consistent with the expected effect of varying pitch and diameter and were in close agreement with BEMT predictions, lending confidence that the dynamometer performed as expected and is dependable for future data collection efforts. The method used in this study is recommended for validating wind tunnel propeller dynamometers, especially for Group 2 unmanned aircraft, to ensure reliable performance data.

Keywords: propeller; propulsion; UAS; dynamometer; thrust; pitch; torque; wind tunnel

Citation: Jdiobe, M.; Rouser, K.; Paul, R.; Rouser, A. Validation of a Wind Tunnel Propeller Dynamometer for Group 2 Unmanned Aircraft. *Appl. Sci.* **2022**, *12*, 8908. https://doi.org/10.3390/app12178908

Academic Editor: Yosoon Choi

Received: 28 June 2022
Accepted: 2 September 2022
Published: 5 September 2022

1. Introduction

Unmanned aircraft systems (UAS) continue to prove their utility in the performance of both missions that were once conducted by manned platforms and those that are entirely novel altogether. In the United States, the current regulatory environment permits the commercial operation of unmanned vehicles weighing less than 55 pounds in the National Airspace System (NAS). Services such as pipeline patrol, communication relay, surveillance, and surveying for agricultural and security purposes are offered for hire by businesses utilizing this rule-set. For-profit entities and the public continue to demand services via UAS. The Federal Aviation Administration (FAA), the regulator in the United States, has been responsive to the demand, as evidenced by recent expansion of the existing Part 107 rules to allow for limited operations over people and nighttime flights. The economic value of expanding UAS operations has been recognized by the agency, including the generation of a road map to expand operations, such as beyond visual line-of-sight flights based on the risk on a risk-management framework proposed under the upcoming Modernization of Special Airworthiness Certificates (MOSAIC) rule-set.

As UAS operations expand and businesses are growing to meet customer demand, the need to optimize mission performance becomes paramount to efficiently and profitably provide services. Propulsion system optimization is among the many areas a designer considers. Group 2 UAS are based on medium-sized unmanned aerial vehicles (UAVs), with 21–55 lbs maximum take-off weight and flying lower than a 3500-foot operational ceiling and under 250 knots cruise airspeed, according to the US Department of Defense. Among Group 2 UAS, common power plants include low-cost electric motors and internal combustion engines. These devices convert stored energy to propulsion with a simple fixed-pitch propeller, as weight and cost constraints most often preclude the use of variable-pitch propeller options. The use of fixed-pitch propellers requires a compromise between climb and cruise performance. Thus, accurately understanding propeller performance is an important factor contributing to the operating envelope and mission capability of the vehicle.

Early in the design cycle for a new platform, performance estimates are developed using low-order models to take advantage of the ability to perform rapid design iterations and mission performance evaluations. As the design matures, an increase in fidelity of the estimates is desired. Commonly, models such as Blade Element Momentum Theory (BEMT) are applied in early design stages to predict propeller performance. However, BEMT models are subject to limitations, particularly at low Reynolds numbers and low advance ratios [1]. Researchers have demonstrated the ability to accurately predict propeller performance using computational fluid dynamics (CFD), even in difficult-to-resolve flow conditions [2,3]. Although such examples commendably replicate performance characteristics based on available data, CFD practitioners still require validation of their modeling results. Besides its use as a validation tool, experimental propeller characterization remains a viable option, especially for the UAS community, due to the smaller wind tunnel facility requirements compared to full-scale aircraft propeller testing.

There is an ever-growing body of knowledge from wind tunnel experiments using different configurations to assess propeller performance; however, there does not appear to be a unifying method to ensure that the different configurations are valid. Czyz et al., 2022, studied the aerodynamic performance of propellers with various pitch in a wind tunnel for electric propulsion applications [4]. Podsedkowski et al., 2020, conducted experimental tests of variable pitch propellers for UAVs [5], the study involved a propeller of 16 inches in diameter and various pitch. Podsedkowski et al. designed and built a measuring station that operated similarly to a propeller dynamometer. Avanzini et al., 2020, developed a test bench for measuring propeller aerodynamic performance and electrical parameters; this involved using measurements of thrust, torque, and electric power to validate models used for preliminary designs of UAVs [6]. Islami and Hartono, 2019, developed a small propeller test bench system; this study involved the use of a rig with loads cells to measure thrust and torque for small propellers (10 inches in diameter) [3]. Experimental measurements were compared to results obtained from CFD and BEMT [3]. These studies [3–8] have formed a basis of knowledge useful to Group 2 UAS; however, they do not specifically address a method for validating a wind tunnel dynamometer, which is essential for credible experimental results. There are many potential sources of experimental artifacts that can affect data and yet not be manifest from an uncertainty analysis. For example, the presence of fluid–structure interaction between the propeller, motor, instrumentation and support structure can influence results in a way that does not effect bias (systematic) or precision (random) error. This current paper proposes a novel method to be adopted as common practice for validating such wind tunnel dynamometers.

1.1. Previous Dynamometer Work

There are many existing wind tunnel propeller dynamometers, which can generally be categorized by scale and configuration. Small-scale dynamometers are typically used to evaluate propellers with up to about 10-inch diameter, and include those at the University of Illinois at Urbana–Champaign (UIUC) and Ohio State University ([9–14]). Brandt and Selig ([9,10]) and Deters et al. ([11,12]) noted the effect of low-Reynolds-number operation on such propellers from a wind tunnel propeller dynamometer, and Dantsker et al. ([13]) reported the performance of small folding propellers. McCrink and Gregory ([14]) compared blade element momentum (BEM) modeling results with wind tunnel experimental data for small propellers operating at low Reynolds numbers. Van Trueren et al. ([15]) evaluated small UAS propellers designed for minimum induced drag using a wind tunnel propeller dynamometer at the United States Air Force Academy. Gamble and Arena ([16]) described automatic dynamic propeller testing at low Reynolds numbers and designed a dynamometer. Bellcock and Rouser ([17]) described the design of a wind tunnel propeller dynamometer at Oklahoma State University (OSU) for evaluating a jet-blowing flow controller on small propellers to suppress boundary layer separation. Figure 1 shows the previous OSU wind tunnel propeller dynamometer design described by Bellcock and Rouser to evaluate a modified 10-inch diameter electric propeller. Morris ([18]) presented a method for validating a mobile propeller dynamometer for UAS applications; however, there has not otherwise been previous work on a method to validate a wind tunnel propeller dynamometer for Group 2 UAS applications.

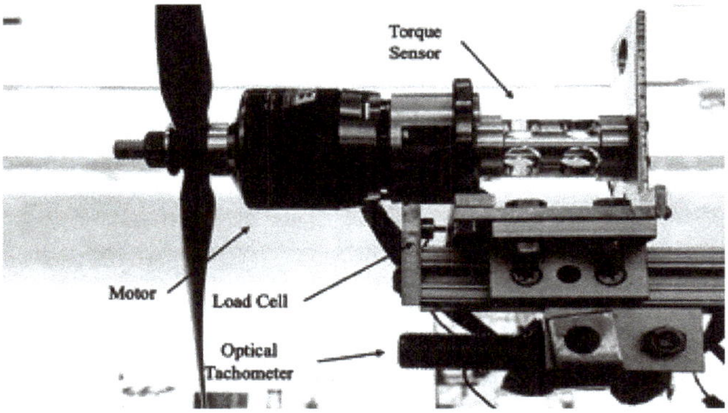

Figure 1. Previous OSU dynamometer for propeller flow control.

Examples of large wind tunnel propeller dynamometers are typically found in government and industry. Boldman et al. ([19]) described a dynamometer used in the United Technologies Technology Research Center: a 10 ft by 15 ft large subsonic wind tunnel used to evaluate an advanced ducted propeller. National Aeronautics and Space Administration (NASA) facilities have been previous described, including a 2000 hp dynamometer at NASA Langley used in a 16 ft, high-speed wind tunnel ([20]), shown in Figure 2, and a 1000 hp dynamometer at NASA Ames used in a 12 ft wind tunnel ([21]), shown in Figure 3. The propeller diameters used in these NASA facilities range from 4 ft to 10 ft and are roughly one half to one third of the test section size. Further, the propellers are located between one half to two diameters ahead of the vertical strut. In order to collect credible propeller performance data, it is important for wind tunnel dynamometers to be designed to reduce fluid–structure interaction between the propeller flow-field and the wind tunnel test section and dynamometer vertical support.

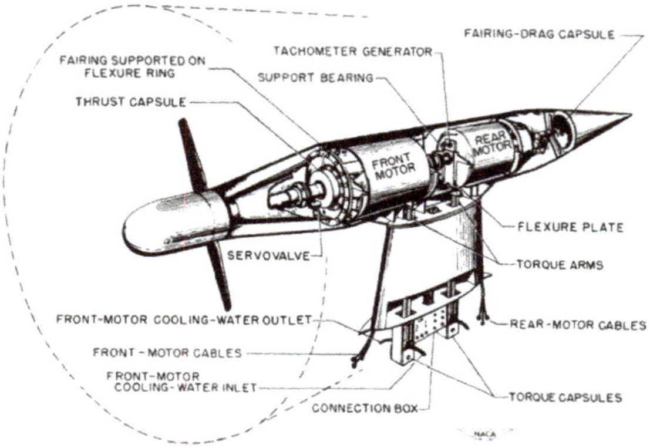

Figure 2. NASA Langley 2000 hp dynamometer schematic [20].

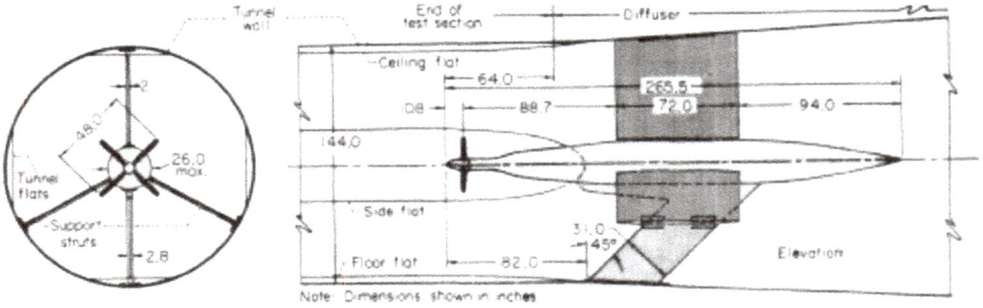

Figure 3. NASA Ames 1000 hp dynamometer schematic [21].

Dynamometer configurations generally can be classified by the means by which they measure thrust and torque. Thrust is typically measured with a linear load cell that is inline or offset from the propeller shaft, or in a moment arm arrangement. The aforementioned OSU dynamometer includes a linear, offset load cell for measuring thrust, which requires accounting for moment created by the offset distance. Alternatively, the NASA dynamometer in Figure 2 includes inline thrust measurement with a pneumatic thrust capsule. Torque is also typically measured inline or by using a moment arm arrangement. Figure 1 shows an example of an inline torque meter integrated into the previous OSU dynamometer, whereas the NASA Langley dynamometer includes torque arms for taking measurements with a moment. The advantages and disadvantages of these measurement approaches are discussed later in the design rationale for our proposed dynamometer.

1.2. Proposed Validation Method

The method includes a comparison of experimental results to BEMT analytical results over a relevant range of test conditions. A validated dynamometer should be able to resolve low-Reynolds-number effects. Furthermore, the method includes comparing experimental results for propellers of at least three different diameters and pitch over the same range of relevant test conditions. A validated dynamometer should distinguish a consistent trend in performance across different diameters and pitch. Finally, the proposed performance figures of merit should at least include thrust coefficient and propeller efficiency, noting that the power coefficient can be derived from those two figures of merit. The motivation for establishing this method is to assist those conducting propeller wind tunnel experiments, especially for Group 2 UAS, to improve the credibility of their results. This, in turn, will improve the confidence of those using propeller wind tunnel data in mission planning and aircraft design.

1.3. Objectives

The wind tunnel propeller dynamometer in this current study is intended to measure the propeller performance of Group 2 UAS. This paper will address the details and rationale for the dynamometer design. The objective of this paper is to present a method to validate the design using BEMT and experimental data from a 3 ft by 3 ft subsonic wind tunnel test section. The study evaluates the performance of three different propeller diameters, ranging from 14 to 18 inches, and three different magnitudes of pitch, ranging from 10 to 14 inches. Tunnel airspeeds range from 25 to 50 ft/s, and shaft speeds range from 1500 to 5500 revolutions per minute (RPM). The propeller dimensions considered in this paper are common and a good representation of Group 2 UAS propellers. However, there is a wide range of propellers in the Group 2 category. The objective of this paper is not to measure or improve propeller performance nor to present or improve dynamometer design (both of these are already well-documented), but rather, it is about a method for validating a propeller dynamometer.

1.4. Propeller Theory

This section provides a brief overview of parameters used to characterize propeller performance, and then presents the methodology for the BEMT code implemented over the course of this research to provide comparison data to contrast with the experimental results to validate our proposed propeller dynamometer.

1.4.1. Performance Characterization

Propellers are characterized by the amount of torque and thrust they produce at a given shaft speed, and by the ratio of the power transferred to the air versus the mechanical power supplied, known as propeller efficiency [10,11,14]. As is typical in aerodynamics applications, the dimensional thrust and power are not typically specified; rather, non-dimensional coefficients are presented to allow the end-user of the data to adapt the results to their application (i.e., operating with a different atmospheric density or at a different velocity). Unlike aircraft wing aerodynamics, which are non-dimensionalized using freestream velocity, propeller performance coefficients are based in the propeller frame of reference, using chord-wise velocity at a given radial location as a function of both freestream and rotational velocities.

Reynolds number is defined as the ratio of momentum force to viscous shear force. For propellers, Reynolds number is based on chord length (c), relative velocity (V_{rel}), air density (ρ), and dynamic viscosity (μ). In order to satisfy the objectives of this research for validating a wind tunnel propeller dynamometer for Group 2 UAS, testing was conducted at low Reynolds numbers.

$$Re = \frac{\rho c V_{rel}}{\mu} \tag{1}$$

Propeller characteristics are typically cataloged as a function of the ratio between freestream and angular velocity to allow for translation to arbitrary operating speeds. This ratio is known as the advance ratio (J), and is shown symbolically in Equation (2), where V is freestream velocity, n is the rotational speed in revolutions per second, and D is propeller diameter.

$$J = \frac{V}{nD} \tag{2}$$

Thrust coefficient, defined as shown in Equation (3), is a non-dimensional quantity that relates thrust produced (T) to the rotational velocity (n) and propeller diameter (D), where ρ is the density of the air the propeller is acting on.

$$C_T = \frac{T}{\rho n^2 D^4} \tag{3}$$

Similarly, power and torque coefficients are non-dimensional quantities that relate power (P) and torque (Q), respectively, to the rotational velocity and propeller diameter, as in Equation (4).

$$C_p = \frac{P}{\rho n^3 D^5} \tag{4}$$

$$C_Q = \frac{Q}{\rho n^2 D^5} \tag{5}$$

Finally, propeller efficiency (η_p) is the ratio of power transferred to the air by the propeller to the mechanical power required to turn the propeller, as shown in Equation (6).

$$\eta_p = \frac{J C_T}{C_p} \tag{6}$$

1.4.2. Blade Element Momentum Theory

Blade Element Momentum Theory (BEMT) is a common methodology for predicting propeller performance in terms of the coefficients defined in Section 1.4.1. BEMT requires only a few inputs. The code implemented for this research is described succinctly by the flowchart presented as Figure 4, and is similar to examples found in [22–25].

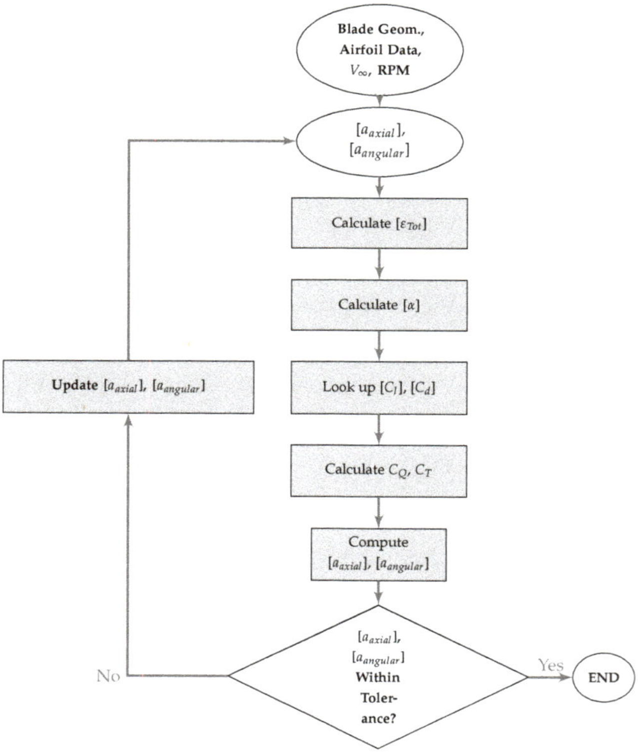

Figure 4. BEMT solution process.

The first step in the BEMT solution process is to discretize propeller geometry for analysis. Input files catalog propeller twist and the local airfoil profile for n radial segments, specified by distance from the hub (r), each of length dr, from the hub to the tip. The measurements describing propeller geometry specification are shown as Figure 5.

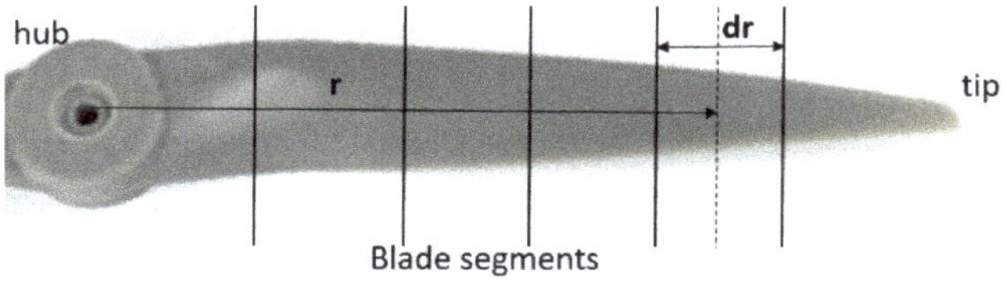

Figure 5. Propeller discretized into blade segments.

In addition to propeller geometry data, the propeller operating condition is input by specifying freestream velocity (V_∞) and RPM for a given run of the BEMT code.

After a run case begins, the code takes on assumed values for the axial and angular inflow factors, a_{axial_i} and $a_{angular_i}$, respectively, for each propeller segment of length dr_i. The initial assumed values for a_{axial_i} and $a_{angular_i}$ are 0.1 and 0.01, respectively. These terms are induction factors describing the axial and angular velocity components, V_{axial_i} and V_{Θ_i}, respectively, within an annular streamtube containing dr_i. Due to the propeller rotation, the fluid within streamtube i acquires the velocity components modeled as

$$V_{axial_i} = [a_{axial_i}]V_\infty$$

and

$$V_{\Theta_i} = [a_{angular_i}]\omega r_i$$

which are accounted for during application of momentum conservation equations.

Subsequently, the total downwash angle is computed for each blade segment. The local flow geometry and definitions for force directions for a blade segment are shown in Figure 6.

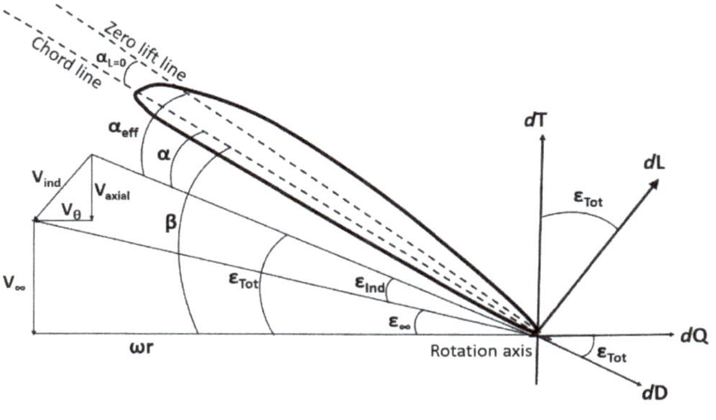

Figure 6. Velocities and force directions on propeller blade segment i.

Given the freestream velocity, rotational velocity at the radial location under consideration, and the induced velocities due to the propeller motion, the total downwash angle at segment i is computed as shown in Equation (7).

$$\varepsilon_{Tot_i} = \tan^{-1}\left(\frac{V_\infty + V_{axial_i}}{\omega r_i - V_{\Theta_i}}\right) \quad (7)$$

Next, with the total downwash angle defined, the local lift and drag coefficients for the airfoil sections can be determined. For each blade segment, the effective angle of attack

α_{eff_i} is the sum of the geometric angle of attack (AOA), α_i and zero-lift AOA, $\alpha_{L=0_i}$. The geometric AOA is defined in Equation (8), where β_i is the geometric pitch angle

$$\alpha_i = \beta_i - \varepsilon_{Tot_i}. \tag{8}$$

Given the effective angle of attack, α_{eff_i}, the C_l and C_d for each section is straightforward to determine from tables of 2-D aerodynamic data. As the APC (Advanced Precision Composites) propellers studied experimentally are predominately made up of National Advisory Committee for Aeronautics (NACA) 4412 airfoils [14], this cross-section was assumed for each propeller segment in the BEMT code. In this work, the 2-D input aerodynamic data are developed from XFOIL [26] analysis at the Reynolds number computed based on the vector sum of the freestream and rotational velocity and chord at the 75% radial location, as is common in propeller aerodynamics [11].

Then, the total thrust and torque the propeller is producing are estimated. For each blade segment, the incremental thrust and torque are shown as Equations (9) and (10), respectively.

$$dT_i = q_i c_i [C_{l_i} \cos(\varepsilon_{Tot_i}) - C_{d_i} \sin(\varepsilon_{Tot_i})] A dr_i \tag{9}$$

$$dQ_i = q_i c_i r_i [C_{l_i} \sin(\varepsilon_{Tot_i}) + C_{d_i} \cos(\varepsilon_{Tot_i})] A dr_i \tag{10}$$

where dynamic pressure at radial location i is defined as shown in Equation (11).

$$q_i = \frac{1}{2}\rho\left[(V_\infty + V_{axial_i})^2 + (\omega r_i - V_{\Theta_i})^2\right] \tag{11}$$

The total thrust and torque produced by the propeller are estimated by integrating the incremental thrust and torque contributions along the blade span, and multiplying by the number of blades (N) on the propeller. The total power of the propeller is obtained by multiplying angular velocity with total torque of the propeller ($P = \omega Q$) [22].

Finally, in order to determine if the conservation of axial and angular momentum is satisfied by the current solution, the induction factors a_{axial_i} and $a_{angular_i}$ are computed for each radial section using Equations (12) and (13) and the incremental thrust and torque found previously using Equations (9) and (10).

$$dT_i = 4\pi r_i \rho V_\infty^2 (1 + [a_{axial_i}])[a_{axial_i}] dr_i \tag{12}$$

$$dQ_i = 4\pi r_i^2 \rho \omega V_\infty (1 + [a_{axial_i}])[a_{angular_i}] dr_i \tag{13}$$

If the induction factors match the values at the beginning of the solution procedure within a user-defined tolerance, outputs are stored for the flow condition under consideration. Otherwise, the induction factors are updated with an average of the newly calculated and initial inflow factor guess, and the solution procedure is repeated until convergence is achieved; the solution is considered converged when the new a_{axial_i} and $a_{angular_i}$ are less than 1×10^{-5}.

2. Materials and Methods

2.1. Propeller Dynamometer Design

The scale of the dynamometer components is dictated by size of the wind tunnel test section (3 ft by 3 ft) and max propeller diameter (18 inch) such that the propeller diameter is half that of the wind tunnel (consistent with dynamometer designs noted in the previous work in Section 1.1). A typical highly loaded APC 18 in propeller is expected to draw about 4 kW of power at 6000 RPM and low airspeeds. Therefore, a 4 kW Magna-Power direct current (DC) power supply is selected.

To avoid overloading the dynamometer motor, a 5 kW Great Planes Rimfire 50 cc electric motor is selected to drive the propeller. The dynamometer drive motor has a max voltage of 55 V, which is higher than the DC power supply's 32 V range, avoiding the potential for the supply to over-volt the drive system. The motor has a 230 kV rating, which limits max shaft speed to 7360 RPM at 32 V, well within the dyno motor limit of about 12,500 RPM.

A Castle Creations Phoenix Edge 160 HV electronic speed controller (ESC) is selected, as its 50 V and 160 A range is greater than the DC power supply output. The ESC is placed outside the dynamometer cowling such that freestream air and propeller wake provide adequate cooling flow. The ESC receives a pulse-width modulation (PWM) throttle signal from a GT Power Professional Digital Servo Tester that is powered by a 7.4 V to 12 V DC input and provides a 4.8 V output. Table 1 includes a summary of the dynamometer electrical and instrumentation components.

Table 1. Dynamometer electrical and instrumentation components.

Component	Manufacturer	Model	Specifications
Drive Motor	Great Plains	Rimfire 50CC	5 kW, 55 V, 230 kV
DC Power Supply	Magna-Power	SL32-125/208 +LXI	4 kW, 32 V, 125 A
Electronic Speed Controller	Castle Creations	Phoenix Edge HV160	50 V, 160 A
Throttle Controller	GT Power	Pro Digital Servo Tester	7.4 V to 12 V DC input; 4.8V output
Hall-Effect Sensor	Honeywell	SS460S	1.5 micro-sec rise–fall
Thrust–Torque Load Cell	Futek	MBA500	50 lb, 50 in-lb; Error 0.25% RO

A Honeywell SS460S Hall-effect sensor is epoxied inside the motor to detect shaft speed. A Futek MBA500 torque and thrust bi-axial load cell is mounted between and inline with the drive motor and dynamometer horizontal support, using custom-designed and 3D-printed cowling components, as shown in Figures 7 and 8. The inline arrangement is an improvement over the previous OSU dynamometer design, minimizing the effect of vibrations that can be experienced with an offset, moment-arm arrangement. The load cell has a 50 lb thrust limit and 50 in-lb torque limit with an error of 0.25% of read-out.

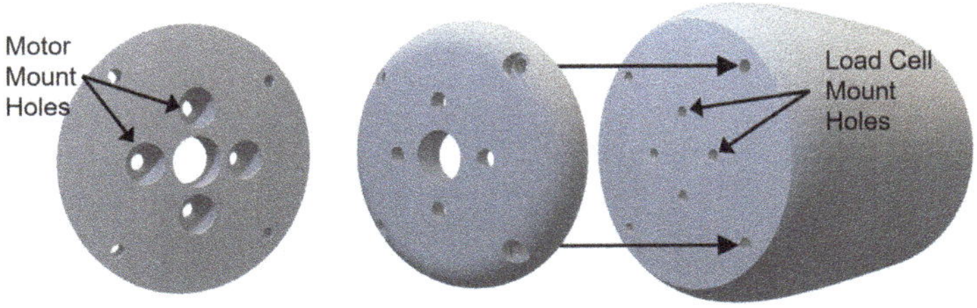

Figure 7. Motor mount backside (**left**) and cowling assembly (**right**).

Figure 8. Dynamometer load cell arrangement (**left**) with cowling (**right**).

The dynamometer support structure is fabricated from 2 in by 2 in quad-rail, t-slot aluminum extrusion. The horizontal support is shrouded in a 3 in diameter polyvinyl chloride (PVC) pipe, as shown in Figure 9. The space between the rail and pipe is filled with sand to damp vibrations induced by fluid–structure interactions. The length of the vertical support is such that the horizontal support is in the center-line of the wind tunnel when mounted to a 2.5 in thick, 6 ft long, 2 ft wide optical breadboard that rests on the bottom of the wind tunnel test section.

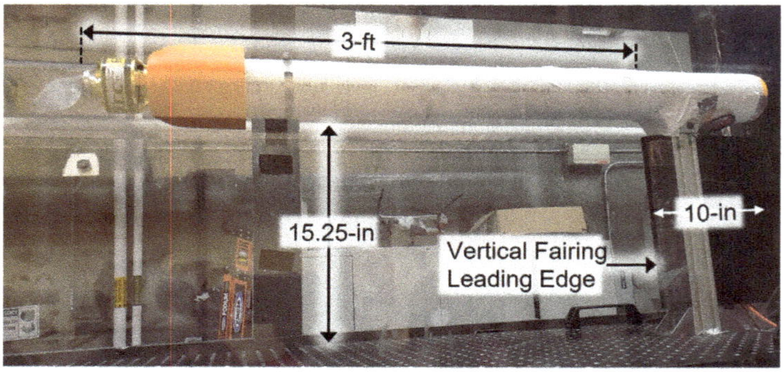

Figure 9. Dynamometer dimensions in the wind tunnel.

The vertical support includes symmetric airfoil fairing pieces that were 3D printed from polylactic acid (PLA) filament and inserted into the quad rail slots. The airfoil leading edge extends 1.5 inches ahead of the quad rail, and the trailing edge extends 6.5 inches behind, such that the total chord length of the vertical support is 10 inches. The distance between the propeller plane of rotation and the leading edge of the vertical support fairing is 36 inches, equal to twice the distance of the maximum 18 in propeller diameter, consistent with that of NASA designs noted in previous work, and also an improvement over the previous OSU design.

2.2. Wind Tunnel and Data Acquisition System

The dynamometer is in operation at Oklahoma State University in the Advanced Technology Research Center (ATRC). The wind tunnel has a 125 hp draw down drive motor. The test section has a 3 ft by 3 ft area. The wind tunnel has a pitot-static probe positioned at the entrance of the test section, 18 in from the bottom of the test section. The pitot-static probe is 3 ft from the propeller rotational plane on the propeller dynamometer, and it is plumbed to an Omega differential pressure transducer with a 0.072 psi range. The pressure transducer is driven by a 24 V, 10 A National Instruments (NI) power supply. The transducer signal passes through a Phoenix Contact interface module that converts the wired signal to a D-SUB port. The signal is then sent into an NI analog input module. This analog input module is attached to an NI 8-slotted chassis that compiles the signals received and transmits the data to a Dell Precision Tower 5810 computer. This computer also uses the same NI chassis for sending signals to drive the wind tunnel fan through an NI analog output module and a corresponding Phoenix Contact D-SUB interface.

The wires from the dynamometer Hall-effect sensor are connected to an Arduino Uno to compute RPM measurements. The Arduino Uno sends this RPM data to the Dell computer through a USB cable. The dynamometer Futek thrust and torque load cell is connected to the Dell computer by two USB connectors corresponding to each measurement, as shown in Figure 10 and Table 2.

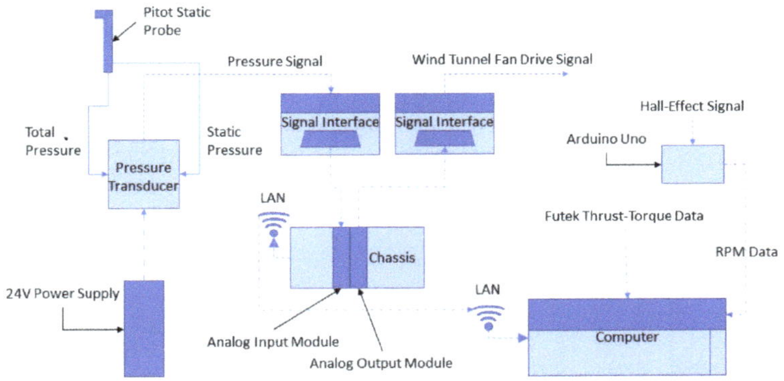

Figure 10. Wind tunnel data acquisition schematic.

Table 2. Wind tunnel data acquisition instruments.

Instruments	Manufacturer	Model	Specifications
Differential Pressure Transducer	Omega Engineering	PX653 -02D5V	Range 0.072 psi, +/−0.3% of full scale
Power Supply	National Instruments	NI PS-16	24 V, 10 A, 240 W
Signal Interface to D-SUB	Phoenix Contact	2281212	37-pole
Signal Management Chassis	National Instruments	NI cDAQ-9188	8 slots for modules
Analog Signal Input Module	National Instruments	NI 9220	16-bit, +/−10 V, 16 channels
Analog Signal Output Module	National Instruments	NI 9264	16-bit, +/−10 V, 4 mA 16 channels
RPM Signal Processor	Arduino	Uno Rev3	Operates at 5 V, Clock Speed 16 MHz
Computer DAQ	Dell	Precision 5810 Tower	64-bit, 32 GB RAM, 3.6 GHz processor

2.3. Experimental Procedures

Experiments in this study obtained data for five APC propellers, as depicted in Table 3. Data include wind tunnel air speed (ranging from 25 ft/s to 50 ft/s); propeller RPM (ranging from 1500 to 5500), thrust, and torque; and power supply voltage and current. Airspeed, RPM, and power supply data were obtained by visually reading measurement displays. Thrust and torque data were recorded using Sensit software. The wind tunnel utilizes a closed-loop speed controller to maintain airspeed at a desired value. The procedure used in this study for obtaining propeller data is as follows:

1. Open Arduino software for displaying propeller RPM; the Arduino measures the RPM at 4 Hz.
2. Open Sensit software to tare instruments and adjust settings for autonomous testing to record thrust and torque.
3. Turn on wind tunnel fan drive motor power and set test section speed to 25 ft/s.
4. Set propeller speed to 1500 RPM using servo tester and Arduino display.
5. Visually read and manually record all displays, averaging five measurements for propeller RPM and power supply voltage and current.

6. Run Sensit software autonomous recorder for 10 s at 100 samples per second.
7. Repeat steps 5 through 8 at propeller speeds ranging from 1500 to 5500 RPM.
8. Repeat steps 5 through 9 for wind tunnel air speeds ranging from 25 to 50 ft/s.

Table 3. Propeller test matrix to study effects of diameter and pitch.

		Pitch	
Diameter	10	12	14
18	x	x	x
16		x	
14		x	

3. Results

The method proposed to validate the dynamometer is to first show proper application of the BEMT code to match existing experimental data for propellers with geometry similar to those used in this study. Then, the BEMT code is used to validate experimental propeller performance from the dynamometer used here. Equation (1) is used to estimate the range of Reynolds number conditions for each propeller tested. Reynolds numbers for the study stay between 28,000 to 94,000 for the 14-inch diameter propeller and between 68,000 to 230,000 for 18-inch propellers. The low-Reynolds-number conditions are associated with low freestream velocity and low angular velocity.

3.1. Manufacturer-Published Propeller Data and BEMT Results

Figure 11, is a plot of BEMT results including thrust coefficient and propeller efficiency versus the advance ratio for an 18 × 12E APC propeller. The plot includes results from blade element momentum theory (BEMT) and APC published data. The published APC data are obtained analytically according to the APC database website, and no further information is provided regarding data methodology. The BEMT results cover a range of propeller speeds from 1500 to 5500 RPM, whereas the APC data range is from 1000 to 6000 RPM. Each line on the plot represents a different RPM. The plot indicates that thrust coefficient decreases as advance ratio increases. Initially, propeller efficiency increases as advance ratio increases, then rapidly decreases for advance ratios greater than 0.63. APC results extend to a maximum advance ratio greater than 0.7, but BEMT results in this study are less than 0.7. Increasing RPM results in both higher thrust coefficient and propeller efficiency. The BEMT propeller efficiency peaks are lower than those from APC and occur at lower advance ratios. Thrust coefficient results from BEMT are lower than those from APC; however, they are similar in slope.

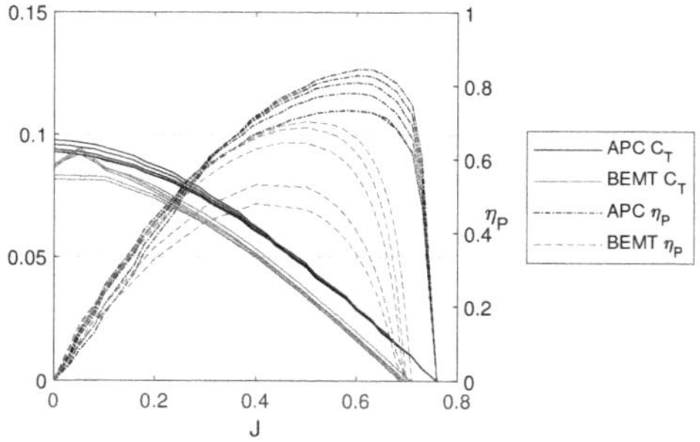

Figure 11. BEMT and APC C_T, η_P of a 18 × 12 APC propeller. Each line represents a constant RPM increasing from bottom to top of the plot.

Figure 12 shows experimental results from UIUC for a 14 × 12E APC propeller compared to those from the BEMT code. The plot for BEMT and UIUC experimental results includes thrust coefficient and propeller efficiency as a function of advance ratio at 3500 RPM [27]. The results are in close agreement in terms of propeller efficiency up to an advance ratio of 0.6. The BEMT code under-predicts thrust coefficient by as much as 15% for advance ratios below 0.3. In general, the BEMT results are more reliable than the APC published performance in the previous figure. Though the BEMT results are only reliable for validating performance over advance ratios of 0.3 to 0.6, they capture the general trends beyond that range, including the slope of the thrust coefficient for advance ratios between 0.6 and 0.8 and the rapid drop in propeller efficiency at an advance ratio of about 0.8. The other main take-away is that the BEMT code used in this study is indeed properly applied, acknowledging that the analytical model is not expected to capture complicated viscous flow effects at high and low advance ratios where the blade experiences very low and high relative angles of attack. Discrepancies may also possibly result from experimental uncertainty and airfoil aerodynamic data that does not capture three-dimensional flow effects.

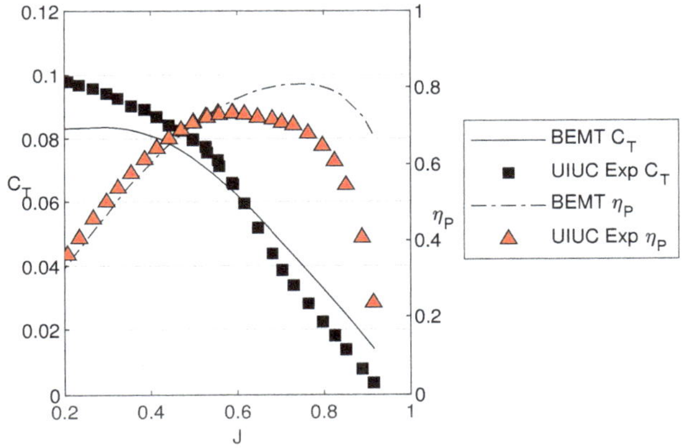

Figure 12. BEMT and UIUC C_T, η_P of a 14 × 12E APC propeller at 3500 RPM [27].

3.2. Experimental Results Compared to Blade Element Momentum Theory (BEMT) Results

Figure 13 shows the experimental and BEMT results for 18 × 10E, 18 × 12E, and 18 × 14 APC propellers. The first two propellers are of a comparable thin, electric type, and the third propeller is classified as a sport propeller. The plotted results include thrust coefficient and propeller efficiency as functions of the advance ratio. All of the plots indicate that the thrust coefficient decreases as the advance ratio increases. Initially, propeller efficiency increases as advance ratio increases, then rapidly decreases at high advance ratios. Observations from Plots A and B in Figure 13 indicate the experimental peak efficiency for propellers 18 × 10E and 18 × 12E occurs at higher advance ratios as pitch increases. The experimental thrust coefficient lines increase with pitch for the 18 × 10E and 18 × 12E propellers, which is expected. The BEMT results are generally consistent with experimental results for advance ratios between 0.3 and 0.6, such that the dynamometer appears to produce valid performance. The BEMT results have less agreement with the 18 × 14 sport propeller than with the thin electric propellers. It also appears that the dynamometer is able to show that performance trends are not consistent across the 18 × 12E thin electric and 18 × 14 sport propellers with increasing pitch.

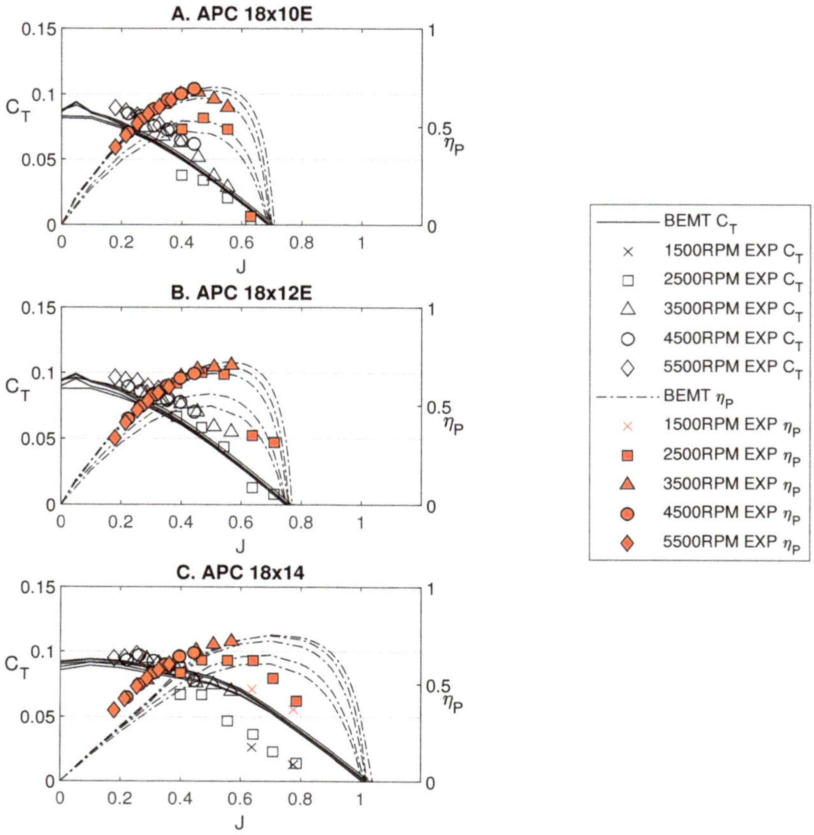

Figure 13. BEMT and experimental C_T and η_P for APC propellers 18 × 10E, 18 × 12E, and 18 × 14.

Figure 14 shows experimental and BEMT results in plots A, B, and C for 18 × 12E, 16 × 12E, and 14 × 12 APC propellers, respectively. The first two propellers are more comparable, both being of a thin electric type. Consistent with aforementioned results for all the propellers, efficiency initially increases as advance ratio increases, then rapidly decreases, and thrust coefficient decreases with increasing advance ratio. Results from plots D and E in Figure 14 indicate peak efficiency occurs at a lower advance ratio with decreasing diameter, which is expected. The slope of the thrust coefficient for both experimental and BEMT decreases as the propeller diameter decreases. The BEMT results are in good agreement with experimental results over advance ratios from 0.3 to 0.6 for the thin electric propellers, but under-predict performance for the 14 × 12 sport propeller. The BEMT code appears to be better for validating dynamometer data from thin electric propellers, and the propeller dynamometer appears to be able to resolve differences between propeller types: thin electric and sport.

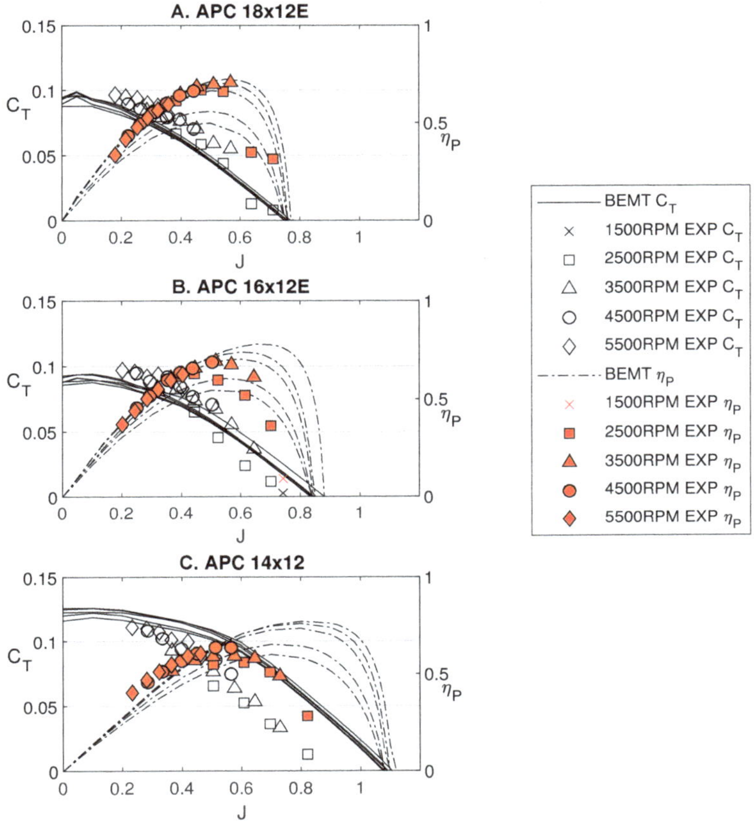

Figure 14. BEMT and experimental C_T and η_P for APC propellers 18 × 12E, 16 × 12, and 14 × 12.

4. Discussion

4.1. Comparison of BEMT and Experimental Results

As indicated in Figure 11, there is a significant difference between the BEMT code and APC results, likely due to different analytical methods and application of airfoil data. The APC published data have been found inconsistent by other studies: Alves, 2014 [28] and Trevor's master thesis, 2009 [29]. Figure 12 furthermore shows that the BEMT code produces more reliable data for validating dynamometer experimental data, which is likely due to the treatment of Reynolds number effects. Increasing the Reynolds number results in increased thrust coefficient and propeller efficiency because an increase in the Reynolds number increases the sectional lift coefficient and decreases the blade drag coefficient [11]. The BEMT analysis captures these effects by incorporating airfoil data, rendering a more conservative prediction for both propeller efficiency and thrust coefficient.

As the advance ratio increases to a magnitude of about 0.6 at either low airspeed or low rotational speeds, the propeller is expected to encounter a sufficiently low Reynolds number that it is susceptible to boundary layer separation. The resulting effect is a sharp decline in propeller efficiency and near zero thrust coefficient as the flow relative to the propeller may render a negative angle of attack. Under this condition, flow is expected to separate around the bottom (pressure side) of the propeller. The experimental data from the dynamometer is consistent with this expectation, having better agreement with the BEMT results than those of APC. Thus, the BEMT code is shown to be accurate and useful to validate dynamometer performance over a range of advance ratios from about 0.3 to 0.6.

4.2. Effect of Pitch and Diameter

Increasing propeller pitch from 10 to 12 inches increases the susceptibility of boundary layer separation. This is apparent in Figure 13, where experimental results for efficiency peak at lower advance ratios with increasing pitch, particularly for thin electric propellers. The dynamometer is able to resolve the difference between thin electric and sport propellers, though the trend in pitch is not comparable across the propeller types. Therefore, the BEMT code appears to be more effective for validating a dynamometer with thin electric propellers, and a reliable dynamometer should be able to indicate a trend in pitch and difference in propeller types.

Likewise, as propeller diameter decreases from 18 to 16 inches, peak propeller efficiency shifts to lower advance ratios due to low operating Reynolds numbers, which is apparent from experimental data in Figure 14. At low Reynolds numbers, the flow is dominated by viscous forces [11], hence increasing the sectional drag coefficient and flow separation; therefore hindering propeller efficiency and thrust coefficient, as manifested in the experimental results. Thus, a reliable dynamometer should also be able to resolve the effects of propeller diameter.

4.3. Uncertainty Analysis

Instrument bias error in this study is summarized in Table 4. The thrust–torque load cell measured the thrust and torque produced by the propeller, the pressure transducer measured the dynamic pressure, and the Hall-effect sensor measured the rotational speed of the propeller. Bias error is sufficiently low in this study to support conclusions.

Table 4. Instrumentation error

Instrument	Bias Error
Thrust–Torque load cell	0.25% RO
Temperature probe	0.05
Pressure transducer	0.3%
Hall-effect sensor	1.5 ms

In this study, an uncertainty analysis was performed on 18 × 10 propeller data. The standard deviation of measurements from the torque and thrust sensors are computed and used to determine precision error for thrust coefficient and propeller efficiency calculations. Figure 15 includes a plot of error for an 18 × 10 APC propeller. The plot of precision error shows how insignificant the error contribution is in the measurement data, as it is indistinguishable from the actual data measurement points. Therefore, the error in experimental data is sufficiently low.

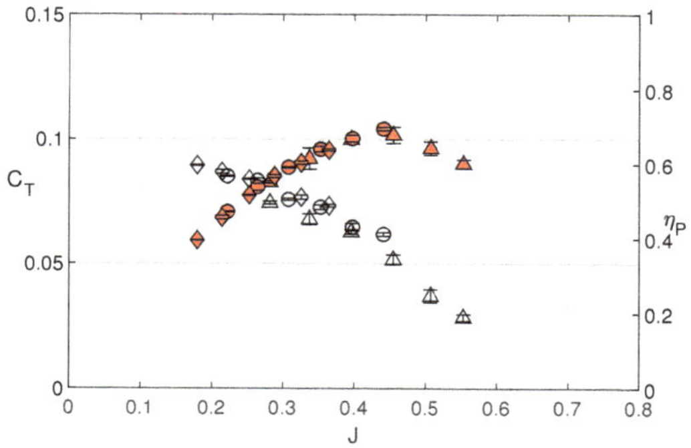

Figure 15. Plot of error on an 18 × 10 APC propeller.

Table 5 includes a breakdown of precision error contributions from the thrust and torque at a rotational speed of 3500 RPM and various wind tunnel speeds from 25 to 50 ft/s. Percent error for thrust and propeller efficiency is calculated using the standard deviation of thrust and propeller efficiency divided by average thrust and propeller efficiency values. The percent error is higher for thrust coefficient than propeller efficiency but does not exceed 7.7%.

Table 5. Thrust and propeller efficiency error at 3500 RPM for an 18 × 10 APC propeller.

Speed (ft/s)	C_T	η_P	C_T Error	η_P Error	% C_T Error	% η_P Error
25	0.0744	0.5523	0.0006	0.0029	0.7812	0.5255
30	0.0678	0.6147	0.0009	0.0055	1.3342	0.8993
35	0.0627	0.6661	0.0025	0.0181	4.0577	2.7155
40	0.0515	0.6777	0.0013	0.0113	2.5679	1.6712
45	0.0370	0.6424	0.0020	0.0219	5.3265	3.4038
50	0.0284	0.6009	0.0022	0.0283	7.6710	4.7056

5. Conclusions

Group 2 unmanned aircraft represent a large and continually growing segment of aerospace operations and businesses that demand optimal mission performance enabled by propulsion systems. It is critical that reliable experimental propeller performance data are available to UAS designers and mission planners, especially when progressing from low-order models to validated, higher-fidelity estimates. Wind tunnel propeller dynamometer designs have been well-documented for a range of propeller sizes, and the principles have been applied to the dynamometer design in this study. However, it is important to have a proper method to validate dynamometer performance, which is particularly challenging for propellers at the low-Reynolds-number operating conditions often associated with Group 2 UAS. Because there is a lack of validated wind tunnel performance data for this particular scale, an approach to validating such wind tunnel propeller dynamometers is presented here. The method includes using BEMT code and experimental results to authenticate a dynamometer.

The proper application of the BEMT code was shown by comparing results to existing propeller data of a smaller scale (14 × 12E), revealing less than 10% difference between BEMT and experimental results over a range of advance ratios from 0.3 to 0.6. The BEMT code was then applied to larger-scale propellers to predict performance with a wind tunnel dynamometer at airspeeds relevant to Group 2 UAS. BEMT and experimental results were in good agreement, particularly for thin electric propellers, up to advance ratios of about 0.6, above which Reynolds number effects become problematic such that BEMT predicted propeller efficiency increases as thrust coefficient approaches zero.

The validation method proposed here also involved experimentally demonstrating expected effects of propeller diameter and pitch. Results showed that a reliable dynamometer should resolve that peak efficiency occurs at higher advance ratios with increasing pitch, showing peak efficiency at about a 25% higher advance ratio when increasing pitch from an 18 × 10E to 18 × 12E propeller. This effect was particularly noticeable for thin electric propellers. Peak efficiency also shifts to lower advance ratios as propeller diameter decreases. Peak efficiency occured at a 10% lower advance ratio from an 18 × 12E to 16 × 12E propeller. Furthermore, a dynamometer should be able to resolve differences in propeller type, as shown by results for thin electric and sport propellers, particularly apparent when comparing 18 × 12E thin electric propeller results to those of an 18 × 14 sport propeller.

Use of the method presented here is recommended for validating wind tunnel propeller dynamometers for Group 2 UAS. It is important to apply it to advance ratios between about 0.3 and 0.6 to ensure reliable propeller performance data. A validated dynamometer should produce thrust coefficient and propeller efficiency results within 10% of the results from BEMT analysis. Furthermore, a validated dynamometer should be able to resolve performance effects associated with varying propeller diameter and pitch, as well as propeller type. Future work related to this study is recommended to show the effects of novel flow control methods to mitigate degraded propeller performance due to low-Reynolds-number operating conditions. Results from such a study will be enabled with a validated propeller dynamometer.

Author Contributions: M.J.—Primary author, developer of BEMT code, experimental data post-processing. K.R.—Developed propeller dyno, writing and editing. R.P.—Writing and editing, assisting with BEMT code. A.R.—Assisting with experimental data collection. All authors have read and agreed to the published version of the manuscript.

Funding: The work described in this article received no external funding.

Data Availability Statement: Tabular data are included as Appendixes A and B.

Acknowledgments: The authors acknowledge the effort expended by Brian Pizana on integration of the LabView data display system. Additionally, David Kelley's load cell shroud design, build, and installation into the test facility is gratefully acknowledged. Finally, Thomas Rannock's assistance mounting the hall-effect sensor to collect RPM data is acknowledged.

Conflicts of Interest: The authors declare no conflict of interest.

Appendix A. Propeller Modelling

The BEM code requires inputs of (1) propeller geometry and (2) 2-D sectional aerodynamic characteristics along the span. The data used in the BEM analysis in this paper are presented in this Appendix.

Appendix A.1. Propeller Geometry

The blade element model requires a geometric description of the propeller geometry to specify the twist distribution and airfoil profile along the length of the blade. Beta is the measured geometric pitch angle between the chord line and fixed plane of rotation. The tables below capture the inputs to the BEM code used to generate the theoretical data in Figures 11–14. As documented in the narrative, since the vast majority of the propeller blade was reported to feature the NACA 4412 cross-section, the BEM results for all radial stations used airfoil data from this profile at the appropriate Reynolds number.

Table A1. APC 14 × 12 propeller geometry.

r/R	c/R	Beta
0.08	0.134	33.34
0.15	0.136	43.37
0.23	0.147	50.88
0.30	0.146	46.08
0.37	0.148	40.23
0.44	0.153	34.88
0.51	0.157	31.33
0.58	0.157	28.22
0.65	0.154	25.52
0.73	0.147	23.64
0.80	0.132	21.06
0.87	0.110	18.89
0.94	0.076	16.25

Table A2. APC 16 × 12 propeller geometry.

r/R	c/R	Beta
0.06	0.105	25.96
0.12	0.102	33.08
0.19	0.118	49.51
0.25	0.136	47.22
0.31	0.151	40.52
0.37	0.161	34.33
0.44	0.163	30.69
0.50	0.161	27.12
0.56	0.153	23.56
0.62	0.142	20.87
0.69	0.130	19.29
0.75	0.113	18.19
0.81	0.098	16.78
0.87	0.084	15.80
0.94	0.071	14.66
1.00	0.056	13.21

Table A3. APC 18 × 10 propeller geometry.

r/R	c/R	Beta
0.07	0.118	17.45
0.12	0.113	23.32
0.18	0.122	39.95
0.24	0.138	39.84
0.29	0.150	34.43
0.35	0.158	30.05
0.40	0.162	26.22
0.46	0.161	22.68
0.51	0.155	19.17
0.57	0.146	18.71
0.62	0.136	15.91
0.68	0.121	15.76
0.74	0.107	14.77
0.79	0.092	14.12
0.85	0.076	13.54
0.90	0.064	13.29
0.96	0.036	11.60

Table A4. APC 18 × 12 propeller geometry.

r/R	c/R	Beta
0.07	0.116	20.10
0.12	0.108	26.44
0.18	0.119	41.44
0.24	0.135	45.10
0.29	0.146	38.65
0.35	0.158	33.53
0.40	0.162	27.55
0.46	0.162	25.11
0.51	0.157	23.11
0.57	0.150	20.21
0.62	0.139	19.13
0.68	0.128	17.51
0.74	0.114	15.91
0.79	0.098	14.50
0.85	0.085	13.07
0.90	0.072	12.63
0.96	0.060	12.43

Table A5. APC 18 × 14 propeller geometry.

r/R	c/R	Beta
0.10	0.166	25.85
0.15	0.160	31.05
0.21	0.164	36.02
0.26	0.168	41.86
0.32	0.161	39.68
0.37	0.154	36.01
0.43	0.145	33.35
0.49	0.137	31.61
0.54	0.127	29.28
0.60	0.117	27.61
0.65	0.105	25.28
0.71	0.094	23.91
0.76	0.081	21.55
0.82	0.071	19.34
0.87	0.058	19.07
0.93	0.045	16.94

Appendix A.2. 2-D Sectional Aerodynamic Characteristics for APC Propeller Airfoils

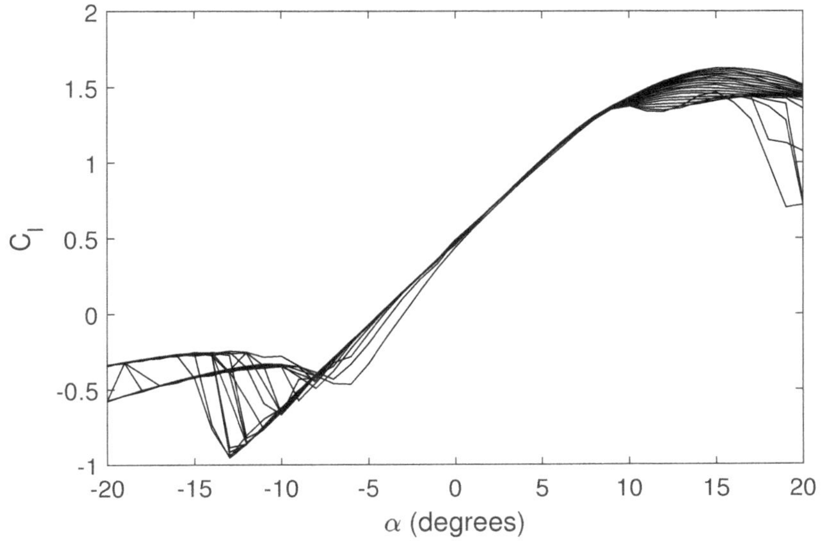

Figure A1. Lift coefficient from XFOIL at different Reynolds numbers ranging from 2×10^4 to 1×10^6.

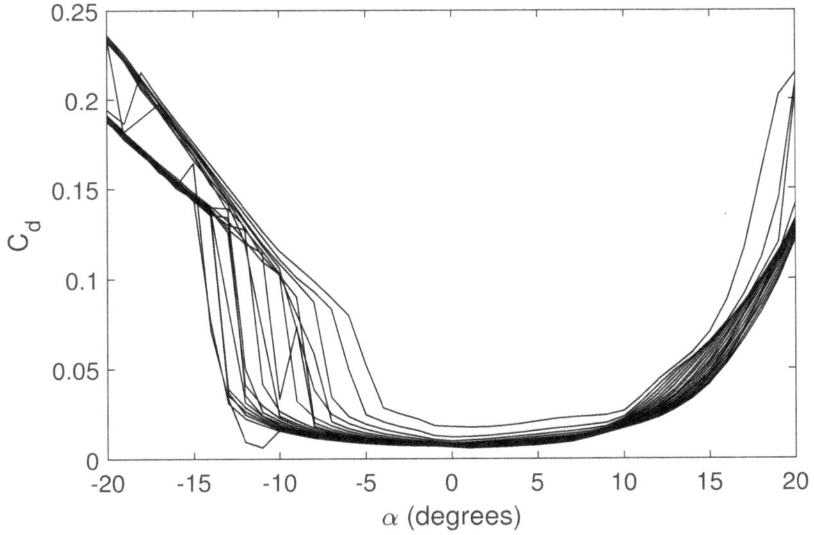

Figure A2. Drag coefficient from XFOIL at different Reynolds numbers ranging from 2×10^4 to 1×10^6.

Appendix B. Experimental Data

Table A6. APC 14 × 12 propeller wind tunnel raw data.

Speed (ft/s)	RPM	Prop Power (ft-lb/s)	Thrust (lb)	Torque (ft-lb)
25	1517	2.12	−0.0026	0.0133
25	2548	22.46	0.4938	0.0842
25	3504	64.05	1.3175	0.1745
25	4507	139.06	2.5406	0.2946
25	5512	241.54	3.9042	0.4184
30	1553	0.26	−0.0954	0.0016
30	2537	20.95	0.3907	0.0789
30	3497	66.53	1.2598	0.1817
30	4611	147.11	2.5109	0.3047
30	5464	242.12	3.7959	0.4231
35	1506	−1.77	−0.1895	−0.0112
35	2585	19.12	0.2780	0.0706
35	3570	67.26	1.1316	0.1799
35	4520	137.48	2.2219	0.2905
35	5558	262.62	3.8355	0.4512
40	1532	−2.81	−0.2715	−0.0175
40	2503	12.73	0.0892	0.0486
40	3571	63.73	0.9422	0.1704
40	4572	143.93	2.1730	0.3006
40	5639	271.58	3.7189	0.4599
45	1757	−5.30	−0.4801	−0.0288
45	2597	9.19	−0.0488	0.0338
45	3586	61.78	0.7965	0.1645
45	4517	143.36	2.0220	0.3031
45	5511	265.88	3.4931	0.4607
50	1817	−6.93	−0.6062	−0.0364
50	2545	2.18	−0.2460	0.0082
50	3525	48.38	0.4729	0.1311
50	4554	141.67	1.7964	0.2971
50	5559	260.34	3.1404	0.4472

Table A7. APC 14 × 12 propeller wind tunnel processed data.

J	Cp	Cq	Ct	eta
0.8475	0.0269	0.0043	−0.0010	−0.0305
0.5046	0.0604	0.0096	0.0657	0.5496
0.3669	0.0662	0.0105	0.0928	0.5143
0.2853	0.0675	0.0107	0.1081	0.4567

Table A7. *Cont.*

J	Cp	Cq	Ct	eta
0.2333	0.0641	0.0102	0.1111	0.4041
0.9935	0.0031	0.0005	−0.0342	−10.8215
0.6081	0.0570	0.0091	0.0525	0.5594
0.4412	0.0692	0.0110	0.0890	0.5681
0.3346	0.0667	0.0106	0.1021	0.5121
0.2824	0.0660	0.0105	0.1099	0.4703
1.1952	−0.0230	−0.0037	−0.0722	3.7498
0.6963	0.0492	0.0078	0.0360	0.5089
0.5042	0.0657	0.0105	0.0767	0.5888
0.3982	0.0662	0.0105	0.0940	0.5656
0.3239	0.0680	0.0108	0.1073	0.5112
1.3428	−0.0348	−0.0055	−0.1000	3.8612
0.8219	0.0361	0.0057	0.0123	0.2803
0.5761	0.0622	0.0099	0.0639	0.5913
0.4499	0.0670	0.0107	0.0899	0.6039
0.3648	0.0673	0.0107	0.1011	0.5477
1.3172	−0.0434	−0.0069	−0.1344	4.0793
0.8911	0.0233	0.0037	−0.0062	−0.2387
0.6454	0.0596	0.0095	0.0535	0.5801
0.5124	0.0692	0.0110	0.0857	0.6347
0.4199	0.0706	0.0112	0.0994	0.5912
1.4152	−0.0513	−0.0082	−0.1587	4.3749
1.0104	0.0059	0.0009	−0.0328	−5.6469
0.7295	0.0491	0.0078	0.0329	0.4887
0.5647	0.0667	0.0106	0.0749	0.6340
0.4626	0.0674	0.0107	0.0878	0.6031

Table A8. APC 16 × 12 propeller wind tunnel raw data.

Speed (ft/s)	RPM	Prop Power (ft-lb/s)	Thrust (lb)	Torque (ft-lb)
25	1517	3.3576	0.0123	0.0211
25	2536	32.8465	0.8271	0.1237
25	3514	94.5394	2.0389	0.2569
25	4493	207.2060	3.7727	0.4404
25	5558	397.2475	5.9101	0.6825
30	1562	−0.3397	−0.1378	−0.0021
30	2582	30.0892	0.5983	0.1113
30	3512	98.4049	2.0062	0.2676
30	4581	212.5770	3.6594	0.4431
30	5527	392.3271	5.7619	0.6778

Table A8. *Cont.*

Speed (ft/s)	RPM	Prop Power (ft-lb/s)	Thrust (lb)	Torque (ft-lb)
35	1567	−1.9920	−0.2394	−0.0121
35	2563	20.8815	0.3098	0.0778
35	3531	96.5214	1.8132	0.2610
35	4450	210.3278	3.5873	0.4513
35	5522	384.5077	5.5179	0.6649
40	1583	−4.0567	−0.3500	−0.0245
40	2567	16.5423	0.1505	0.0615
40	3494	93.6156	1.6205	0.2559
40	4543	216.9924	3.4430	0.4561
40	5577	416.5673	5.7364	0.7133
45	1558	−5.7658	−0.4778	−0.0353
45	2550	8.5515	−0.0995	0.0320
45	3562	92.5229	1.3869	0.2480
45	4624	221.8607	3.2377	0.4582
45	5555	418.4753	5.5473	0.7194
50	1591	−7.2435	−0.5630	−0.0435
50	2614	3.1477	−0.2578	0.0115
50	3491	71.9359	0.8795	0.1968
50	4470	202.9432	2.7901	0.4335
50	5560	405.0720	5.0691	0.6957

Table A9. APC 16 × 12 propeller wind tunnel processed data.

J	Cp	Cq	Ct	eta
0.7416	0.0219	0.0035	0.0027	0.0916
0.4436	0.0459	0.0073	0.0652	0.6295
0.3201	0.0497	0.0079	0.0837	0.5392
0.2504	0.0521	0.0083	0.0947	0.4552
0.2024	0.0528	0.0084	0.0969	0.3719
0.8643	−0.0020	−0.0003	−0.0286	12.1662
0.5229	0.0399	0.0063	0.0455	0.5965
0.3844	0.0518	0.0082	0.0824	0.6116
0.2947	0.0504	0.0080	0.0884	0.5164
0.2443	0.0530	0.0084	0.0956	0.4406
1.0051	−0.0118	−0.0019	−0.0494	4.2061
0.6145	0.0283	0.0045	0.0239	0.5192
0.4460	0.0500	0.0080	0.0737	0.6575
0.3539	0.0544	0.0087	0.0918	0.5970
0.2852	0.0521	0.0083	0.0917	0.5023

Table A9. *Cont.*

J	Cp	Cq	Ct	eta
1.1371	−0.0233	−0.0037	−0.0708	3.4508
0.7012	0.0223	0.0035	0.0116	0.3639
0.5152	0.0500	0.0080	0.0673	0.6924
0.3962	0.0528	0.0084	0.0845	0.6347
0.3228	0.0548	0.0087	0.0935	0.5508
1.2997	−0.0348	−0.0055	−0.0997	3.7291
0.7941	0.0118	0.0019	−0.0078	−0.5235
0.5685	0.0467	0.0074	0.0554	0.6745
0.4379	0.0512	0.0081	0.0767	0.6567
0.3645	0.0557	0.0089	0.0911	0.5965
1.4142	−0.0410	−0.0065	−0.1127	3.8860
0.8607	0.0040	0.0006	−0.0191	−4.0950
0.6445	0.0386	0.0061	0.0366	0.6113
0.5034	0.0518	0.0082	0.0708	0.6874
0.4047	0.0537	0.0086	0.0831	0.6257

Table A10. APC 18 × 10 propeller wind tunnel raw data.

Speed (ft/s)	RPM	Prop Power (ft-lb/s)	Thrust (lb)	Torque (ft-lb)
25	1493	1.7466	−0.0729	0.0112
25	2502	38.1768	0.7454	0.1457
25	3550	134.2274	2.9654	0.3611
25	4516	290.6071	5.4911	0.6145
25	5577	555.7216	8.8060	0.9515
30	1587	0.2933	−0.2048	0.0018
30	2545	38.1715	0.6941	0.1432
30	3563	132.8820	2.7229	0.3561
30	4545	303.2770	5.4487	0.6372
30	5605	566.9247	8.6415	0.9659
35	1491	−2.9645	−0.3939	−0.0190
35	2530	29.5088	0.4109	0.1114
35	3537	130.3511	2.4809	0.3519
35	4549	293.3680	4.9574	0.6158
35	5548	553.0162	8.1742	0.9518
40	1528	−5.4990	−0.5608	−0.0344
40	2540	17.9976	0.0193	0.0677
40	3521	119.0318	2.0167	0.3228
40	4540	296.0171	4.7365	0.6226
40	5554	586.3313	8.3267	1.0081
45	1636	−7.4761	−0.7072	−0.0436

Table A10. *Cont.*

Speed (ft/s)	RPM	Prop Power (ft-lb/s)	Thrust (lb)	Torque (ft-lb)
45	2540	7.0606	−0.3134	0.0265
45	3550	103.2212	1.4735	0.2777
45	4531	282.4474	4.1983	0.5953
45	5533	553.6882	7.4286	0.9556
50	1958	−11.9256	−0.8949	−0.0582
50	2599	−0.6172	−0.5372	−0.0023
50	3614	97.6277	1.1733	0.2580
50	4538	289.6495	4.0235	0.6095
50	5491	547.6228	6.9882	0.9523

Table A11. APC 18 × 10 propeller wind tunnel processed data.

J	Cp	Cq	Ct	eta
0.6698	0.0066	0.0011	−0.0103	−1.0433
0.3997	0.0308	0.0049	0.0377	0.4881
0.2817	0.0380	0.0060	0.0744	0.5523
0.2214	0.0399	0.0064	0.0852	0.4724
0.1793	0.0405	0.0065	0.0896	0.3961
0.7561	0.0009	0.0001	−0.0257	−20.9487
0.4715	0.0293	0.0047	0.0339	0.5455
0.3368	0.0372	0.0059	0.0678	0.6147
0.2640	0.0409	0.0065	0.0834	0.5390
0.2141	0.0407	0.0065	0.0870	0.4573
0.9390	−0.0113	−0.0018	−0.0561	4.6505
0.5534	0.0231	0.0037	0.0203	0.4873
0.3958	0.0373	0.0059	0.0627	0.6661
0.3078	0.0394	0.0063	0.0758	0.5914
0.2523	0.0410	0.0065	0.0840	0.5173
1.0471	−0.0195	−0.0031	−0.0760	4.0794
0.6299	0.0139	0.0022	0.0009	0.0428
0.4544	0.0345	0.0055	0.0515	0.6777
0.3524	0.0400	0.0064	0.0727	0.6400
0.2881	0.0433	0.0069	0.0854	0.5681
1.1002	−0.0216	−0.0034	−0.0836	4.2569
0.7087	0.0055	0.0009	−0.0154	−1.9971
0.5070	0.0292	0.0046	0.0370	0.6424
0.3973	0.0384	0.0061	0.0647	0.6689
0.3253	0.0414	0.0066	0.0768	0.6037
1.0215	−0.0201	−0.0032	−0.0738	3.7520

Table A11. *Cont.*

J	Cp	Cq	Ct	eta
0.7695	−0.0004	−0.0001	−0.0252	43.5244
0.5534	0.0262	0.0042	0.0284	0.6009
0.4407	0.0392	0.0062	0.0618	0.6945
0.3642	0.0419	0.0067	0.0733	0.6380

Table A12. APC 18 × 12 propeller wind tunnel raw data.

Speed (ft/s)	RPM	Prop Power (ft-lb/s)	Thrust (lb)	Torque (ft-lb)
25	1465	3.2814	−0.0176	0.0214
25	2605	58.2540	1.4280	0.2135
25	3495	155.3620	3.2658	0.4245
25	4477	326.9132	5.6701	0.6973
25	5563	701.1023	9.4475	1.2035
30	1528	2.0179	−0.0675	0.0126
30	2565	54.5721	1.2151	0.2032
30	3565	163.1383	3.1777	0.4370
30	4537	338.4828	5.5851	0.7124
30	5578	673.1802	9.3279	1.1524
35	1555	−0.3309	−0.2151	−0.0020
35	2582	49.1241	0.9251	0.1817
35	3514	160.9492	2.9766	0.4374
35	4551	347.4308	5.4896	0.7290
35	5512	658.7973	8.9966	1.1413
40	1545	−4.1078	−0.4108	−0.0254
40	2511	29.4174	0.2568	0.1119
40	3531	161.4175	2.7666	0.4365
40	4534	342.2056	5.1365	0.7207
40	5535	660.1781	8.6858	1.1390
45	1594	−7.6263	−0.5733	−0.0457
45	2535	23.2764	0.1630	0.0877
45	3533	150.9018	2.3357	0.4079
45	4543	356.3566	5.0688	0.7490
45	5571	685.1968	8.6133	1.1745
50	1620	−10.5793	−0.7594	−0.0624
50	2511	4.8097	−0.3416	0.0183
50	3525	153.1912	2.1657	0.4150
50	4518	341.6694	4.5298	0.7221
50	5584	672.1857	7.9985	1.1495

Table A13. APC 18 × 12 propeller wind tunnel processed data.

J	Cp	Cq	Ct	eta
0.6826	0.0132	0.0021	−0.0026	−0.1341
0.3839	0.0417	0.0066	0.0666	0.6128
0.2861	0.0460	0.0073	0.0846	0.5255
0.2234	0.0461	0.0073	0.0895	0.4336
0.1798	0.0515	0.0082	0.0966	0.3369
0.7853	0.0072	0.0011	−0.0091	−1.0029
0.4678	0.0409	0.0065	0.0584	0.6680
0.3366	0.0456	0.0073	0.0791	0.5844
0.2645	0.0459	0.0073	0.0858	0.4950
0.2151	0.0491	0.0078	0.0948	0.4157
0.9003	−0.0011	−0.0002	−0.0281	22.7538
0.5422	0.0361	0.0057	0.0439	0.6591
0.3984	0.0469	0.0075	0.0763	0.6473
0.3076	0.0466	0.0074	0.0838	0.5530
0.2540	0.0498	0.0079	0.0937	0.4780
1.0356	−0.0141	−0.0022	−0.0544	3.9998
0.6372	0.0235	0.0037	0.0129	0.3492
0.4531	0.0464	0.0074	0.0702	0.6856
0.3529	0.0465	0.0074	0.0790	0.6004
0.2891	0.0493	0.0078	0.0897	0.5263
1.1292	−0.0238	−0.0038	−0.0714	3.3827
0.7101	0.0181	0.0029	0.0080	0.3152
0.5095	0.0433	0.0069	0.0592	0.6965
0.3962	0.0481	0.0077	0.0777	0.6401
0.3231	0.0501	0.0080	0.0878	0.5657
1.2346	−0.0315	−0.0050	−0.0915	3.5890
0.7965	0.0038	0.0006	−0.0171	−3.5512
0.5674	0.0443	0.0070	0.0551	0.7069
0.4427	0.0469	0.0075	0.0702	0.6629
0.3582	0.0488	0.0078	0.0811	0.5950

Table A14. APC 18 × 14 propeller wind tunnel raw data.

Speed (ft/s)	RPM	Prop Power (ft-lb/s)	Thrust (lb)	Torque (ft-lb)
25	1568	10.7621	0.2042	0.0655
25	2505	59.5548	1.3286	0.2270
25	3490	174.4010	3.6346	0.4772
25	4529	349.5821	6.0447	0.7371
25	5593	638.6751	9.3844	1.0904
30	1548	7.8084	0.0967	0.0482
30	2551	66.4363	1.3760	0.2487

Table A14. *Cont.*

Speed (ft/s)	RPM	Prop Power (ft-lb/s)	Thrust (lb)	Torque (ft-lb)
30	3589	186.9023	3.5863	0.4973
30	4586	389.4933	6.4469	0.8110
30	5620	672.2617	9.5368	1.1423
35	1508	4.0276	−0.0445	0.0255
35	2515	52.8140	0.9359	0.2005
35	3599	194.3804	3.4965	0.5157
35	4523	376.0632	5.9869	0.7940
35	5520	677.5513	9.4613	1.1721
40	1557	1.1586	−0.1782	0.0071
40	2494	46.1492	0.7157	0.1767
40	3544	179.5197	3.0255	0.4837
40	4600	396.5264	5.9353	0.8231
40	5549	678.3220	8.9954	1.1673
45	1509	−3.3940	−0.3962	−0.0215
45	2545	39.6254	0.4673	0.1487
45	3533	188.1003	2.9451	0.5084
45	4552	404.3274	5.7631	0.8482
45	5514	686.6947	8.7378	1.1892
50	1450	−3.1609	−0.4205	−0.0208
50	2548	34.1172	0.2830	0.1279
50	3514	188.8147	2.7235	0.5131
50	4504	378.1394	4.9877	0.8017
50	5537	711.8156	8.5555	1.2276

Table A15. APC 18 × 14 propeller wind tunnel processed data.

J	Cp	Cq	Ct	eta
0.6378	0.0353	0.0056	0.0263	0.4745
0.3992	0.0479	0.0076	0.0670	0.5577
0.2865	0.0519	0.0083	0.0944	0.5210
0.2208	0.0476	0.0076	0.0932	0.4323
0.1788	0.0462	0.0074	0.0949	0.3673
0.7752	0.0266	0.0042	0.0128	0.3715
0.4704	0.0506	0.0081	0.0669	0.6213
0.3344	0.0512	0.0081	0.0881	0.5756
0.2617	0.0511	0.0081	0.0970	0.4966
0.2135	0.0479	0.0076	0.0955	0.4256
0.9284	0.0149	0.0024	−0.0062	−0.3863
0.5567	0.0420	0.0067	0.0468	0.6202
0.3890	0.0528	0.0084	0.0854	0.6296
0.3095	0.0514	0.0082	0.0926	0.5572

Table A15. *Cont.*

J	Cp	Cq	Ct	eta
0.2536	0.0510	0.0081	0.0982	0.4887
1.0276	0.0039	0.0006	−0.0233	−6.1519
0.6415	0.0376	0.0060	0.0364	0.6204
0.4515	0.0510	0.0081	0.0762	0.6741
0.3478	0.0515	0.0082	0.0887	0.5987
0.2883	0.0502	0.0080	0.0924	0.5305
1.1928	−0.0125	−0.0020	−0.0550	5.2532
0.7073	0.0304	0.0048	0.0228	0.5306
0.5095	0.0540	0.0086	0.0746	0.7046
0.3954	0.0542	0.0086	0.0880	0.6414
0.3264	0.0518	0.0082	0.0909	0.5726
1.3793	−0.0131	−0.0021	−0.0633	6.6513
0.7849	0.0261	0.0042	0.0138	0.4147
0.5692	0.0551	0.0088	0.0698	0.7212
0.4440	0.0524	0.0083	0.0778	0.6595
0.3612	0.0531	0.0084	0.0883	0.6010

Table A16. BEMT 14 × 12 APC propeller data.

	Run at 1500 RPM				Run at 2500 RPM		
J	Ct	Cp	eta	J	Ct	Cp	eta
0	0.116505	0.095371	0	0	0.120413	0.09114	0
0.1	0.1191	0.096204	0.1238	0.1	0.122606	0.092654	0.1323
0.2	0.116987	0.09698	0.2413	0.2	0.120389	0.092897	0.2592
0.3	0.112323	0.096158	0.3504	0.3	0.114882	0.091208	0.3779
0.4	0.107912	0.096696	0.4464	0.4	0.109803	0.091389	0.4806
0.5	0.10123	0.096886	0.5224	0.5	0.102563	0.091209	0.5622
0.6	0.091524	0.095327	0.5761	0.6	0.092471	0.089293	0.6214
0.7	0.076785	0.089886	0.598	0.7	0.077674	0.083682	0.6497
0.8	0.057768	0.079462	0.5816	0.8	0.058664	0.073121	0.6418
0.82	0.053774	0.076932	0.5732	0.82	0.054675	0.070565	0.6353
0.84	0.049737	0.074271	0.5625	0.84	0.05064	0.067875	0.6267
0.86	0.045665	0.071482	0.5494	0.86	0.046572	0.065058	0.6156
0.88	0.041559	0.068563	0.5334	0.88	0.04247	0.062111	0.6017
0.9	0.037418	0.065511	0.514	0.9	0.038327	0.059025	0.5844
0.92	0.033237	0.062322	0.4906	0.92	0.034157	0.055811	0.5631
0.94	0.029032	0.059004	0.4625	0.94	0.029952	0.05246	0.5367
0.96	0.024804	0.055558	0.4286	0.96	0.025726	0.048981	0.5042
0.98	0.020557	0.051986	0.3875	0.98	0.021481	0.045377	0.4639
1	0.016275	0.048275	0.3371	1	0.017208	0.041638	0.4133
1.02	0.011962	0.044424	0.2747	1.02	0.012898	0.037753	0.3485
1.04	0.007627	0.040439	0.1961	1.04	0.008566	0.033736	0.2641
1.06	0.003261	0.036314	0.0952	1.06	0.004208	0.02958	0.1508
1.08	−0.00112	0.032055	0	1.08	−0.00018	0.025285	0

Table A17. BEMT 14 × 12 APC propeller data.

	Run at 3500 RPM				Run at 4500 RPM		
J	Ct	Cp	eta	J	Ct	Cp	eta
0	0.125598	0.087058	0	0	0.126318	0.087174	0
0.1	0.126155	0.091201	0.1383	0.1	0.126552	0.091189	0.1388
0.2	0.124468	0.09092	0.2738	0.2	0.125037	0.090954	0.2749
0.3	0.11977	0.087061	0.4127	0.3	0.120573	0.087146	0.4151
0.4	0.114512	0.086613	0.5288	0.4	0.115748	0.086531	0.5351
0.5	0.10644	0.085837	0.62	0.5	0.108108	0.08555	0.6318
0.6	0.095132	0.083141	0.6865	0.6	0.09678	0.082499	0.7039
0.7	0.079304	0.076609	0.7246	0.7	0.080408	0.075532	0.7452
0.8	0.06029	0.065795	0.7331	0.8	0.061124	0.064404	0.7593
0.9	0.040036	0.051493	0.6998	0.9	0.040953	0.050017	0.7369
0.92	0.035885	0.048236	0.6844	0.92	0.036821	0.046742	0.7247
0.94	0.031698	0.04484	0.6645	0.94	0.032652	0.043328	0.7084
0.96	0.027491	0.041316	0.6388	0.96	0.028464	0.039786	0.6868
0.98	0.023266	0.037666	0.6053	0.98	0.024257	0.036117	0.6582
1	0.019013	0.03388	0.5612	1	0.020024	0.032312	0.6197
1.02	0.014723	0.029948	0.5015	1.02	0.015755	0.028361	0.5666
1.04	0.010412	0.025882	0.4184	1.04	0.011465	0.024276	0.4912
1.06	0.006077	0.021678	0.2971	1.06	0.007151	0.020053	0.378
1.08	0.001714	0.017333	0.1068	1.08	0.00281	0.015687	0.1935
1.1	−0.0027	0.012822	0	1.1	−0.00158	0.011156	0

Table A18. BEMT 14 × 12 APC propeller data.

	Run at 5500 RPM		
J	Ct	Cp	eta
0	0.123014	0.088978	0
0.1	0.123362	0.092334	0.1336
0.2	0.122795	0.092177	0.2664
0.3	0.119054	0.088335	0.4043
0.4	0.115502	0.087775	0.5264
0.5	0.109183	0.087113	0.6267
0.6	0.099042	0.084519	0.7031
0.7	0.083177	0.077686	0.7495
0.8	0.063967	0.066415	0.7705
0.9	0.043895	0.051963	0.7603
0.92	0.039779	0.048681	0.7518
0.94	0.035629	0.045264	0.7399
0.96	0.03146	0.041722	0.7239
0.98	0.027273	0.038057	0.7023
1	0.023057	0.034257	0.6731
1.02	0.018808	0.030316	0.6328
1.04	0.014528	0.026235	0.5759
1.06	0.010231	0.022026	0.4924
1.08	0.005904	0.017674	0.3608
1.1	0.001552	0.013183	0.1295
1.12	−0.00282	0.008563	0

Table A19. BEMT 16 × 12 APC propeller data.

	Run at 1500 RPM				Run at 2500 RPM		
J	Ct	Cp	eta	J	Ct	Cp	eta
0	0.085876	0.054212	0	0	0.087976	0.051102	0.0001
0.1	0.087824	0.053416	0.1644	0.05	0.09141	0.048734	0.0938
0.2	0.084148	0.055068	0.3056	0.1	0.089942	0.050272	0.1789
0.3	0.077795	0.055587	0.4199	0.15	0.088294	0.051207	0.2586
0.4	0.069433	0.055175	0.5034	0.2	0.085906	0.051731	0.3321
0.5	0.056947	0.052283	0.5446	0.25	0.081713	0.051391	0.3975
0.6	0.041507	0.046349	0.5373	0.3	0.078507	0.051681	0.4557
0.62	0.038166	0.044778	0.5284	0.35	0.074719	0.051577	0.507

Table A19. *Cont.*

J	Run at 1500 RPM Ct	Cp	eta	J	Run at 2500 RPM Ct	Cp	eta
0.64	0.034772	0.04309	0.5165	0.4	0.069887	0.051096	0.5471
0.66	0.031309	0.041305	0.5003	0.45	0.064014	0.049936	0.5769
0.68	0.027831	0.03939	0.4804	0.5	0.057398	0.048138	0.5962
0.7	0.02431	0.037359	0.4555	0.55	0.049997	0.045554	0.6036
0.72	0.020756	0.035212	0.4244	0.6	0.041963	0.042133	0.5976
0.74	0.017167	0.032947	0.3856	0.61	0.040309	0.041362	0.5945
0.76	0.013543	0.030561	0.3368	0.62	0.038625	0.040549	0.5906
0.78	0.009882	0.028051	0.2748	0.63	0.036933	0.039711	0.5859
0.8	0.006194	0.025422	0.1949	0.64	0.035235	0.038847	0.5805
0.82	0.00247	0.022665	0.0894	0.65	0.033523	0.037953	0.5741
0.84	−0.00129	0.019769	0	0.66	0.031774	0.037046	0.5661
				0.67	0.030038	0.036094	0.5576
				0.68	0.028295	0.035116	0.5479
				0.69	0.026543	0.034108	0.537
				0.7	0.024779	0.033069	0.5245
				0.71	0.023005	0.032001	0.5104
				0.72	0.021225	0.030905	0.4945
				0.73	0.019437	0.029781	0.4765
				0.74	0.017641	0.028626	0.456
				0.75	0.015833	0.027439	0.4328
				0.76	0.014018	0.026222	0.4063
				0.77	0.012195	0.024977	0.376
				0.78	0.010363	0.023699	0.3411
				0.79	0.008521	0.022388	0.3007
				0.8	0.006676	0.021051	0.2537
				0.81	0.004818	0.019679	0.1983
				0.82	0.002953	0.018275	0.1325
				0.83	0.001087	0.016846	0.0535
				0.84	−0.0008	0.015364	0

Table A20. BEMT 16 × 12 APC propeller data.

J	Run at 3500 RPM Ct	Cp	eta	J	Run at 4500 RPM Ct	Cp	eta
0	0.091625	0.046855	0.0001	0	0.092669	0.046612	0.0001
0.05	0.093724	0.048012	0.0976	0.05	0.093815	0.047934	0.0979
0.1	0.093444	0.046553	0.2007	0.1	0.09422	0.046474	0.2027
0.15	0.091655	0.047061	0.2921	0.15	0.092592	0.046807	0.2967
0.2	0.088928	0.047665	0.3731	0.2	0.089905	0.047511	0.3785
0.25	0.084502	0.047117	0.4484	0.25	0.085613	0.046844	0.4569
0.3	0.080939	0.047187	0.5146	0.3	0.082206	0.046732	0.5277
0.35	0.076508	0.046817	0.572	0.35	0.077914	0.046143	0.591
0.4	0.07113	0.04603	0.6181	0.4	0.072388	0.04536	0.6384
0.45	0.06509	0.04474	0.6547	0.45	0.066287	0.04388	0.6798
0.5	0.058375	0.042832	0.6814	0.5	0.059333	0.041852	0.7088
0.55	0.050927	0.040152	0.6976	0.55	0.051622	0.038981	0.7284
0.6	0.042914	0.03667	0.7022	0.6	0.043537	0.035409	0.7377
0.65	0.0345	0.032427	0.6916	0.65	0.035138	0.031122	0.7339
0.66	0.032756	0.031507	0.6862	0.66	0.033384	0.030202	0.7295
0.67	0.031026	0.030542	0.6806	0.67	0.031661	0.029231	0.7257
0.68	0.029288	0.029549	0.674	0.68	0.029928	0.028231	0.7209
0.69	0.027542	0.028528	0.6661	0.69	0.028187	0.027202	0.715
0.7	0.025783	0.027476	0.6569	0.7	0.026435	0.026143	0.7078
0.71	0.024016	0.026394	0.646	0.71	0.024673	0.025053	0.6992
0.72	0.022241	0.025284	0.6334	0.72	0.022905	0.023936	0.689
0.73	0.02046	0.024145	0.6186	0.73	0.021131	0.02279	0.6769
0.74	0.01867	0.022976	0.6013	0.74	0.019346	0.021612	0.6624
0.75	0.016869	0.021775	0.581	0.75	0.017552	0.020404	0.6452
0.76	0.01506	0.020544	0.5571	0.76	0.01575	0.019165	0.6246
0.77	0.013244	0.019283	0.5289	0.77	0.013941	0.017896	0.5998
0.78	0.011419	0.01799	0.4951	0.78	0.012123	0.016596	0.5697
0.79	0.009583	0.016665	0.4543	0.79	0.010293	0.015262	0.5328
0.8	0.007745	0.015312	0.4047	0.8	0.008463	0.013902	0.487
0.81	0.005895	0.013925	0.3429	0.81	0.006619	0.012506	0.4287

Table A20. *Cont.*

	Run at 3500 RPM				Run at 4500 RPM		
J	Ct	Cp	eta	J	Ct	Cp	eta
0.82	0.004037	0.012506	0.2647	0.82	0.004768	0.011079	0.3529
0.83	0.002179	0.011061	0.1635	0.83	0.002918	0.009626	0.2516
0.84	0.000295	0.009563	0.0259	0.84	0.001042	0.008119	0.1078
0.85	−0.0016	0.008024	0	0.85	−0.00085	0.006572	0

Table A21. BEMT 16 × 12 APC propeller data.

	Run at 5500 RPM		
J	Ct	Cp	eta
0	0.088353	0.049098	0.0001
0.05	0.088759	0.047726	0.093
0.1	0.089333	0.049054	0.1821
0.15	0.088744	0.049311	0.27
0.2	0.087214	0.049639	0.3514
0.25	0.08391	0.048822	0.4297
0.3	0.082031	0.048793	0.5044
0.35	0.079658	0.048579	0.5739
0.4	0.076248	0.04843	0.6298
0.45	0.071152	0.047193	0.6785
0.5	0.06465	0.04516	0.7158
0.55	0.057235	0.042263	0.7448
0.6	0.049323	0.038625	0.7662
0.65	0.041045	0.034279	0.7783
0.7	0.032459	0.029283	0.7759
0.71	0.030723	0.028197	0.7736
0.72	0.028979	0.027085	0.7703
0.73	0.027225	0.025944	0.7661
0.74	0.025463	0.024775	0.7606
0.75	0.023689	0.023574	0.7536
0.76	0.021909	0.022348	0.7451
0.77	0.020121	0.021093	0.7345
0.78	0.018327	0.01981	0.7216
0.79	0.01652	0.018495	0.7056
0.8	0.014708	0.017153	0.686
0.81	0.012889	0.015782	0.6615
0.82	0.011058	0.014378	0.6306
0.83	0.009225	0.012949	0.5913
0.84	0.007388	0.011487	0.5403
0.85	0.005552	0.009998	0.472
0.86	0.003711	0.00848	0.3764
0.87	0.001873	0.00694	0.2347
0.88	-4.6×10^{-5}	0.005306	0

Table A22. BEMT 18 × 10 APC propeller data.

	Run at 1500 RPM				Run at 2500 RPM		
J	Ct	Cp	eta	J	Ct	Cp	eta
0	0.081598	0.04384	0.0001	0	0.083143	0.040721	0.0001
0.1	0.081126	0.043859	0.185	0.1	0.082665	0.040736	0.2029
0.2	0.073919	0.045084	0.3279	0.2	0.075251	0.041802	0.36
0.3	0.063474	0.044402	0.4289	0.3	0.063844	0.040675	0.4709
0.4	0.049834	0.041687	0.4782	0.4	0.050198	0.037909	0.5297
0.5	0.033647	0.036335	0.463	0.5	0.034014	0.032501	0.5233
0.52	0.030238	0.03495	0.4499	0.52	0.030608	0.031105	0.5117
0.54	0.02679	0.033463	0.4323	0.54	0.02716	0.029606	0.4954
0.56	0.023282	0.031862	0.4092	0.56	0.023654	0.027993	0.4732
0.58	0.019742	0.030144	0.3799	0.58	0.020115	0.026262	0.4442
0.6	0.016161	0.02831	0.3425	0.6	0.016536	0.024416	0.4064
0.62	0.012532	0.026357	0.2948	0.62	0.012912	0.022451	0.3566
0.64	0.008862	0.024284	0.2335	0.64	0.009242	0.020365	0.2904
0.66	0.005141	0.022084	0.1536	0.66	0.005524	0.018152	0.2008
0.68	0.001368	0.019753	0.0471	0.68	0.001752	0.015806	0.0754
0.7	−0.00244	0.017301	0	0.7	−0.00205	0.013342	0

Table A23. BEMT 18 × 10 APC propeller data.

	Run at 3500 RPM				Run at 4500 RPM		
J	Ct	Cp	eta	J	Ct	Cp	eta
0	0.086271	0.035999	0.0001	0	0.087591	0.035367	0.0001
0.05	0.093724	0.031641	0.1481	0.05	0.094369	0.031521	0.1497
0.1	0.085242	0.03613	0.2359	0.1	0.086356	0.035581	0.2427
0.15	0.0816	0.036736	0.3332	0.15	0.082798	0.036001	0.345
0.2	0.076543	0.036784	0.4162	0.2	0.077649	0.036109	0.4301
0.25	0.07042	0.03607	0.4881	0.25	0.071552	0.035151	0.5089
0.3	0.064699	0.035419	0.548	0.3	0.065684	0.034424	0.5724
0.35	0.058239	0.034283	0.5946	0.35	0.059006	0.033202	0.622
0.4	0.050991	0.032524	0.6271	0.4	0.051596	0.031346	0.6584
0.45	0.043119	0.030091	0.6448	0.45	0.043614	0.028823	0.6809
0.5	0.034826	0.027004	0.6448	0.5	0.035329	0.02571	0.6871
0.55	0.026242	0.023257	0.6206	0.55	0.026764	0.021929	0.6712
0.56	0.024487	0.022424	0.6115	0.56	0.025012	0.02109	0.6642
0.57	0.022724	0.021561	0.6008	0.57	0.023254	0.02022	0.6555
0.58	0.020956	0.020668	0.5881	0.58	0.02149	0.01932	0.6451
0.59	0.019178	0.019748	0.573	0.59	0.019717	0.018392	0.6325
0.6	0.017386	0.018796	0.555	0.6	0.017929	0.017433	0.617
0.61	0.015586	0.017817	0.5336	0.61	0.016133	0.016447	0.5984
0.62	0.013772	0.016806	0.5081	0.62	0.014324	0.015429	0.5756
0.63	0.011946	0.015764	0.4774	0.63	0.012503	0.014379	0.5478
0.64	0.010112	0.014693	0.4405	0.64	0.010675	0.013301	0.5137
0.65	0.008266	0.01359	0.3954	0.65	0.008834	0.01219	0.471
0.66	0.006404	0.012452	0.3394	0.66	0.006977	0.011045	0.4169
0.67	0.004527	0.01128	0.2689	0.67	0.005105	0.009864	0.3468
0.68	0.002643	0.010078	0.1783	0.68	0.003227	0.008655	0.2535
0.69	0.000758	0.00885	0.0591	0.69	0.001348	0.007419	0.1254
0.7	−0.00114	0.007586	0	0.7	−0.00055	0.006146	0

Table A24. BEMT 18 × 10 APC propeller data.

	Run at 5500 RPM		
J	Ct	Cp	eta
0	0.087393	0.036192	0.0001
0.05	0.091741	0.03353	0.1368
0.1	0.085971	0.036425	0.236
0.15	0.083188	0.036832	0.3388
0.2	0.078903	0.037062	0.4258
0.25	0.073194	0.036112	0.5067
0.3	0.067477	0.035377	0.5722
0.35	0.060822	0.034089	0.6245
0.4	0.05343	0.032173	0.6643
0.45	0.045496	0.029608	0.6915
0.5	0.037198	0.026445	0.7033
0.55	0.028659	0.022638	0.6963
0.56	0.026914	0.021795	0.6915
0.57	0.02516	0.020921	0.6855
0.58	0.023402	0.020019	0.678
0.59	0.021633	0.019089	0.6686
0.6	0.019852	0.018129	0.657
0.61	0.018059	0.01714	0.6427
0.62	0.016255	0.016122	0.6251
0.63	0.01444	0.015073	0.6036
0.64	0.012614	0.013993	0.5769
0.65	0.010779	0.012885	0.5438
0.66	0.008928	0.011741	0.5018
0.67	0.00706	0.010563	0.4478
0.68	0.005187	0.009356	0.377
0.69	0.003315	0.008126	0.2815
0.7	0.001422	0.006856	0.1451
0.71	−0.0005	0.005541	0

Table A25. BEMT 18 × 12 APC propeller data.

	Run at 1500 RPM				Run at 2500 RPM		
J	Ct	Cp	eta	J	Ct	Cp	eta
0	0.08776	0.050578	0.0001	0	0.089605	0.047186	0.0001
0.1	0.08784	0.050651	0.1734	0.05	0.095373	0.041851	0.1139
0.2	0.080957	0.051864	0.3122	0.1	0.089451	0.047136	0.1898
0.3	0.072641	0.051939	0.4196	0.15	0.086755	0.047852	0.2719
0.4	0.060556	0.05	0.4845	0.2	0.08314	0.048283	0.3444
0.5	0.045184	0.045335	0.4983	0.25	0.077679	0.047746	0.4067
0.6	0.027835	0.037834	0.4414	0.3	0.073076	0.047604	0.4605
0.62	0.024188	0.035973	0.4169	0.35	0.067486	0.046881	0.5038
0.64	0.020514	0.033992	0.3862	0.4	0.061007	0.045602	0.5351
0.66	0.016806	0.03189	0.3478	0.45	0.053593	0.043561	0.5536
0.68	0.013054	0.029665	0.2992	0.5	0.045632	0.040865	0.5583
0.7	0.009261	0.027314	0.2373	0.55	0.037162	0.037461	0.5456
0.72	0.005429	0.024838	0.1574	0.56	0.035427	0.036692	0.5407
0.74	0.001563	0.022235	0.052	0.57	0.033661	0.035884	0.5347
0.76	−0.00239	0.019469	0	0.58	0.031881	0.035047	0.5276
				0.59	0.030091	0.034183	0.5194
				0.6	0.028289	0.03329	0.5099
				0.61	0.026472	0.032366	0.4989
				0.62	0.024642	0.031412	0.4864
				0.63	0.022808	0.030428	0.4722
				0.64	0.020972	0.029416	0.4563
				0.65	0.019123	0.028371	0.4381
				0.66	0.017263	0.027296	0.4174
				0.67	0.015394	0.026191	0.3938
				0.68	0.013515	0.025055	0.3668
				0.69	0.011624	0.023887	0.3358
				0.7	0.009726	0.022689	0.3001
				0.71	0.007818	0.021459	0.2587
				0.72	0.005894	0.020194	0.2102
				0.73	0.003967	0.0189	0.1532
				0.74	0.00203	0.017573	0.0855
				0.75	0.000099	0.016226	0.0046
				0.76	−0.00192	0.014788	0

Table A26. BEMT 18 × 12 APC propeller data.

	Run at 3500 RPM				Run at 4500 RPM		
J	Ct	Cp	eta	J	Ct	Cp	eta
0	0.093187	0.042393	0.0001	0.0	0.094435	0.041946	0.0001
0.05	0.098665	0.039479	0.125	0.05	0.099102	0.039411	0.1257
0.1	0.092652	0.042449	0.2183	0.1	0.093781	0.042008	0.2232
0.15	0.089629	0.043048	0.3123	0.15	0.090865	0.042478	0.3209
0.2	0.085564	0.04342	0.3941	0.2	0.086862	0.042666	0.4072
0.25	0.079649	0.042695	0.4664	0.25	0.080847	0.04206	0.4805
0.3	0.074447	0.042297	0.528	0.3	0.075664	0.041459	0.5475
0.35	0.068598	0.041406	0.5799	0.35	0.069765	0.040511	0.6027
0.4	0.061936	0.039986	0.6196	0.4	0.062859	0.03894	0.6457
0.45	0.054503	0.037875	0.6476	0.45	0.055261	0.036743	0.6768
0.5	0.046549	0.035117	0.6628	0.5	0.047212	0.033904	0.6963
0.55	0.038095	0.031649	0.662	0.55	0.038667	0.030352	0.7007
0.56	0.036363	0.030867	0.6597	0.6	0.029826	0.026073	0.6864
0.57	0.034602	0.030046	0.6564	0.61	0.028018	0.025128	0.6801
0.58	0.032826	0.029196	0.6521	0.62	0.026198	0.024153	0.6725
0.59	0.03104	0.028318	0.6467	0.63	0.024375	0.023149	0.6634
0.6	0.029242	0.027412	0.6401	0.64	0.022549	0.022114	0.6526
0.61	0.02743	0.026475	0.632	0.65	0.020711	0.021049	0.6396
0.62	0.025605	0.025507	0.6224	0.66	0.018861	0.019951	0.6239
0.63	0.023776	0.024509	0.6112	0.67	0.017002	0.018824	0.6052
0.64	0.021945	0.023483	0.5981	0.68	0.015135	0.017666	0.5826
0.65	0.020102	0.022424	0.5827	0.69	0.013255	0.016475	0.5552
0.66	0.018247	0.021334	0.5645	0.7	0.011369	0.015254	0.5217
0.67	0.016383	0.020215	0.543	0.71	0.009472	0.014001	0.4803
0.68	0.01451	0.019064	0.5175	0.72	0.007563	0.012714	0.4283
0.69	0.012625	0.017881	0.4872	0.73	0.005647	0.011395	0.3618

Table A26. *Cont.*

J	Run at 3500 RPM Ct	Cp	eta	J	Run at 4500 RPM Ct	Cp	eta
0.7	0.010732	0.016668	0.4507	0.74	0.003722	0.010045	0.2742
0.71	0.00883	0.015423	0.4065	0.75	0.001805	0.008673	0.1561
0.72	0.006913	0.014143	0.352	0.76	−0.0002	0.007211	0
0.73	0.004992	0.012833	0.284				
0.74	0.003061	0.011491	0.1971				
0.75	0.001137	0.010128	0.0842				
0.76	−0.00087	0.008674	0				

Table A27. BEMT 18 × 12 APC propeller data.

J	Run at 5500 RPM Ct	Cp	eta
0	0.09419	0.042857	0.0001
0.05	0.096246	0.041393	0.1163
0.1	0.093672	0.042954	0.2181
0.15	0.091632	0.043485	0.3161
0.2	0.088347	0.043764	0.4037
0.25	0.082518	0.043005	0.4797
0.3	0.077552	0.042407	0.5486
0.35	0.071746	0.041423	0.6062
0.4	0.064867	0.03979	0.6521
0.45	0.057305	0.037541	0.6869
0.5	0.049281	0.034658	0.711
0.55	0.040735	0.031063	0.7213
0.6	0.031922	0.026765	0.7156
0.61	0.03012	0.025819	0.7116
0.62	0.028306	0.024843	0.7064
0.63	0.026489	0.023838	0.7001
0.64	0.024667	0.022803	0.6923
0.65	0.022836	0.021739	0.6828
0.66	0.020991	0.020642	0.6712
0.67	0.019137	0.019516	0.657
0.68	0.017275	0.01836	0.6398
0.69	0.0154	0.017171	0.6188
0.7	0.013518	0.015954	0.5931
0.71	0.011625	0.014703	0.5613
0.72	0.009725	0.013424	0.5216
0.73	0.007811	0.012109	0.4709
0.74	0.005892	0.010765	0.405
0.75	0.003962	0.009387	0.3166
0.76	0.002019	0.007975	0.1925
0.77	0.000053	0.006518	0.0063
0.78	−0.00193	0.005018	0

Table A28. BEMT 18 × 14 APC propeller data.

J	Run at 1500 RPM Ct	Cp	eta	J	Run at 2500 RPM Ct	Cp	eta
0	0.085415	0.06742	0	0	0.087765	0.064712	0
0.1	0.089085	0.067858	0.1313	0.1	0.091317	0.065966	0.1384
0.2	0.087219	0.068008	0.2565	0.2	0.089502	0.065514	0.2732
0.3	0.082829	0.067216	0.3697	0.3	0.084572	0.064372	0.3941
0.4	0.079115	0.067328	0.47	0.4	0.080618	0.064143	0.5027
0.5	0.074976	0.067589	0.5546	0.5	0.076067	0.064241	0.592
0.6	0.066401	0.066564	0.5985	0.6	0.067051	0.06292	0.6394
0.7	0.053511	0.062076	0.6034	0.7	0.05402	0.058262	0.649
0.8	0.036623	0.053115	0.5516	0.8	0.037146	0.049226	0.6037
0.82	0.033178	0.050991	0.5335	0.82	0.033704	0.047087	0.5869
0.84	0.029707	0.048761	0.5118	0.84	0.030235	0.04484	0.5664
0.86	0.026205	0.046422	0.4855	0.86	0.026738	0.042487	0.5412
0.88	0.022669	0.043968	0.4537	0.88	0.023206	0.040016	0.5103
0.9	0.019139	0.04143	0.4158	0.9	0.019677	0.03746	0.4728
0.92	0.015588	0.038788	0.3697	0.92	0.016132	0.034804	0.4264

Table A28. *Cont.*

	Run at 1500 RPM				Run at 2500 RPM		
J	Ct	Cp	eta	J	Ct	Cp	eta
0.94	0.012009	0.036035	0.3133	0.94	0.012558	0.032034	0.3685
0.96	0.008412	0.033177	0.2434	0.96	0.008963	0.029158	0.2951
0.98	0.004779	0.030196	0.1551	0.98	0.005338	0.026163	0.1999
1	0.001143	0.027123	0.0421	1	0.001703	0.02307	0.0738
1.02	−0.00251	0.023945	0				

Table A29. BEMT 18 × 14 APC propeller data.

	Run at 3500 RPM				Run at 4500 RPM		
J	Ct	Cp	eta	J	Ct	Cp	eta
0	0.091282	0.062056	0	0	0.092083	0.062174	0
0.1	0.093465	0.06533	0.1431	0.1	0.094021	0.065304	0.144
0.2	0.092049	0.064838	0.2839	0.2	0.09276	0.06488	0.2859
0.3	0.087734	0.062328	0.4223	0.3	0.08871	0.062469	0.426
0.4	0.084411	0.060945	0.554	0.4	0.086386	0.059778	0.578
0.5	0.079281	0.060692	0.6531	0.5	0.08081	0.06004	0.673
0.6	0.069364	0.058886	0.7068	0.6	0.07049	0.05796	0.7297
0.7	0.055083	0.053283	0.7237	0.7	0.055803	0.051971	0.7516
0.8	0.03828	0.044125	0.694	0.8	0.039027	0.042769	0.73
0.82	0.034853	0.041959	0.6811	0.82	0.035614	0.040587	0.7195
0.84	0.031397	0.039685	0.6646	0.84	0.032173	0.038296	0.7057
0.86	0.027915	0.037304	0.6435	0.86	0.028706	0.035899	0.6877
0.88	0.024398	0.034806	0.6169	0.88	0.025204	0.033383	0.6644
0.9	0.020886	0.032222	0.5834	0.9	0.021707	0.030781	0.6347
0.92	0.017357	0.029536	0.5406	0.92	0.018195	0.028078	0.5962
0.94	0.013799	0.026737	0.4851	0.94	0.014653	0.02526	0.5453
0.96	0.010222	0.023831	0.4118	0.96	0.011093	0.022336	0.4768
0.98	0.006614	0.020806	0.3115	0.98	0.007503	0.019293	0.3811
1	0.002997	0.017681	0.1695	1	0.003905	0.016149	0.2418
				1.02	0.000297	0.012905	0.0235
				1.04	−0.00357	0.009725	0

Table A30. BEMT 18 × 14 APC propeller data.

	Run at 5500 RPM		
J	Ct	Cp	eta
0	0.09007	0.063308	0
0.1	0.091743	0.066001	0.139
0.2	0.090891	0.065724	0.2766
0.3	0.087228	0.063431	0.4125
0.4	0.085463	0.06057	0.5644
0.5	0.080695	0.060761	0.664
0.6	0.071717	0.059175	0.7272
0.7	0.05761	0.053402	0.7552
0.8	0.040953	0.044162	0.7419
0.82	0.037555	0.041971	0.7337
0.84	0.034124	0.039671	0.7226
0.86	0.030666	0.037262	0.7078
0.88	0.027178	0.034743	0.6884
0.9	0.023694	0.032139	0.6635
0.92	0.02019	0.029433	0.6311
0.94	0.016663	0.026619	0.5884
0.96	0.013116	0.0237	0.5313
0.98	0.009539	0.020664	0.4524
1	0.005948	0.017526	0.3394
1.02	0.002353	0.014293	0.1679
1.04	−0.00151	0.01112	0

References

1. Gur, O.; Rosen, A. Propeller performance at low advance ratio. *J. Aircr.* **2005**, *42*, 435–441. [CrossRef]
2. Kutty, H.A.; Rajendran, P. 3D CFD simulation and experimental validation of small APC slow flyer propeller blade. *Aerospace* **2017**, *4*, 10. [CrossRef]
3. Islami, Z.S.; Hartono, F. Development of small propeller test bench system. *IOP Conf. Ser. Mater. Sci. Eng.* **2019**, *645*, 012017. [CrossRef]
4. Czyż, Z.; Karpiński, P.; Skiba, K.; Wendeker, M. Wind Tunnel Performance Tests of the Propellers with Different Pitch for the Electric Propulsion System. *Sensors* **2022**, *22*, 2. [CrossRef] [PubMed]
5. Podsędkowski, M.; Konopiński, R.; Obidowski, D.; Koter, K. Variable Pitch Propeller for UAV-Experimental Tests. *Energies* **2020**, *13*, 5264. [CrossRef]
6. Avanzini, G.; Nisio, A.D.; Lanzolla, A.; Stigliano, D. A test-bench for battery-motor-propeller assemblies designed for multirotor vehicles. In Proceedings of the 2020 IEEE 7th International Workshop on Metrology for AeroSpace (MetroAeroSpace), Pisa, Italy, 22–24 June 2020; pp. 600–605. [CrossRef]
7. Scanavino, M.; Vilardi, A.; Guglieri, G. An Experimental Analysis on Propeller Performance in a Climate-controlled Facility. *J. Intell. Robot. Syst.* **2020**, *100*, 505–517. [CrossRef]
8. Speck, S.; Herbst, S.; Kim, H.; Stein, F.G.; Hornung, M. Development, Startup Operations and Tests of a Propeller Wind Tunnel Test Rig. In Proceedings of the 33rd AIAA Applied Aerodynamics Conference, Dallas, TX, USA, 22–26 June 2015. [CrossRef]
9. Brandt, J.B. Small-Scale Propeller Performance at Low Speeds. Master's Thesis, University of Illinois at Urbana-Champaign, Champaign, IL, USA, 2005.
10. Brandt, J.; Selig, M. Propeller Performance Data at Low Reynolds Numbers. In Proceedings of the 49th AIAA Aerospace Sciences Meeting Including the New Horizons Forum and Aerospace Exposition, Orlando, FL, USA, 4–7 January 2011. [CrossRef]
11. Deters, R.W.; Ananda Krishnan, G.K.; Selig, M.S. Reynolds number effects on the performance of small-scale propellers. In Proceedings of the 32nd AIAA Applied Aerodynamics Conference, Atlanta, GA, USA, 16–20 June 2014; p. 2151. [CrossRef]
12. Deters, R.W. *Performance and Slipstream Characteristics of Small-Scale Propellers at Low Reynolds Numbers*; University of Illinois at Urbana-Champaign: Champaign, IL, USA, 2014.
13. Dantsker, O.; Caccamo, M.; Deters, R.W.; Selig, M.S. Performance Testing of Aero-Naut CAM Folding Propellers. In Proceedings of the AIAA AVIATION 2020 FORUM, Virtual Event, 15–19 June 2020; p. 2762.
14. McCrink, M.H.; Gregory, J.W. Blade Element Momentum Modeling of Low-Re Small UAS Electric Propulsion Systems. In Proceedings of the 33rd AIAA Applied Aerodynamics Conference 2015, Dallas, TX, USA, 22–26 June 2015.
15. Van Treuren, K.; Sanchez, R.; Bennett, B.; Wisniewski, C. Testing UAS Propellers Designed for Minimum Induced Drag. In Proceedings of the AIAA AVIATION 2021 FORUM, Virtual Event, 2–6 August 2021. [CrossRef]
16. Gamble, D.; Arena, A. Automated Dynamic Propeller Testing at Low Reynolds Numbers. In Proceedings of the 48th AIAA Aerospace Sciences Meeting Including the New Horizons Forum and Aerospace Exposition, Orlando, FL, USA, 4–7 January 2010; p. 853.
17. Bellcock, A.; Rouser, K. Design of Vortex Generator Jets for Small UAS Propellers at Low Reynolds Number Operation. In Proceedings of the 2018 AIAA Information Systems-AIAA Infotech @ Aerospace, Kissimmee, FL, USA, 8–12 January 2018. [CrossRef]
18. Lowe, T.E. Mobile Propeller Dynamometer Validation. Master's Thesis, Oklahoma State University, Stillwater, AK, USA, 2013. Available online: http://argo.library.okstate.edu/login?url=https://www.proquest.com/dissertations-theses/mobile-propeller-dynamometer-validation/docview/1517984259/se-2?accountid=4117 (accessed on 27 June 2022).
19. Boldman, D.R.; Iek, C.; Hwang, D.P.; Larkin, M.; Schweiger, P. Effect of a Rotating Propeller on the Separation Angle of Attack and Distortion in Ducted Propeller Inlets. In Proceedings of the 31st Aerospace Sciences Meeting, Reno, NV, USA, 11–14 January 1993.
20. Corson, B.; Maynard, J. The Langley 2000-Horsepower Propeller Dynamometer and Tests at High Speed of an NACA 10-(3)(08)-03 Two-Blade Propeller; NACA TN 2859. 1952. Available online: https://ntrs.nasa.gov/citations/19930083637 (accessed on 27 June 2022).
21. Reynolds, R.M.; Samonds, R.I.; Walker, J.H. An Investigation of Single- and Dual-Rotation Propellers at Positive and Negative Thrust, and in Combination with an NACA 1-series D-Type Cowling at Mach Numbers up to 0.84; NACA TR 1336. 1957. Available online: https://ntrs.nasa.gov/citations/19930092325 (accessed on 27 June 2022).
22. Theodorsen, T. *Theory of Propellers*; McGraw-Hill: New York, NY, USA, 1948; Chapter 7, p. 14.
23. Bangga, G. Comparison of Blade Element Method and CFD Simulations of a 10 MW Wind Turbine. *Fluids* **2018**, *3*, 73. [CrossRef]
24. Plaza, B.; Bardera, R.; Visiedo, S. Comparison of BEM and CFD results for MEXICO rotor aerodynamics. *J. Wind. Eng. Ind. Aerodyn.* **2015**, *145*, 115–122. [CrossRef]
25. Abdelhamid, B.; Smaïli, A.; Guerri, O.; Masson, C. Comparison of BEM and Full Navier-Stokes CFD Methods for Prediction of Aerodynamics Performance of HAWT Rotors. In Proceedings of the 2017 International Renewable and Sustainable Energy Conference (IRSEC), Tangier, Morocco, 4–7 December 2017. [CrossRef]
26. Drela, M. XFOIL: An Analysis and Design System for Low Reynolds Number Airfoils. In *Low Reynolds Number Aerodynamics*; Springer: Berlin/Heidelberg, Germany, 1989.
27. Selig, M.S. *UIUC Airfoil Data Site*; Department of Aeronautical and Astronautical Engineering University of Illinois at Urbana-Champaign: Champaign, IL, USA, 2007.

28. Alves, P.J.F. Low Reynolds Number Propeller Performance Measurement in Wind Tunnel Test Rig. Master's Thesis, Universidade da Beira Interior, Covilhã, Portugal, 2014. Available online: https://ubibliorum.ubi.pt/bitstream/10400.6/6454/1/3828_7604.pdf (accessed on 27 June 2022).
29. Lowe, T.E. Development of a Microsoft Excel Based Uav Propeller Design and Analysis Tool. Master's Thesis, Oklahoma State University, Stillwater, OK, USA, 2009. Available online: https://shareok.org/bitstream/handle/11244/48995/Lowe_okstate_0664M_14425.pdf?sequence=1 (accessed on 27 June 2022).

applied sciences

MDPI

Article

Multirotor Unmanned Aerial Vehicle Configuration Optimization Approach for Development of Actuator Fault-Tolerant Structure

Yisak Debele, Ha-Young Shi, Assefinew Wondosen, Jin-Hee Kim and Beom-Soo Kang *

Department of Aerospace Engineering, Pusan National University, Busan 46241, Korea; yisaktol@pusan.ac.kr (Y.D.); shy621@pusan.ac.kr (H.-Y.S.); wondebly@pusan.ac.kr (A.W.); rlawls129@pusan.ac.kr (J.-H.K.)
* Correspondence: bskang@pusan.ac.kr; Tel.: +82-51-510-2310

Featured Application: The proposed approach can be utilized to support the design of novel actuator fault-tolerant multirotor configurations capable of performing desired maneuvers.

Abstract: Presently, multirotor unmanned aerial vehicles (UAV) are utilized in numerous applications. Their design governs the system's controllability and operation performance by influencing the achievable forces and moments produced. However, unexpected causalities, such as actuator failure, adversely affect their controllability, which raises safety concerns about their service. On the other hand, their design flexibility allows further design optimization for various performance requirements, including actuator failure tolerance. Thus, this study proposed an optimization framework that can be employed to design a novel actuator fault-tolerant multirotor UAV configuration. The approach used an attainable moment set (AMS) to evaluate the achievable moment from a multirotor configuration; similarly, standard deviation geometries (SDG) were employed to define performance requirements. Therefore, given a UAV configuration, actuator fault situation, and SDG derived from the designed mission requirement, the suggested optimization framework maximizes the scaling factor of SDG and fits it into the AMS by adjusting the design parameters up to a sufficient margin. The framework is implemented to optimize selected parameters of the Hexacopter-type of parcel delivery multirotor UAV developed by the PNU drone, and a simulation was conducted. The result showed that the optimized configuration of the UAV achieved actuator fault tolerance and operation-performing capability in the presence of a failed actuator.

Keywords: fault-tolerant configuration; multirotor UAV; attainable moment set; required moment

Citation: Debele, Y.; Shi, H.-Y.; Wondosen, A.; Kim, J.-H.; Kang, B.-S. Multirotor Unmanned Aerial Vehicle Configuration Optimization Approach for Development of Actuator Fault-Tolerant Structure. *Appl. Sci.* **2022**, *12*, 6781. https://doi.org/10.3390/app12136781

Academic Editor: Wei Huang

Received: 9 June 2022
Accepted: 1 July 2022
Published: 4 July 2022

1. Introduction

Nowadays, unmanned aerial vehicles (UAVs) are widely used in civilian and military applications. They are used for tactical reconnaissance, territory surveillance, target placement, and other military operations, as well as mapping, field monitoring, meteorological exploration, highway inspection, package delivery, and other civil applications. Their rangy applicability is due to their excellent design, which makes them efficient and cost-effective. They are also renowned for flying at varying speeds, hovering over locations, maintaining a stable position, and performing sophisticated maneuvers. Unfortunately, unanticipated events, such as actuator and sensor failures, can negatively impact their performance and raise safety concerns. Especially in multirotor UAVs, which use merely spinning rotors for thrust generation, actuator failure is a severe issue. Such causality potentially results in flight troubles, leading to a vehicle accident, resulting in a catastrophe and injuries to civilians.

An effective way to mitigate this problem is to develop a fault-tolerant system that can endure a failure and continue to operate without significant performance degradation.

The article by Fourlas et al. [1] presents a complete survey on UAV fault-tolerant systems. Generally, two main components make up active fault-tolerant schemes. The first component is the fault diagnosis unit responsible for detecting, isolating, and identifying the fault. A second unit is a reconfiguration unit that employs an appropriate methodology that can compensate for the appearance of faults so that the UAV continues its flight mission or lands safely [2]. However, the reconfigurability of multirotor UAVs is possible whenever the UAV is designed so that it allows alternative actuator distribution to compensate for failed actuators.

Researchers suggest several configurations of multirotor UAV layout to address the issue of actuator failure. The use of servomotors to convert the vehicle to reconfigurable ones by tilting rotors [3], changing the spinning direction of unidirectional rotors [4], the use of bidirectional rotors [5], and actuator redundancy that results in a bigger structure [6] are among suggested solutions. Although these solutions could regain control for the considered fault condition, post-failure mission execution capacities are limited to indoor and controlled environments. Howbeit, in densely populated areas where landing is impracticable, recovery operations are usually put through autonomous, obstacle-free, and time-optimal path planning to prerecord location and guidance by or landing on a moving vehicle by the vision-based detection technology of markers [7]. In such a situation, the UAV should be feasible for outdoor applications of such landing site searching operations that may require excellent maneuverability in flight with high perturbation. Taking the design flexibility of multirotor UAVs, appropriately arranging actuators at the design level allows compensation for failed actuators.

Durham et al. [8] proposed a method of determining an aircraft's capability to perform the desired maneuver in a nominal case. The authors represented the required moment as a time history of moments and directly overlaid it into an attainable moment set (AMS) envelope that shows the aircraft's maximum moment-producing capability. As a result, they infer that the existence of requested moment points outside the envelope indicates the inability to conduct the intended operation. However, an attempt involving improving the shroud and including the outside points is not mentioned. Hence, this work contributes to filling the gap by proposing a framework that can optimize a given multirotor UAV configuration to be actuator fault-tolerant and capable of performing desired recovery operation maneuvers. Hence, it provides flexibility in designing advanced failsafe operations that meet the environmental factors.

This paper presents a methodology that is used to evaluate previously treated alternative solutions in the literature [3–6] and optimize a given design of multirotor UAVs to tolerate actuator failure and perform maneuvers required by post-failure missions. The needed moment force to track a predefined mission trajectory is denoted as a time history of required moments that can be obtained from simulation and analytically converting the desired course into control input. The system requirement that imprints these required moments derived from the designed mission and disturbance rejection was geometrically represented as standard deviation geometry (SDG) [9–11]. Similarly, the maximum capacity of a given multirotor configuration in generating moment force is represented by the attainable moment set (AMS) as a convex polytope whose shape is influenced by design parameters, such as the number of actuators, position, orientation, and propeller-related parameters. For a system to be capable of fulfilling its task, the AMS should inscribe sufficiently scaled-up SDG to ensure the system requirement is below the system capability. Therefore, the proposed approach focuses on formulating the optimization problem that considers actuator health status and a related algorithm to evaluate the enclosure of required moment points within the AMS up to the enforced marginal requirement. The proposed method was applied on a Hexarotor type of UAV designed for urban parcel delivery and developed by a PNU drone to optimize its actuator tilting angle and arm installation angle and grant the system actuator fault tolerance. Furthermore, the model of the selected UAV employing an active tilting mechanism was simulated for its fault tolerance at hovering and following a preplanned path.

A brief structure of the paper is given here: Section 2 comprises a theoretical and mathematical overview of multirotor UAVs as well as an introduction to the assessment tools and the assessment of the effect of actuator failure on the system; Section 3 elaborates controllability criteria and their geometrical representation of system requirements; Section 4 introduces an overview of the approbation and mathematical formulation of the optimization problem and the Point-in-AMS checking algorithm; Section 5 discusses implementation details; Section 6 comprises the results and discussion; and Section 7 briefly concludes the paper.

2. Overview of Multirotor UAVs

2.1. Multirotor UAV Configuration

Multirotor UAVs are aerial vehicles that employ more than two rotors with fixed pitch spinning blades, so-called propellers. The spinning of each propeller through the air produces aerodynamic forces that are proportional to the square of their rotation rate ω. The thrust force f acts along the propeller's axis, where the drag moment τ_d acts about the propeller's axis [12].

The thrust force of the ith propeller is modeled as:

$$f_i = k_t \omega_i^2, \tag{1}$$

where k_t is thrust coefficient defined by propeller geometric characteristics.

The drag moment that is generated in reaction to the air resistance around the propeller is given as:

$$\tau_{i,d} = k_d \omega_i^2, \tag{2}$$

where k_d is a constant of drag coefficient defined by propeller geometric characteristics.

The rotors' number, geometrical distribution, and orientation characterize multirotor UAV configurations, as shown in Figure 1. The convectional design has single propellers arranged with an even number and alternating spinning directions to balance out the drag moment generated about the vertical axis of the airframe plane. However, according to design requirements, such as power consumption, size, weight, control ease, payload, and growing application in tasks requiring long flight time and complex maneuvers, various configurations of multirotor UAVs have been constructed. The limitation of the conventional design was resolved by introducing unconventional designs characterized by overlapping propellers and the nonparallel arrangement of propellers.

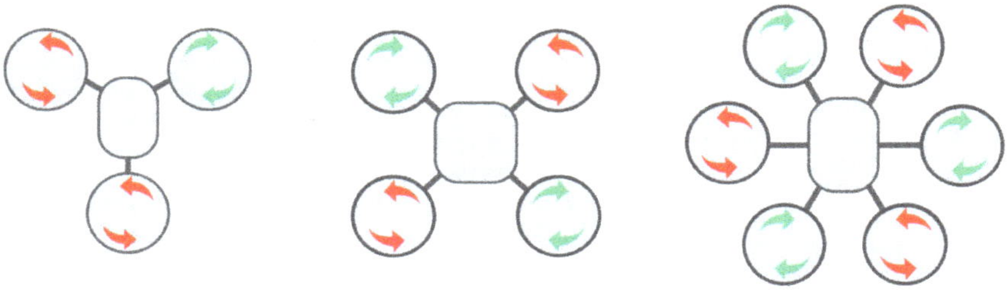

Figure 1. Various layouts of multirotor UAVs.

In all multirotor configurations, the generalized effect of aerodynamic forces generated from each propeller on the overall airframe is governed by the propeller's position and orientation. Therefore, it is necessary to define the propeller's position and orientation relative to the origin of the body frame.

The position of ith propeller x_i can be given as:

$$x_i = \begin{bmatrix} \cos\theta_i \\ \sin\theta_i \\ 0 \end{bmatrix} \times l_i \tag{3}$$

where θ_i is the ith propeller position angle about the Z_b axis in the horizontal $(X_b Y_b)$ aircraft plane, which is formed by the arm with a length l_i and X_b of the right-hand body coordinate.

The orientation of the ith propeller can be given as:

$$q_i = \boldsymbol{R}_{Z_b}(\theta_i)\boldsymbol{R}_{Y_p}(\alpha_{i,y})\boldsymbol{R}_{x_p}(\alpha_{i,x})e_3 \tag{4}$$

where $\boldsymbol{R}_{Z_b}(\cdot)$ is the rotation matrix for the arm rotation θ_i about Z_b the axis of the body frame; $\boldsymbol{R}_{Y_p}(\cdot)$ is the rotation matrix for the propeller rotation $\alpha_{i,y}$ about Y_p of the propeller coordinate; $\boldsymbol{R}_{x_p}(\cdot)$ is the rotation matrix for the propeller rotation $\alpha_{i,x}$ about Y_p of propeller coordinate; and e_3 is a unit vector. The detailed computations and descriptions of the rotation matrixes $\boldsymbol{R}$, the generalized propeller's position matrix x, and the orientation matrix q are presented in Appendix A.

A vertically orientated propeller ($\alpha_{i,y} = \alpha_{i,x} = 0$) applies all its generated force to lift the vehicle, and tilting the propellers results in the vectorization of vertical thrust into lateral force along the plane of the airframe. Moment force about the body frame is generated by virtue of the propellers being positioned some distance from the center of mass. As a result, the steady-state model of thrust and drag induced, as well as its relationship with propeller orientation and location, is expressed as:

$$\boldsymbol{F} = \sum_{i=1}^{3} q_i f_i \tag{5}$$

$$\tau = \sum_{i=1}^{3} x_i \times f_i + \tau_{di} \tag{6}$$

where $\boldsymbol{F}$ is the generalized force generated in the $[x,\ y,\ z]^T$ direction of the airframe, whereas τ gives the generalized moment generated about $[x,\ y,\ z]^T$ direction of the airframe, which results in 6-D force and moment space $\boldsymbol{R}^k$.

Generally, the above formulation can be written compactly by using the effectiveness matrix, $\boldsymbol{B} \,\epsilon\, \boldsymbol{R}^{k\times n}$, which maps actuator space to moment space $\boldsymbol{R}^n \longrightarrow \boldsymbol{R}^k$ as:

$$\begin{bmatrix} \boldsymbol{F} \\ \tau \end{bmatrix} = \boldsymbol{B} \begin{bmatrix} \omega_i^2 \\ \vdots \\ \omega_n^2 \end{bmatrix} \tag{7}$$

As a result, a multirotor UAV system's potential to generate force can be assessed and characterized using configuration parameters.

2.2. AMS Based Multirotor Configuration Assessment

An AMS is a powerful method to assess and understand the system's maximum potential in generating moment force [13]. In a multirotor UAV, the achievable moment force produced from a system using admissible control input is called an attainable moment and is affected by design parameters. Thus, the set of all attainable moments in its three axes is denoted by the AMS, $\Lambda \in \mathbb{R}^3$, as follows:

$$\Lambda = \left\{ m \in \mathbb{R}^3 \middle| m = B_{sub}u,\ u_{min} < u < u_{max} \right\} \tag{8}$$

where $B_{sub} \in \mathbb{R}^{3\times n}$ is the effectiveness matrix that takes rows corresponding to the three moment directions from the original $\boldsymbol{B}$ given in Equation (7); it is characterized by a set of design parameters, such as propeller position, orientation, and constant coefficients, and maps the actuator control input to moment space, where u is the control input constrained between the operational range of the actuators.

Similarly, Equation (7) can be represented geometrically as a higher dimension convex polytope, which is expressed as the following:

$$\Omega = \left\{ \boldsymbol{m} \in \mathbb{R}^3 \middle| \boldsymbol{B}_{sub}^{+} \boldsymbol{m} \le \boldsymbol{u},\ \boldsymbol{B}_{sub}^{+} \in \mathbb{R}^{n \times 3},\ \boldsymbol{u} \in \mathbb{R}^n \right\} \tag{9}$$

where B_{sub}^{+} denotes the pseudo inverse of B_{sub}.

Therefore, the AMS convex polytope can be calculated given a feasible control set (FCS) and the effectiveness matrix B by evaluating the moment produced at the extremes of control inputs. The polytope vertex and facet are defined using a convex hull algorithm. In this work, a MATLAB function *convhull* was employed.

2.3. Multirotor UAV Configuration with Failed Actuator

In multirotor UAVs, the failure of an actuator results in the loss of ability to generate a moment required to control and stabilize the system. The unintentional damage of one or more actuators from a systematically arranged configuration results in an unbalance in their contributing direction.

Similarly, by replacing the effectiveness matrix B in Equations (8) and (9) with a modified effectiveness matrix B_f, the effect of the failed actuator can be treated as follows:

$$B_f = B f_i \tag{10}$$

where f_i is the fault indicator $n \times n$ identity matrix $f = I(n)$, whose i^{th} column corresponding to the failed actuator is zero.

As a result, this section emphasizes that multirotor UAV behavior and controllability are influenced by their design and actuator health.

3. Controllability Criteria

3.1. Null Controllability

In the event of an actuator failure, it is essential to employ an emergent hovering to regain control before the decision to continue following the mission path or performing an emergency landing [14]. An emergent hovering is guaranteed if the system is null controllable, which describes the possibility of driving the UAV state to its hovering state in a finite time with admissible control $\mho$. Thus, it necessitates the resultant attainable moment set Λ origin to have neighborhood moment points with radius r.

Hence, the distribution of moment points around the origin o, where $m = 0$ and radius r are represented by sphere geometry g_s as follows:

$$g_s = \{\mathcal{O} + u \mid \|u\|_2 \le r\} \tag{11}$$

where $\|\cdot\|_2$ denotes the Euclidean norm, i.e., $\|u\|_2 = \left(u^T u\right)^2$.

In doing so, Equation (11) depicts that having a large radius r around the origin o clearly illustrates the UAV's capability to produce adequate control moments to reject disturbance and stabilize the system to hover at a location.

3.2. Maneuverability Requirement

Recalling the previous discussion, setting the UAV at an emergent hovering mode and landing may not handle the causality in some situations. Nowadays, efficient, safe landing searching algorithms autonomously plan routes that need complex and precise maneuvers. To fully implement these algorithms, the system should have the ability to produce all the moments required to meet the designed mission profile and disturbance rejection. A given UAV system is said to be capable of performing the maneuver when the requirement lies below the maximum capability of the system.

The designed mission trajectories can be converted into a sequence of control commands analytically or obtained from simulation and represented as the time history of moments (THM). Based on the nature of the operation, some maneuvers may not have the same relative control authority requisite in different moment directions. This work utilizes a statistical tool, standard deviation geometry (SDG), to define the weakest and strongest direction and geometrically characterize the required moment. If equal control is required in all directions, such as in one of the situations considered in the previous section, the geometry term indicates spheroid. In contrast, if weighted control authority is desired, standard deviation ellipsoid (SDE) would be indicated.

Suppose $X \in \mathbb{R}^3$ is the trivariate Gaussian time history of moment data. By taking each point of the moment time series as an observation, the mean of the desired moment data can be calculated as:

$$\overline{X} = \frac{1}{n-1}\sum_{i=1}^{n} X_i \tag{12}$$

The covariance matrix of trivariate data X is expressed as:

$$c = \frac{1}{n-1}\sum_{i=1}^{n}(X_i - \overline{X})(X_i - \overline{X})^T \tag{13}$$

where $\overline{X}$ is the mean value, and $c \in \mathbb{R}^{3\times 3}$ is the symmetric and positive semi-definite matrix.

A corresponding ellipsoid can be constructed with the inverse square root of eigenvalues, $\lambda_1 > \lambda_2 > \lambda_3$, to be its principal semi-axes oriented by the corresponding eigenvectors.

We can parameterize the ellipsoid as the image of the unit ball under an affine transformation as:

$$g_e = \{\mathcal{O} + Wu \mid u_2 \leq 1\} \tag{14}$$

where $W = c^{1/2}$ is the symmetric and positive semi-definite matrix.

In addition, SDG can be extended to assess the probability of randomly scattered moment points falling inside the scaled ellipsoid and its corresponding magnification factor. In this work, an efficient computation algorithm for the confidence level analysis of SDG is used from the work of [9]. As shown in Figure 2, the 3D data example shows the underlying idea of how SDG and confidence level analysis can be applied to later formulations of optimization problems.

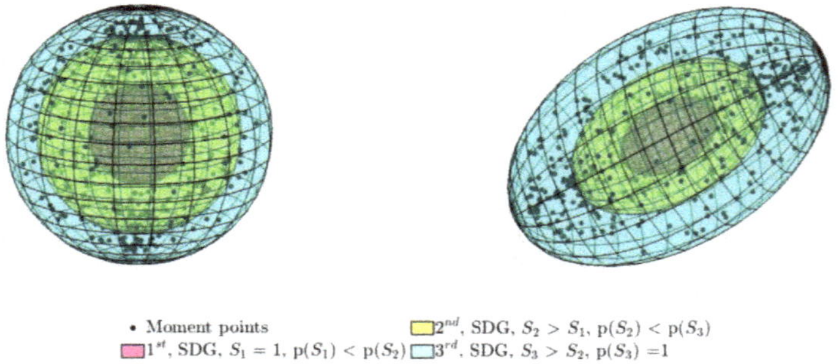

Figure 2. Visualization of standard deviation geometry scaling and corresponding confidence level.

4. Optimization

4.1. Optimization Framework

In Section 3 discussions, the secret behind the variation in multirotor UAVs configuration is elaborated, and a powerful tool is introduced to quantify their moment force generation capability. Furthermore, the effect of actuators' complete failure in controllability and possible ways of alleviating the issue are described. Consequently, this section proposes an optimization framework that can assist the structural design of multirotor UAVs that considers their future control in nominal and actuator failure situations.

The proposed optimization technique aims to find design parameters that give a multirotor UAV system actuator fault-tolerant capability. As shown in Figure 3, it evaluated the AMS from the initial design parameters and specified the actuator effectiveness value.

Firstly, the distribution of moment points is evaluated, and the relative control authority demand is represented as SDG. It checks for the fulfillment of controllability criteria stated in Section 3 by overlaying each required moment point needed to produce the designed mission inside the AMS envelope. The inclusion of all points inside the AMS guarantees the fulfillment of the necessary performance. However, if points exist outside the AMS envelope, the framework maximizes the envelope to include the points. This can be accomplished through fitting and maximizing the SDG to find the largest possible magnification of SDE and the achievable controllability margin by updating design parameters, such as the actuator tilting angle, considering actuator health conditions. Therefore, the optimization outcome will be a set of design parameters that grant actuator fault tolerance. This parameter can be stored in lookup tables and used to reconfigure a system.

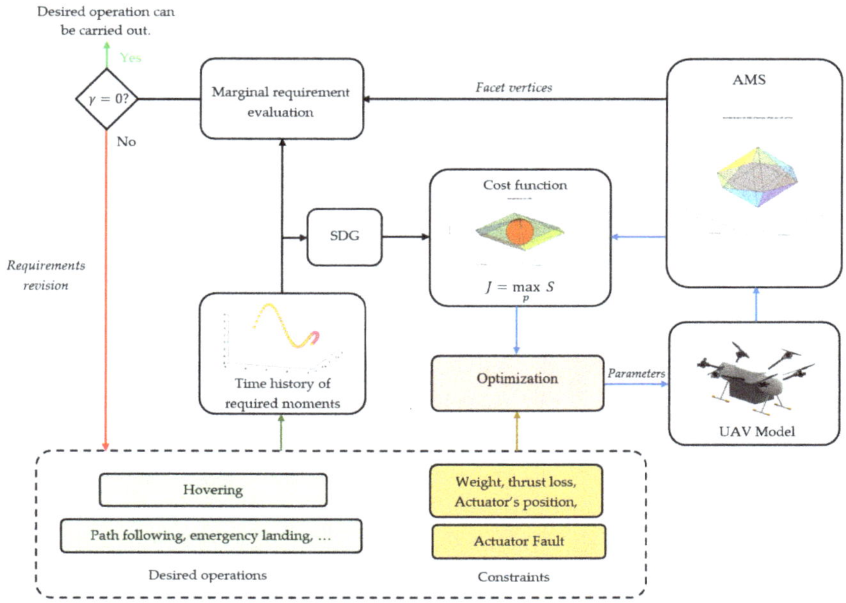

Figure 3. Optimization framework.

4.2. Optimization Formulation

For a given set of design parameter p that describes the UAV configuration, the set of actuator failure possibilities ξ, and the defined mission requirement, the optimization problem was formulated as the fitting geometry of the mission profile moment requirement into an AMS convex polytope.

As shown in Figure 4, a 2D example of moment data points of various mission requirements demonstrates the formulation visually. The first mission demands equal control authority in all moment directions; in contrast, the second data set requires higher strength in one of its directions, resulting in weighted control authority requests. Both data distribution are represented geometrically as a circle and an ellipse using Equations (11)–(14), respectively, and the concentric geometries portray different levels of their magnification. Similarly, the violet polygon signifies the AMS, whereas the concentric convex polytope (broken line) shows the marginal constraint. Recalling the properties of the AMS and controllability criteria, sufficient magnification, and fitting of these geometries into the AMS by adjusting design parameters ensure the enclosure of the required moment point within the geometries and inside the AMS.

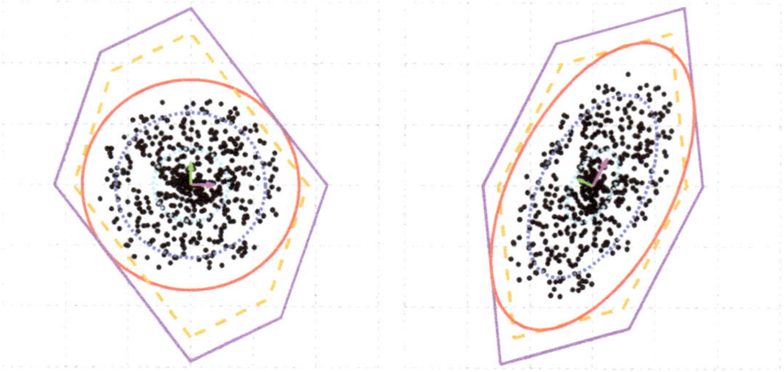

Figure 4. 2D visualization of the optimization problem formulation of fitting SDG into the AMS.

For the first case, the above statement can be formulated mathematically by using Equations (9)–(11) as a problem of fitting and maximizing directly the radius of spheroid subjected to an inequality equation that describes the AMS polytope:

$$\begin{aligned} &\textit{maximize } S \\ &\text{Subject to, } S||B_{fi}^{+}||_2 + B_{fi}^{+T} o \leq u_i \quad \text{for } i = 1,2,\ldots,\text{n} \\ &u_{i,min} < u_i < u_{i,max} \\ &S > 0 \end{aligned} \tag{15}$$

The effect of actuator failure was considered through a modified effectiveness matrix that features the actuator health status indicator in Equation (10):

$$B_f^{+} = (B_f(\xi))^{+} \tag{16}$$

where B_f^{+} is the pseudo inverse of B_f subjected to a set of actuator failure possibilities ξ.

For the second case, where the required moments are directionally distributed, the problem is modified by Equation (14):

$$\begin{aligned} &\textit{maximize } \log\det(SW) \\ &\text{Subject to, } S||WB_{fi}^{+}||_2 + B_{fi}^{+T} o \leq u_i \quad \text{for } i = 1,2,\ldots,\text{n} \\ &u_{i,min} < u_i < u_{i,max} \\ &S > 0 \end{aligned} \tag{17}$$

Note that the formulation can be verified by computing the confidence level $\mathfrak{p}$, corresponding to scale factor s, which defines the probability of randomly scattered required moment data points falling inside the magnified geometry, as shown in Table 1. For a three-dimensional SDE, a scaling factor $S \geq 5$ gives a confidence level of 1.

Table 1. Confidence level of scaled SDE for different scaling factors and dimensions [9].

Dimensionality (n)	**Scale Factor S**					
	1	**2**	**3**	**4**	**5**	**6**
1	0.6827	0.9545	0.9973	0.9999	1.0000	1.0000
2	0.3935	0.8647	0.9889	0.9997	1.0000	1.0000
3	0.1987	0.7385	0.9707	0.9989	1.0000	1.0000
4	0.0902	0.5940	0.9389	0.9970	0.9999	1.0000

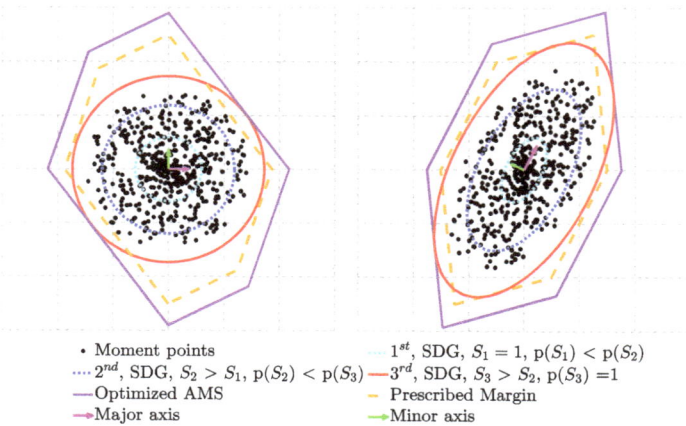

4.3. Inside-AMS-Point Check

In this section, Algorithm 1 is proposed to check the orientation of required moment points relative to the AMS and address the issue of marginal requirement. In a convex polytope analysis, each facet is a hyperplane that divides a space into half-spaces. As shown in Figure 5, conventionally, the normal vector of a convex polytope facet is supposed to be oriented to the exterior [15]. On the other hand, the signed distance between an arbitrary point x_i and a plane tells the orientation of the point relative to that plane. The positive distance indicates the existence of a point x_i on the same side of the facet normal vector $\hat{n}$, and negative if it is on the opposite side [16]. Therefore, if the distance of each required moment data point from all facets of the AMS is negative, it shows the existence of all points inside the AMS envelope.

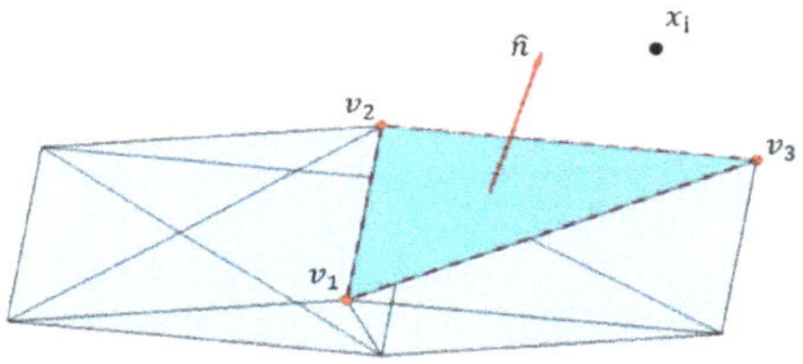

Figure 5. Norm vector and vertex of an AMS facet.

$\mathcal{F}_i$ is a triangular facet of an AMS with the vertex $v_j = [v_{jx}\ v_{jy}\ v_{jz}]$, where $i = 1, 2, \ldots, 2C_2^m$ and j = 1, 2, 3 number of vertexes.

The normal unit vector to a facet of an AMS can be given as follows:

$$\hat{n} = \frac{(v_2 - v_1) \times (v_3 - v_1)}{|(v_2 - v_1) \times (v_3 - v_1)|} \tag{18}$$

The signed distance d_j between an arbitrary point x_0 and a facet of an AMS can be calculated as for all vertices on the facet:

$$d_j = \hat{n} \cdot (x_0 - v_j) \tag{19}$$

If all vertices lay on the same plane, the signed distance should be:

$$d_1 = d_2 = d_3 = d \tag{20}$$

The determination of the point orientation relative to the AMS can be summarized based on the sign d as follows:

$$\text{f } d = \begin{cases} < 0 & \text{the point is inside of AMS} \\ > 0 & \text{the point is outside of AMS} \\ = 0 & \text{on the boundary of AMS} \end{cases}$$

In case marginal requirement $\zeta \in \mathbb{R}^+$ is prescribed, the criteria can be modified as follows:

$$\text{if } d = \begin{cases} < -\zeta & \text{the point is inside AMS upto specified margin} \\ > -\zeta & \text{the point is outside from specified margin} \\ = -\zeta & \text{on specified margin} \end{cases}$$

The pseudo-code below describes the procedures involved in determining the orientation of moment points about the AMS.

Algorithm 1 Inside-AMS-point check

x_i for $i = 1, 2, \ldots, n$ required moment with n number of points
$\mathcal{F}_j$ is a facet from AMS for $j = 1, 2 \ldots 2C_2^m$
v_k^j for $k = 1, 2, 3$ vertices of AMS facet
ζ marginal requirement
for all i
    for all j
        for all k
            n = **norm** $(\mathcal{F}_j)$ //norm vector for each facet
            d = **dot** $(n, (x_i - v_k^j)$ //signed distance between each facet and points in a moment's history
    If $d > -\zeta$
        outside point= h_i //h_i is outside of the AMS
    Else if $d < -\zeta$
        inside point= h_i //h_i is inside the AMS
If z* = **size**(outside point)! = 0
        performance requirement not fulfilled
If z* = **size**(outside point) = 0
        performance requirement fulfilled

Our proposed optimization framework uses this algorithm to assess whether performance criteria are met for specified marginal requirements. Furthermore, the number of points residing outside of the margin of the AMS for an arbitrary S can be quantified using by exclusion ratio γ, as expressed:

$$\gamma = \frac{Z^*}{Z} \tag{21}$$

where $0 \leq \gamma \leq 1$ and it is defined as the ratio of the set of points outside the margin of the AMS Z^* to the set of all points of the required moment Z. $\gamma = 0$ indicates the existence of all points inside the AMS, whereas $\gamma = 1$ implies the existence of all points outside the AMS envelope.

5. Implementation

The proposed method was implemented on parcel delivery Hexarotor UAV developed by a PNU drone to optimize its actuator tilting angle and arm installation angle. This implementation aimed to validate the presented approach and show functional application practices of the computed parameters through a simulation of the assumed UAV.

The preliminary design of the assumed UAV had a standard coplanar configuration, as shown in Figure 6. The output of the proposed method for possible actuator's complete failure one at a time and desired post-failure operations were computed. The possible practice of deploying this optimized tilting angle for reorienting the actuators is using an active tilting mechanism, as shown in Figure A1. These situations were demonstrated with a simulation in its hovering and path-following mission.

Figure 6. Preliminary design of urban package delivery drone developed by a PNU drone.

5.1. Plant Modeling and Simulation

In this work, a simulation of an assumed UAV was presented. Although the detailed modeling process of the system is beyond the scope of this study, a subjective description of the level of abstraction related to actuator failure and reconfigurability mechanisms is elaborated in this section.

Nowadays, the advancements of modeling software and efficient computers enable the simulation of highly abstracted models. Multibody modeling tools allow the development of high-fidelity simulation models without getting into the complexity of the mathematical modeling of a system [17]. In this modeling process, SOLIDWORKS 3D CAD modeling software was employed to model the digital copy of the UAV structure with its inertial parameters. In contrast, physical models, such as D.C. motors, R.C. servo motors, and other relevant components, were modeled with Simscape Multibody™. It is an extension of MATLAB/Simulink. It has tools to simulate a mechanical system with multiple degrees of freedom which allows modeling the individual components and their integration, including their energy interaction [18]. The library contains all the blocks required to define physical systems, such as bodies, joints, actuators, and sensors. The solver simulates the dynamics of the physical system by developing and solving differential equations [19].

The block diagram of the UAV HFM developed in Simscape Multibody™ is shown in Figure 7. Inside the UAV block, the inertial properties of the UAV were defined by a body block that contained a CAD file of the UAV airframe. Based on the XML file generated from CAD, the relative position and orientation of components were specified by transformation block models, whereas the relative motion constraints were modeled in the joint block. The propulsion system was composed of two central units. The first unit was responsible for generating thrust, and it had a D.C. motor model block and propeller model block, while the second unit was responsible for vectorizing the generated thrust, and it had an R.C. servo motor model block and tilting mechanism model.

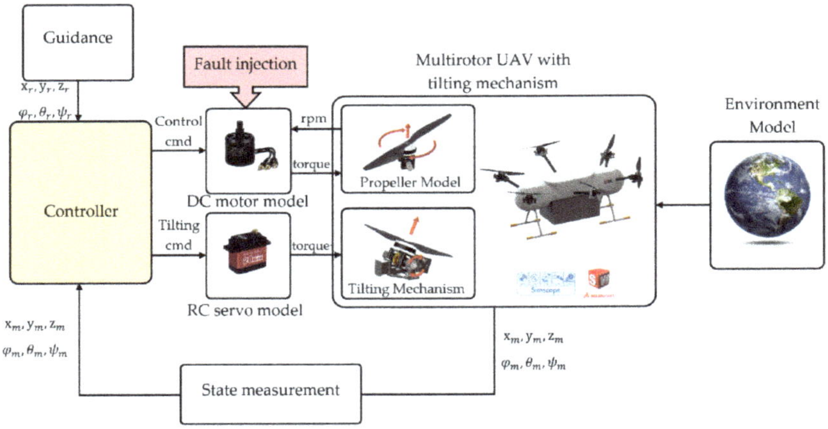

Figure 7. UAV system modeling and simulation block diagram.

Furthermore, these actuation blocks allow fault injection at a specified simulation time. The environmental model block was applied to define the gravitational force and model contact between the UAV and CAD modeled ground and obstacles. The translation and rotational state of the UAV with respect to the world reference were measured by transform sensor block. The propellers' angular rate and tilting angles were measured by sensor option on the respective joint block during the simulation. The model blocks are configured according to the manufacturer's datasheet of selected components.

The controller block receives the position setpoint from the waypoint-based trajectory generator and the state of the UAV from the state measurement block. It outputs the control signal to the actuator block. The stabilization and control of the plant were implemented in the control block, which uses the cascaded closed-loop PID position and attitude control. The precomputed tilting angles combined with the fault tag are stored in the lookup table to reconfigure the UAV [20–22]. In this test platform, by assuming the presence of a perfect actuator fault detection and isolation system, fault signals were generated automatically, and corresponding reconfiguration parameters were selected after some detection time. Thus, each tilting mechanism servo received an actuation signal and executed structural reconfiguration.

5.2. Parameter Selection

The preliminary design of the proposed UAV had six equally spaced propellers on the same plane. The propellers were arranged in alternating order of their spinning direction. The propellers' counterclockwise (CCW) rotation about the Z-axis of the propeller coordinate was taken as positive rotation, whereas the clockwise rotation was assumed as negative, and the thrust generated by the propellers was directed parallel to both the airframe and propeller coordinate Z-direction. Even though it was not fully controllable, this arrangement fulfilled the minimum number of propellers required to provide actuator fault tolerance [23].

Recalling the discussion in Section 1, actuator failure causes the loss of force and moment unbalance, which results in an incapability to maintain entire attitude and altitude control. A typical scheme for solving this situation is scarifying control of one or more DOF, usually yaw motion to control rolling and pitching motion independently [24]. Vectorizing thrust by tilting the propeller was another technique many researchers presented. The inward, sideways, or combined tilting of propellers proved to enhance the multirotor UAVs' fault tolerance and maneuverability [25–27].

Thus, as shown in Figure 8, actuators 3–6 were established to tilt inward and outward about the axis perpendicular to the arm axis. In contrast, actuators 1 and 2 were situated to make sideways tilting about the arm axis. The additional parameter β controls the deviation between the lateral thrust vector produced by the vectorization of the thrust produced by tilting the propellers and the arm axis. Angle β results in offsetting two symmetric and opposite propellers' lateral thrust. Figure 8c,d shows the modified configuration of the preliminary design shown in Figure 8a,b. The green line represents the direction of the lateral thrust offsetting by β, whereas the grey line represents the preliminary design arm axis.

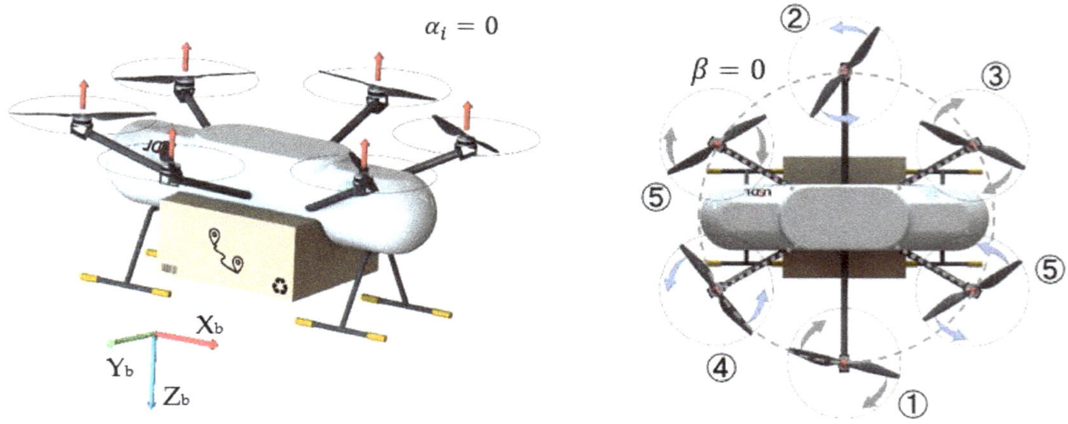

(a) (b)

Figure 8. *Cont.*

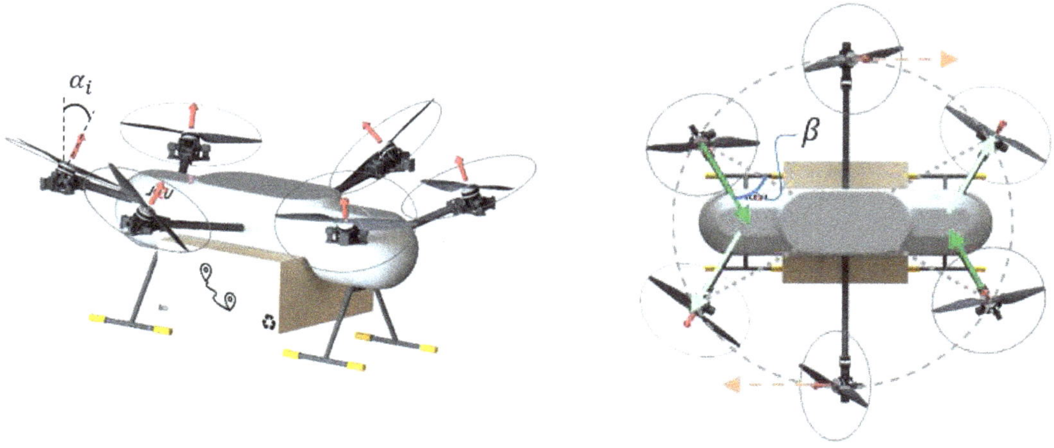

(**c**) (**d**)

Figure 8. Comparison of proposed UAV preliminary design and UAV with tilting mechanism (**a**) preliminary design where all propellers tilting angles are zero (coplanar) (**b**) preliminary design with axis offsetting angle zero (**c**) α_i tilted propeller configuration (**d**) configuration with axis offset angle β.

The design parameters to be optimized were chosen as:

$$p = [\beta, \alpha_1, \alpha_2, \alpha_3, \alpha_4, \alpha_5, \alpha_6]^T \quad (22)$$

where α_i is the propeller's tilting angle, and β is the lateral thrust offsetting angle; the outward tilting angle was taken as a positive tilting angle.

6. Result and Discussion

6.1. Optimization Result

The proposed framework's verification by optimizing parameters in p for each actuator failure possibility in the platform and chosen post-failure operation performance requirement is presented. The two common operations, hovering at the location and following an obstacle-free trajectory to return home, are considered. The required moment data to accomplish these operations and reject the associated disturbance in the nominal condition were logged from the simulation and used as a performance requirement for optimization in faulty conditions. If the framework is implemented correctly, the parameters must converge to a value that gives a maximum cost function and the least exclusion ratio for a given marginal demand. If this is violated, the parameters should not be accepted as optimum, and we recommend that the operation and parameter constraints be revised. In order to limit the maximum vertical thrust loss due to tilting to 5% and consider installation constraints, the domain of parameters is defined as follows:

$$\mathcal{D} = \{\mathrm{p} | 0 < \beta < 30, -20 < \alpha_{\mathrm{i}} < 20, \mathrm{i} = 1, 2 \ldots 6\}$$

To perform the optimization, the particle swarm optimization (PSO) algorithm was implemented to search for a combination of parameters that maximizes the cost function. The algorithm used randomly distributed population sizes of 500 and 400 iterations.

6.1.1. Null Controllability

This section presents the optimization result of the assumed UAV towards achieving actuator fault-tolerant capabilities in a single actuator total failure while hovering. The framework used Equation (15) to maximize and fit the sphere described in Equation (11) into the AMS, and Figure 9 shows the result as a plot of parameters and cost function against the number of iterations for actuator-1 total failure. The result showed that the parameters were converged to values that maximize the cost function within their constraint limits. The initial values, optimal values, and the resulting cost function computed by the optimization framework for each actuator's possible failure are listed in Table 2.

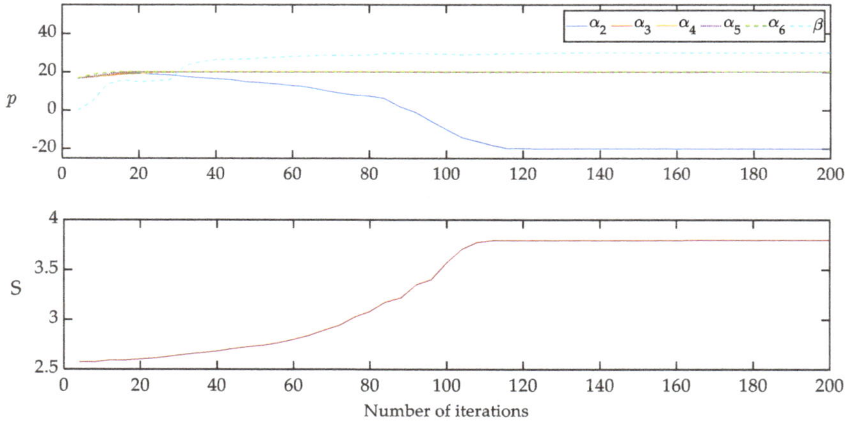

Figure 9. Optimization output: plot of actuator's tilting angle and cost function trend against the number of iterations for actuator 1 failure and the null controllability performance requirement.

Table 2. Parameter optimizations result in null controllability, single actuator failure at a time.

Fault Condition	Parameters (Angles in Degree)		Cost Function
	Initial Value	Optimization	
Actuator 1 failed	$\begin{bmatrix} 0 \\ 15 \\ 15 \\ 15 \\ 15 \\ 15 \\ 15 \end{bmatrix}$	$\begin{bmatrix} 30 & 0 & -20 & 20 & 20 & 20 & 20 \end{bmatrix}^T$	3.8862
Actuator 2 failed		$\begin{bmatrix} 30 & 20 & 0 & 20 & 20 & 20 & 20 \end{bmatrix}^T$	3.8862
Actuator 3 failed		$\begin{bmatrix} 30 & 20 & 13.5 & 0 & 19 & 20 & 20 \end{bmatrix}^T$	3.5078
Actuator 4 failed		$\begin{bmatrix} 30 & 20 & 13 & -16 & 0 & 20 & 20 \end{bmatrix}^T$	3.4718
Actuator 5 failed		$\begin{bmatrix} 30 & 20 & 5 & 20 & 19 & 0 & 20 \end{bmatrix}^T$	3.6263
Actuator 6 failed		$\begin{bmatrix} 30 & 20 & 13 & 20 & 20 & -20 & 0 \end{bmatrix}^T$	3.5029

A comparison of the preliminary designs of the AMS (yellow) and the configuration augmented with optimum parameters (aqua) for each actuator's failure is presented in Figure 10. In preliminary design, actuator-1 total failure results in an inability to produce a negative yaw moment and a negative roll moment simultaneously, and actuators-2 total failure results in an inability to produce a positive yaw moment and a positive roll moment. Likewise, the complete failure of actuators-3–4–5–6 degrades the system's controllability, so the system loses its attitude control. In contrast, owing to the vectorization of the vertical thrust force into the lateral force via optimum angle tilting and arm installation angle of the produced lateral force from symmetrically located actuators, the yaw moment was produced independently with a slight loss of roll moment in the optimal configuration. As a result, sufficient control was produced around the origin of the AMS, as shown on the optimized configuration AMS by origin-centered sphere geometry.

The marginal evaluation result for actuator 1 failure optimization is depicted as shown in Figure 11. The actuator-1 failure in the preliminary design results in $S = 0$ and $\gamma = 0.462$, which indicates 46.2% of the required moment points outside the AMS envelope, as shown in Figure 11a. Given the marginal value of $\zeta = 1$, the coverage of all points within the prescribed margin was ensured through the magnification of g_s by $S = 3.26$. Furthermore,

Figure 11b shows the maximum achievable scaling factor $S = 3.886$ and the corresponding marginal value of $\zeta = 1.316$.

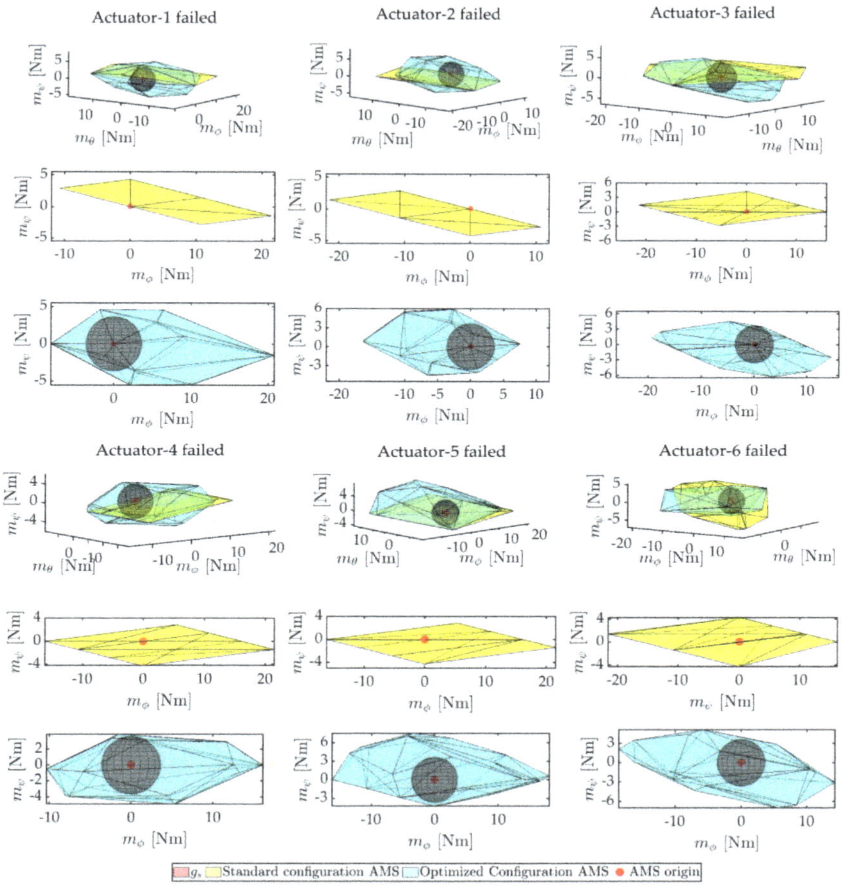

Figure 10. AMS comparison of preliminary configuration with optimized configuration for each actuator failure.

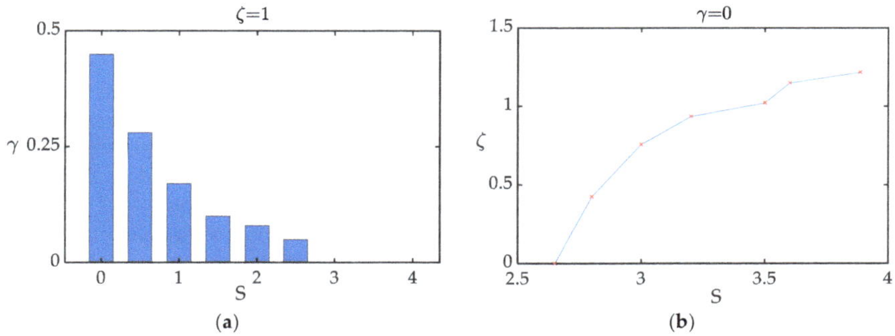

Figure 11. The marginal evaluation results. (**a**) Exclusion ratio for sampled scaling factors at marginal value $\zeta = 1$ (**b**) Achievable marginal requirement.

6.1.2. Maneuver Requirement

In this case, the proposed framework was used to find the optimum design parameters that would allow the system to execute its assigned mission in the event of an actuator failure. The required moment data to track mission trajectory were obtained from the assumed UAV model simulation at nominal conditions. The distribution of moment data points in its three directions of moment space $\mathbb{R}^3$ was portrayed geometrically by constructing the SDE using Equations (12)–(14). The framework used Equation (17) to maximize and fit the SDE described by Equation (14) into the AMS, and Figure 12 shows the result as a plot of parameters and cost function against the number of iterations. Similarly, the parameters were converged to values that maximize the cost function within their constraint limits.

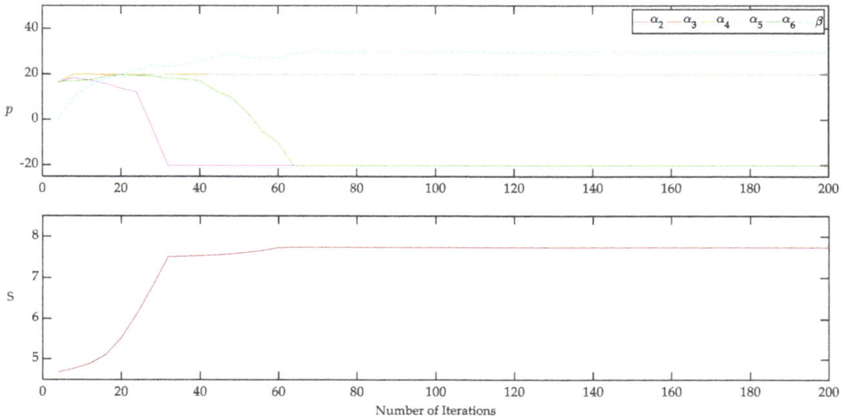

Figure 12. Optimization output: plot of actuator's parameters and cost function trend against the number of iterations for actuator 1 failure and prescribed maneuver performance requirement.

Similarly, a comparison of the preliminary design of the AMS (yellow) and the configuration augmented with optimum parameters (aqua) for each actuator's failure is presented in Figure 13. Unlike the hovering operation, the maneuver requires different control authorities in all moment directions in this operation. In this case, the optimization fits Equation (14), which describes the required moments to meet the assigned maneuver into the AMS using the formulation in Equation (17). Similarly, the results demonstrated that sufficient control authority was obtained in all directions, based on their relative weight. The initial values, optimal values, and the resulting cost function computed for each actuator's possible failure are listed in Table 3.

Using Algorithm 1 given in Section 4, the orientation of points can be defined using the exclusion ratio γ and confidence level $p_r(S)$ given marginal value ζ for an arbitrary value of scale factor S, as shown in Figure 14a. In the preliminary design, failure in actuator one results in a loss of controllability in one of the directions; hence, the geometry will have zero radii that result in $S = 0$ and the corresponding $p_r(S) = 0$. In this circumstance, about 1/3 of the moments required to perform the needed operation were present outside of the AMS envelope. As S increases, the number of points flowing into the AMS polytope increases, whereas the number of points outside the envelopes decreases, as indicated by decreasing of γ. At $S = 5$ the confidence level reaches a maximum $p_r(S) = 1$, which shows the existence of all points within the ellipsoid and hence in the AMS envelope. However, 5.2% of points reside outside the AMS's prescribed margin. Further magnification of the ellipsoid results in the enlargement of the AMS and crossing of the remaining points across the specified margin inside the AMS. At $S = 6.35$ all points were orientated inside of the requested margin.

Moreover, the maximum marginal value that can be imposed is depicted in Figure 14b. At $\zeta = 0$ all points are orientated inside the AMS polytope without marginal specification. For $\zeta > 0$, the polytope must be enlarged to keep $\gamma = 0$. As a result of imposed constraint on the parameter, the maximum marginal value that can be achieved was $\zeta = 2.05$, which corresponds to the maximum scale factor ($S = 7.747$). Therefore, the computed parameters can be used to reconfigure the UAV to tolerate the considered fault and perform the desired maneuver.

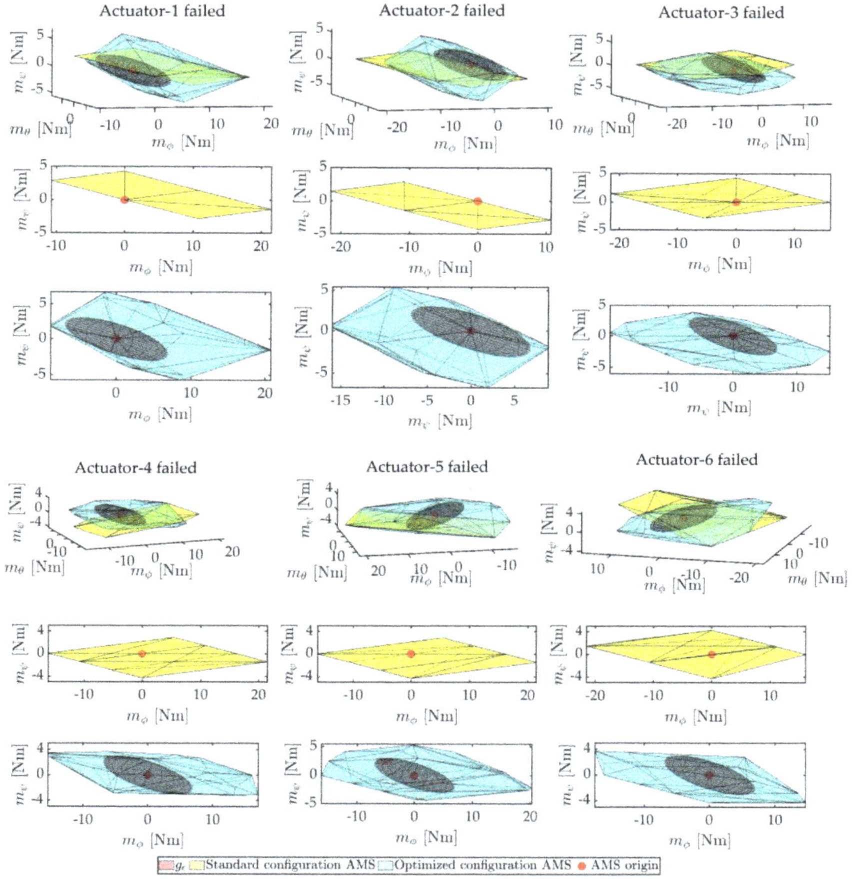

Figure 13. AMS comparison of preliminary configuration with optimized configuration for each actuator failure and prescribed maneuver performance requirement.

Table 3. Mission-based parameter optimization results for single actuator failure at a time.

Fault Condition	Parameters (Angles in Degree) Initial Value	Parameters (Angles in Degree) Optimization	Cost Function
Actuator 1 failed	$\begin{bmatrix}0\\15\\15\\15\\15\\15\\15\end{bmatrix}$	$\begin{bmatrix}30 & 0 & -20 & 20 & 20 & 20 & -20\end{bmatrix}^T$	7.747
Actuator 2 failed		$\begin{bmatrix}30 & 20 & 0 & 20 & -20 & 20 & 10\end{bmatrix}^T$	8.416
Actuator 3 failed		$\begin{bmatrix}30 & 20 & 20 & 0 & 20 & 20 & -12\end{bmatrix}^T$	8.595
Actuator 4 failed		$\begin{bmatrix}30 & 20 & 20 & -20 & 0 & 20 & 20\end{bmatrix}^T$	8.101
Actuator 5 failed		$\begin{bmatrix}30 & 20 & -5 & -20 & 20 & 0 & -10\end{bmatrix}^T$	8.173
Actuator 6 failed		$\begin{bmatrix}30 & 20 & 17 & 20 & 20 & -20 & 0\end{bmatrix}^T$	6.751

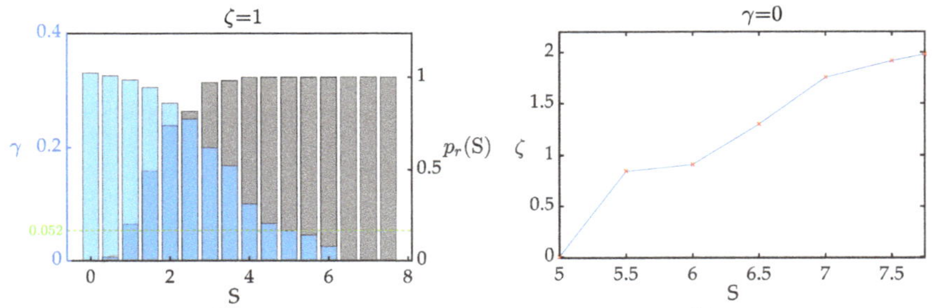

(**a**) (**b**)

Figure 14. The marginal evaluation results for designed mission profile (**a**) Trend of exclusion ratio and confidence level against cost function at a different stage of design parameter optimization is plotted at $\zeta = 1$; (**b**) Achievable marginal requirement.

6.2. Simulation Result

6.2.1. Scenario 1

The assumed UAV model simulation was used to prove the optimized configuration's ability to survive specified actuator failure while hovering at the target as shown in simulation Video S1. As shown in Figure 15, the UAV with the preliminary actuator orientation was commanded to take off to six meters and hover. While hovering, the fault was injected into actuator1 at a simulation time of 20 s, and the propellers were steered to tilt after sufficient detection time. The simulation result demonstrated that the optimum configuration compensated for the lost control after some perpetuation and stabilized towards its hovering state, as shown in Figure 16.

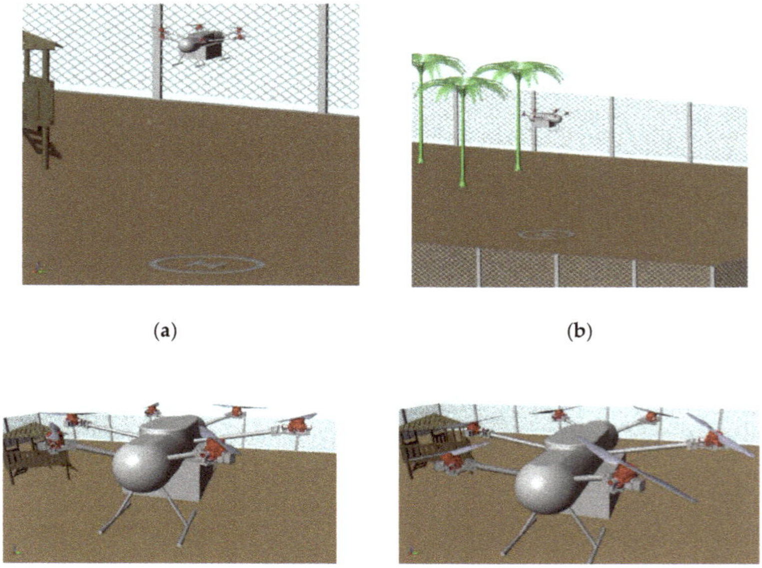

(**a**) (**b**) (**c**) (**d**)

Figure 15. Simulation for hovering flight (**a**) Hovering at a given height in the nominal situation. (**b**) Right-side view of hovering at a given height in the presence of actuator failure. (**c**) Close-up view of hovering flight before actuator failure. (**d**) Close-up view of actuator's reorientation after actuator failure at recovered hovering.

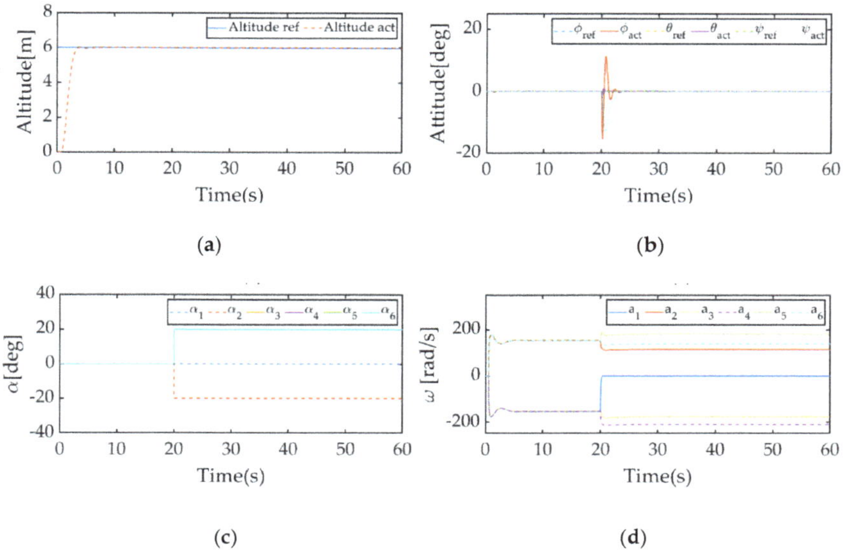

Figure 16. Hovering test result for fault injected at t = 20 s on propeller (**a**) Altitude of the vehicle (**b**) Attitude response (**c**) Propeller tilting angle (**d**) Propeller's rotation rate.

6.2.2. Scenario 2

In this scenario, the ability of a configuration with optimum design parameters to navigate via waypoints was evaluated in the event of a single actuator failure. The waypoints are positioned so that they reflect the tasks that are carried out to avoid static barriers that may be encountered in real-world applications. The B-spline trajectory generating technique established in [28] was used to combine the waypoints as shown on Figure 17.

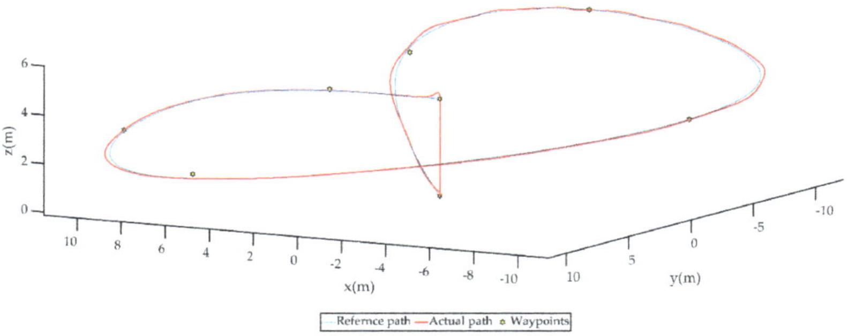

Figure 17. The path followed by the UAV.

As shown in Figure 18, the possible environmental confrontation is depicted as windows at different heights, trees, and a house. The first window was placed in such a way that it allowed the UAV to pass at a lower altitude below two meters, whereas the second window was placed at the height of six meters. Following the mission profile, the UAV was ordered to take off to the altitude of four meters (Figure 18a) pitch forward about ten meters, and follow the curved path to the first and second window obstacles while rolling, pitching, and descending to the height of two meters simultaneously (Figure 18c). Then it had to ascend simultaneously to an altitude of six meters (Figure 18d) to pass through the opening, and finally land at the depicted landing pad (Figure 18b). Therefore, in this flight path, the performance of the optimized configuration during a single actuator failure was conducted to fulfill the specified operation.

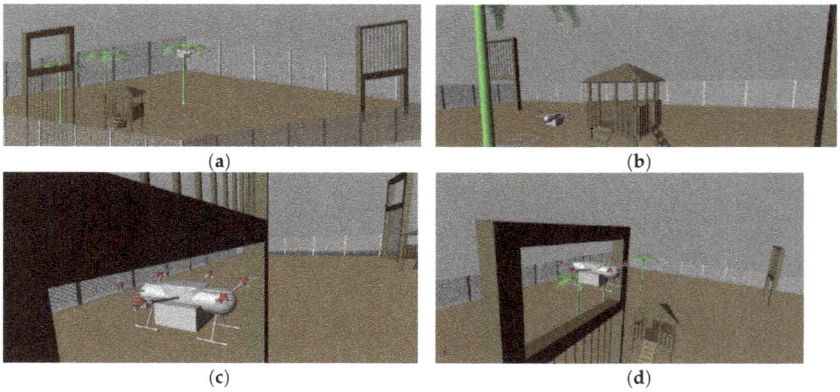

Figure 18. Trajectory following simulation result with actuator-1 failed (**a**) Take-off (**b**) Landing (**c**) Passing through obstacle 1 (**d**) Passing through obstacle 2.

The UAV was reconfigured to the optimum propeller tilting and offset angle listed in Table 3 corresponding to the actuator-1 failure. As shown in Figure 19, the result showed that the desired operation is fulfilled while the actuator-1 failed with optimized parameters.

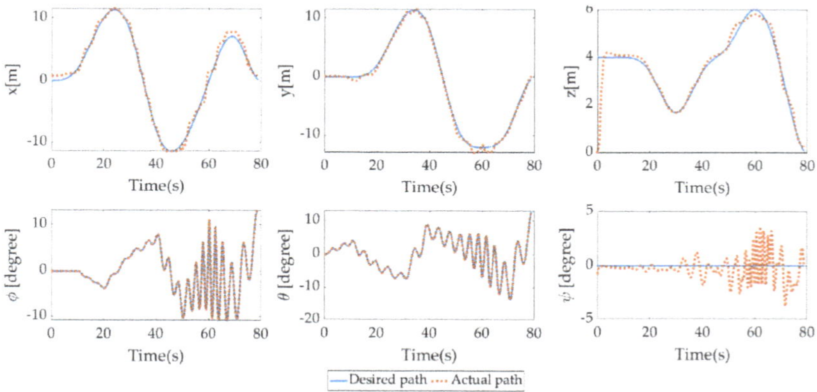

Figure 19. Simulation result of the optimized configuration in tolerating actuator 1 failure and performing maneuvers.

7. Conclusions

This work proposed a reliable optimization strategy that can be employed to design actuator fault-tolerant multirotor UAV configuration. The framework considers the required moment data derived from the designed mission profile and disturbance rejection requirement. Given the required moment as a geometry that describes its distribution and the actuator's health status indicator, the optimizer aims to maximize the scaling factor of the geometry and fit into the AMS, such that the requirements lay under the system capability in the presence of a failed actuator. An efficient marginal evaluation algorithm is proposed to quantify the extent of capability margin. The framework is applied to the delivery drone concept developed by the PNU drone with six rotors. The assumed UAV is modified with a one-direction rotor active tilting mechanism to allow the system to reconfigure itself in the event of failure and recovery. Firstly, the strategy is verified by a multivariable optimization of selected design parameters for performing a given task under fault conditions, and the resulting trend of the cost function and parameter was plotted. The optimization result shows that the proposed approach maximizes the AMS to enclose requirements under system capability, and the resulting cost function is clearly plotted against the exclusion ratio to show the orientation of points relative to the AMS. The author believes that this work is a fundamental and essential step in designing fail-safe operations, such as obstacle-free trajectory, safe landing site search, etc.

Supplementary Materials: The following supporting information can be downloaded at: https://www.mdpi.com/article/10.3390/app12136781/s1, Video S1: Simulation of actuator fault tolerant multirotor UAV with tilting actuators.

Author Contributions: Conceptualization, Y.D.; methodology, Y.D.; software, Y.D. and H.-Y.S.; validation, Y.D. and A.W.; investigation, Y.D. and H.-Y.S.; writing original draft preparation, Y.D.; writing review and editing, Y.D., H.-Y.S. and A.W.; visualization, Y.D. and J.-H.K.; supervision, B.-S.K. All authors have read and agreed to the published version of the manuscript.

Funding: This work was supported by the National Research Foundation of Korea (NRF) grant funded by the Korean government (MIST) (No. 2012R1A5A1048294).

Conflicts of Interest: The authors declare no conflict of interest.

Appendix A. Position and Orientation Matrix Derivation

The relationship between the design parameters and the force and moment generated can be summarized as follows:

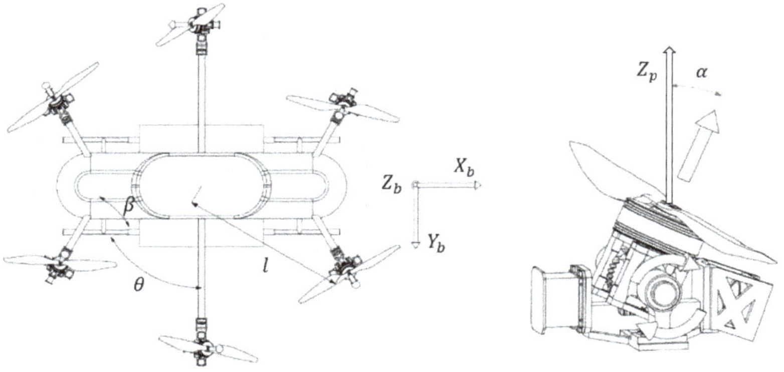

Figure A1. Structural layout of proposed UAV.

The propeller's position can be described with rotation about the body frame Z_b axis by angle θ as shown in Equation (3) in Section 2. The generalized position matrix x of the assumed Hexarotor UAV preliminary configuration shown is defined as:

$$x = \begin{bmatrix} 0 & 0 & \mathrm{s}(\theta) & -\mathrm{s}(\theta) & \mathrm{s}(\theta) & -\mathrm{s}(\theta) \\ \mathrm{c}\theta & -\mathrm{c}\theta & -\mathrm{c}(\theta) & \mathrm{c}(\theta) & \mathrm{c}(\theta) & -\mathrm{c}(\theta) \\ 0 & 0 & 0 & 0 & 0 & 0 \end{bmatrix}$$

Referring from Section 5, for optimization purposes, the orientation matrix was modified with offsetting angle β as:

$$\hat{x} = \begin{bmatrix} 0 & 0 & \mathrm{s}(\theta-\beta) & -\mathrm{s}(\theta-\beta) & \mathrm{s}(\theta-\beta) & -\mathrm{s}(\theta-\beta) \\ \mathrm{c}(\theta-\beta) & -\mathrm{c}(\theta-\beta) & -\mathrm{c}(\theta-\beta) & \mathrm{c}(\theta-\beta) & \mathrm{c}(\theta-\beta) & -\mathrm{c}(\theta-\beta) \\ 0 & 0 & 0 & 0 & 0 & 0 \end{bmatrix}$$

The orientation of each propeller can be computed as the successive rotation of the arm with an angle θ about the body frame Z_b axis and about propellers coordinate axis y_p and x_p with angles α_y and α_x, respectively.

$$R_{Z_b}(\theta) = \begin{bmatrix} c_\theta & s_\theta & 0 \\ -s_\theta & c_\theta & 0 \\ 0 & 0 & 1 \end{bmatrix}$$

$$R_{y_p}(\alpha_y) = \begin{bmatrix} c_{\alpha_y} & 0 & -s_{\alpha_y} \\ 0 & 1 & 0 \\ s_{\alpha_y} & 0 & c_{\alpha_y} \end{bmatrix}$$

$$R_{x_p}(\alpha_x) = \begin{bmatrix} 1 & 0 & 0 \\ 0 & c_{\alpha_x} & s_{\alpha_x} \\ 0 & -s_{\alpha_x} & c_{\alpha_x} \end{bmatrix}$$

From Equation (4), the generalized propellers orientation matrix is given as:

$$q = \begin{bmatrix} \mathrm{s}(\alpha_1) & -\mathrm{s}\alpha_2 & -\mathrm{s}(\theta)\mathrm{s}\alpha_3 & -\mathrm{s}(\theta)\mathrm{s}\alpha_4 & -\mathrm{s}(\theta)\mathrm{s}\alpha_5 & -\mathrm{s}(\theta)\mathrm{s}\alpha_6 \\ 0 & 0 & -\mathrm{c}(\theta)\mathrm{s}\alpha_3 & -\mathrm{c}(\theta)\mathrm{s}\alpha_4 & -\mathrm{c}(\theta)\mathrm{s}\alpha_5 & -\mathrm{c}(\theta)\mathrm{s}\alpha_6 \\ \mathrm{c}\alpha_1 & \mathrm{c}\alpha_2 & \mathrm{c}\alpha_3 & \mathrm{c}\alpha_4 & \mathrm{c}\alpha_5 & \mathrm{c}\alpha_6 \end{bmatrix}$$

Recalling from Section 2 for opmization purposes, the orientation matrix can be modified with offsetting angle β as:

$$\hat{q} = \begin{bmatrix} \mathrm{s}\alpha_1 & -\mathrm{s}\alpha_2 & -\mathrm{s}(\theta-\beta)\mathrm{s}\alpha_3 & -\mathrm{s}(\theta-\beta)\mathrm{s}\alpha_4 & -\mathrm{s}(\theta-\beta)\mathrm{s}\alpha_5 & -\mathrm{s}(\theta-\beta)\mathrm{s}\alpha_6 \\ 0 & 0 & -\mathrm{c}(\theta-\beta)\mathrm{s}\alpha_3 & -\mathrm{c}(\theta-\beta)\mathrm{s}\alpha_4 & -\mathrm{c}(\theta-\beta)\mathrm{s}\alpha_5 & -\mathrm{c}(\theta-\beta)\mathrm{s}\alpha_6 \\ \mathrm{c}\alpha_1 & \mathrm{c}\alpha_2 & \mathrm{c}\alpha_3 & \mathrm{c}\alpha_4 & \mathrm{c}\alpha_5 & \mathrm{c}\alpha_6 \end{bmatrix}$$

References

1. Fourlas, G.K.; Karras, G.C. A Survey on Fault Diagnosis and Fault-Tolerant Control Methods for Unmanned Aerial Vehicles. *Machines* **2021**, *9*, 197. [CrossRef]
2. Baldini, A.; Felicetti, R.; Freddi, A.; Longhi, S.; Monteriù, A. Actuator Fault-Tolerant Control Architecture for Multirotor Vehicles in Presence of Disturbances. *J. Intell. Robotics Syst.* **2020**, *99*, 859–874. [CrossRef]
3. Pose, C.D.; Giribet, J.I.; Mas, I. Fault Tolerance Analysis for a Class of Reconfigurable Aerial Hexarotor Vehicles. *IEEE/ASME Trans. Mechatron.* **2020**, *25*, 1851–1858. [CrossRef]
4. Schneider, T.; Bouabdallah, S.; Rudin, G.D.K. Fault-Tolerant Multirotor Systems. Master's Thesis, ETH Zurich, Zürich, Switzerland, 2011.
5. Michael, A.; Klaus, M.D.; Danie, G.; Jan, S. Design of a Multi Rotor MAV with regard to Efficiency, Dynamics and Redundancy. In Proceedings of the AIAA 2012-4779, AIAA Guidance, Navigation, and Control Conference, Minneapolis, MN, USA, 13–16 August 2012.
6. Sanjuan, A.; Nejjari, F.; Sarrate, R. Reconfigurability Analysis of Multirotor UAVs under Actuator Faults. In Proceedings of the 2019 4th Conference on Control and Fault Tolerant Systems (SysTol), Casablanca, Morocco, 18–20 September 2019; pp. 26–31. [CrossRef]
7. Yang, T.; Li, P.; Zhang, H.; Li, J.; Li, Z. Monocular Vision SLAM-Based UAV Autonomous Landing in Emergencies and Unknown Environments. *Electronics* **2018**, *7*, 73. [CrossRef]
8. Wayne, D.; Kenneth, A.B.; Roger, B. *Aircraft Control Allocation*; Wiley: Chichester, UK, 2017.
9. Wang, B.; Shi, W.; Miao, Z. Confidence Analysis of Standard Deviational Ellipse and Its Extension into Higher Dimensional Euclidean Space. *PLoS ONE* **2015**, *10*, e0118537. [CrossRef] [PubMed]
10. Wolfgang, K.H.; Léopold, S. *Applied Multivariate Statistical Analysis*; Springer: Berlin/Heidelberg, Germany, 2015. [CrossRef]
11. Gong, J. Clarifying the Standard Deviational Ellipse. *Geogr. Anal.* **2002**, *34*, 155–167. [CrossRef]
12. Jiang, G.; Richard, M.V. A nonparallel hexrotor UAV with faster response to disturbances for precision position keeping. In Proceedings of the 2014 IEEE International Symposium on Safety, Security, and Rescue Robotics, Hokkaido, Japan, 27–30 October 2014; pp. 1–5.
13. Zhang, J.; Max, S.; Florian, H. Attainable Moment Set Optimization to Support Configuration Design: A Required Moment Set Based Approach. *Appl. Sci.* **2021**, *11*, 3685. [CrossRef]
14. Wen, F.H.; Hsiao, F.Y.; Shiau, J.K. Analysis and Management of Motor Failures of Hexacopter in Hover. *Actuators* **2021**, *10*, 48. [CrossRef]
15. Haicheng, L.; Rodney, T.; Peter, V.O.; Martijn, M. Executing convex polytope queries on nD point clouds. *Int. J. Appl. Earth Obs. Geoinf.* **2021**, *105*, 102625. [CrossRef]

16. Weisstein, E.W. Point-Plane Distance. From MathWorld—A Wolfram Web Resource. Available online: https://mathworld.wolfram.com/Point-PlaneDistance.html (accessed on 9 May 2022).
17. Denery, T.; Ghidella, J.; Mosterman, P.; Shenoy, R. Creating Flight Simulator Landing Gear Models Using Multidomain Modeling Tools. In Proceedings of the AIAA Meeting Papers on Disc, (2006-6821), Keystone, CO, USA, 21–24 August 2006; American Institute of Aeronautics and Astronautics: Reston, VA, USA, 2006.
18. MathWorks Inc. *Simscape MultibodyTM*; MathWorks Inc.: Natick, MA, USA, 2020.
19. Khurrum, M.; Norilmi, A.I. Application of multibody simulation tool for dynamical analysis of tethered aerostat. *J. King Saud Univ. Eng. Sci.* **2022**, *34*, 209–216. [CrossRef]
20. Sun, W.; Zhang, X.; Lin, G.; Wang, H.; Han, J. Extreme Maneuvering Control and Planning of Multi-Rotor UAV for High-Speed Invading Target Avoidance. In Proceedings of the 2021 IEEE International Conference on Real-Time Computing and Robotics (RCAR), Xining, China, 15–19 July 2021; pp. 387–392. [CrossRef]
21. Chen, C.; Zhang, J.; Zhang, D.; Shen, L. Control and flight test of a tiltrotor unmanned aerial vehicle. *Int. J. Adv. Robot. Syst.* **2017**, *14*, 172988141667814. [CrossRef]
22. Hussein, H. Fault-Tolerant Control of a Multirotor Unmanned Aerial Vehicle under Hardware and Software Failures. Ph.D. Thesis, Université Libanaise, Beirut, Lebanon, 2020.
23. Vey, D.; Lunze, J. Structural reconfigurability analysis of multirotor UAVs after actuator failures. In Proceedings of the 2015 54th IEEE Conference on Decision and Control (CDC), Osaka, Japan, 15–18 December 2015; pp. 5097–5104. [CrossRef]
24. Merheb, A.; Noura, H.; Bateman, F. Emergency Control of AR Drone Quadrotor UAV Suffering a Total Loss of One Rotor. *IEEE/ASME Trans. Mechatron.* **2017**, *22*, 961–971. [CrossRef]
25. Giribet, J.I.; Sanchez, R.S.; Ghersin, A.S. Analysis and design of a tilted rotor hexacopter for fault tolerance. *IEEE Trans. Aerosp. Electron. Syst.* **2016**, *52*, 1555–1567. [CrossRef]
26. Giribet, J.I.; Pose, C.D.; Ghersin, A.S.; Mas, I. Experimental Validation of a Fault-Tolerant Hexacopter with Tilted Rotors. *Int. J. Electr. Electron. Eng. Telecommun.* **2018**, *7*, 58–65. [CrossRef]
27. Michieletto, G.; Ryll, M.; Franchi, A. Control of statically hoverable multi-rotor aerial vehicles and application to rotor-failure robustness for hexacopters. In Proceedings of the 2017 IEEE International Conference on Robotics and Automation (ICRA), Singapore, 29 May–3 June 2017; pp. 2747–2752. [CrossRef]
28. Gauthier, R.; Cristina, S.M.; Sihem, T.; Mathieu, B.; Nicolas, M. Minimum-time B-spline trajectories with corridor constraints. Application to cinematographic quadrotor flight plans. *Control Eng. Pract.* **2019**, *89*, 190–203. [CrossRef]

applied sciences

MDPI

Article

A Small UAV Optimized for Efficient Long-Range and VTOL Missions: An Experimental Tandem-Wing Quadplane Drone

Michał Okulski * and Maciej Ławryńczuk

Institute of Control and Computation Engineering, Warsaw University of Technology, ul. Nowowiejska 15/19, 00-665 Warsaw, Poland; maciej.lawrynczuk@pw.edu.pl
* Correspondence: pl.micas.pro@gmail.com

Abstract: Most types of Unmanned Aerial Vehicle (UAV, drone) missions requiring Vertical-Take-Off-and-Landing (VTOL) capability could benefit if a drone's effective range could be extended. Example missions include Search-And-Rescue (SAR) operations, a remote inspection of distant objects, or parcel delivery. There are numerous research works on multi-rotor drones (e.g., quadcopters), fixed-wing drones, VTOL quadplanes, or tilt-motor/tilt-wing VTOLs. We propose a unique compact VTOL UAV optimized for long hover and long-range missions with great lifting capacity and manoeuvrability: a tandem-wing quadplane with fixed motors only. To the best of our knowledge, such a drone has not yet been researched. The drone was designed, built, and tested in flight. Construction details, its advantages, and issues are discussed in this research.

Keywords: UAV; quadcopter; quadplane; multicopter; multirotor; VTOL; tandem-wing; long-range

Citation: Okulski, M.; Ławryńczuk, M. A Small UAV Optimized for Efficient Long-Range and VTOL Missions: An Experimental Tandem-Wing Quadplane Drone. *Appl. Sci.* **2022**, *12*, 7059. https://doi.org/10.3390/app12147059

Academic Editors: Sylvain Bertrand and Hyo-sang Shin

Received: 20 May 2022
Accepted: 7 July 2022
Published: 13 July 2022

1. Introduction

Currently, the Unmanned Aerial Vehicle (UAV) market is snowballing—thanks to many successful civilian and military applications, including, but not limited to photogrammetry [1], remote inspection [2], parcel delivery [3], disaster recovery [4,5], etc. Most UAV missions will benefit from extending the drone range as much as possible while maintaining the Vertical-Take-Off-and-Landing (VTOL) capability. We may consider medical supplies [6], Search-And-Rescue (SAR) operations [7,8], and even non-critical missions such as ordinary remote inspection of distant objects (e.g., wind turbines) or simple parcel delivery.

In this paper, we make the following contribution: we investigate a rare type of UAV (Figure 1), designed for a mission where a small-sized, agile, long-range VTOL drone is capable of precise hovering over a distant target. Many aspects of the drone are discussed, including, but not limited to test flight results (Figures 2 and 3). We define the size and weight constraints as follows: a ready-to-fly drone should fit into a square of 1×1 m and weigh less than 3 kg (without payload). It should be able to carry at least 300 g of payload and stay airborne for at least 15 min. This research aims to design, build, and test a UAV with maximized range and hovering time.

There are many different types of UAVs, each having unique features. Figure 4 presents the most common configurations:

(a) A flying wing [9,10]—typically used for long-endurance missions, e.g., photogrammetry or aerial photography. It is not a VTOL drone.
(b) A fixed-wing plane [11,12]—similar applications as for a flying wing. Fixed-plane UAVs are usually bigger and can carry more payload. If the plane has landing gear, it can operate from a runway. Again, it is not a VTOL drone.
(c) A helicopter [13,14]—can fulfil VTOL missions when a heavy payload is required. Complex mechanical design, many moving parts, and inefficient due to the energy required for the tail rotor, which does not contribute to the lifting force.

(d) A hexacopter [15,16]—similar applications to the helicopter. Especially popular in aerial photography, the film industry, and crop spraying. Less complex design; typically, it has no moving parts except the motors and propellers. It can survive if one motor or propeller is lost.

(e) A quadcopter [12,17–19]—the most popular type of non-professional UAV. It is very robust and agile; it can be tiny, as well as bigger and more powerful. Usually, it has a limited range because its electric propulsion always comprises lifting force (static thrust) and manoeuvrability (pitch-speed).

(f) A co-axial helicopter [20,21]—a rare type of UAV. It basically has the same features as a helicopter, except there is no tail rotor; thus, it is more energy-efficient. However, double main rotors mounted on a long shaft make it prone to wind gusts.

(g) A quadplane [22–24]—combines the benefits of a VTOL quadcopter with all features of a fixed-wing plane. The drawbacks include higher take-off mass and increased drag due to extra motors and motor holders or additional motor tilting mechanics. The wings and the tailplane make this type of drone less agile in hover and more prone to wind gusts.

(h) A tandem-wing quadplane—combines essential advantages of VTOL and fixed-wing UAVs. There are a few full-scale designs, and prototypes [25,26] mainly use multiple motors to hover (e.g., 6, 8, 12, or more). Existing tandem-wing drones mostly use a tilt-motor [22,27,28] or tilt-wing design [24,29]. We decided to select this type of UAV because it has no extra motor holder beams (the two wings support the motors), and it can have a smaller wingspan than a quadplane with a comparable wing surface. It is agile in hover because it acts almost like a regular quadcopter. Thanks to the wings and the pusher motor, it can fly fast in a level flight without using significant energy. Furthermore, all its electric propulsion can be optimized for a single role: four main motors for hover conditions (high static thrust) and the pusher motor for fast level flight (high pitch-speed).

Figure 1. The Elka1Q—an experimental tandem-wing quadplane UAV.

Figure 2. The Elka1Q drone in one of its test flights.

Figure 3. View from the on-board video camera; altitude ca. 55 m above the airfield.

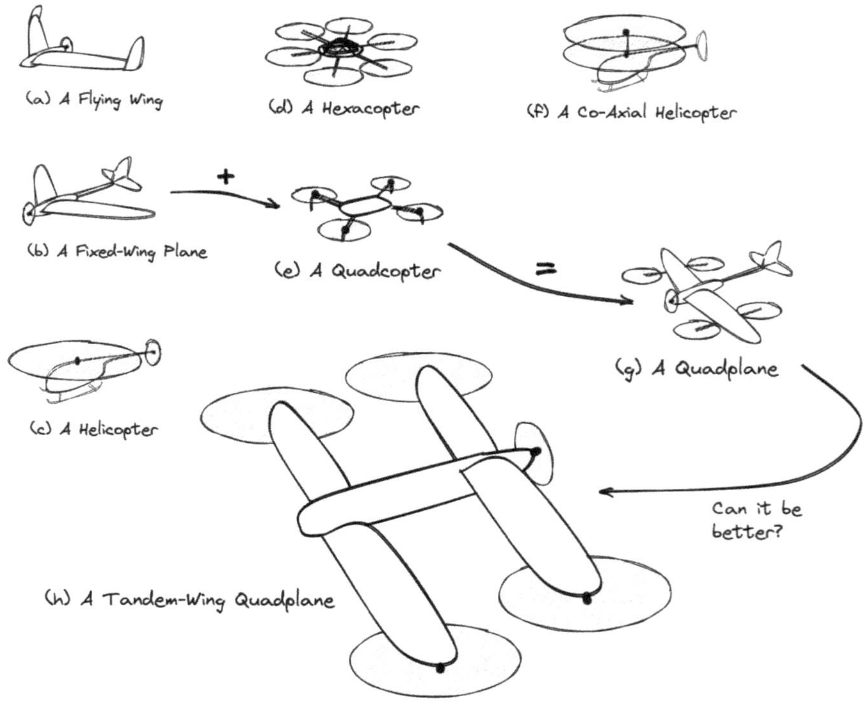

Figure 4. An overview of popular UAV types.

2. Materials and Methods

2.1. Selecting the Drone Type and Its Design Details

The need to start and land vertically implies choosing one of the rotorcraft vehicle types. A multirotor, especially a **quadcopter**, seems to be a better choice than a classic helicopter. A helicopter has many more complex mechanical elements (e.g., main rotor head, a swashplate, pushrods). In contrast, the quadcopter has just four motors with fixed propellers and no moving parts at all. There are two ways of reaching a distant target by a quadcopter: fly fast and drain the battery quickly or stay mid-air longer, but fly at a slower speed. In both cases, flying forward causes more significant electric energy consumption than hovering. However, adding wings and a propulsion system dedicated (and optimized) for a forward flight might increase the drone's range (a drawback: adding some extra weight). We wanted to keep the drone small and agile. We finally decided to research a rare **tandem-wing configuration** rather than a typical quadplane. Such a design allowed us to increase the wing area without exceeding size constraints and minimize structural support elements—the wings could become the quadcopter motors' holders. The tandem-wing (i.e., a lifting-tail) airplane can have its centre of gravity more aft than a regular tail plane. That is beneficial from the quadcopter point of view because the main motors are loaded more evenly.

We decided to use a pusher motor as the propulsion system for the level flight to avoid drag produced by the propeller wake. The four main propellers' wake could introduce strong turbulence over the wingtips; therefore, we moved the motors below the wings.

2.2. Electric Propulsion

A helpful rule of thumb is that one starts with the electric propulsion design first because it is easier to match the mechanical design of the fuselage to known components' sizes and weights. That is especially important for the battery pack—usually the heaviest and largest element.

At the time of writing, the eCalc [30] seems to be the best and most accurate tool, including, but not limited to finding the best motor–propeller–battery setup (see Figures 5 and 6). Another great tool, especially for predicting electric RC plane performance, is MotoCalc [31]. We used it for detailed pusher propulsion performance prediction, e.g., to estimate top speed in level flight and find the stall speed for various wing setups (see Figure 7). Considering the general assumptions mentioned in Section 1, we decided to look for high-D/P-ratio propellers (diameter-to-pitch ratio), a high-voltage setup, and lithium-ion batteries rather than lithium-polymer ones. A propeller with a high D/P ratio provides great static thrust (a lift force in the case of a drone), but the thrust drops suddenly when the airspeed increases. That phenomenon, however, may impact drone manoeuvrability. Typically, fast and agile drones (e.g., racing quadcopters) use the D/P ratio in the range [1 . . . 2], where propellers with D/P close to 1 (e.g., 5 × 4) allow high-speed flights, but consume much energy in hover. A high-voltage setup decreases the current needed (assuming a constant total output power)—lower current benefits in thinner electric wires, and smaller Electronic Speed Controllers (ESCs), thus a more lightweight setup. The Li-ion batteries usually offer greater capacity than Li-poly batteries of the same weight, but cannot work in high-current load conditions. The final setup of the quadcopter's electric propulsion is presented in Figure 5 and Table 1, and the predicted performance of the drone (and its range) can be found in Figure 6. Predicted stall speed, top speed, and the total flight time can be found in Figure 7. A summary of the predicted performance is presented in Table 2.

Table 1. The final electric propulsion setup.

Main Quadcopter Motors	4× T-Motor MN3110 KV470
Main Quadcopter Propellers	4× T-Motor Carbon-Fibre (CF) 12 × 4
Main Quadcopter ESCs	4× Hobbywing Micro 35A 3-6S BLHeli
Battery Type	Li-Ion, 18× Sony US18650VTC5 cell, custom-made battery pack
Battery Setup	6S3P
Battery Capacity	7500 mAh
Battery Voltage	21.6 V nominal, 18–25.2 V full range
Battery Weight	936 g
Pusher Motor	1× T-Motor F60 PRO II KV1750
Pusher Propeller	1× APC-E 5 × 5 (2-blade propeller) or 1× HQProp Ethix S5 5 × 4 × 3 (3-blade propeller)
Pusher Motor ESC	1× Beatles 50 A
Total Drive Weight	ca. 1800 g

Table 2. Predicted performance of the electric drive setup; assumed 90% max allowed discharge and 300 g payload (e.g., a video camera with a companion single-board computer).

All-up Weight	2900 g (2600 g + 300 g of payload)
Hover Flight Time	ca. 20.4 min
Top Speed (plane mode, using pusher motor) [1]	**ca. 34–42 m/s (122–144 km/h)**
Top Speed (quadcopter mode only)	ca. 8.9–12.5 m/s (32–45 km/h)
Max Range (VTOL, plane mode) [1]	**ca. 25 km**
Max Range (quadcopter mode only)	ca. 4.6 km
Stall Speed (plane mode)	ca. 11.1–13.9 m/s (40–50 km/h)

[1] The drone's range and top speed are much higher compared to a typical quadcopter.

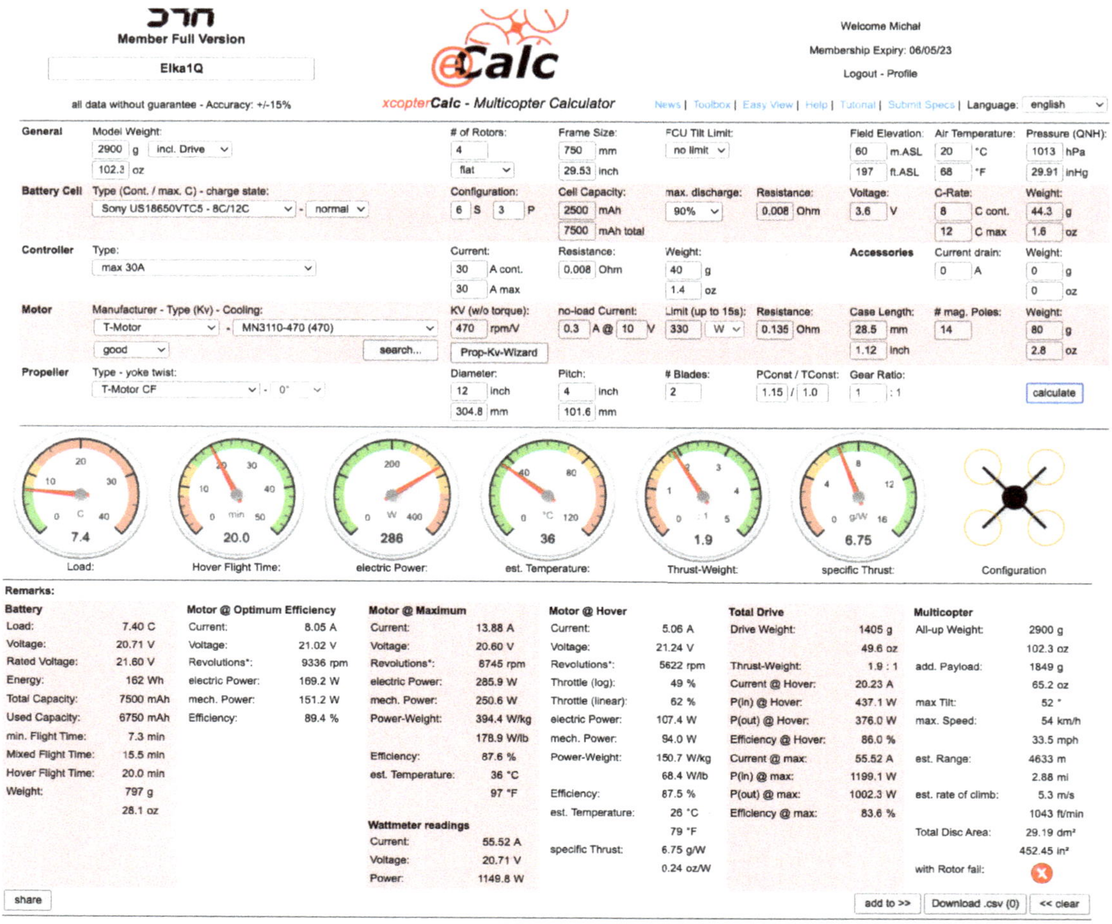

Figure 5. The eCalc [30] tool helped us find the best electric propulsion setup for the Elka1Q drone ; disclaimer from eCalc: * *The manufacturer limitation is NOT monitored* (relates to motor revolutions).

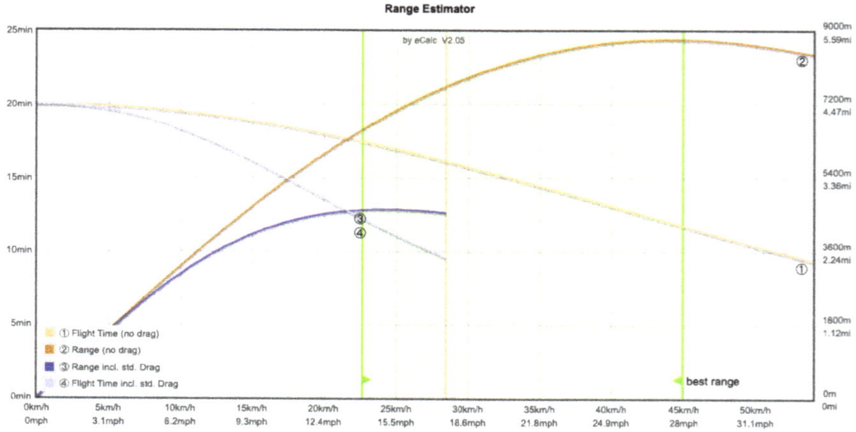

Figure 6. The eCalc [30] tool estimates the drone range and its general performance.

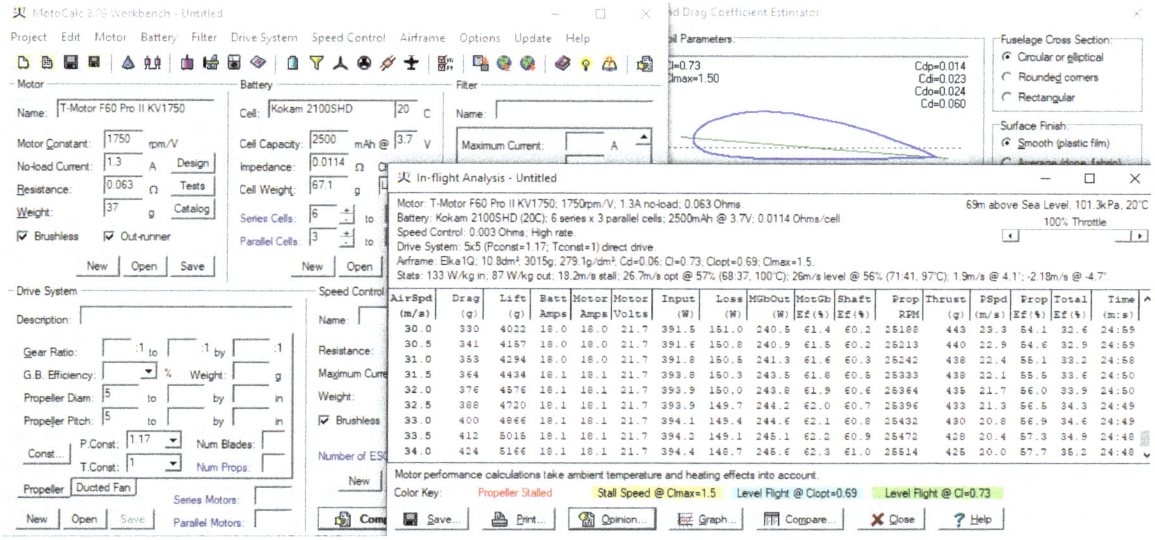

Figure 7. The MotoCalc 8.09 [31] workbench; the tool helped us estimate the drone performance in level flight (plane mode).

2.3. Mechanical Design

The overall shape of the drone (as seen in Figures 8 and 9) is a compromise among the general assumptions (described in Section 1), size and weight of significant components (such as the battery pack), and smart usage of available materials.

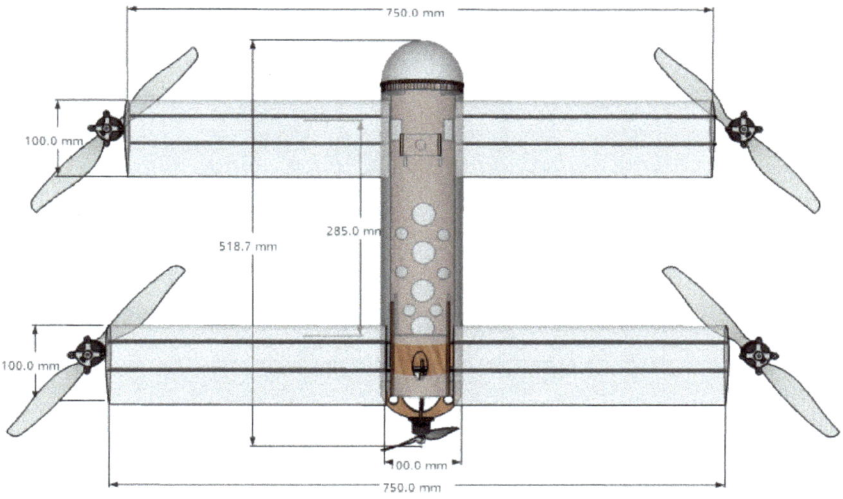

Figure 8. An overview of the Elka1Q drone dimensions—top view.

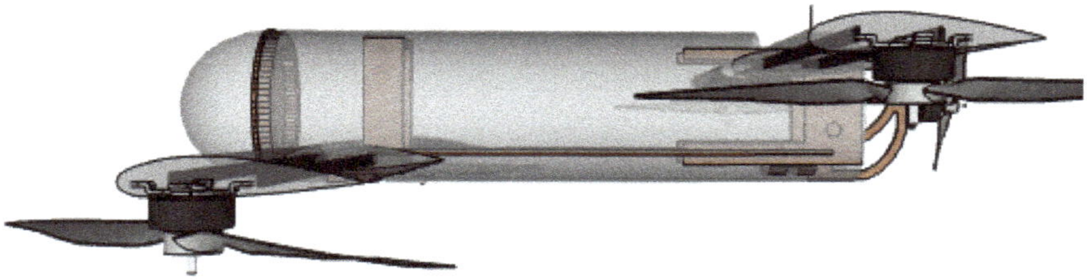

Figure 9. An overview of the Elka1Q drone—side view.

2.3.1. Wings

Typically, drone arms are made of carbon-fibre tubes because they are very stiff and lightweight at the same time. However, such a single tube could have a too big a diameter to fit into the drone's wing. Instead, we decided to use double 6 × 2 mm carbon-fibre flat bars as wing spars. Additionally, the space between them forms a convenient tunnel for electric wires. The wings are built of two matching full-balsa wood elements: a bottom and a top half, both CNC 3D milled and glued together. The leading and trailing edges of a wing are usually prone to accidental damage (especially a very thin trailing edge); therefore, both edges are reinforced with carbon-fibre 4 × 1 mm flat bars. The carbon-fibre wing spars at the wingtips support the main motor holders (CNC milled from a 3mm-thick aluminium sheet). The two elements of the holders are screwed together to catch protruding wing spars tightly. Finally, the surface of the wing is covered by Oracover [32] film. The wing construction proves to be light and very durable. We could say it is a perfect balance between stiffness and elasticity. Initially, we chose a wing profile (an airfoil) optimized for high-speed flight: the P-51D tip (BL215) airfoil (see Figure 10). Generally speaking, high-speed airfoils have low drag, but, on the other hand, have a low lift coefficient, which results in a high stall speed, and that means the plane has to maintain high enough speed to stay airborne in a level flight. That should not be an issue if the pusher motor can accelerate the drone to that speed. Due to safety reasons, we decided to modify the original wings—we made them much thicker (see Figure 11). Such a thick airfoil (thickness increased from 12% to 25% of the airfoil chord) gives us a much higher lift coefficient (resulting in a lower stall speed) at the cost of lowering the top speed. Nevertheless, lower stall speed means we could perform the in-flight experiments of switching between quadcopter and plane mode at lower (i.e., safer) speed, and we could do that in a less spacious airfield.

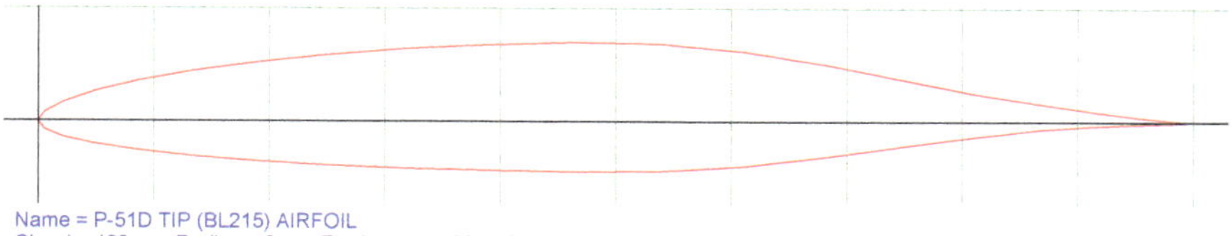

Figure 10. The original airfoil for both drone wings; generated by Airfoil Plotter [33].

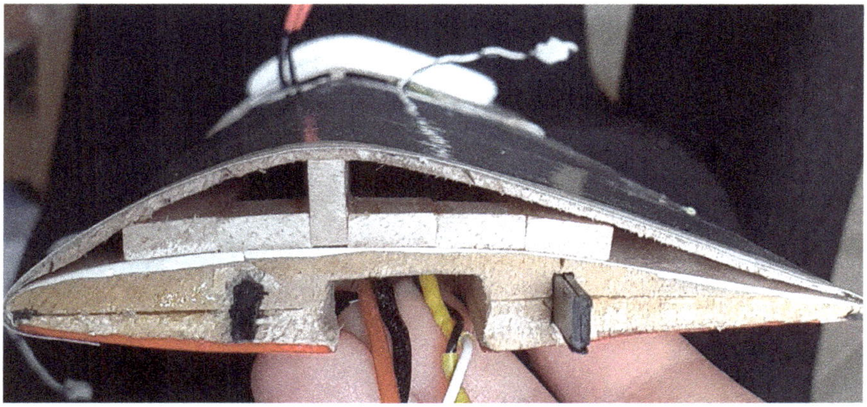

Figure 11. An actual final wing airfoil—the original P-51D tip (BL215) airfoil is still visible underneath an extra top wing surface. The original wing was full-balsa construction, while the modification was based on a few balsa wood strips and a 1.5 mm-thick balsa wood covering.

The wing configuration used in the drone is called a "tandem-wing" or sometimes a "lifting-tail plane". Those names refer to the fact that the aft wing is not just a horizontal stabilizer, like in a classic "tailplane" configuration, but it contributes to the total lift force produced by the plane. It is a rare configuration due to possible stability and controllability issues [34,35]. Sometimes, quite the opposite statements can be found—tandem-wing planes are easier to pilot because of safer stall behaviour [36]. However, there were at least a few successful tandem-wing planes, e.g., Quickie designed by Elbert Leander "Burt" Rutan (and later QAC Quickie Q2) [36,37] and the Proteus [38] built by Scaled Composites (Rutan's company). Another famous tandem-wing plane is the "Flying Flea" (French name: "Pou du Ciel"), designed by Henri Mignet in 1933. A thorough study of many more historical and modern tandem-wing planes and UAVs, as well as their aerodynamic and stability studies, can be found in [34].

A wing that produces lift force also generates a downwash, i.e., the airflow direction behind the trailing edge of the wing is deflected down by the aerodynamic action of the wing. That phenomenon changes the effective Angle of Attack (AoA) of the rear wing in the tandem-wing configuration. Most tandem-wing planes have the front wing mounted lower than the rear wing to minimize the downwash effect of the front wing [34,35]. Additionally, it is recommended to set a higher AoA of the front wing than the aft wing—such a wing setup affects the stall behaviour of the tandem-wing plane. The front wing with a higher AoA will stall first while the aft wing still produces lift force—that situation will cause the plane to pitch down, increase the speed, and ultimately, end the front wing's stall (bring back its lift force) [36]. Following the suggestions, the front wing of the Elka1Q drone was mounted at ca. 4° AoA and the aft wing at ca. 2° AoA.

Finally, there is at least one more critical aspect of every aircraft having wings: Centre of Gravity (CG, CoG). It is crucial to keep the longitudinal stability of an aircraft. We used a CG calculator from the eCalc toolset [30]. The results of the calculation are presented in Figure 12.

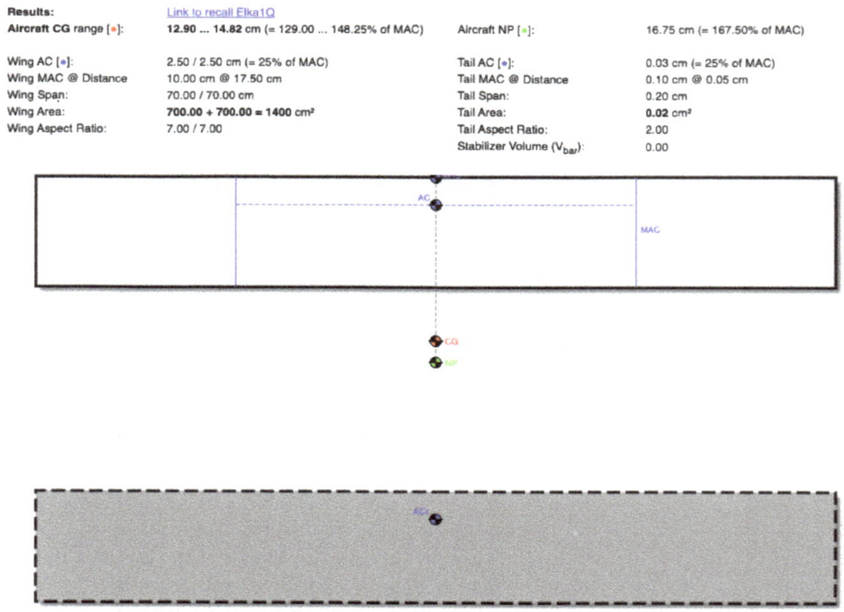

Figure 12. Centre of Gravity (CG) calculated for the Elka1Q drone by the eCalc [30] tool.

2.3.2. Fuselage

The early design of the fuselage (Figure 13) was based on carbon-fibre components (CNC milled from a 2.5 mm-thick CF plate). After a few flight tests, we discovered an issue: the fuselage was twisting about the longitudinal axis, as seen in Figure 14.

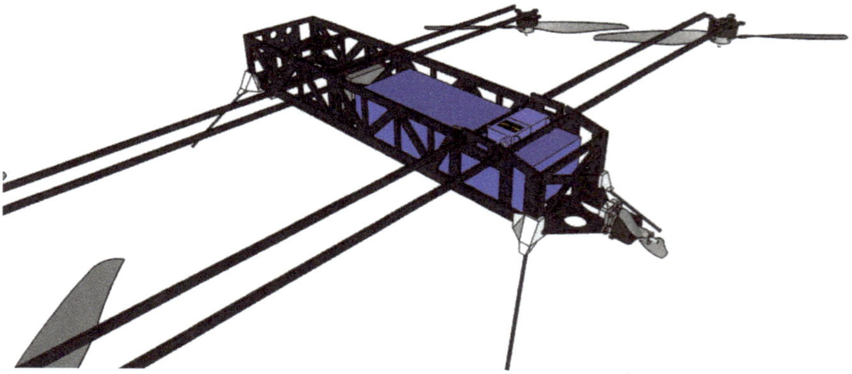

Figure 13. The early Elka1Q fuselage mainly was made of carbon-fibre elements, but it was still not rigid enough.

Figure 14. Flight tests revealed that the early Elka1Q fuselage was twisting about the longitudinal axis (see the arrows).

The final fuselage design was based on a rigid PVC tube (100 mm diameter and 1 mm wall) and a lighter, but still solid plywood structure (Figures 15–17). The PVC tube acts similarly to a monocoque structure, eliminating the twisting about the longitudinal axis.

The landing gear is non-retractable—we made four fixed legs of 3 mm spring steel wire supported by pinewood blocks at the bottom of the fuselage.

The overall structure of the wings and the fuselage proved to be very rigid and robust, surviving a few serious crash landings. The most significant disadvantage of such a compact construction is complicated maintenance of internal components, e.g., access to electronic boards, wires, and connectors.

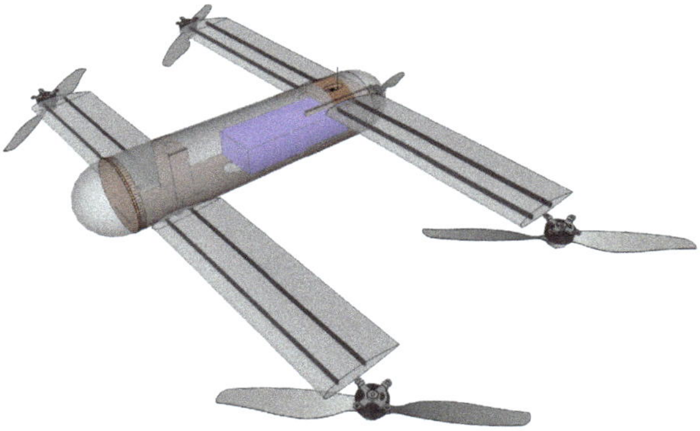

Figure 15. The final design of Elka1Q fuselage—a PVC tube with internal plywood structure.

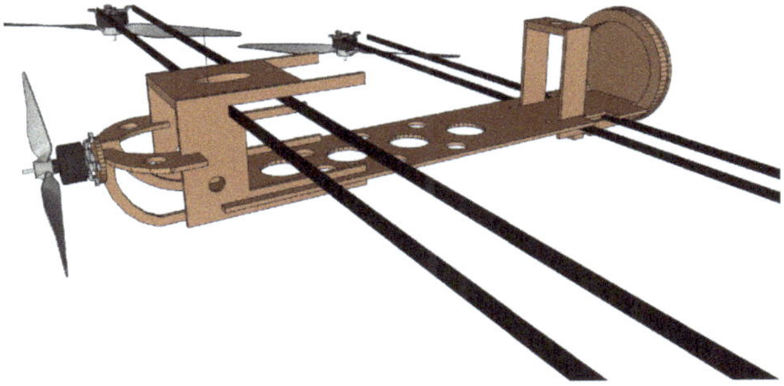

Figure 16. The final design of the Elka1Q fuselage—the internal plywood structure.

Figure 17. The final design of Elka1Q fuselage—a PVC tube with internal plywood structure.

2.4. Electronic Systems

The complete diagram of all electronic components installed in the drone and wirings among them are presented in Figure 18. We chose the Holybro Kakute F7 AIO [39] as the central Flight Controller (FC) board—mainly due to its compact size and efficient primary microcontroller (STM32F745). The FC board was attached to the fuselage through vibration dampers—which is crucial for correct onboard Inertial Measurement Unit (IMU) readings.

The wiring diagram (Figure 18) reveals many connections across far drone sections. To simplify the maintenance of the electronic components, we designed a dedicated connector board (Figure 19). The board exposes all signal sockets and separates the voltage supply for the servos. A separate DC/DC converter provides a 5 V supply for the servos and video camera. The servos could generate dangerous voltage spikes that could interfere, e.g., with the FC or other crucial components. The FC board supplies other components through a built-in 5 V DC/DC converter, which should be free of any voltage spikes. The diagram shows that all power lines go through the FC board because the FC has a built-in current sensor (up to 120 A). It is worth mentioning that the FC has only 6 Pulse Width Modulation (PWM) output channels: four of them are used by the four main quadcopter motor ESCs, and the two elevon servos use the remaining two. Because of the lack of another PWM output, the pusher motor ESC is connected directly to the Radio Control (RC) receiver. We used a Futaba T14SG transmitter and a Futaba R7008SB receiver to pilot the drone. The receiver has the S.Bus output—a single connector to send all 14 channels to the FC board. A 3DR radio transceiver (433 MHz) was used for telemetry and autopilot commands from the Ground Station (GS). A GPS and compass were placed in a compact module mounted outside the fuselage—to improve GPS signal reception and move the compass away from substantial magnetic field interference induced by the power wires. The FC can communicate its status via programmable LEDs (WS2812)—two such LEDs are mounted in the front section, on both sides of the fuselage. The FC has a built-in barometer for altitude reading. However, the barometer's accuracy is limited. We planned to install a down-facing rangefinder for a smooth auto-landing feature. Initially, an ultrasonic sensor was tested, but we decided to replace it with a tiny, affordable, and very reliable Time-of-Flight (ToF) laser rangefinder (Pololu VL53L1X [40]). With its range up to 400cm, the auto-landing mode works perfectly smoothly.

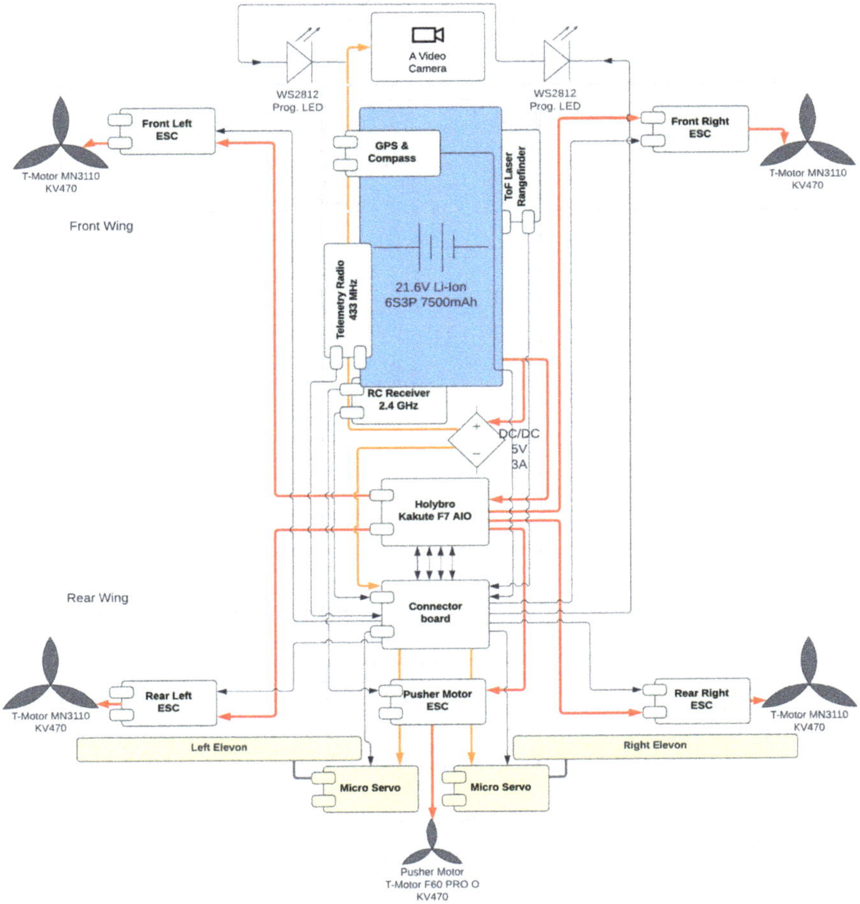

Figure 18. A diagram explaining the wiring of all electronic components of the Elka1Q drone.

We used good-quality 18 AWG power wires with XT60 and XT30 connectors. For the signal connectors, we used Ninigi NXG-02 [41] (2mm raster connectors), 2–6 pins, depending on a particular component to connect.

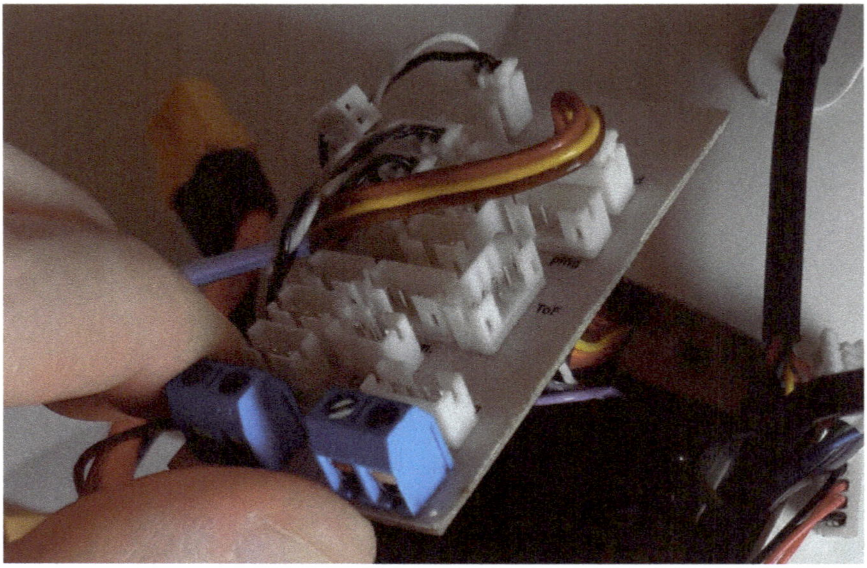

Figure 19. A dedicated connector board improves the maintenance because the flight controller board is mounted deep inside the drone's fuselage.

2.5. Control Principles

Figure 20 explains the forces acting on the drone's body and primary axes of rotation. The drone dynamics can be analysed from a quadcopter and a plane point of view. The forces and moments equations can be derived from Euler's equations for rigid body dynamics—this is thoroughly explained in [12,17,18]. We also explained in our previous paper [42] how the quadcopter control forces Fq (see Figure 20) are mixed to obtain desired moments for the roll, pitch, and yaw axes. Although the quadcopter-like control always actuates all four motors, the control task can be decomposed into isolated controllers for each rotation axis separately [42]. The Fp force in Figure 20 is the forward-directed force produced by the pusher propeller (when the pusher motor is active).

The trailing edge of the rear wing was converted into full-length elevons, i.e., control surfaces that act as an elevator when deflected in the same direction and as ailerons when deflected differentially [12]. The drone has no rudder, which means that while flying in the plane mode (a level flight using pusher motor and wings' lift force, quadcopter motors shut down), only rotation around the roll and pitch axes is possible. We decided that such a simplified control should be enough to maintain the horizontal flight and perform basic turns.

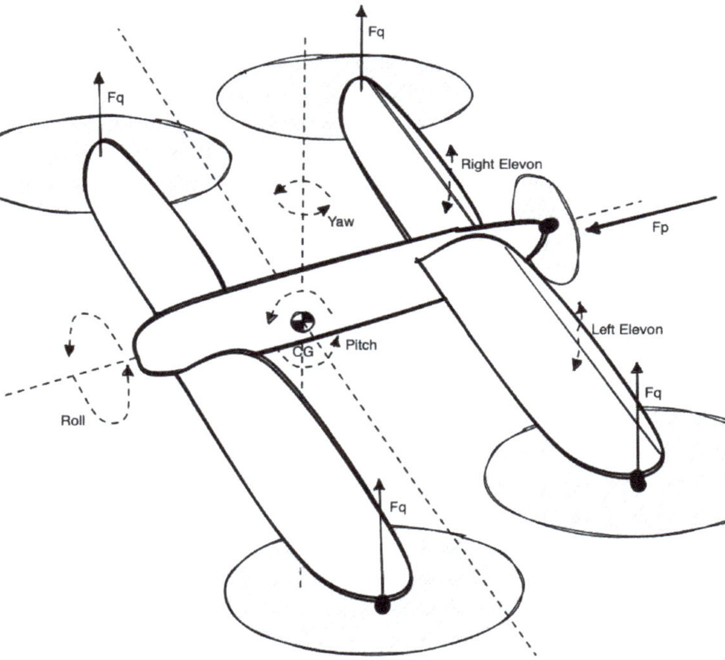

Figure 20. Axes of rotation and control forces produced by the drone's motors and elevons (at the rear wing).

2.6. Flight Controller Software

We chose the Kakute F7 AIO as the primary FC because the ArduPilot [43] software can be compiled and installed on that board. The ArduPilot is a leading open-source autopilot software. It is well tested and greatly supported by a broad community of UAV enthusiasts. We compiled the ArduCopter subset of the ArduPilot stack. Figures 21 and 22 show a high-level overview of the ArduCopter software components. The structure of ArduCopter attitude controllers is shown in Figure 23. The original attitude controller is a cascade of Proportional (P) and Proportional–Integral–Derivative (PID) controllers [44] (for each axis separately). Although the original diagram mentions a Feed-Forward (FF) component, it was eventually disabled in our build of the ArduCopter software. We investigated the overall agility of the attitude controller and eventually implemented a novel variant of the Model Predictive Controller (MPC) [44,45] as described in [46]. The control law of the attitude PID controller (see Figure 23) is simple enough that the embedded microcontroller can compute it efficiently in the main control loop at 400 Hz. However, our MPC controller was computationally heavy and barely fit into the main control loop. An interesting control scheme optimization (event-triggered control scheme) was proposed in [47,48]—future research could check how much processor time could be saved without losing attitude stability.

We implemented two new flight modes [49]: the FixedTestTrajectory and the Elka1Q mode. The former was helpful for model identification and controller tuning (described in detail in our other papers [42,50]), and the latter was entirely dedicated to the plane-like flight phase.

Let us consider the flight dynamics of a fixed-wing plane with symmetric elevon control surfaces. When elevons deflect, the plane starts to rotate over its roll or pitch axis, depending on the direction of the elevons' deflection (to recap: the same direction of deflection leads to pitch rotation, and differential deflection leads to roll rotation). A simplified fixed-wing model can safely assume that the angle of the elevons' deflection is proportional to the plane rotation speed. Based on that statement, we eventually simplified the Elka1Q mode. When activated, the main quadcopter motors are slowed down to a constant idle speed and no longer actuated. The pitch and roll control signals from the rate controllers (see Figure 23) are directly translated into the desired elevons' deflection angles. Such a simplification has advantages: the ArduCopter stabilize mode implements the attitude position and rate (i.e., rotation speed) controllers. Still, it allows control of the target roll and pitch angles via the human pilot's RC transmitter. Since the drone has not been tested in a wind tunnel, we were not sure how big we should allow for the deflection angles to be for controlling the drone safely. Due to that concern, we implemented a live-tuning (via RC transmitter knobs) of scale coefficients for the output pitch and roll signals, which were fed into the elevons' mixer.

We decided to intentionally not implement any transition phase—a flip of an RC transmitter switch turns the Elka1Q mode on and off immediately. In our reasoning, we assumed it is safe to accelerate the drone in the regular quadcopter mode because the ArduCopter's stabilize mode maintains the drone's attitude in the air (i.e., it keeps the roll, pitch, and yaw tilt angles fixed). When the drone flies fast enough (faster than the estimated stall speed), disabling the quadcopter motors and enabling the elevons should (at least in theory) let it keep roughly the same attitude in the air in an actual on-wing level flight. Similarly, going back to quadcopter mode in a fast forward flight should not be a problem. The autopilot will just use different actuators to maintain the target attitude—the quadcopter motors instead of elevons. The only predicted side effect could be related to some rate of climbing (increasing altitude) because, for a moment, there will be two lift force sources: the wings and the quadcopter motors.

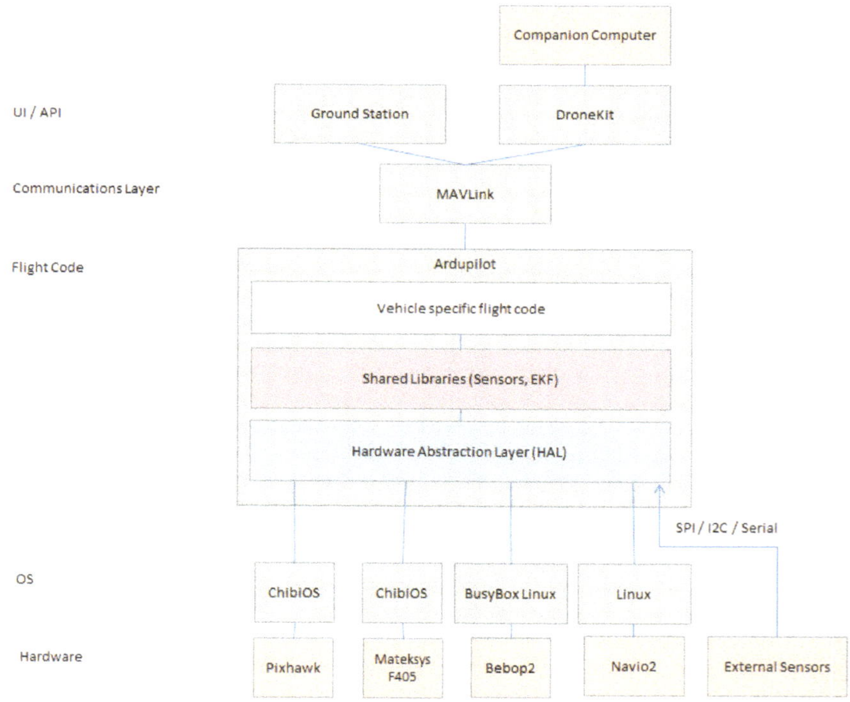

Figure 21. A high-level architecture overview of the ArduPilot software; image source: https://ardupilot.org/dev/docs/apmcopter-code-overview.html, accessed on 1 May 2022; published under the CC BY-SA 3.0 license: https://creativecommons.org/licenses/by-sa/3.0/legalcode, accessed on 1 May 2022.

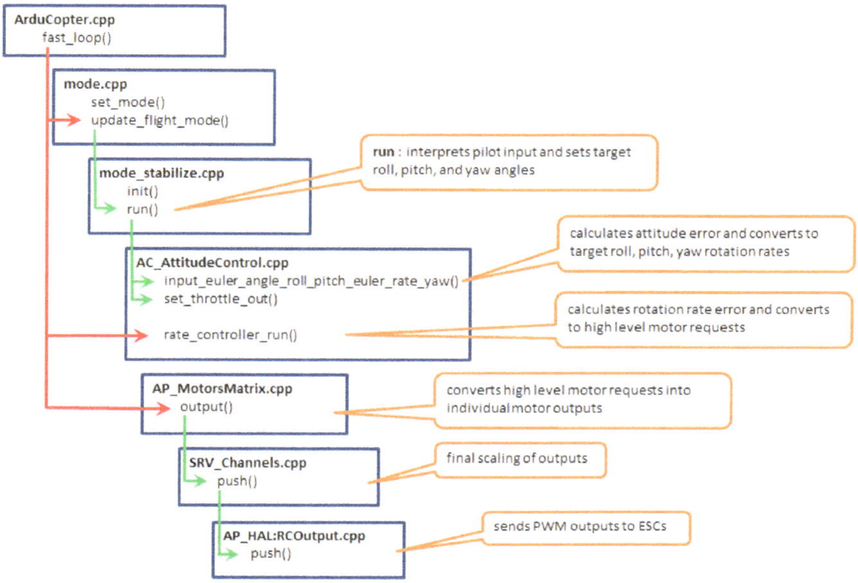

Figure 22. A high-level architecture overview of the ArduCopter software—a sequence of actions in a manually controlled flight mode; image source: https://ardupilot.org/dev/docs/apmcopter-code-overview.html, accessed on 1 May 2022 published under the CC BY-SA 3.0 license: https://creativecommons.org/licenses/by-sa/3.0/legalcode, accessed on 1 May 2022.

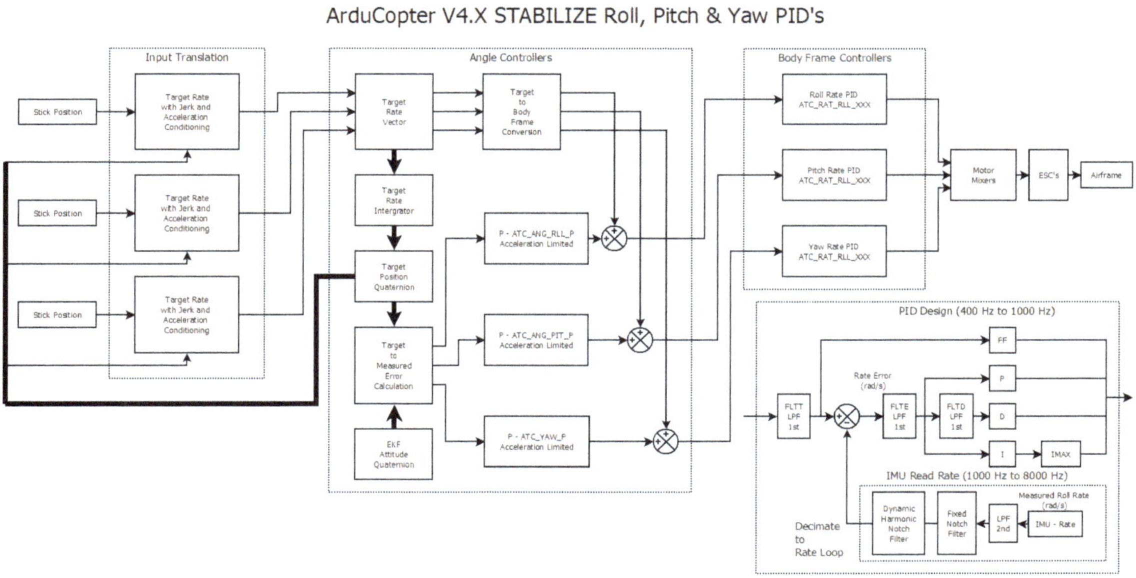

Figure 23. The structure of ArduCopter attitude controllers. Image source: https://ardupilot.org/dev/docs/apmcopter-programming-attitude-control-2.html, accessed on 1 May 2022; published under the CC BY-SA 3.0 license: https://creativecommons.org/licenses/by-sa/3.0/legalcode, accessed on 1 May 2022.

3. Results

3.1. Experiments with Limited Flight Freedom

A series of initial tests was performed in a custom-made limited-freedom test harness, which allows tilting the drone in a single axis only while holding it firmly and locking the remaining axes of rotation (Figure 24). Such a safe test environment was facilitative when we worked on the model identification and implementation of the attitude controllers [42,46]. The final Elka1Q mode was tested in this environment as well.

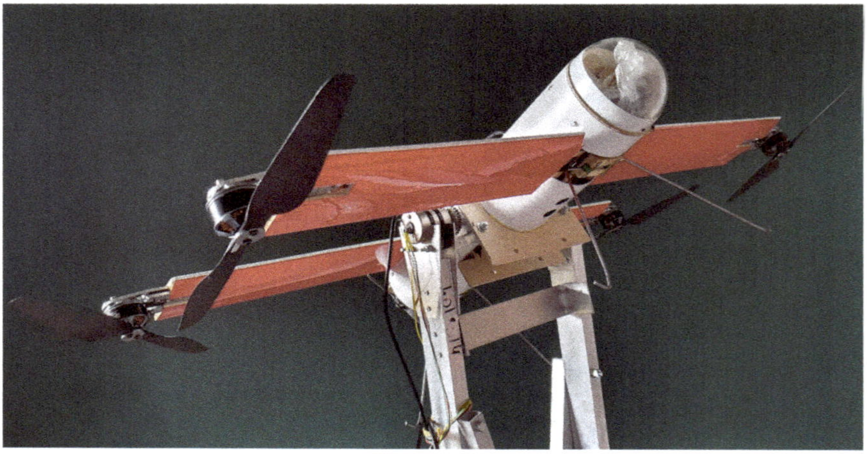

Figure 24. The drone locked in a limited-freedom test harness.

3.2. Real Flight Experiments

The in-flight experiments (Figure 25) were divided into two stages:

1. VTOL (quadcopter) mode flights and onboard systems check: stabilize mode, hover, GPS position hold mode, GPS return-to-land mode, smooth auto-landing using the laser rangefinder, and low-speed forward flight with the pusher motor.
2. In-flight mode transition test and plane mode performance analysis.

The first stage of test flights went without any serious issues. We discovered some instability in the onboard barometer readings—most likely due to pressure changes inside the fuselage caused by wind gusts. The barometer is not crucial for the flights; GPS provides a redundant coarse altitude value. The laser rangefinder provides precise and accurate low altitude readings necessary for the auto-landing feature. The drone behaves correctly in hover, responds sharply to pilot commands, and proved to be quite resistant to wind gusts. Crosswind gusts induced some mild roll oscillations due to the presence of wing surfaces. Low-speed horizontal flight using the pusher motor did not reveal any stability issues. The drone maintained its attitude correctly.

The second stage of test flights revealed a few issues, some severe enough to lead to a few crash landings eventually. Figure 26 shows a telemetry log from an experimental flight when at about the 120th second, the mode was changed from quadcopter to Elka1Q. The drone dived rapidly. Immediate manual intervention (switching back to the quadcopter mode) rescued the drone while it still stayed airborne. The situation repeated in a few other attempts, sometimes flipping the drone by 180° (a half-loop). We concluded that the drone could have misplaced its CG. The eCalc CG tool states that the calculation results should be carefully examined in test flights due to numerous limitations, including, but not limited to "fat" fuselage effects and aerodynamic performance analysis. The overall static and dynamic plane stability will also depend on the control surface effectiveness and the attitude controller robustness. Eventually, we fixed that issue by moving the CG by ca. 10 mm forward.

Figure 25. Flight experiments.

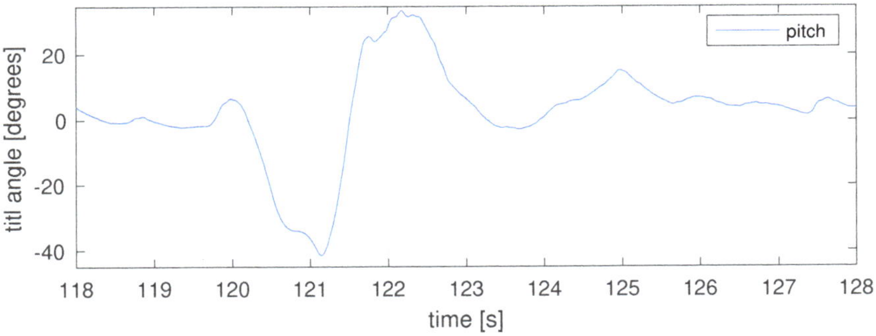

Figure 26. Telemetry log from an in-flight experiment: a sudden pitch down (a dive) after the transition into plane mode because of misplaced CG—eventually rescued thanks to an immediate switch back to quadcopter mode.

Further flights showed that the pitch controller still is not always stable enough—in some flights, after accelerating the drone and switching it to plane mode, the drone became extremely pitch-unstable (Figure 27). Fast horizontal flights in quadcopter mode (using the pusher motor to accelerate) also revealed some pitch instability (oscillations in Figure 28 starting from the 111th second).

Figure 29 presents power usage in quadcopter mode (<55 s), during the transition phase (55–65 s; the drone accelerates with pusher motor; quadcopter motors still work) and plane mode (>65 s; only pusher motor works; other motors were shut down). It is worth mentioning that the electric power needed during the hover phase and the level flight in plane mode is nearly identical (total current drawn from the battery was ca. 20 A). The key difference is related to the horizontal speed of the drone in plane mode (above 100 km/h), while any non-zero forward flight speed in quadcopter mode requires more power than in hover. The eCalc predicts (see Figures 5 and 6) that in quadcopter mode, the practical maximum speed will be less than 35 km/h at maximum main motor power (current readings over 60 A!). That proves the plane mode's efficiency—the drone can fly 3× faster while consuming 3× less electric energy!

Due to safety reasons, we performed the top speed test only partially. Figure 30 shows that we measured ca. 105 km/h top speed in a level flight.

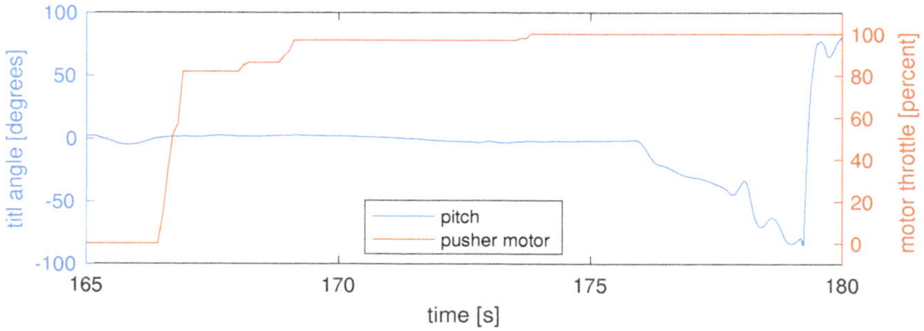

Figure 27. Telemetry log from an in-flight experiment: pitch instability in plane mode flight.

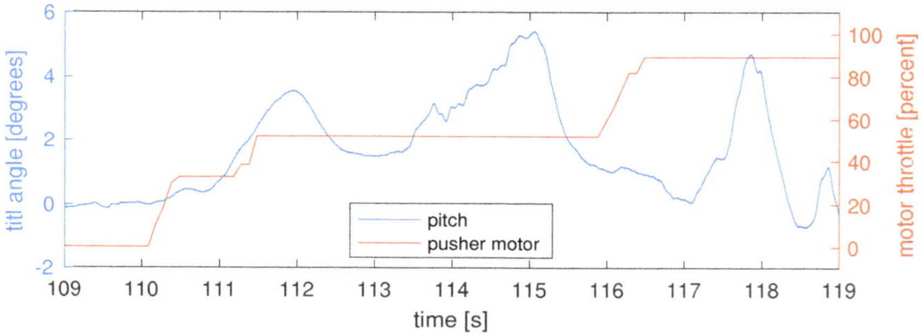

Figure 28. Telemetry log from an in-flight experiment: a close-up on pitch instability observed in quadcopter mode when the pusher motor increased the drone's speed in a horizontal flight.

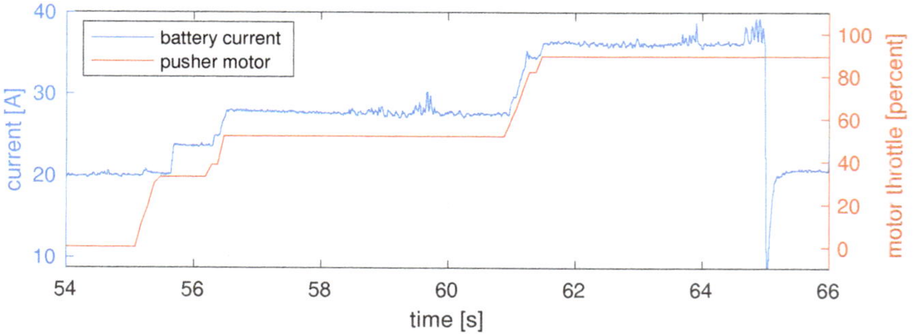

Figure 29. Telemetry log from an in-flight experiment: battery current readings in quadcopter mode (up to 55th second), accelerating with pusher motor (up to 65th second) and quadcopter motors shut down when fully transitioned into plane flight mode.

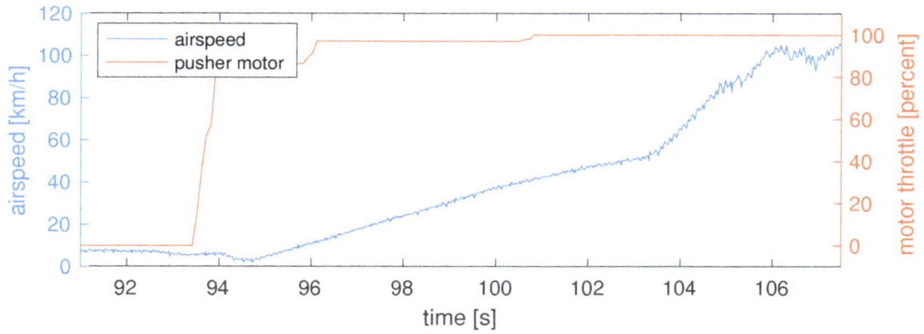

Figure 30. Telemetry log from an in-flight experiment: top speed test in a horizontal flight in Elka1Q mode.

4. Discussion

The mechanical construction of the drone proved to be successful. It was lightweight and highly robust and rigid. The only disadvantage of the construction was related to the complicated maintenance of internal components.

The complete electronic components used in the drone worked with no issues. Especially, the Kakute F7 AIO Flight Controller was a good choice—it was tiny, yet highly reliable and robust.

Thanks to excellent documentation and community support, working with the ArduCopter software stack was a pleasure. Implementing new flight modes was eventually relatively easy.

The VTOL test flights revealed no issues, which means the optimized quadcopter electric propulsion provides significant lift force and still guarantees good manoeuvrability. The motors and propellers were designed to carry 300 g of payload, while the drone could easily lift an additional 1500 g of payload—or hover longer with a nominal payload.

The pusher motor also performed well—the acceleration was instant, and the high pitch-speed of the propeller surely could accelerate the drone to an even higher top speed than the measured 105 km/h.

Building wings around the quadcopter motor holders also proved a good idea. However, the aerodynamic configuration of a tandem wing turned out to be surprising and challenging. The first serious problem was related to the correct CG localization. The calculated value seemed to be placed too far aft. That mistake caused one of the crash landings.

Another issue was related to pitch instability. It was caused (most likely) by too small (and thus inefficient) elevons. However, the findings based on telemetry logs were sometimes inconclusive in this matter.

The drone finally proved it could fly fast using its tandem-wing and the pusher motor only. Due to airfield restrictions, we could not perform a full top speed test and long-endurance flights. Nevertheless, we measured the electric power usage in both flight phases. Based on that, we could estimate the drone range.

5. Conclusions

This research aimed to design, build, and test a unique UAV in flight: optimized for efficient hover, but able to fly in long-range missions. We chose a tandem-wing configuration because it is the most compact variant of a VTOL drone compared to a fixed-wing one. We proved in our test flights that the drone could fly very fast (in plane mode) using the same electric power as in hover (in quadcopter mode). That feature lets it cover a longer distance than a typical multi-rotor could do. Additionally, we could optimize motors and propellers for a single purpose, unlike motor-tilting constructions. The tandem-wing configuration is also significantly smaller and lighter than a typical quadplane of the same wing area.

Eventually, we designed, calculated, simulated, built, and tested the Elka1Q drone in static tests and in flight. We also identified a model of the drone dynamics to improve its attitude controller implementation and experiment with custom-made MPC controllers.

The flights proved the wings almost do not affect the drone behaviour in hover and VTOL manoeuvres.

Fast horizontal flights using the pusher motor and transitioning to plane mode proved the drone could cover a significant distance in a short amount of time, i.e., it can operate in long-range missions. We measured the electric power usage and concluded that the prototype drone offers a 5× more extended range than a typical VTOL-only UAV (e.g., a multirotor). The compact mechanical construction of the tandem-wing and the fuselage proved to be extremely robust—it survived some crash landings with very little damage, having a take-off weight of ca. 2900 g!

The most significant disadvantages of the Elka1Q drone are related to its pitch stability issues. Some were caused by a misplaced CG, some plausibly by inefficient elevons, but some remain unexplained. We suggest examining the pitch controller in a wide range of conditions and possibly tuning it further.

Future research could optimize the wings and measure all the physical limits of such a drone, such as the actual top speed and range with a selected mission scenario.

There is a Supplementary Material available—a video summary of the research.

Supplementary Materials: The following supporting information (Video: Elka1Q UAV—a Tandem-Wing Quadplane.) can be downloaded at: https://www.youtube.com/watch?v=2cdPLeVac24, accessed on 19 May 2022.

Author Contributions: Conceptualization, M.O. and M.Ł.; methodology, M.O. and M.Ł.; software, M.O.; validation, M.O.; formal analysis, M.O. and M.Ł.; investigation, M.O.; resources, M.O.; data curation, M.O.; writing—original draft preparation, M.O.; writing—review and editing, M.O. and M.Ł.; visualization, M.O.; supervision, M.Ł.; project administration, M.Ł. All authors have read and agreed to the published version of the manuscript.

Funding: This research received no external funding.

Institutional Review Board Statement: Not applicable.

Informed Consent Statement: Not applicable.

Data Availability Statement: Not applicable.

Conflicts of Interest: The authors declare no conflict of interest.

Abbreviations

The following abbreviations are used in this manuscript:

UAV	Unmanned Aerial Vehicle
SAR	Search-And-Rescue
PID	Proportional–Integral–Derivative
IMU	Inertial Measurement Unit
MPC	Model Predictive Controller
GPC	Generalized Predictive Controller
FC	Flight Controller
CG	Centre of Gravity
AoA	Angle of Attack
ESC	Electronic Speed Controller

References

1. Razi, P.; Sumantyo, J.T.S.; Perissin, D.; Kuze, H.; Chua, M.Y.; Panggabean, G.F. 3D Land Mapping and Land Deformation Monitoring Using Persistent Scatterer Interferometry (PSI) ALOS PALSAR: Validated by Geodetic GPS and UAV. *IEEE Access* **2018**, *6*, 12395–12404. [CrossRef]
2. Máthé, K.; Buşoniu, L. Vision and control for UAVs: A survey of general methods and of inexpensive platforms for infrastructure inspection. *Sensors* **2015**, *15*, 14887–14916. [CrossRef] [PubMed]
3. Heutger, M.; Kückelhaus, M. *Unmanned Aerial Vehicles in Logistics a DHL Perspective on Implications and Use Cases for the Logistics Industry*; DHL Customer Solutions & Innovation: Troisdorf, Germany, 2014.
4. Waharte, S.; Trigoni, N. Supporting search and rescue operations with UAVs. In Proceedings of the 2010 International Conference on Emerging Security Technologies, Canterbury, UK, 6–7 September 2010; pp. 142–147.
5. Aljehani, M.; Inoue, M. Performance Evaluation of Multi-UAV System in Post-Disaster Application: Validated by HITL Simulator. *IEEE Access* **2019**, *7*, 64386–64400. [CrossRef]
6. Scott, J.E.; Scott, C.H. Drone Delivery Models for Medical Emergencies. In *Delivering Superior Health and Wellness Management with IoT and Analytics*; Springer International Publishing: Cham, Switzerland, 2020; pp. 69–85.
7. Alicandro, M.; Dominici, D.; Massimini, V. Surveys with UAV photogrammetry: Case studies in l'Aquila during the post-earthquake scenario. In Proceedings of the EGU General Assembly Conference Abstracts, Vienna, Austria, 12–17 April 2015; p. 14987.
8. Perez-Grau, F.; Ragel, R.; Caballero, F.; Viguria, A.; Ollero, A. Semi-autonomous teleoperation of UAVs in search and rescue scenarios. In Proceedings of the 2017 International Conference on Unmanned Aircraft Systems (ICUAS), Miami, FL, USA, 13–16 June 2017; pp. 1066–1074. [CrossRef]
9. Chung, P.H.; Ma, D.M.; Shiau, J.K. Design, manufacturing, and flight testing of an experimental flying wing UAV. *Appl. Sci.* **2019**, *9*, 3043. [CrossRef]
10. Jensen, A.M.; Baumann, M.; Chen, Y. Low-cost multispectral aerial imaging using autonomous runway-free small flying wing vehicles. In Proceedings of the IGARSS 2008-2008 IEEE International Geoscience and Remote Sensing Symposium, Boston, MA, USA, 8–11 July 2008; Volume 5, pp. V-506.
11. Austin, R. *Unmanned Aircraft Systems: UAVS Design, Development and Deployment*; John Wiley & Sons: Hoboken, NJ, USA, 2011.
12. Valavanis, K.P.; Vachtsevanos, G.J. *Handbook of Unmanned Aerial Vehicles*; Springer: Dordrecht, The Netherlands, 2015; Volume 1.
13. Cai, G.; Chen, B.M.; Peng, K.; Dong, M.; Lee, T.H. Modeling and Control System Design for a UAV Helicopter. In Proceedings of the 2006 14th Mediterranean Conference on Control and Automation, Ancona, Italy, 28–30 June 2006; pp. 1–6. [CrossRef]
14. Peng, K.; Cai, G.; Chen, B.M.; Dong, M.; Lum, K.Y.; Lee, T.H. Design and implementation of an autonomous flight control law for a UAV helicopter. *Automatica* **2009**, *45*, 2333–2338. [CrossRef]
15. Mehmood, H.; Nakamura, T.; Johnson, E.N. A maneuverability analysis of a novel hexarotor UAV concept. In Proceedings of the 2016 International Conference on Unmanned Aircraft Systems (ICUAS), Arlington, VA, USA, 7–10 June 2016; pp. 437–446.
16. Jannoura, R.; Brinkmann, K.; Uteau, D.; Bruns, C.; Joergensen, R.G. Monitoring of crop biomass using true colour aerial photographs taken from a remote controlled hexacopter. *Biosyst. Eng.* **2015**, *129*, 341–351. [CrossRef]
17. Mahony, R.; Kumar, V.; Corke, P. Multirotor aerial vehicles: Modeling, estimation, and control of quadrotor. *IEEE Robot. Autom. Mag.* **2012**, *19*, 20–32. [CrossRef]
18. Yang, H.; Lee, Y.; Jeon, S.Y.; Lee, D. Multi-rotor drone tutorial: Systems, mechanics, control and state estimation. *Intell. Serv. Robot.* **2017**, *10*, 79–93. [CrossRef]
19. Eraslan, Y.; Özen, E.; Oktay, T. The Effect of Change in Angle Between Rotor Arms on Trajectory Tracking Quality of a PID Controlled Quadcopter. In Proceedings of the EJONS 10th International Conference on Mathematics, Engineering, Natural Medical Sciences, Batumi, Georgia, 15–17 May 2020.
20. Koehl, A.; Rafaralahy, H.; Boutayeb, M.; Martinez, B. Aerodynamic modelling and experimental identification of a coaxial-rotor UAV. *J. Intell. Robot. Syst.* **2012**, *68*, 53–68. [CrossRef]
21. Dzul, A.; Hamel, T.; Lozano, R. Modeling and nonlinear control for a coaxial helicopter. In Proceedings of the IEEE International Conference on Systems, Man and Cybernetics, Toronto, ON, Canada, 9–12 October 2002; Volume 6, p. 6.
22. Govdeli, Y.; Muzaffar, S.M.B.; Raj, R.; Elhadidi, B.; Kayacan, E. Unsteady aerodynamic modeling and control of pusher and tilt-rotor quadplane configurations. *Aerosp. Sci. Technol.* **2019**, *94*, 105421. [CrossRef]
23. Flores, G.R.; Escareño, J.; Lozano, R.; Salazar, S. Quad-tilting rotor convertible mav: Modeling and real-time hover flight control. *J. Intell. Robot. Syst.* **2012**, *65*, 457–471. [CrossRef]
24. Muraoka, K.; Okada, N.; Kubo, D. Quad tilt wing vtol uav: Aerodynamic characteristics and prototype flight. In Proceedings of the AIAA Infotech@ Aerospace Conference and AIAA Unmanned... Unlimited Conference, Seattle, WA, USA, 6–9 April 2009; p. 1834.
25. ATEA Concept Design. 2022. Available online: https://www.ascendance-ft.com/products/ (accessed on 10 May 2022).
26. eMagic One—An eVTOL That Really Works. 2022. Available online: https://emagic-aircraft.com/ (accessed on 10 May 2022).
27. Alba-Maestre, J.; Prud'homme van Reine, K.; Sinnige, T.; Castro, S.G. Preliminary propulsion and power system design of a tandem-wing long-range eVTOL aircraft. *Appl. Sci.* **2021**, *11*, 11083. [CrossRef]
28. OKTAY, T.; Metin, U.; ÇELİK, H.; KONAR, M. PID based hierarchical autonomous system performance maximization of a hybrid unmanned aerial vehicle (HUAV). *Anadolu Univ. J. Sci. Technol. A-Appl. Sci. Eng.* **2017**, *18*, 554–562. [CrossRef]

29. Çetinsoy, E.; Dikyar, S.; Hançer, C.; Oner, K.; Sirimoglu, E.; Unel, M.; Aksit, M. Design and construction of a novel quad tilt-wing UAV. *Mechatronics* **2012**, *22*, 723–745. [CrossRef]
30. eCalc—Reliable Electric Drive Simulations. 2022. Available online: https://www.ecalc.ch/index.htm (accessed on 18 May 2022).
31. MotoCalc Electric Flight Performance Prediction Software. 2022. Available online: http://www.motocalc.com/ (accessed on 18 May 2022).
32. Oracover Film. 2022. Available online: https://www.oracover.de/film (accessed on 18 May 2022).
33. Airfoil Plotter. 2022. Available online: http://airfoiltools.com/plotter/index?airfoil=p51dtip-il (accessed on 18 May 2022).
34. Minardo, A. *The Tandem Wing: Theory, Experiments and Practical Realisations*; Politecnico DI Milano: Milan, Italy, 2014.
35. Brinkworth, B. On the aerodynamics of the Miles Libellula tandem-wing aircraft concept, 1941–1947. *J. Aeronaut. Hist. Pap.* **2016**, *2*, 10–58.
36. Tandemwings—Nest of Dragons. 2022. Available online: https://www.nestofdragons.net/weird-airplanes/tandemwings/ (accessed on 18 May 2022).
37. The Quickie Q2. 2022. Available online: https://www.eaa.org/eaa/news-and-publications/eaa-news-and-aviation-news/bits-and-pieces-newsletter/03-14-2017-the-quickie-q2 (accessed on 18 May 2022).
38. Linehan, D. *Burt Rutan's Race to Space: The Magician of Mojave and His Flying Innovations*; Zenith Press: Singapore, 2011.
39. Kakute F7 AIO V1.5—Holybro. 2022. Available online: http://www.holybro.com/product/kakute-f7-aio-v1-5/ (accessed on 18 May 2022).
40. VL53L1X Time-of-Flight Distance Sensor Carrier with Voltage Regulator, 400cm Max. 2022. Available online: https://www.pololu.com/product/3415 (accessed on 18 May 2022).
41. NXG-02, Raster Signal Connectors. 2022. Available online: https://ninigi.com/gb/en/product/ninigi/raster-signal-connectors/nxg-02/112945/ (accessed on 18 May 2022).
42. Okulski, M.; Ławryńczuk, M. A Novel Neural Network Model Applied to Modeling of a Tandem-Wing Quadplane Drone. *IEEE Access* **2021**, *9*, 14159–14178. [CrossRef]
43. ArduPilot. 2020. Available online: http://ardupilot.org/about/ (accessed on 1 September 2020).
44. Tatjewski, P. *Advanced Control of Industrial Processes: Structures and Algorithms*; Springer Science & Business Media: Berlin/Heidelberg, Germany, 2007.
45. Camacho, E.F.; Bordons, C. *Model Predictive Control*; Springer: London, UK, 1999.
46. Okulski, M.; Ławryńczuk, M. How Much Energy Do We Need to Fly with Greater Agility? Energy Consumption and Performance of an Attitude Stabilization Controller in a Quadcopter Drone: A Modified MPC vs. PID. *Energies* **2022**, *15*, 1380. [CrossRef]
47. Tabuada, P. Event-triggered real-time scheduling of stabilizing control tasks. *IEEE Trans. Autom. Control* **2007**, *52*, 1680–1685. [CrossRef]
48. Yan, S.; Gu, Z.; Park, J.H.; Xie, X. Adaptive Memory-Event-Triggered Static Output Control of TS Fuzzy Wind Turbine Systems. *IEEE Trans. Fuzzy Syst.* **2021**, 1–11. [CrossRef]
49. Adding a New Flight Mode to Copter. 2021. Available online: https://ardupilot.org/dev/docs/apmcopter-adding-a-new-flight-mode.html (accessed on 11 November 2021).
50. Okulski, M.; Ławryńczuk, M. Identification of Linear Models of a Tandem-Wing Quadplane Drone: Preliminary Results. In *Advanced, Contemporary Control*; Springer: Cham, Switzerland, 2020; pp. 219–228.

applied sciences

MDPI

Article

Robust Approach to Supervised Deep Neural Network Training for Real-Time Object Classification in Cluttered Indoor Environment

Bedada Endale [1], Abera Tullu [2], Hayoung Shi [1] and Beom-Soo Kang [1,*]

[1] Department of Aerospace Engineering, Pusan National University, Busan 46241, Korea; endale@pusan.ac.kr (B.E.); shy621@pusan.ac.kr (H.S.)
[2] Department of Aerospace Engineering, Sejong University, 209, Neungdong-ro, Gwangjin-gu, Seoul 05006, Korea; tuab@sejong.ac.kr
* Correspondence: bskang@pusan.ac.kr; Tel.: +82-51-510-2310

Abstract: Unmanned aerial vehicles (UAVs) are being widely utilized for various missions: in both civilian and military sectors. Many of these missions demand UAVs to acquire artificial intelligence about the environments they are navigating in. This perception can be realized by training a computing machine to classify objects in the environment. One of the well known machine training approaches is supervised deep learning, which enables a machine to classify objects. However, supervised deep learning comes with huge sacrifice in terms of time and computational resources. Collecting big input data, pre-training processes, such as labeling training data, and the need for a high performance computer for training are some of the challenges that supervised deep learning poses. To address these setbacks, this study proposes mission specific input data augmentation techniques and the design of light-weight deep neural network architecture that is capable of real-time object classification. Semi-direct visual odometry (SVO) data of augmented images are used to train the network for object classification. Ten classes of 10,000 different images in each class were used as input data where 80% were for training the network and the remaining 20% were used for network validation. For the optimization of the designed deep neural network, a sequential gradient descent algorithm was implemented. This algorithm has the advantage of handling redundancy in the data more efficiently than other algorithms.

Keywords: object classification; deep learning; convolutional neural network; network architecture

Citation: Endale, B.; Tullu, A.; Shi, H.; Kang, B.-S. Robust Approach to Supervised Deep Neural Network Training for Real-Time Object Classification in Cluttered Indoor Environment. *Appl. Sci.* **2021**, *11*, 7148. https://doi.org/10.3390/app11157148

Academic Editors: Sylvain Bertrand, Hyo-sang Shin and Seong-Ik Han

Received: 9 June 2021
Accepted: 28 July 2021
Published: 2 August 2021

1. Introduction

The emergence of artificial intelligence and computer vision technologies bring forth a wide range of applications. As a result, various unmanned systems are being deployed in both civilian and military domains. Equipped with these technologies, self-driving cars [1–4] and autonomously navigating UAVs [5–9] are being integrated into our daily life. All of these and other important applications of integrated artificial intelligence and computer vision technologies rely, in one way or another, on training neural networks, which is crucial for the classification of objects in images taken by visual sensors.

Moreover, the ability of a computing machine to autonomously detect and classify objects leveraged the autonomous navigation of unmanned aerial vehicles in cluttered environments. This capability further incites a wide range of applications of UAVs. UAV missions, such as door-to-door package delivery, search and rescue of victims in a collapsed building, indoor first aid, and target tracking in urban environments, demand that the UAV has environmental perception. To this end, training a companion computer onboard the UAV is mandatory. Training computing machines to perceive the surrounding environment is a state-of-the-art technology, generally known by the name "machine learning": a sub-discipline of artificial intelligence. For the process of machine learning, a network of mathematical abstracts (neurons) are layered in structured way. The depth of the network is determined by its number of layers and training this deep network to hierarchically extract desired features in input data is referred to as "deep learning". Various definitions of deep learning were reviewed by Zhang et al. [10]. A comprehensive review of deep learning and its variants was presented by Yann et al. [11] and Jurgen Schmidhuber [12]. A survey of the wide range of applications of deep learning as well as its challenges and future directions was reported by Laith et al. [13]

There are various deep learning approaches where supervised deep learning is one of the well known and widely used approaches. In this type of neural network learning approach, the dataset is labeled manually for training the network. The three commonly known neural networks to which supervised deep learning is applied are the convolutional neural network (CNN), artificial neural network (ANN), and recurrent neural network (RNN). Ever since its conception, supervised deep learning is being implemented in various areas such as the autonomous navigation of both ground [14] and aerial vehicles [15], speech and pattern recognition [16], and medical image analysis [17].

The commonly utilized network that is often implemented in deep learning for object detection and classification is CNN [18–22]. This CNN is proved to outperform other networks on various tasks [23]. There are many variants of CNN and their differences are dictated by the network parameters, such as number of layers, number of intra-layer neurons and the types of inter-layer connections in the architecture. These parameters greatly affect the efficiency of the performance of a given CNN. An extensive survey of the recent architectures of deep CNN was reported by Asiffullah et al. [24].

Neha et al. [25] analyzed three variants of CNN—AlexNet, GoogleNet and ResNet50—and reported that the number of layers in a network architecture affects the performance of CNN. Karen et al. [26] found that the increase in the number of layers of CNN enhances the performance accuracy of the network. Christian et al. [27] also proposed that increasing the depth of the network remarkably improves the performance of the network. In their study on training deep convolutional neural networks using huge image data, Alex et al. [28] concluded that a very large network, such as ImageNet LSVRC-2010, is essential to achieving good results.

Indefinite increment in network depth, however, incurs a tiresome training process, requires high storage and computing capacities, and includes the difficulties of network architectural design that lead to performance degradation. Tiresome pre-training processes, such as input data labeling and the need for high performance computers, are common setbacks for deep network implementation. A companion computer on-board a UAV has to have enough memory to store huge amounts of activations and weights of the deep network and needs to conduct resource intensive image processing in real-time for safe navigation.

Challenges related to large network training were presented and explained by Michael [29]. Soumya et al. [30] also discussed the challenges of training CNNs. Many researchers suggested remedies to the challenges. Hugo et al. [31] suggested a deep neural network training procedure that leverages the performance of the network. Kaiming et al. [32] proposed a learning framework, named residual learning, that resolves the difficulty of training large networks and performance degradation. Stephan et al. [33] proposed a stability training approach that enhances network tolerance to small perturbations in input image data.

In addition to training difficulty, the work of selecting and designing a particular deep CNN architecture is not simple. This is because the optimization algorithms have to be gauged from the point of view of specific deep learning problems. There is, however, a most appropriate optimization for a particular problem which is the result of the "no free lunch theorem" of mathematical optimization reported by Tamás et al. [34]. Ivana et al. [35] took the aforementioned network parameters and parameters such as number of filters per layer and filter size in order to perform parameter optimization and obtain a network architecture with the best performance. Gao et al. [36] introduced DenseNets, where each layer is connected to every other layer. They reported that this network reduces some of the problems that come with increasing the depth of CNN architecture.

The remainder of this work is organized into sections. Section 2 describes the objective of this research and the methodologies followed. Problem specific input data augmentation techniques are listed and augmentation strategies are explained in Section 3. The designed network architecture and its training procedure are presented in Section 4. Results and discussions are presented in Section 5 and finally the conclusion is drawn in Section 6.

2. Problem Statement and Methodology

The objective of this research is to enable a quadcopter UAV engaged in search and rescue operations to acquire the capability of classifying objects in a wreckage of collapsed building. For this to be realized, a companion computer onboard the quadcopter has to be well trained with plausible indoor objects. For the first phase of this work, we randomly picked plausible indoor and surrounding objects, such as windows, doors, walls, columns, pipes, poles, tables, fans, nets, and trees.

For training, we considered supervised deep learning of the convolutional neural network approach. The architecture of this CNN is problem specific. Therefore, we designed a custom CNN architecture specific to the aforementioned problem. Often, network training requires big input data. There are no big data for the stated scenario. In cases when the available input data are scarce, a common method to enhance the number input data is to augment the available data through various methods. Connor et al. [37] conducted an extensive survey on augmentation methods of input image data. In our case, we used mission specific augmentation methods.

A mission specific augmented input data generation approach is pertinent to the quadcopter UAV engaged in search and rescue operations, in a collapsed building. Under such an operation, the quadcopter has to enter the collapsed building through any hole available and avoid collision with obstacles as it searches for targets. The quadcopter will possibly encounters environments with variable brightness as well as laid or inverted objects. While the quadcopter is taking images of the scene, the images can be blurred due to the vibration of the quadcopter. If images are taken during the rolling of the quadcopter, objects in the images might appear rotated. If images are taken at a glance, objects in the images appear sheared. These possibilities are taken into consideration during image data augmentation. In designing the neural network model, data cleaning and code development took most of the time and computing resources. Data training frameworks and code development are carried out with Python libraries. Image preparation and manipulation is performed with the Image Processing Toolbox™ from MathWorks®. Units' dropout is randomly implemented in a ratio of 0.1 to 0.4 on the fully connected (dense) layer of applied layers.

3. Data Preparation and Pre-Processing Methodology

A quadcopter UAV based mission for the search and rescue of a target in a collapsed building was taken into consideration during augmentation of the input image data. We used a stereo camera to collect the 3D image information in SVO format. Sample images are presented in Figure 1. The image information is then converted to 2D stereo images for training the supervised deep neural network.

Before the data were analyzed, they were organized into an appropriate form. The raw image data were prepared in such a way that they retain complete information about the environment they represent: the aspect of the environment that the data are describing. Random sample image inputs of the actual experimental indoor environment are shown in Figure 2. The integrity, completeness, validity, consistency and uniqueness of the dataset are maintained as much as possible.

The whole preparation process contains tasks such as refining, integrating and transforming input data. For our small dataset, we applied an image augmentation technique to train the network.

(a) Setup

(b) puricare (fan)

(c) window

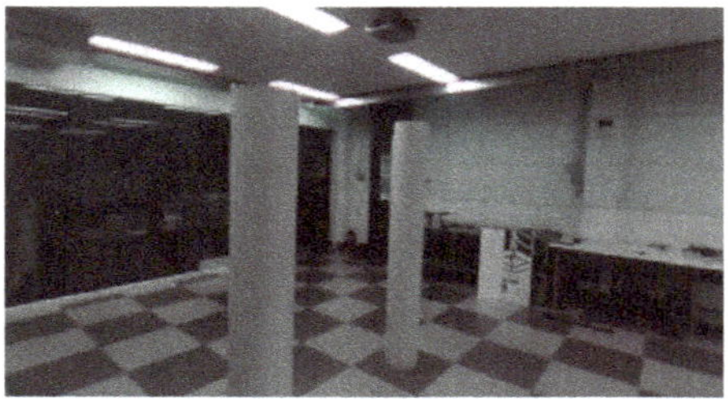

(d) pole

Figure 1. Experimental setup and sample stereo images of objects used to train the network.

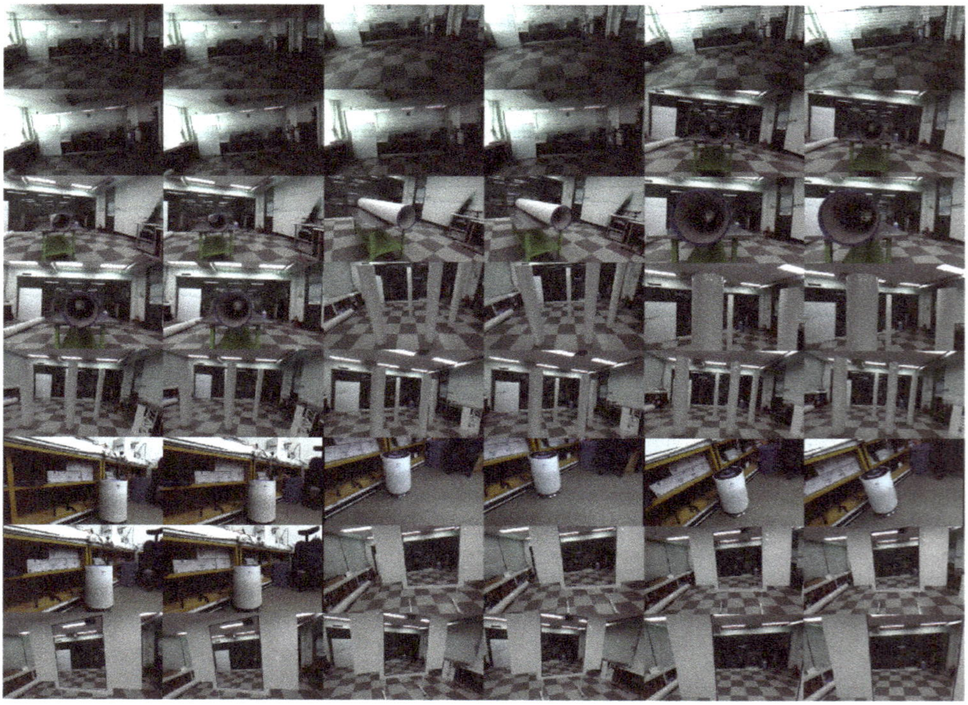

Figure 2. Sample random stereo images used to train our deep neural network.

This augmentation technique enhances the size of input data by zooming, shearing, rotating, blurring, flipping and changing the color of already existing scarce input data. The technique performs transformations to yield believable-looking images in the scene. The network model makes use of this technique to perceive wider aspects of the input data. Deliberately introducing imperfections into our dataset was essential to making our model more resilient to the harsh realities it will encounter in real world situations. Degrading image quality by applying Gaussian blurring was one we applied to our images for this purpose. Figure 3 shows a random sample of the blurred images used for training.

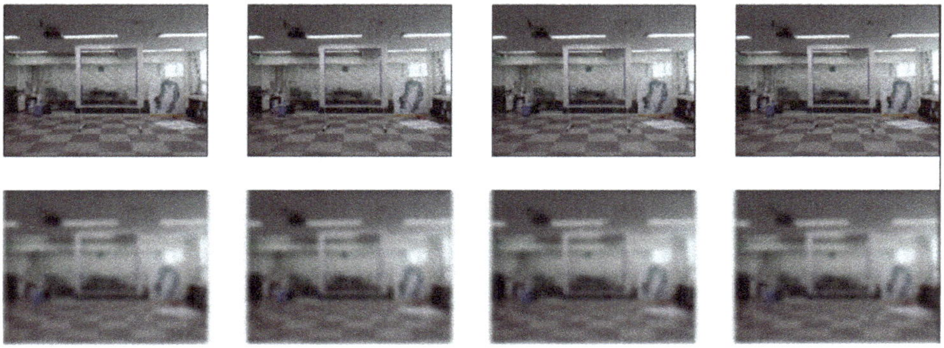

Figure 3. Original (**top**) and blurred (**bottom**) sample images.

Many types of imperfections can make their way into an image: blur, poor contrast, noise, joint photographic experts group (JPEG) compression, and more. Of these, blurring is among the most detrimental to image classification. Blurring an image is taking neighboring pixels and averaging them, in effect reducing detail and creating what can be perceived as blur. When we implement different amounts of blur, we are determining how many neighboring pixels to include. We measured this spread from a single pixel as the standard deviation in both the horizontal and vertical directions. The larger the standard deviation, the more blur an image receives. In this work, we filtered the image with a Gaussian filter of 4 standard deviations. Image flipping is one of the commonly used approaches in image augmentation techniques. Flipping images vertically or horizontally does not alter the classes of objects in the images. Sample horizontally flipped (left and right) images used in the model training are displayed in Figure 4a.

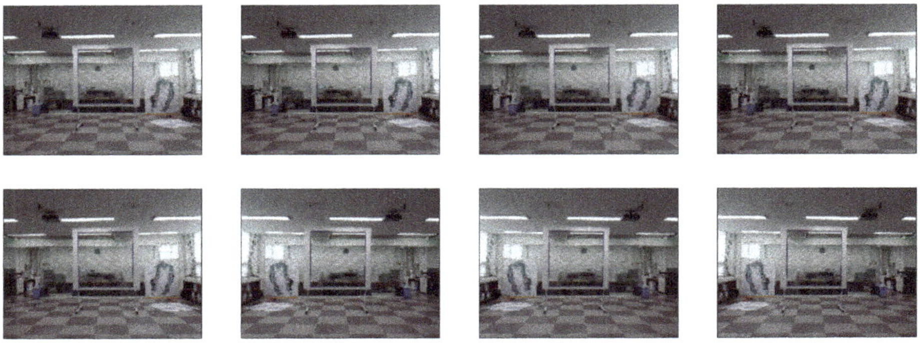

(**a**) left and right flipped

Figure 4. *Cont.*

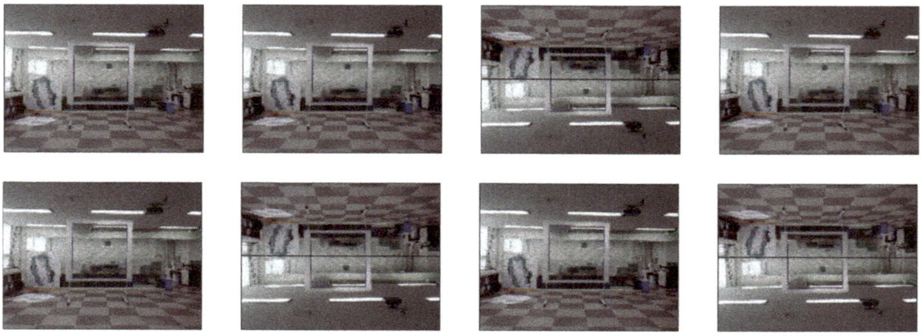

(b) top and bottom flipped

Figure 4. Left and right and top and bottom flipped images used for training.

Even though it is not as common as horizontal flipping, we flipped input images vertically (up-side-down) and the sample is displaced in Figure 4b. Both of these horizontal and vertical flipping techniques align with the scenario of objects under a collapsed building.

To make the network model invariant to the change in positions of objects in images, input images were randomly cropped. Each input image was randomly cropped from 10% to 100% of its original area, and the ratio of width to height of the region was randomly selected between 0.5 and 2. Sample cropped images are shown in Figure 5a with the width and height scaled to 180 pixels.

Varying the colors of input images is another image augmentation technique. We varied the brightness, contrast, saturation and hue of input images. Brightness is randomly varied between 50% to 150% of the original image and sample images are as shown in Figure 5b. Similarly, the hues of the input images were randomly varied. Sample images are displayed in Figure 6a.

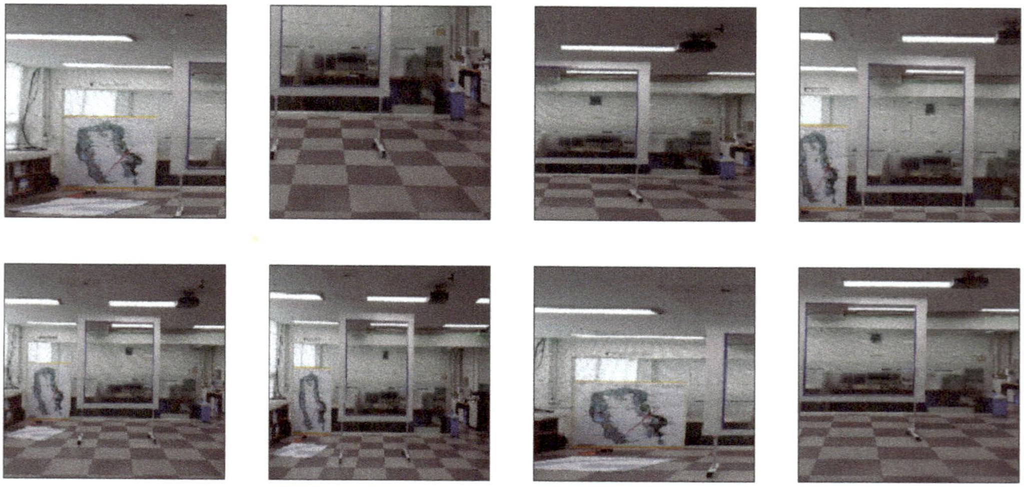

(a) random cropped

Figure 5. *Cont.*

(b) random brightness

Figure 5. Randomly cropped and brightened images used for training.

We also created random color jitter instances and randomly changed the brightness, contrast, saturation and hue of the images as shown in Figure 6b. Finally, a multiple image augmentation technique was applied. All the aforementioned techniques were overlaid and applied to each input image used in the model training. Figure 6c shows the different augmentation techniques on sample images.

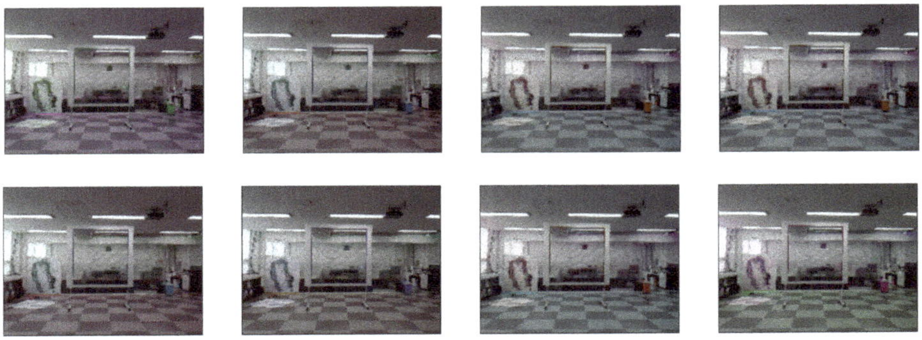

(a) random hue

(b) random color jitters

Figure 6. *Cont.*

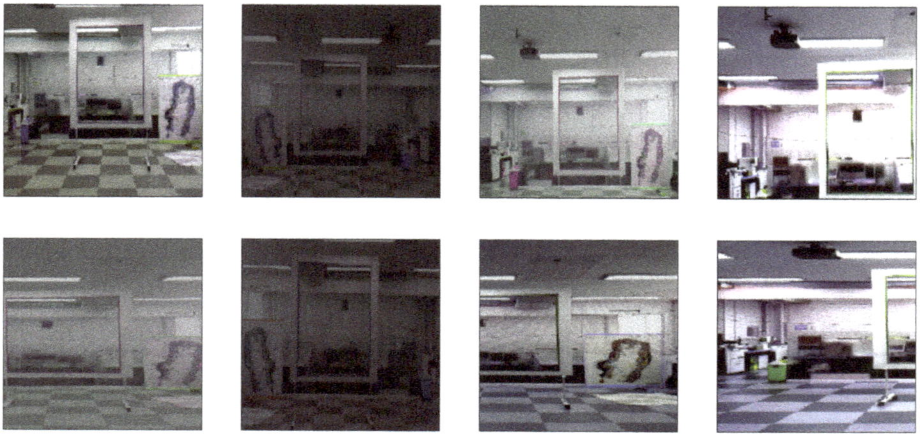

(**c**) multiple augmentation

Figure 6. Images with random hue, color jitters and multiple augmentation used for training.

4. Network Architecture Design

There are two components in artificial neural network models. These are filters and network architecture. The filter defines the weight parameters and the network architecture defines parameters such as how many layers, how many neurons per layer, type of intra-layer and inter-layer neurons connections in the network. Network architecture design is problem specific. What is accurate for one problem may not perform well for the others. Therefore, designing a problem specific deep neural network architecture is a common practice with the intention of reducing computational burdens and enhancing the accuracy or speed of the network for the objective it is designed for.

In this section, our designed deep neural network architecture for object classification is presented. The designed architecture is based on CNN. As an objective, since we have very few classes of objects to classify, it is better to have very few but an effective number of layers in the network to make it light-weight for fast object classification to ensure the real-timeness of the classification processes. In this light-weight CNN design, we followed a common CNN architecture design trend [38], where the sequence of layers are with few convolutional layers and activation functions followed by pooling layers as shown in Figure 7.

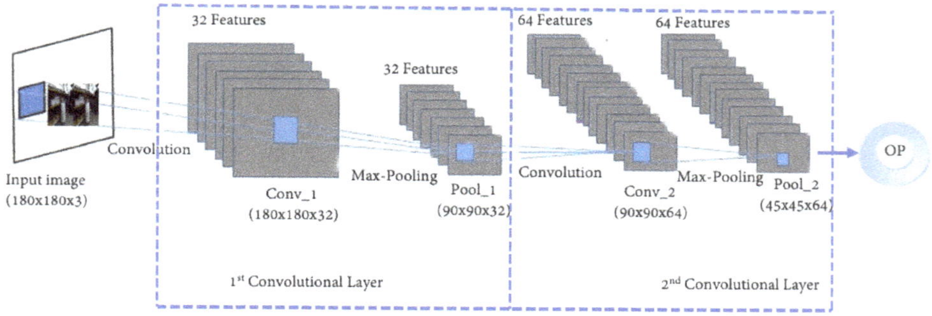

Figure 7. Data flow in the first two layers of a convolutional neural network (CNN).

Deep neural learning networks and deep learning are popular algorithms. Most of the outcomes of these algorithms depend on their cautious architectural design and the choice of appropriate activation functions. Commonly deployed activation functions in neural network design are shown in Figure 8.

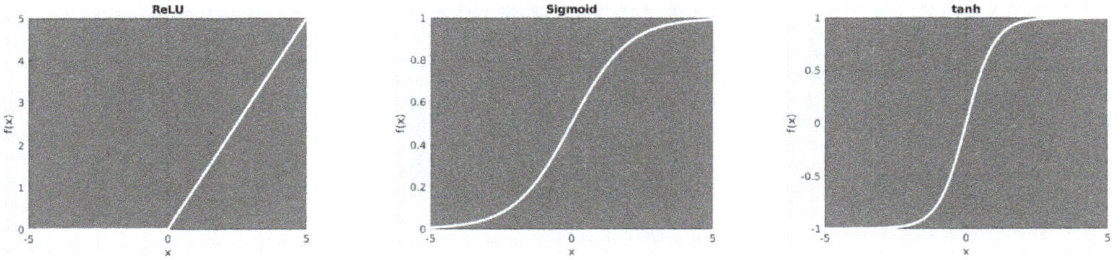

(**a**) rectified linear unit (**b**) sigmoid (**c**) tanhyperbolic

Figure 8. Different activation functions.

In Figure 7, a susceptive area of a part with a specified weight vector (a filter) is transformed bit by bit over a two-dimensional arrangement of process parameters, which are constituent of an image. The resulting arrangement of following activation incidents of this part then provide parameters to higher-level parts and so forth. The basic parts in each convolutional block are a convolutional layer, a rectified linear unit (ReLU) activation function, and a succeeding max-pooling operation.

Each layer has a particular goal. The layers may be replicated with inconsistent variables as part of the convolutional network. The types of layers we deployed in our artificial neural network design are listed as follows:

- image input layer
- convolution 2D layer
- batch normalization layer
- relu layer
- max pooling 2D layer
- fully connected layer
- softmax layer
- classification layer

There can be various layers of each kind of layer. Some convolutional nets have hundreds of layers. Convolution is the process of highlighting expected features in an image. This layer applies sliding convolutional filters to an image to extract features. A batch normalization layer normalizes each input channel across a mini-batch. It automatically divides up the input channel into batches. This reduces the sensitivity to the initialization. Relu layer is a layer that uses the rectified linear unit activation function.

Maxpooling 2D Layer creates a layer that breaks the 2D input into rectangular pooling regions and outputs the maximum value of each region. The input pool size specifies the width and height of a pooling region. Pool size can have one element (for square regions) or two for rectangular regions. The fully connected layer connects all of the inputs to the outputs with weights and biases. Softmax finds a maximum of a set of values using the logistic function. A classification layer computes the cross-entropy loss for multi-class classification problems with mutually exclusive classes.

The layers of the network are summarized in the Table 1. The neural network design contains of three convolutional structures each of which has a layer that calculates the maximum value. There is a fully connected layer with 256 units. The neural network model uses a rectified linear unit (relu) activation function. The image-quantity is a variable quantity of configuration (32, 180, 180, 3). That is a quantity of 32 images of the configuration of $180 \times 180 \times 3$, where the last feature represents the colors red, blue and green channels.

Table 1. Some layers of the neural network used for object classification training.

Layer (Type)	Output Shape	Parameter
Rescaling	(None, 180, 180, 3)	0
conv1 (Conv2D)	(None, 180, 180, 32)	896
max-pooling1 (MaxPooling2D)	(None, 90, 90, 32)	0
conv2 (Conv2D)	(None, 90, 90, 64)	18,496
max-pooling2 (MaxPooling2D)	(None, 45, 45, 64)	0
conv3 (Conv2D)	(None, 45, 45, 128)	73,856
max-pooling3 (MaxPooling2D)	(None, 22, 22, 128)	0
Flatten	(None, 61,952)	0
dense (Dense)	(None, 256)	15,859,968
dense-1 (Dense)	(None, 5)	1285

4.1. Training the Neural Network

The task of training a deep neural network is the biggest setback of all the processes therein. This setback worsens as the depth of the network increases. Xavier et al. [39] discussed the difficulty of training deep neural networks and suggested the work around procedures to be taken during training. In this section, different methods are explored to optimize the network output. The general network training flowchart is shown in Figure 9.

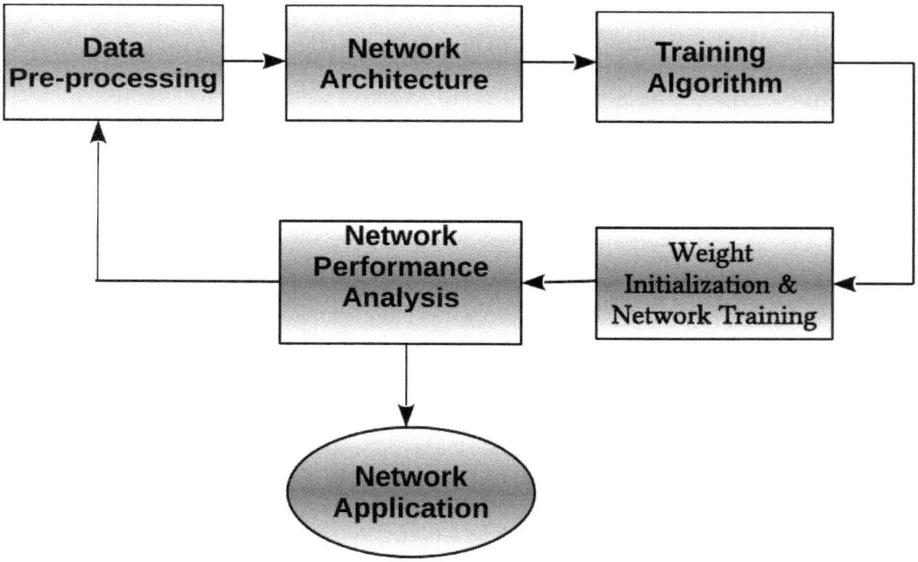

Figure 9. Flowchart of neural network training, analysis and application process.

Network training is all about determining the optimal values of network parameters for which the network performs best. Neural networks are a common category of parametric nonlinear functions of the form that transforms a vector **u** of input variables to a vector **v** of output variables. A simple way to obtain network parameters is to make the similarity of curve fitting, and thus minimize the sum of squares error function. Given a training set containing a set of input vectors $\mathbf{u}_n$, where $n = 1, \ldots, N$, together with a corresponding set of target vectors $\top_n$, we calculate the minimum of the error function.

$$\xi(\boldsymbol{\Omega}) = \frac{1}{2}\sum_{n=1}^{N} \|\mathbf{v}(\mathbf{u}_n, \boldsymbol{\Omega}) - \top_n\|^2. \tag{1}$$

Considering a single target variable k that can take any real value and that k has a Gaussian distribution with an x-dependent mean, which is given by the output of the neural network, so that

$$p(k|\mathbf{u}, \boldsymbol{\Omega}) = \mathbb{N}(\top|v(\mathbf{u}, \boldsymbol{\Omega}), \alpha^{-1}), \tag{2}$$

where α is the precision of Gaussian noise. For N independent data $\boldsymbol{\varnothing} = \{\mathbf{u}_1, ..., \mathbf{u}_N\}$ with corresponding target values $\top = \{\top_1, ..., \top_N\}$, we can build the consistent probability function.

$$p(\top|\boldsymbol{\varnothing}, \boldsymbol{\Omega}, \alpha) = \prod_{n=1}^{N} p(\top_n|\mathbf{u}_n, \boldsymbol{\Omega}, \alpha). \tag{3}$$

We can evaluate the error function by calculating the negative logarithm of the probability function.

$$\frac{\alpha}{2}\sum_{n=1}^{N} \{v(\mathbf{u}_n, \boldsymbol{\Omega}) - \top_n\}^2 - \frac{N}{2}\ln\alpha + \frac{N}{2}\ln(2\pi). \tag{4}$$

Equation (4) is used to learn parameters $\boldsymbol{\Omega}$ and α. In neural networks design literature, it is customary to take into account the minimization of the error function more than the maximization of the probability function. We are following the same approach here. First we obtain $\boldsymbol{\Omega}$ by maximizing the probability function (or minimizing the sum-of-squares error function).

$$\xi(\boldsymbol{\Omega}) = \frac{1}{2}\sum_{n=1}^{N} \{v(\mathbf{u}_n, \boldsymbol{\Omega}) - \top_n\}^2, \tag{5}$$

where operational constants are rejected. The value of $\boldsymbol{\Omega}$ obtained by minimizing $\xi(\boldsymbol{\Omega})$ represented by $\boldsymbol{\Omega}_{mxl}$ corresponds to the maximum probability solution. Having obtained $\boldsymbol{\Omega}_{mxl}$, α can be found by minimizing the negative logarithm of probability

$$\frac{1}{\alpha_{mxl}} = \frac{1}{N}\sum_{n=1}^{N} \{v(\mathbf{u}_n, \boldsymbol{\Omega}_{mxl}) - \top_n\}^2. \tag{6}$$

The practically local maximum of the probability may be obtained that corresponds to the local minimum of the error function. If there are multiple target variables, which are assumed to be independent depending on $\mathbf{u}$ and $\boldsymbol{\Omega}$ with common α, then the conditional distribution of the target values is obtained by:

$$p(\top|\mathbf{u}, \boldsymbol{\Omega}) = \mathbb{N}(\top|\mathbf{v}(\mathbf{u}, \boldsymbol{\Omega}), \alpha^{-1}\mathbf{I}). \tag{7}$$

With the same argument as for a single target variable, we obtain the maximum likelihood weights by minimizing the sum-of-squares error function. The noise precision is given by

$$\frac{1}{\alpha_{mxl}} = \frac{1}{NK}\sum_{n=1}^{N} \|\mathbf{v}(\mathbf{u}_n, \boldsymbol{\Omega}_{mxl}) - \top_n\|^2, \tag{8}$$

where K is the number of target variables. There is a real matching of the error function and the output unit activation function. In regression, the network can be viewed as having an output activation function that is the unitary, so that $v_k = b_k$. The corresponding sum-of-squares error function can be expressed as

$$\frac{\partial\xi}{\partial b_\top} = v_k - \top_k, \tag{9}$$

which is used for error backpropagation. In the case of binary classification, in which there is a single target variable k so that k = 1 represents class $\mathtt{CL}_1$ and k = 0 denotes class $\mathtt{CL}_2$. Assuming a network to have a single output whose activation function is a logistic sigmoid,

$$v = \sigma(b) = \frac{1}{1 + exp(-b)}. \tag{10}$$

In this instance, $0 \leq v(\mathbf{u}, \mathbf{\Omega}) \leq 1$. $v(\mathbf{u}, \mathbf{\Omega})$ is explained as the conditional probability $p(\mathtt{CL}_1|\mathbf{u})$ given by $1 - v(\mathbf{u}, \mathbf{\Omega})$. The conditional distribution of targets given inputs is then a Bernoulli distribution with the form

$$p(\top|\mathbf{u}, \mathbf{\Omega}) = v(\mathbf{u}, \mathbf{\Omega})^{\top}\{1 - v(\mathbf{u}, \mathbf{\Omega})\}^{1-\top}. \tag{11}$$

Considering a training set of unconventional results, then the error function, which is given by the negative logarithm probability, is a *measure of the performance the classifier or log loss* error function which is given by the form

$$\xi(\mathbf{\Omega}) = -\sum_{n=1}^{N}\{\top_n \ln v_n + (1 - \top_n)\ln(1 - v_n)\}, \tag{12}$$

where v_n denotes $v(\mathbf{u}, \mathbf{\Omega})$. Using the log loss error function in place of the sum-of-squares for a classifier improves training and generalization. If there are K separate binary classifications to be done, we can use a network that has K outputs, each of which has a logistic sigmoid activation function. Associated with each output is a binary class label $k_t \in \{0, 1\}$, where $t = 1, \ldots, \top$. Assuming that the class labels are individualistic, given the input vectors, the conditional distribution of the target is

$$p(\top|\mathbf{u}, \mathbf{\Omega}) = \prod_{k=1}^{K}(\mathbf{u}, \mathbf{\Omega})^{\top_k}[1 - v_{\top}(\mathbf{u}, \mathbf{\Omega})]^{1-\top_k}. \tag{13}$$

Evaluating the negative logarithm of the corresponding probability function then gives the following error function

$$\xi(\mathbf{\Omega}) = -\sum_{n=1}^{N}\sum_{k=1}^{K}\{k_{n\top} \ln v_{n\top} + (1 - k_{n\top})\ln(1 - v_{n\top})\}, \tag{14}$$

where $v_{n\top}$ represents $v_{\top}(\mathbf{u}_n, \mathbf{\Omega})$. Considering the standard multi-class classification problem in which each input is assigned to one of the $\top$ mutually exclusive classes. The binary target variables $k_{\top} \in \{0, 1\}$ have a 1-of-$\top$ coding strategy showing the class, and the network outputs are explained as $v_{\top}(\mathbf{u}, \mathbf{\Omega}) = p(k_{\top} = 1|\mathbf{u})$, which results in the following error function

$$\xi(\mathbf{\Omega}) = -\sum_{n=1}^{N}\sum_{\top=1}^{K} k_{\top n} \ln v_{\top}(\mathbf{u}_n, \mathbf{\Omega}). \tag{15}$$

The return unitary activation function, which correlates with the canonical link, is then represented by the softmax function

$$v_{\top}(\mathbf{u}, \mathbf{\Omega}) = \frac{exp(b_t(\mathbf{u}, \mathbf{\Omega}))}{\sum_j exp(b_j(\mathbf{u}, \mathbf{\Omega}))}, \tag{16}$$

which fulfills $0 \leq v_{\top} \leq 1$ and $\sum_{\top} v_{\top} = 1$. In general, there is a simple option of both output part activation function and complement error function, in accordance with the type of solution being sought. In the case of regression, linear outputs and a sum-of-squares error are used. For multiple unconstrained duplicate classifications we use logistic sigmoid outputs with a log-loss error function. In the case of multi-class classification we implement softmax outputs with the comparable multi-class log-loss error function. For a classifier problem concerning two classes, a single logistic sigmoid output can be used, or optionally network with two outputs having a softmax output activation function can be implemented.

Parameter Optimization

We will find a weight vector $\mathbf{\Omega}$ that decreases the selected function $\xi(\mathbf{\Omega})$. If we make a small pace in weight span from $\mathbf{\Omega}$ to $\mathbf{\Omega} + \delta\mathbf{\Omega}$, then the pace in the error function is $\delta\xi \simeq \delta\mathbf{\Omega}^T \nabla\xi(\mathbf{\Omega})$, where the vector $\nabla\xi(\mathbf{\Omega})$ shows in the direction of the largest rate of the pace of the error function. Because the error is an even, continuous function of $\mathbf{\Omega}$, its least value will happen at a point in weight span such that the slope of the error function disappears, therefore

$$\nabla\xi(\mathbf{\Omega}) = 0, \tag{17}$$

or else we could make a small pace in the direction of $-\nabla\xi(\mathbf{\Omega})$ and thereby further decrease the error. Points at which the slope disappears are known as static points, and may be categorized into minimum, maximum, and saddle points. Our aim is to find a vector $\mathbf{\Omega}$ such that $\xi(\mathbf{\Omega})$ takes its least value. Nevertheless, the error function always has an extremely nonlinear dependence on the weights and bias parameters. So there will be many points in weight span where the slope disappears (or is numerically very small). For any point $\mathbf{\Omega}$ that is a local minimum, there will be other points in the weight span that are similar minima.

Additionally, there will always be many different static points and in particular multiple different minima. A minimum that corresponds to the least value of the error function for any weight vector is called a global minimum. Any other minima corresponding to larger values of the error function are called local minima. For a better application of neural networks, it may not be important to find the global minimum but it may be good to compare many local minima to find an acceptably good result. because there is no expectation of obtaining an analytical result to the equation $\nabla\xi(\mathbf{\Omega}) = 0$, we seek iterative numerical methods. Most numerical methods include selecting some initial value $\mathbf{\Omega}^0$ for the weight vector and then operating through the weight span in sequence of pace of the following form,

$$\mathbf{\Omega}^{(\lambda+1)} = \mathbf{\Omega}^{(\lambda)} + \Delta\mathbf{\Omega}^{(\lambda)}, \tag{18}$$

where λ labels the iteration step. Separate algorithms use many alternatives for the weight vector improvement $\Delta\mathbf{\Omega}^{(\lambda)}$. There are methods that make use of slope detail and hence need, after each improvement, the value of $\nabla\xi(\mathbf{\Omega})$ to be calculated at the new weight vector $\mathbf{\Omega}^{(\lambda+1)}$. It is viable to calculate the slope of an error function effectively by means of the backpropagation procedure. The use of this gradient detail can lead to better improvements in the speed with which the minima of the error function can be detected.

The simplest way to use gradient detail is to choose the weight update in Equation (18) to include a small pace in the direction of the negative slope so that

$$\mathbf{\Omega}^{(\lambda+1)} = \mathbf{\Omega}^{(\lambda)} - \mu\nabla\xi(\mathbf{\Omega}^{(\lambda)}), \tag{19}$$

where the specification $\mu > 0$ is called the learning rate. Following each such improvement, the slope is re-calculated for the new weight vector and the computation is repeated. The error function is explained regarding a training set, and so each pace needs all the training sets to be computed in order to estimate $\nabla\xi$. Methods that use the whole dataset at once are named batch methods. At each pace, the weight vector is moved in the direction of the largest rate of decrease of the error function and so this approach is called the gradient descent or steepest descent. Even though this approach might look sensible, in fact it turns out to be a poor method. For batch optimization, there are more effective methods, such as conjugate gradients and quasi-Newton methods, which are stronger and faster than the simple gradient descent. Far from gradient descent, these algorithms have the effects that the error function always diminishes at each iteration except for the weight vector has reached a local or global minimum.

To find an acceptably good minimum, it may be mandatory to run a gradient-based algorithm many times, each time using a different randomly selected starting point, and comparing the resulting performance on an independent validation set. However, there is an on-line sort of gradient descent that has shown convenience in practice for training neural networks. Error functions based on maximum probability for a set of independent observations include a sum of terms, one for every data point,

$$\xi(\mathbf{\Omega}) = \sum_{n=1}^{N} \xi_n(\mathbf{\Omega}). \tag{20}$$

On-line gradient descent, also known as *sequential gradient descent* or *stochastic gradient descent*, makes an update to the weight vector based on one data point at a time so that

$$\mathbf{\Omega}^{(\lambda+1)} = \mathbf{\Omega}^{(\lambda)} - \mu \nabla \xi_n(\mathbf{\Omega}^{(\lambda)}). \tag{21}$$

In this way, the update is redone by repeating through the data either in succession or by random selection of points and replacing. The power of on-line methods handle superfluity in the data much more effectively in contrast to the batch methods.

One of the challenges with training networks with a limited number of input data is over fitting which greatly degrades the performance of the network. A recommended way of reducing this overfitting is through randomly dropping a certain number of neurons from layers [40]. We implemented this regularization method and were able to achieve an improvement in the performance of our network.

5. Results and Discussion

The datasets used for this study are problem specific and a neural network is designed for this specific purpose that does not rely on the use of other pre-trained networks and transfer learning. So network training was done from scratch. Network training was performed on NVIDIA® AGX Xavier Developer Kit for 50 epochs. The accuracies for mini-batch, validation and the corresponding losses are summarized in Table 2 for some selected epochs.

Using sequential gradient descent (SGD) with momentum (0.9) and a piece wise adjustment to the learning rate schedule, training was run on a 512-core Volta GPU with Tensor Cores. For GPUs with less memory, it may be necessary to reduce the batch size. The choice of SGD optimization has advantages for small datasets. Based on the SGD optimization scheme, ten classes of indoor objects were used for training the model. The validation precision of the architecture to predict an object which was not contained in training data was 76.5%. Furthermore, the performance of the network was improved by fine tuning the weights (parameters) and learning rate. More importantly, the performance of the network improved by about 10.5% using data augmentation and dropout techniques.

As we observe from our experimental results, training and validation precision are separated by a large boundary in the case where data augmentation was not implemented. The model in this case attained only around 68.5% precision on the validation dataset. Shown in Figure 10, the training precision improves linearly over time, but the validation precision stalls around 68.5% in the learning process. Moreover, the difference in precision between training and validation losses is clearly observed which is an indication of overfitting.

Table 2. Network training on 512-core Volta GPU result summary.

Epoch	Mini-Batch Accuracy	Vald. Accuracy	Mini-Batch Loss	Vald. Loss
1	37.54%	57.22%	1.48	1.10
4	68.53%	67.03%	0.82	0.94
8	74.08%	70.44%	0.68	0.74
12	79.09%	72.62%	0.55	0.68
16	82.30%	75.20%	0.50	0.69
20	83.55%	73.98%	0.41	0.75
24	88.20%	74.52%	0.40	0.77
27	88.09%	74.39%	0.32	0.84
31	91.62%	74.39%	0.26	0.78
35	92.04%	76.84%	0.23	0.83
39	93.08%	76.29%	0.18	1.06
43	94.56%	72.62%	0.16	1.14
47	93.89%	76.98%	0.18	0.90
50	94.48%	76.16%	0.15	1.09

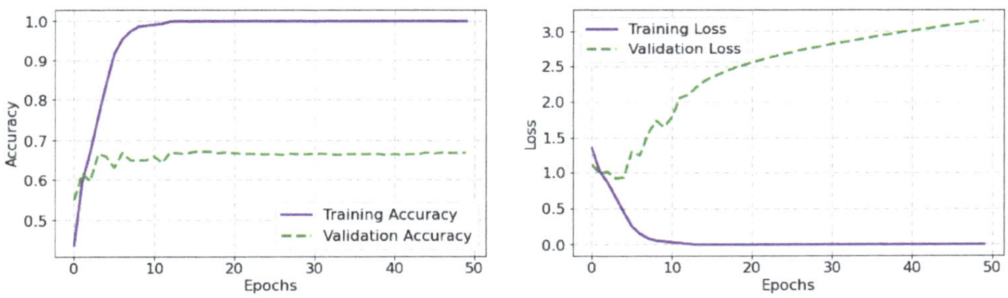

Figure 10. Model performance without data augmentation.

When there is a small number of training datasets, the model sometimes learns from unwanted details from the training dataset that negatively impact the performance of the model on new datasets. This means that the model will not be able to generalize on a new dataset. To reduce the problem of overfitting, we implemented image augmentation and dropout techniques. Image augmentation produces more training images from our existing image by augmenting them using random transformations that yield believable-looking images. This helps expose the model to more aspects of the image and generalize better. Image augmentation as depicted in Figure 11 improved the model performance to well above 74.5%. At the same time, the loss during model training is much reduced, indicating a better model performance.

In the dropout scheme, we randomly decreased the number of output units in the dense layer in a range of 10% to 40%. Despite the fact that there was not much noticeable improvement as a result of dropout in training and validation accuracy, there was a progressive improvement of the validation and loss functions as shown in Figure 12a–d. Increasing the number of dropouts from the output units again is not a guarantee for better model performance as it may degrade model performance. In this study, the best model performance is achieved for both validation precision and validation loss at 30% dropout of the output units.

The performance of the trained model network is evaluated using validation datasets. Validation data are not used to train the network. Our experimental results reveal that the model classified the data in the training dataset with a success rate of 94.5% and validated results with a success rate of 76.5%. As shown in Figure 12, the training loss reveals how well the model is fitting the training data, while the validation loss reveals how well the model fits new data. Very successful rates on both training and validation loss are achieved at a 30% dropout, which are well below 0.3 and 0.8, respectively.

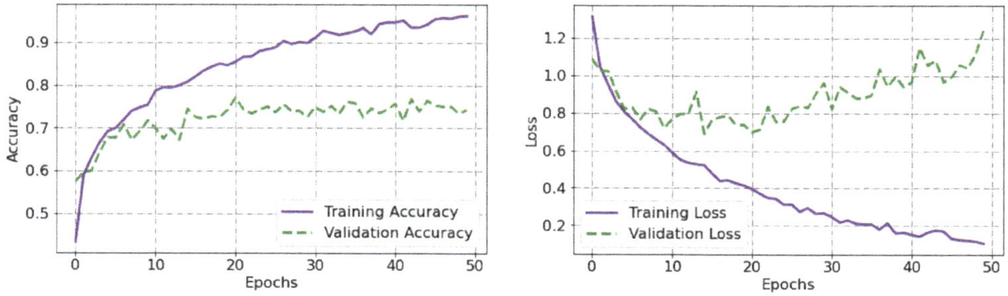

Figure 11. Model performance with data augmentation.

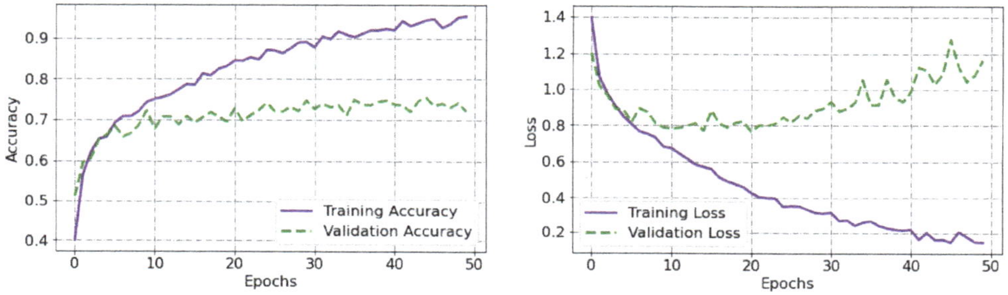

(**a**) 10% dropout

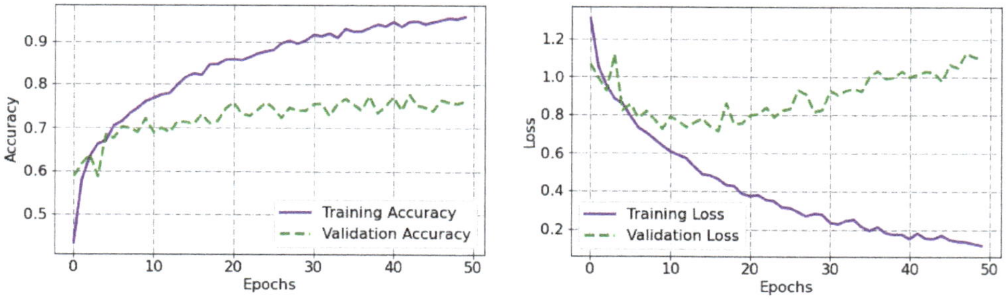

(**b**) 20% dropout

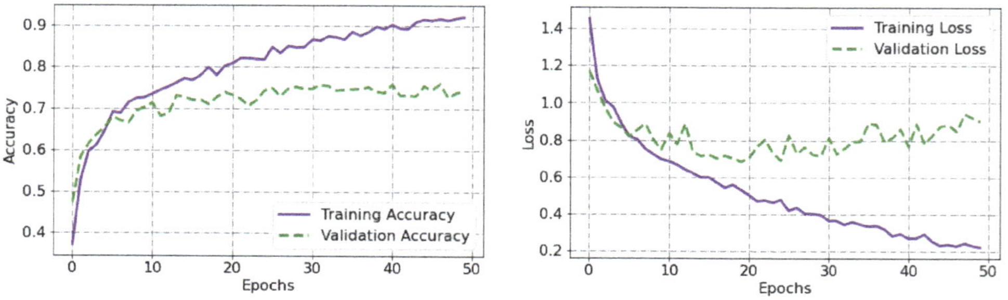

(**c**) 30% dropout

Figure 12. *Cont.*

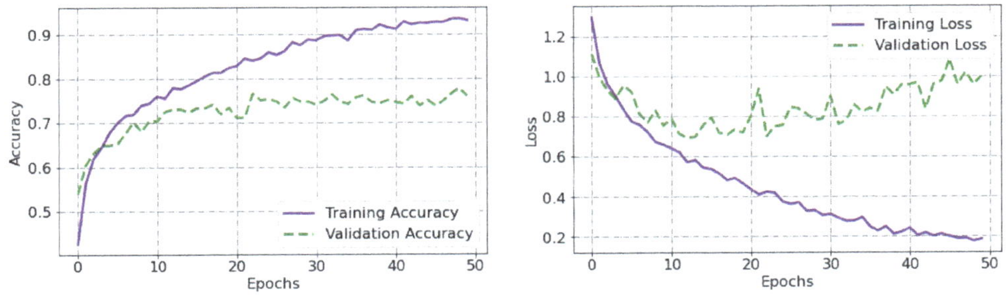

(**d**) 40% dropout

Figure 12. Improvement in validation accuracy and loss function of the model architecture.

To study the precision over various label groups, the confusion matrix (error matrix) is plotted, which is shown in Figure 13. The label imbalance noted in the training set is an issue in the classification accuracy. The confusion chart (matrix) illustrates higher precision and recall for walls and confuses most tables with doors. Even if there are very few miss-predicted objects, the overall performance of the model to predict an object that was not contained in the training or validation sets is successful. Since the purpose of this study is to demonstrate a basic classification network training approach with raw data, possible next steps that could be taken to improve classification performance, such as re-sampling the training set or achieving better label balance or using a loss function more robustly to label imbalance (e.g., weighted cross-entropy,) will be explored in future studies.

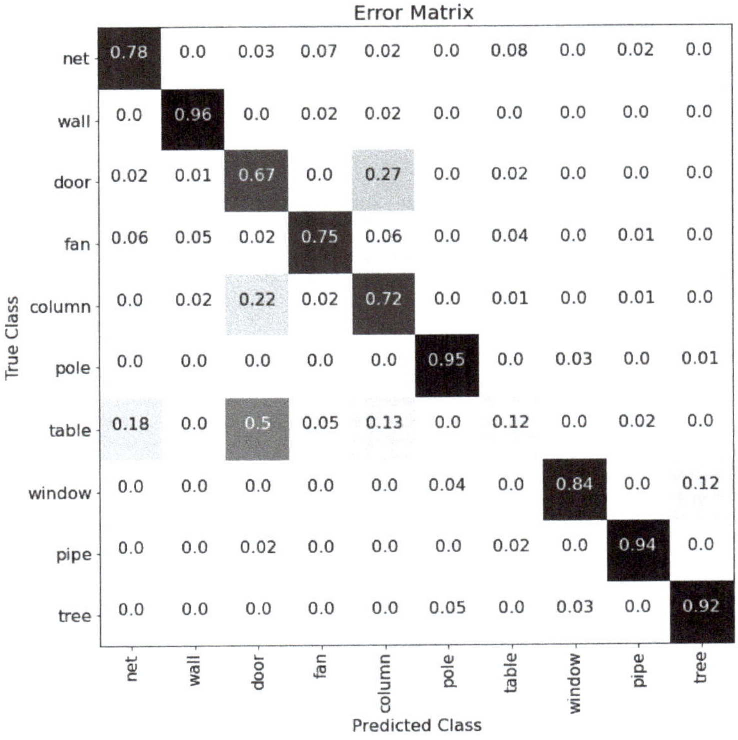

Figure 13. Confusion chart illustrating precision for each class.

6. Conclusions

A wide diversity of design and optimization techniques are used to design neural networks and to analyze their performance. The goal of these novel approaches is not to restrain the expertise architect but is rather to provide hands-on ways and mechanisms that can help obtain less complex, more robust and high-performance designs. In this study, we designed and presented a supervised deep neural network architecture for accurate real-time classification of objects in cluttered indoor environments. We have performed an in-depth analysis of augmentation in image classification for the case where there is not a large dataset. Our experimental results revealed that model performance is enhanced by enforcing different augmentation techniques provided that the level of noise remains reasonable.

To overcome data constraints in the development of the model that require large datasets, we implemented a data augmentation technique for each class of objects. Moreover, we implemented a dropout technique to further improve the performance of the model and reduce the validation loss. In this case, the model achieved a success rate of 94.5% and 76.5% for the training and validation datasets, respectively. Experimental results reveal that the model architecture performs very well in predicting new data that are not included in either the training or validation datasets. In our future work, we will deploy our model on different problems related to images and explore optimal ways to hasten the computational time.

Author Contributions: In this manuscript, model design for the object classifier, hardware integration for collecting data, and model validation and visualization were done by B.E. The overall conceptual design of the methodology, advisory work on pre-training data augmentation and manuscript editing were done by A.T. Data collection, profiling, labeling and arrangement was done by H.-Y.S. and B.-S.K. was responsible for the technical layout of the manuscript and the supervision of the research work in progress. All authors have read and agreed to the published version of the manuscript.

Funding: This research was funded by the Korea Institute for Advancement of Technology (KIAT) grant funded by the Korean Government (MOTIE) (N0002431, The Competency Development Program for Industry Specialist).

Institutional Review Board Statement: Not applicable.

Informed Consent Statement: Not applicable.

Conflicts of Interest: Authors mentioned in this manuscript have strongly involved in this study from the start to the end. The manuscript were thoroughly reviewed by the authors before it's submission to journal of Applied Science. This manuscript has not been submitted to another journal for publication.

References

1. Naghavi, S.H.; Avaznia, C.; Talebi, H. Integrated real-time object detection for self-driving vehicles. In Proceedings of the 10th Iranian Conference on Machine Vision and Image Processing, Isfahan, Iran, 22–23 November 2017.
2. Liu, D.; Cui, Y.; Chen, Y.; Zhang, J.; Fan, B. Video object detection for autonomous driving: Motion-aid feature calibration. *Neurocomputing* **2020**, *409*, 1–11. [CrossRef]
3. Wang, Y.; Liu, D.; Jeon, H.; Chu, Z.; Matson, E. End-to-end learning approach for autonomous driving: A convolutional neural network model. In Proceedings of the International Conference on Agents and Artificial Intelligence 2019, Prague, Czech Republic, 19–21 February 2019.
4. Saloni, W. The Role of Autonomous Unmanned Ground Vehicle Technologies in Defense Applications. Aerospace & Defense Technology Magazine. 2020. Available online: https://www.aerodefensetech.com/component/content/article/adt/features/articles/37888 (accessed on 1 April 2021).
5. Niu, H.; Gonzalez-Prelcic, N.; Heath, R.W. A uav-based traffic monitoring system-invited paper. In Proceedings of the IEEE 87th Vehicular Technology Conference, Porto, Portugal, 3–6 June 2018.
6. Sarthak, B.; Sujit, P. UAV Target Tracking in Urban Environments Using Deep Reinforcement Learning. *arXiv* **2020**. arXiv:2007.10934.
7. Zhen, J.; Balasuriya, A.; Subhash, C. Autonomous vehicles navigation with visual target tracking: Technical approaches. *Algorithms* **2008**, *1*, 153–182.

8. Aguilar, W.G.; Luna, M.A.; Moya, J.F.; Abad, V.; Parra, H.; Ruiz, H. Pedestrian detection for UAVs using cascade classifiers with meanshift. In Proceedings of the IEEE 11th International Conference on Semantic Computing, San Diego, CA, USA, 30 January–1 February 2017.
9. Gageik, N.; Benz, P.; Montenegro, S. Obstacle detection and collision avoidance for a uav with complementary low-cost sensors. *IEEE Access* **2015**, *3*, 599–609. [CrossRef]
10. Zhang, W.J.; Yang, G.; Lin, Y.; Gupta, M.M.; Ji, C. On the definition of deep learning. In Proceedings of the 2018 World Automation Congress (WAC), Stevenson, WA, USA, 3–6 June 2018.
11. Yann, L.; Yoshua, B.; Geoffrey, H. Deep learning. *Nature* **2015**, *521*, 436–444.
12. Jurgen, S. Deep learning in neural networks: An overview. *Neural Netw.* **2015**, *61*, 85–117.
13. Alzubaidi, L.; Zhang, J.; Humaidi, A.J.; Al-Dujaili, A.; Duan, Y.; Al-Shamma, O.; Santamaría, J.; Fadhel, M.A.; Al-Amidie, M.; Farhan, L. Review of deep learning: Concepts, CNN architecture, challenges, applications, future directions. *J. Big Data* **2021**, *8*, 1–74. [CrossRef]
14. Jefferson, S.; Gustavo, P.; Fernando, O.; Denis, W. Vision-Based Autonomous Navigation Using Supervised Learning Techniques. In Proceedings of the 12th Engineering Applications of Neural Networks and 7th Artificial Intelligence Applications and Innovations, Corfu, Greece, 15–18 September 2011.
15. Fabrice, R.N. Machine Learning and a Small Autonomous Aerial Vehicle Part 1: Navigation and Supervised Deep Learning. *Tech. Rep.* **2018**. [CrossRef]
16. Wang, D.; Chen, J. Supervised Speech Separation Based on Deep Learning: An Overview. Available online: https://arxiv.org/pdf/1708.07524.pdf (accessed on 1 April 2021).
17. Yang, X.; Kwitt, R.; Niethammer, M. Fast predictive image registration. In *Deep Learning and Data Labeling for Medical Applications*; Springer: Cham, Switzerland, 2016; pp. 48–57.
18. Louati, H.; Bechikh, S.; Louati, A.; Hung, C.C.; Said, L.B. Deep convolutional neural network architecture design as a bi-level optimization problem. *Neurocomputing* **2021**, *439*, 44–62. [CrossRef]
19. Bayot, R.; Gonalves, T. A Survey on Object Classfication Using Convolutional Neural Networks. 2015. Available online: https://core.ac.uk/download/pdf/62473376.pdf (accessed on 1 April 2021).
20. Thumu, K.; Gurrala, N.R.; Srinivasan, N. Object Classification and Detection using Deep Convolution Neural Network Architecture. *Int. J. Recent Technol. Eng.* **2020**. [CrossRef]
21. Rikiya, Y.; Mizuho, N.; Richard, K.G.D.; Kaori, T. Convolutional neural networks: An overview and application in radiology. *Insights Imaging* **2018**, *9*, 611–629.
22. Richard, L.; Roberto, F. Space Object Classification Using Deep Convolutional Neural Networks. In Proceedings of the 19th International Conference on Information Fusion, Big Sky, MT, USA, 3–10 March 2018.
23. Bai, S.; Kolter, J.Z.; Koltun, V. An empirical evaluation of generic convolutional and recurrent networks for sequence modeling. *arXiv* **2018**, arXiv:1803.01271.
24. Asifullah, K.; Anabia, S.; Umme, Z.; Aqsa, S.Q. A survey of the Recent Architectures of Deep Convolutional Neural Networks. *Artif. Intell. Rev.* **2020**, *53*, 5455–5516.
25. Neha, S.; Vibhor, J.; Anju, M. An Analysis Of Convolutional Neural Networks For Image Classification. *Procedia Comput. Sci.* **2018**, *132*, 377–384.
26. Karen, S.; Andrew, Z. Very Deep Convolutional Networks for Large-Scale Image Recognition. *arXiv* **2015**, arXiv:1409.1556.
27. Christian, S.; Wei, L.; Yangqing, J.; Pierre, S.; Scott, R.; Dragomir, A.; Dumitru, E.; Vincent, V.; Andrew, R. Going Deeper with Convolutions. *arXiv* **2015**, arXiv:1409.4842.
28. Alex, K.; Ilya, S.; Geoffrey, E.H. ImageNet Classification with Deep Convolutional Neural Networks. *Imagenet Compet.* **2012**, *25*, 1097–1105.
29. Nielsen, M. Neural Networks and Deep Learning. Free Online Book, Michael Nielsen, 2019; Chapter 5. Available online: http://neuralnetworksanddeeplearning.com/about.html (accessed on 1 April 2021).
30. Soumya, J.; Dhirendra, K.V.; Gaurav, S.; Amit, P. Issues in Training a Convolutional Neural Network Model for Image Classification. *Adv. Comput. Data Sci.* **2019**, doi10.1007/978-981-13-9942-8_27. [CrossRef]
31. Hugo, L.; Yoshua, B.; Jerome, L.; Pascal, L. Exploring Strategies for Training Deep Neural Networks. *J. Mach. Learn. Res.* **2009**. [CrossRef]
32. He, K.; Zhang, X.; Ren, S.; Sun, J. Deep Residual Learning for Image Recognition. Computer Vision and Pattern Recognition. *arXiv* **2015**, arXiv:1512.03385.
33. Stephan, Z.; Yang S.; Thomas, L.; Ian G. Improving the Robustness of Deep Neural Networks via Stability Training. *arXiv* 2016, arXiv:1604.04326.
34. Tamás, O.; Anton, R.; Ants, K.; Pedro, A.; David, P.; Jan, K.; Pavel, K. Robust Design Optimization and Emerging Technologies for Electrical Machines: Challenges and Open Problems. *Appl. Sci.* **2020**, *10*, 6653.
35. Ivana, S.; Eva, T.; Nebojsa, B.; Miodrag, Z.; Marko, B.; Milan, T. Designing Convolutional Neural Network Architecture by the Firefly Algorithm. *Int. Young Eng. Forum* **2019**. [CrossRef]
36. Gao, H.; Zhuang, L.; van der Laurens, M.; Kilian, Q.W. Densely Connected Convolutional Networks. *arXiv* **2016**, arXiv:1608.06993.
37. Connor, S.; Taghi, K.M. A survey on Image Data Augmentation of Deep Learning. *J. Big Data* **2019**, *6*, 1–48.

38. Keiron, O.; Ryan, N. An Introduction to Convolutional Neural Networks. Neural and Evolutionary Computing. *arXiv* **2015**, arXiv:1511.08458.
39. Xavier, G.; Yoshua, B. Understanding the difficulty of training deep feedforward neural networks. In Proceedings of Machine Learning Research, Sardinia, Italy, 13–15 May 2010.
40. Geoffrey, E.H.; Nitish, S.; Alex, K.; Ilya, S.; Ruslan, R.S. Improving neural networks by preventing co-adaptation of feature detectors. *arXiv* **2012**, arXiv:1207.0580.

applied sciences

MDPI

Article

Swarm Intelligence with Deep Transfer Learning Driven Aerial Image Classification Model on UAV Networks

Saud S. Alotaibi [1], Hanan Abdullah Mengash [2], Noha Negm [3,4], Radwa Marzouk [2], Anwer Mustafa Hilal [5,*], Mohamed A. Shamseldin [6], Abdelwahed Motwakel [5], Ishfaq Yaseen [5], Mohammed Rizwanullah [5] and Abu Sarwar Zamani [5]

1 Department of Information Systems, College of Computing and Information System, Umm Al-Qura University, Mecca 24382, Saudi Arabia; ssotaibi@uqu.edu.sa
2 Department of Information Systems, College of Computer and Information Sciences, Princess Nourah Bint Abdulrahman University, P.O. Box 84428, Riyadh 11671, Saudi Arabia; hamengash@pnu.edu.sa (H.A.M.); rmmarzouk@pnu.edu.sa (R.M.)
3 Department of Computer Science, College of Science & Art at Mahayil, King Khalid University, Abha 62529, Saudi Arabia; nabdelhamid@kku.edu.sa
4 Department of Mathematics and Computer Science, Faculty of Science, Menoufia University, Menoufia 32511, Egypt
5 Department of Computer and Self Development, Preparatory Year Deanship, Prince Sattam Bin Abdulaziz University, Al-Kharj 16278, Saudi Arabia; asmaeil@psau.edu.sa (A.M.); iyasen@psau.edu.sa (I.Y.); mrizwan@psau.edu.sa (M.R.); azmani@psau.edu.sa (A.S.Z.)
6 Department of Mechanical Engineering, Faculty of Engineering and Technology, Future University in Egypt, New Cairo 11835, Egypt; mohamed.abelbar@fue.edu.eg
* Correspondence: a.hilal@psau.edu.sa

Citation: S. Alotaibi, S.; Abdullah Mengash, H.; Negm, N.; Marzouk, R.; Hilal, A.M.; Shamseldin, M.A.; Motwakel, A.; Yaseen, I.; Rizwanullah, M.; Zamani, A.S. Swarm Intelligence with Deep Transfer Learning Driven Aerial Image Classification Model on UAV Networks. *Appl. Sci.* **2022**, *12*, 6488. https://doi.org/10.3390/app12136488

Academic Editors: Sylvain Bertrand and Hyo-sang Shin

Received: 25 April 2022
Accepted: 22 June 2022
Published: 26 June 2022

Abstract: Nowadays, unmanned aerial vehicles (UAVs) have gradually attracted the attention of many academicians and researchers. The UAV has been found to be useful in variety of applications, such as disaster management, intelligent transportation system, wildlife monitoring, and surveillance. In UAV aerial images, learning effectual image representation was central to scene classifier method. The previous approach to the scene classification method depends on feature coding models with lower-level handcrafted features or unsupervised feature learning. The emergence of convolutional neural network (CNN) is developing image classification techniques more effectively. Due to the limited resource in UAVs, it can be difficult to fine-tune the hyperparameter and the trade-offs amongst computation complexity and classifier results. This article focuses on the design of swarm intelligence with deep transfer learning driven aerial image classification (SIDTLD-AIC) model on UAV networks. The presented SIDTLD-AIC model involves the proper identification and classification of images into distinct kinds. For accomplishing this, the presented SIDTLD-AIC model follows a feature extraction module using RetinaNet model in which the hyperparameter optimization process is performed by the use of salp swarm algorithm (SSA). In addition, a cascaded long short term memory (CLSTM) model is executed for classifying the aerial images. At last, seeker optimization algorithm (SOA) is applied as a hyperparameter optimizer of the CLSTM model and thereby results in enhanced classification accuracy. To assure the better performance of the SIDTLD-AIC model, a wide range of simulations are implemented and the outcomes are investigated in many aspects. The comparative study reported the better performance of the SIDTLD-AIC model over recent approaches.

Keywords: computer vision; unmanned aerial vehicles; deep transfer learning; object detection; aerial image classification; parameter optimization

1. Introduction

Unmanned aerial vehicles (UAV) are utilized as a cost-efficient and prompt methodology for taking remote sensing (RS) images. The boon of UAV technology involves least cost, small size, security, natural function, and, especially, the fast and on-demand acquisition of images [1]. The developments of UAV technologies have achieved the state that it can offer intense higher resolution RS images encircling lavish contextual and spatial information. This has allowed studies suggesting numerous original applications for UAV image examination, comprising disaster management, vegetation monitoring, object detection, detection and mapping of archaeological sites, oil and gas pipeline monitoring, and urban site analysis [2,3].

Aerial image classification methodologies grant distinct semantic categories that are usually established through exploiting changes in spatial deployments and structural forms for designing scenes [4]. In opposition with object or pixel related classifier methods, scene classification provides localization data from extensive aerial image which has apparent semantic data of the surfaces. Such methodologies are classified into three categories, which are: high level vision information, low level visual features, and mid-level visual representations [5,6]. Aerial scenes are differentiated by low level characteristics use, structural features, texture, spectral, and so on. Subsequently, low level feature vectors are pictorial ascriptions which can be derived globally or locally and are usually utilized for describing aerial scene images [7,8]. The typical low level feature methods are local binary patterns (LBP), Global Invariant Scale Transform (GIST), and color histogram Scale Invariant Feature Transform (SIFT). Mid-level analytical methods try to advance complete scene illustrations through conveying high order statistical outlines which are created by deriving local visual qualities [9]. The common processing pipeline derives local image patches, and they are programmed as local signals; therefore, creating a complete mid-level depiction of the aerial scenes. The familiar mid-level procedure is Bag of Visual Words (BoVW).

Deep learning (DL) procedures like Convolutional Neural Networks (CNNs) were broadly recognized as a notable approach for numerous computer vision applications (classification, image or video recognition, and detection), and have revealed amazing outcomes in various applications [10]. Therefore, there comes numerous advantages to stopping from utilizing DL methods in emergency response and calamity management applications to restore crucial data in a timely manner and permitting superior research and response in the course of time-critical circumstances, and supporting the decision-making processes [11]. Although CNNs were rising successfully at several classification roles via transfer learning (TL), their interpretation speed on implanted platforms, like those discovered on-board UAVs, is hampered by the high computational cost, which may acquire and the model size of these networks is prohibitive from a memory standpoint for these entrenched gadgets [12]. At the same time, most of the earlier works do not consider hyperparameter tuning process into account.

This article focuses on the design of swarm intelligence with deep transfer learning driven aerial image classification (SIDTLD-AIC) model on UAV networks. The presented SIDTLD-AIC model follows a feature extraction module using RetinaNet model, in which the hyperparameter optimization process is performed by the use of salp swarm algorithm (SSA). The SSA is chosen as it avoids the local optimal constraints, thus achieving a smooth balance between exploration and exploitation. In addition, a cascaded long short term memory (CLSTM) model is executed for classifying the aerial images. At last, seeker optimization algorithm (SOA) is applied as a hyperparameter optimizer of the CLSTM model and thereby results in enhanced classification accuracy. To assure the better performance of the SIDTLD-AIC model, a wide range of simulations are executed and the outcomes are investigated in various aspects.

2. Related Works

Haq et al. [13] applied DL based supervised image classification model and images gathered using UAV for the forest region classification. The DL technique based stacked Autoencoder (SAE) has shown remarkable potential with respect to the assessment of forest areas and image classification. The experiment result shows that DL technique provides improved performance than other machine learning approaches. The researchers in [14] address the shortcoming of multi-labeling UAV images, usually considered by a higher level of dataset content, by presenting a novel technology based on CNN. They are employed as a means to produce an accurate representation of the query images that are analyzed afterward sub-dividing them into a grid of tiles. The multi-label classification process is implemented by combining a radial basis function neural network and a multi-labeling layer comprised of threshold operation. The researchers in [15] proposed a DL algorithm for classifying UAV images derived from the location and sensor of earth's surface. Initially, the labelled and unlabelled UAV images are fed to a pre-trained CNN to generate deep feature representation. Next, we learned strong domain-invariant features with a further network comprised of two fully connected layers.

Rajagopal et al. [16] developed a new optimum DL-based scene classification algorithm captured by UAV. The suggested method includes a residual network-based features extraction (RNBFE) that extract feature from the convolutional layer of a DRN system. Furthermore, the various parameters result in configuration errors because of parameter tuning. Hence, self-adoptive global best harmony search (SGHS) approach is applied to tune the parameter of the presented model. The researchers in [17] present a multi-objective optimization algorithm to evolve deep CNN for scene classification that generates the non-dominant solution in an automatic manner at the Pareto front. Then, we used two sets of benchmark data sets for testing the effectiveness of the scene classification algorithm and making an extensive analysis. Pustokhina et al. [18,19] presented an energy-effective cluster-based UAV system using DL based scene classification model. The suggested method includes a clustering with parameter tuned residual network (C-PTRN) system that operates on two primary processes scene classification and cluster construction.

3. The Proposed Model

In this article, an automated SIDTLD-AIC method was established for the proper identification and classification of images into distinct kinds on UAV networks. The presented SIDTLD-AIC model follows a feature extraction module using RetinaNet model in which the hyperparameter optimization process is performed by the use of SSA. Next, the SOA-CLSTM model is applied to classify the aerial images. Figure 1 depicts the block diagram of SIDTLD-AIC technique.

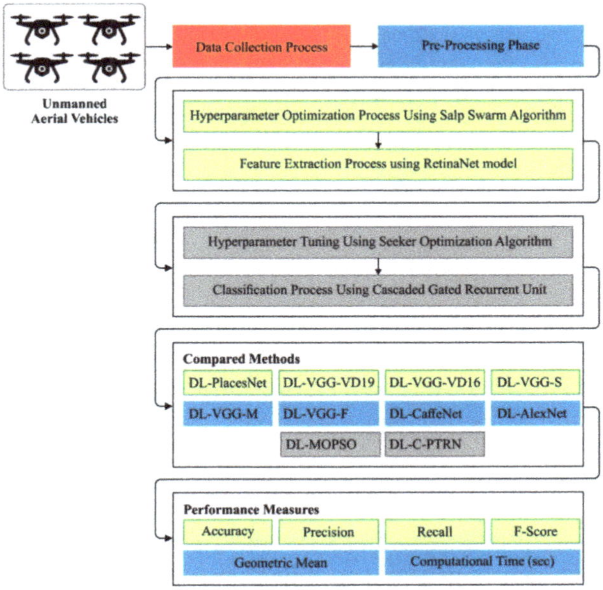

Figure 1. Block diagram of SIDTLD-AIC technique.

3.1. Feature Extraction Using RetinaNet Model

Transfer learning model is applied to enhance the efficiency of the DL model by the use of labeled data. It learns and employs many source processes for enhancing the learning process in relevant domains. It encompasses pre-training approaches which is trained on large scale dataset and is retrained at varying levels of the model on a small training set. The preliminary layer of the pre-training network can be modified upon requirement. The final layer of the model's hyperparameters can be tuned for learning the abilities on new datasets. In this work, the RetinaNet based TL model is applied for deriving feature vectors. An input map was inspired by the individual layer still accomplishing the resulting map. The CNN is designed in an order of layers. Consider $\in R^{h' \times w' \times c'}$ $((h)$: height, w: width, c: channel) are RGB images. Each layer gets X and the set of variables W as input as well as output images $y \in R^{h' \times w' \times c'}$, for example, $y = f(X, W)$. This makes an activation map to demonstrate the reaction of that filter at each spatial region. For calculating the input X with set of filters $W \in R^{\overline{h} x \overline{w} \times \overline{c} \times c'}$ and add a bias $b \in R^{c'}$ as follows.

$$y_{i'j'k'} = f\left(b_{k'} + \sum_{i=1}^{\overline{h}} \sum_{j=1}^{\overline{w}} \sum_{d=1}^{c} W_{ijdk} \times X_{i'+i,j'+j,d'}\right). \quad (1)$$

Next, the max-pooling layer is employed to decrease the computation and parameter with the decreased size of imputing shapes. It evaluates the maximum response of every image channel from $\overline{h} \times W$ sub-windows that implement as sub-sampling function. It can be expressed in the following:

$$y_{i'j'k'} = \max_{\substack{1<i<\overline{h}}}, {}_{1<j<\overline{\omega}} X_{i'+ij'+j,k}. \quad (2)$$

Finally, fully connected (FC) layer is a set of layers that integrate the data extracted by previous layer (feature). This layer gets an input X, processes them, and the last FC layer generates one dimensional vector of size. RetinaNet mainly consists of [20] two fully convolution network (FCNs), ResNet, and feature pyramid network (FPN). The ResNet employs network layer. The widely employed types of network layers are 50_, 101_, and 152_layers. The 101_layer with optimal trained efficacy is chosen. It could eliminate the structure of echocardiography with ResNet and, after, keep them to following sub-network. An FPN is an approach for efficiently eliminating the feature of each dimension from image with a conventional CNN architecture.

Focal loss: it can be improved version of the binary cross entropy (CE), the loss expression is given below:

$$CE\ (p,\ y) = \begin{cases} -\log(p) & if\ y = 1 \\ -\log(l - p) & otherwise, \end{cases} \quad (3)$$

In Equation (3), $y \in [-1, +1]$ indicates the ground truth type and $p \in [0,\ 1]$ represents the prediction probability to type $y = 1$.

$$p_t = \begin{cases} p, & if\ y = l \\ l - p, & otherwise \end{cases} \quad (4)$$

The preceding formula can be abbreviated as follows:

$$CE\ (p,\ y) = CE(p_t) = -\ \log\ (p_t) \quad (5)$$

In order to resolve the problem of data imbalance among the negative and positive samples, the new process is changed into the succeeding process:

$$CE(p_t) = -\alpha_t \log\ (p_t). \quad (6)$$

Among them,

$$\alpha_t = \begin{cases} \alpha, & if\ y = 1, \\ l - \alpha & otherwise, \end{cases} \tag{7}$$

Here, $\alpha \in [0, 1]$ represents the weight factor. In order to resolve the problem, the concentrating variable C was determined to obtain the final process of focal loss:

$$FL(p_t) = -\alpha_t (1 - p_t)^{\gamma} \log (p_t). \tag{8}$$

3.2. Hyperparameter Optimization: SSA

In this work, the hyperparameters of the RetinaNet model such as number of epochs, batch size, learning rate, and momentum are adjusted by the design of SSA. The SSA is simulated from the aggregation performance of salps that procedure a chain of salps and then hunt and move. The salp chain was developed from two kinds of salps, leader and follower [21]. The leader is the salp at the head of chains. Individual salps at the back of chains are followers. During the salp technique, food source F was determined as the individual with optimum fitness amongst every individual. The food source of tth order is $F(t)$. The steps of SSA technique are provided under.

1. Initialization of the population. In order for every individual, the places were arbitrary numbers amongst the upper as well as lower limits. They then compute fitness of every individual and sorted them. An individual with minimal fitness is the food source $F(t)$. $t = 1$, since one iteration was ended.
2. The population place was upgraded. The leader place was upgraded as:

$$x_j^i(t+1) = \begin{cases} F_j(t) + c_1\left[(ub_j - lb_j)c_2 + lb_j\right] & c_3 \geq 0.5 \\ F_j(t) - c_1\left[(ub_j - lb_j)c_2 + lb_j\right] & c_3 < 0.5 \end{cases} \tag{9}$$

where $i = 1$, i.e., the count of leaders is 1. It ranks primary from the populations. $j = 1, 2 \cdots D$. F_j, ub_j, and lb_j are $F(t)$, ub, and lb from the jth dimensional correspondingly. c_2 and c_3 implies the arbitrary numbers from zero and one. c_2 affects the step length of leader movement. c_3 defines if the leader moves forward/backward to food sources. T signifies the maximal number of iterations. c_1 refers the co-efficient of moving length.

$$c_1 = 2e^{-(4t/T)^2} \tag{10}$$

The place of follower is:

$$x_j^i(t+1) = 0.5\left(x_j^{i-1}(t) + x_j^i(t)\right) \tag{11}$$

where $i \geq 2$, and is the sequence of followers from the population. $j = 1, 2 \cdots D$.

3. Compute the fitness of every upgraded individual. The sort of individuals. Upgrade $F(t)$. Improve t by 1.
4. If the iteration accuracy condition was attained or $t = T$, the iteration terminates; or else, go to (2) to remain the iteration.

3.3. Image Classification Using Optimal CLSTM Model

In the final stage, the optimal CLSTM model is utilized to recognize different types of classes that exist in the aerial images [22]. A recurrent neural network (RNN) is a kind of DL method which is depending on existing input and the preceding input. In general, it is suitable for the scenario where the dataset has a consecutive correlation. But while handling a longer series of datasets, there exists an exploiting and vanishing gradient problem. In order to resolve this problem, a long short term memory (LSTM) is utilized that has an internal memory state which adds forget gate. The gate controls the time dependency and the effects of preceding inputs. Bidirectional long short term memory (BiLSTM) and bidirectional RNN (BiRNN) are other variations that reflect preceding input and assume the upcoming input of a certain time frame. This work, can present the BiLSTM RNN and cascaded uni-directional LSTM models. The method comprises the initial layer of bi-directional RNN integrated with uni-directional RNN layer. The bi-directional LSTM comprises forward and backward tracks to learn patterns in two directions.

$$O_n^{f1},\ h_n^{f1},\ i_n^{f1} = L^{f1}\left(i_{n-1}^{f1},\ h_{n-1}^{f1},\ x_n\ :P^{f1}\right), \tag{12}$$

$$O_n^{b1},\ h_n^{b1},\ i_n^{b1} = L^{b1}\left(i_{n-1}^{b1},\ h_{n-1}^{b1},\ x_n : P^{b1}\right), \tag{13}$$

Equations (12) and (13) show the operation of forwarding and backward tracks. Figure 2 depicts the framework of LSTM.

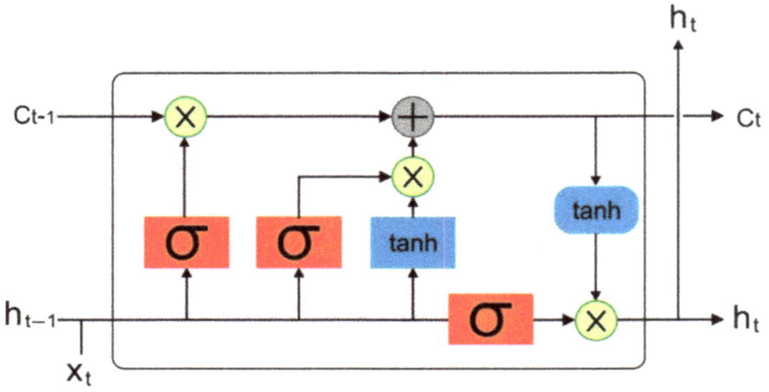

Figure 2. Infrastructure of LSTM.

From the equation, $O_n^{f1}, h_n^{f1}, i_n^{f1}$ and $O_n^{b1}, h_n^{b1}, i_n^{b1}$ indicate the output, the hidden state, and the internal state of the existing state for forwarding and backward LSTM tracks correspondingly. x_n denotes the sequential input, P indicates the LSTM cell variable. The output from these two tracks is integrated as in Equation (14) and forwarded into the next layers.

$$O_n^1 = O_n^{f1} + O_{N-n+1}^{b1}. \tag{14}$$

Bi-directional RNN and uni-directional RNN transform information into an abstract form and assist in learning spatial dependency. The output from the uni-directional layer can be attained by the following equation.

$$O_n^l, h_n^l, i_n^l = \text{LSTM}^l\left(i_{n-1}^l,\ h_{n-1}^l,\ O_n^{l-1}; P^l\right), \tag{15}$$

Now, the output from the lower layer O_n^{l-1} is integrated with preceding internal state i_{n-1}^l and hidden state h_{n-1}^l for obtaining output O_n^l of layer l, and P^l indicates a variable of the LSTM cell. The input dataset comprises a series of instances $(x_1,\ x_2, \ldots,\ x_N)$, while every feature x_n is regarded at time n $(n = 1, 2,\ \ldots,\ N)$. The information is mainly classified into windows of time segment N and fed into the cascading LSTM. We attain predicted score vectors for every time step $(O_1^L,\ O_2^L, \ldots,\ O_N^L)$ at the output. The entire prediction score can be attained by integrating the predictive score vector for the window N. The combination of scores can be implemented by using the sum rule as demonstrated in Equation (16) that implements well than other methodologies. Finally, the predictive score is transformed into probability using a softmax layer over Y.

$$Y = \frac{1}{N}\sum_{(n=1)}^{N} O_n^L. \tag{16}$$

We cascade LSTM to simulate incremental change of n time steps, and every LSTM is utilized for estimating the increment for one time step. In this work, θ-increment learning method learns increment of parameters using the cascaded LSTM network to gain higher frequency approximation, and θ represents the targeted parameter to be calculated.

In order to optimally elect the hyperparameter values of the CLSTM model, the SOA is exploited. In SOA, all the seekers have a central location vector $\vec{c}$, viz., the initial position for finding upcoming solutions, and it is regarded as estimated value Ex. Furthermore, all the seekers have a searching radius $\vec{r}$ regarded as the En', a trust level μ as membership degree, and a searching direction $\vec{d}$. Next, the seeker with some level of trust followed a potential direction and randomly moves towards the second point (novel candidate solution) in some searching radius from their existing location. In every time step t, the search decision-making is carried out for evaluating the four variables, and the seeker moves toward the novel location $\vec{x}(t+1)$. The updating location from the central location can be defined as y-conditional cloud generator [23]:

$$\vec{x}_{ij}(t+1) = \vec{c}_{ij}(t) + \vec{d}_{ij}(t) \times \vec{r}_{ij}(t) \times \sqrt{-\ln(\mu_i)} \tag{17}$$

Here "i" refers to the subscript index of seeker, and "j" indicates the subscript index of parameter dimension. The pseudo-code of the SOA is given in Algorithm 1.

Algorithm 1: Pseudocode of SOA

$t \leftarrow 0$
Initialized generation of S position
$\{x_i(t)|x_i(t) = (x_{i1}, x_{i2}, \ldots, x_{iD}),\ i = 1, \ldots, S,\ t = 0\}$ Uniformly and randomly in the parameters.
Estimate all the seekers: Compute the fitness.
Searching techniques provide search variables involving central location vector, searching direction, searching radius, and trust degree.
Update new location of all the seekers is evaluated.
$t \leftarrow t+1$
When $t < T_{\max}$, then Go to 3; otherwise, End.

Instinctively, central location vector $\vec{c}$ is fixed to existing location $\vec{x}(t)$. Similar to particle swarm optimization (PSO), all the seekers contain a memory stored in its optimal location $\vec{p}$ and a global optimal location g accomplished by communicating with neighboring seekers. Every seeker is categorized into k class in the subscript index, and the seeker in a similar class belongs to virtual neighbors. Therefore, $\vec{g}$ is established in the virtual neighbors.

$$\vec{c} = \vec{x}(t) + r_1\varnothing_1\left(\vec{p}(t) - \vec{x}(t)\right) + r_2\varnothing_2\left(\vec{g}(t) - \vec{x}(t)\right) \tag{18}$$

Now r_1, r_2 indicates the cognitive and social learning rates, correspondingly. $\varnothing_1$ and $\varnothing_2$ denotes the real number randomly and uniformly selected within the range of [0, 1]. In every experiment carried out in the study, $r_1 = 1$, $r_2 = 1$, and $k = 3$.

Generally, all the seekers have four significant directions, named local spacial direction $\vec{d}_{ls}$, local temporal direction $\vec{d}_{lt}$, global spacial direction $\vec{d}_{gs}$, and global temporal direction $\vec{d}_{gt}$, correspondingly.

$$\vec{d}_{lt} = \begin{cases} sign\left(\vec{x}(t) - \vec{x}(t-1)\right) & if\ fit\left(\vec{x}(t)\right) \geq fit\left(\vec{x}(t-1)\right) \\ sign\left(\vec{x}(t-1) - \vec{x}(t)\right) & if\ fit\left(\vec{x}(t)\right) < fit\left(\vec{x}(t-1)\right) \end{cases} \tag{19}$$

$$\vec{d}_{ls} = sign\left(\vec{x}(t) - \vec{x}(t)\right) \tag{20}$$

$$\vec{d}_{gt} = sign\left(\vec{p}(t) - \vec{x}(t)\right) \tag{21}$$

$$\vec{d}_{gs} = sign\left(\vec{g}(t) - \vec{x}(t)\right) \tag{22}$$

From the above equation, $sign\ (\cdot)$ indicates signum function, $\vec{x}'(t)$ represent the location of the seekers with the maximum fitness in a neighbor region, *fit* $\left(\vec{x}(t)\right)$ denotes the fitness function (FF) of $\vec{x}(t)$. Next, searching direction is allocated based on the four directions.

$$\begin{aligned} \vec{d} = sign(\omega\left(sign\left(fit\left(\vec{x}(t)\right) - fit\left(\vec{x}(t-1)\right)\right)\right)\left(\vec{x}(t) - \vec{x}(t-1)\right) \\ +r_1\phi_1\left(\vec{p}(t) - \vec{x}(t)\right) + r_2\phi_2\left(\vec{g}(t) - \vec{x}(t)\right)) \end{aligned} \tag{23}$$

In Equation (7), ω indicates the inertia weight that is fixed to $\omega = (T_{max}\text{-}t)/T_{\max}$. Now, ϕ_1 and ϕ_2 indicates real numbers randomly and uniformly selected within [0, 1].

Search Radius is essential, but challenging, to reasonably provide searching radius. For unimodal optimization problems, the performance is comparatively oblivious to searching radius to some extent. However, for multi-modal problems, various searching radii might lead to various performances of model particularly while handling variety of problems.

The μ variable is considered a quality assessment of location. It is equivalent to the fitness of $\vec{x}(t)$ or the index of ascensive sorting order of the fitness of $\vec{x}(t)$. Especially, the global optimal location has the maximal $\mu_{\max} = 1.0$, when another location has a $\mu < 1.0$.

$$\mu = \mu_{\max} - \frac{S - Sn}{S - 1}(\mu_{\max} - \mu_{\min}) \tag{24}$$

Here, Sn indicates the sequential value of $\vec{x}(t)$ afterward arranging the finesses of neighboring seekers in ascending sequence, $\mu_{\max}$ and $\mu_{\min}$ indicates the maximal and the minimal μ. We adapted $\mu_{\max} = 1.0$, and $\mu_{\min} = 0.2$.

The SOA method develops a fitness function (FF) to accomplish better classification accuracy. It describes a positive integer to characterize the improved performance of the candidate solution. In this work, the reduction of the classification error rate is regarded as the FF, as shown in Equation (25).

$$fitness(x_i) = Classifier\ Error\ Rate(x_i) = \frac{number\ of\ misclassified\ samples}{Total\ number\ of\ samples} * 100 \tag{25}$$

4. Experimental Validation

The performance validation of the proposed model is carried out using the UCM dataset [24]. The dataset contains a total of 2100 images and 21 classes (agricultural, airplane, baseballdiamond, beach, buildings, chaparral, denseresidential, forest, freeway, golfcourse, harbor, intersection, mediumresidential, mobilehomepark, overpass, parkinglot, river, runway, sparseresidential, storagetanks, and tenniscourt). It includes a total of 100 images under each class. The images were manually extracted from large images from the USGS National Map Urban Area Imagery collection for various urban areas around the country. The pixel resolution of this public domain imagery is 1 foot. Each image measures 256 × 256 pixels. For experimental validation, the dataset is split into 70% of training set and 30% of testing set, i.e., 70 images from each class for training and remaining 30 images for testing purposes. Figure 3 showcases the sample images of UCM dataset.

Figure 3. Samples-UCM Dataset.

Figure 4 illustrates the confusion matrices provided by the SIDTLD-AIC model on 70% of UCM datasets as training datasets. The results indicated that the SIDTLD-AIC model has effectually categorized all the 21 classes.

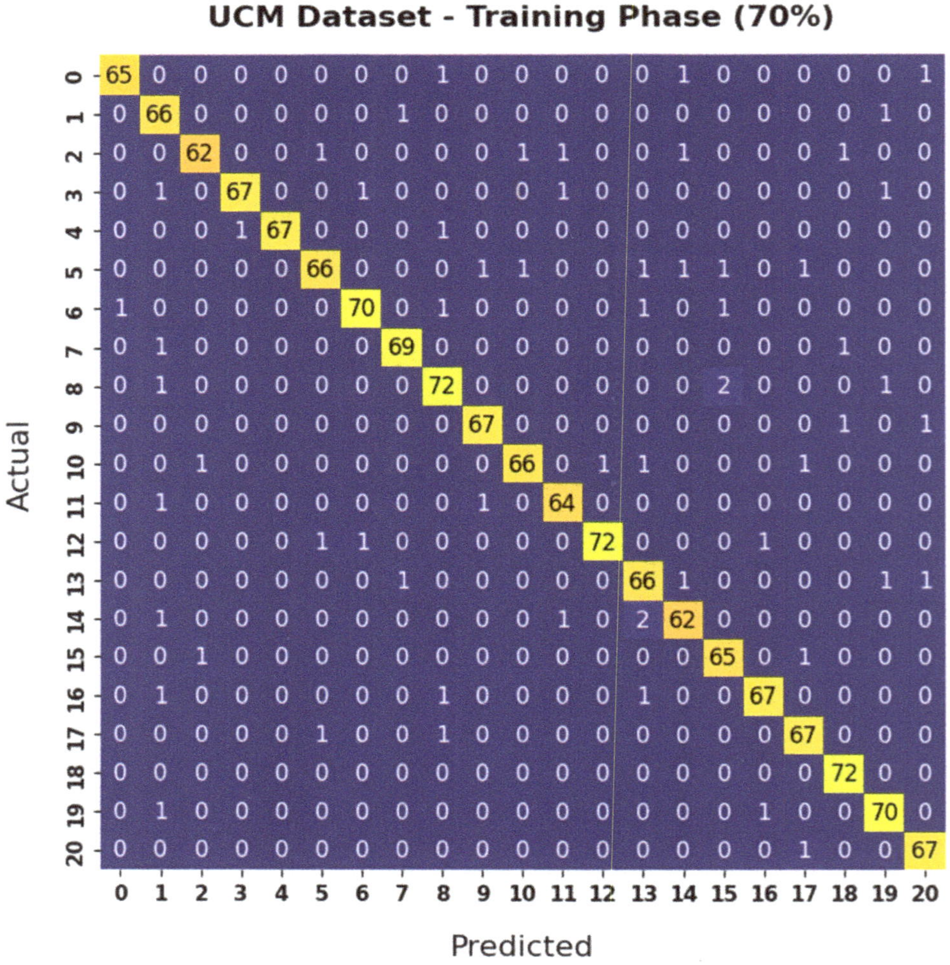

Figure 4. Confusion matrix of SIDTLD-AIC technique on 70% of UCM datasets as training datasets.

Table 1 reports the overall classification outcomes of the SIDTLD-AIC model 70% of UCM datasets as training datasets. The results inferred that the SIDTLD-AIC model has accomplished enhanced classifier outcomes on all class labels. For instance, the SIDTLD-AIC model has recognized class 1 samples with $accu_y$, $prec_n$, $reca_l$, F_{score}, and G_{mean} of 99.39%, 90.41%, 97.06%, 93.62%, and 98.27%, respectively. Along with that, the SIDTLD-AIC method has recognized class 3 samples with $accu_y$, $prec_n$, $reca_l$, F_{score}, and G_{mean} of 99.66%, 98.53%, 94.37%, 96.40%, and 97.11%, correspondingly. Moreover, the SIDTLD-AIC system has recognized class 13 samples with $accu_y$, $prec_n$, $reca_l$, F_{score}, and G_{mean} of 99.32%, 91.67%, 94.29%, 92.96%, and 96.89%, correspondingly. Furthermore, the SIDTLD-AIC approach has recognized class 16 samples with $accu_y$, $prec_n$, $reca_l$, F_{score}, and G_{mean} of 99.66%, 97.10%, 95.71%, 96.40%, and 97.76%, respectively. Lastly, the SIDTLD-AIC method has recognized class 20 samples with $accu_y$, $prec_n$, $reca_l$, F_{score}, and G_{mean} of 99.73%, 95.71%, 98.53%, 97.10%, and 99.16%, correspondingly.

Table 1. Result analysis of SIDTLD-AIC technique with various measures on 70% of UCM datasets as training datasets.

Training Phase (70%)					
Class Labels	**Accuracy**	**Precision**	**Recall**	**F-Score**	**Geometric Mean**
0	99.73	98.48	95.59	97.01	97.73
1	99.39	90.41	97.06	93.62	98.27
2	99.52	96.88	92.54	94.66	96.13
3	99.66	98.53	94.37	96.40	97.11
4	99.86	100.00	97.10	98.53	98.54
5	99.39	95.65	91.67	93.62	95.64
6	99.59	97.22	94.59	95.89	97.19
7	99.73	97.18	97.18	97.18	98.51
8	99.39	93.51	94.74	94.12	97.16
9	99.73	97.10	97.10	97.10	98.47
10	99.59	97.06	94.29	95.65	97.03
11	99.66	95.52	96.97	96.24	98.37
12	99.73	98.63	96.00	97.30	97.94
13	99.32	91.67	94.29	92.96	96.89
14	99.46	93.94	93.94	93.94	96.78
15	99.59	94.20	97.01	95.59	98.36
16	99.66	97.10	95.71	96.40	97.76
17	99.59	94.37	97.10	95.71	98.40
18	99.80	96.00	100.00	97.96	99.89
19	99.59	94.59	97.22	95.89	98.46
20	99.73	95.71	98.53	97.10	99.16
Average	**99.60**	**95.89**	**95.86**	**95.85**	**97.80**

Figure 5 showcases the confusion matrices provided by the SIDTLD-AIC approach on 30% of UCM datasets as testing datasets. The results point out that the SIDTLD-AIC methodology has effectually categorized all the 21 classes.

Table 2 demonstrates the overall classification outcomes of the SIDTLD-AIC method on 30% of UCM datasets as testing datasets. The results exposed that the SIDTLD-AIC model has accomplished higher classifier outcomes on all class labels. For instance, the SIDTLD-AIC method has recognized class 1 samples with $accu_y$, $prec_n$, $reca_l$, F_{score}, and G_{mean} of 99.68%, 100%, 93.75%, 96.77%, and 96.82%, respectively. Next, the SIDTLD-AIC model has recognized class 3 samples with $accu_y$, $prec_n$, $reca_l$, F_{score}, and G_{mean} of 99.52%, 96.43%, 93.10%, 94.74%, and 96.41%, correspondingly. Furthermore, the SIDTLD-AIC system has recognized class 13 samples with $accu_y$, $prec_n$, $reca_l$, F_{score}, and G_{mean} of 99.21%, 93.10%, 90%, 91.53%, and 94.71%, respectively. Moreover, the SIDTLD-AIC methodology has recognized class 16 samples with $accu_y$, $prec_n$, $reca_l$, F_{score}, and G_{mean} of 99.52%, 100%, 90%, 94.74%, and 94.87%, respectively. Finally, the SIDTLD-AIC model has recognized class 20 samples with $accu_y$, $prec_n$, $reca_l$, F_{score}, and G_{mean} of 99.21%, 88.57%, 96.88%, 92.54%, and 98.10%, correspondingly.

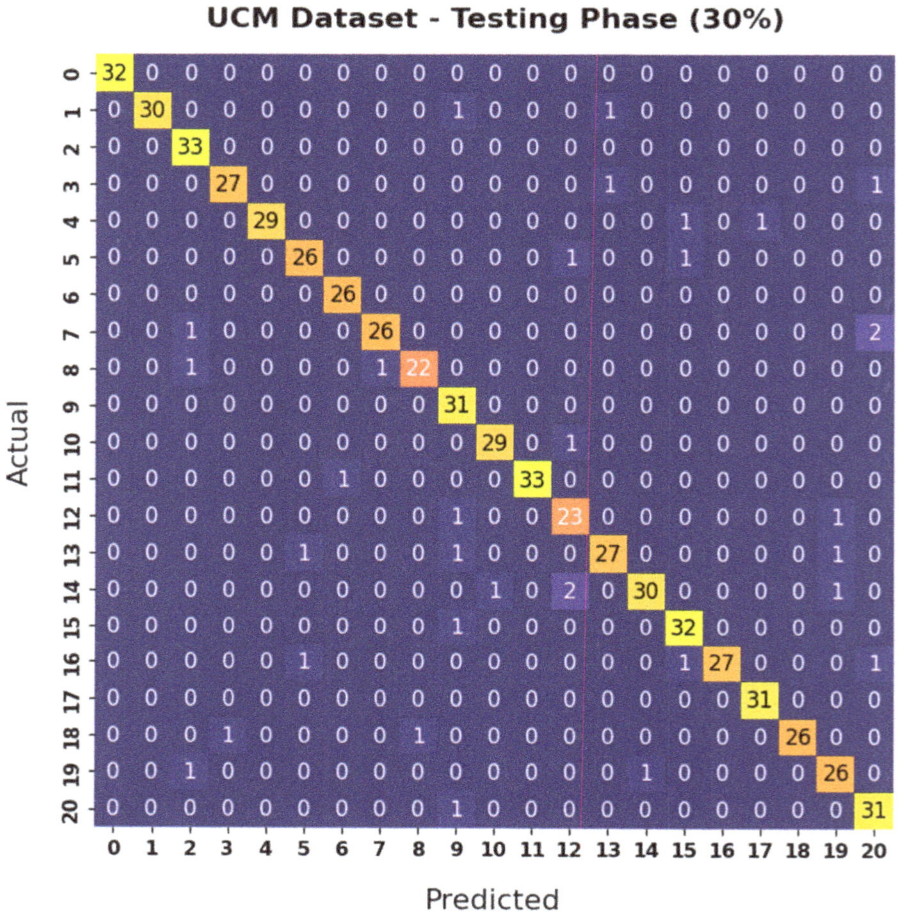

Figure 5. Confusion matrix of SIDTLD-AIC technique on 30% of UCM datasets as testing datasets.

Table 2. Result analysis of SIDTLD-AIC technique with various measures on 30% of UCM datasets as testing datasets.

	Testing (30%)				
Class Labels	**Accuracy**	**Precision**	**Recall**	**F-Score**	**Geometric Mean**
0	100.00	100.00	100.00	100.00	100.00
1	99.68	100.00	93.75	96.77	96.82
2	99.52	91.67	100.00	95.65	99.75
3	99.52	96.43	93.10	94.74	96.41
4	99.68	100.00	93.55	96.67	96.72
5	99.37	92.86	92.86	92.86	96.20
6	99.84	96.30	100.00	98.11	99.92
7	99.37	96.30	89.66	92.86	94.61

Table 2. *Cont.*

	Testing (30%)				
Class Labels	Accuracy	Precision	Recall	F-Score	Geometric Mean
8	99.52	95.65	91.67	93.62	95.66
9	99.21	86.11	100.00	92.54	99.58
10	99.68	96.67	96.67	96.67	98.24
11	99.84	100.00	97.06	98.51	98.52
12	99.05	85.19	92.00	88.46	95.60
13	99.21	93.10	90.00	91.53	94.71
14	99.21	96.77	88.24	92.31	93.85
15	99.37	91.43	96.97	94.12	98.23
16	99.52	100.00	90.00	94.74	94.87
17	99.84	96.88	100.00	98.41	99.92
18	99.68	100.00	92.86	96.30	96.36
19	99.21	89.66	92.86	91.23	96.12
20	99.21	88.57	96.88	92.54	98.10
Average	**99.50**	**94.93**	**94.67**	**94.70**	**97.15**

The training accuracy (TA) and validation accuracy (VA) attained by the SIDTLD-AIC model on UCM dataset is demonstrated in Figure 6. The experimental outcome implied that the SIDTLD-AIC model has gained maximum values of TA and VA. In specific, the VA seemed higher than TA.

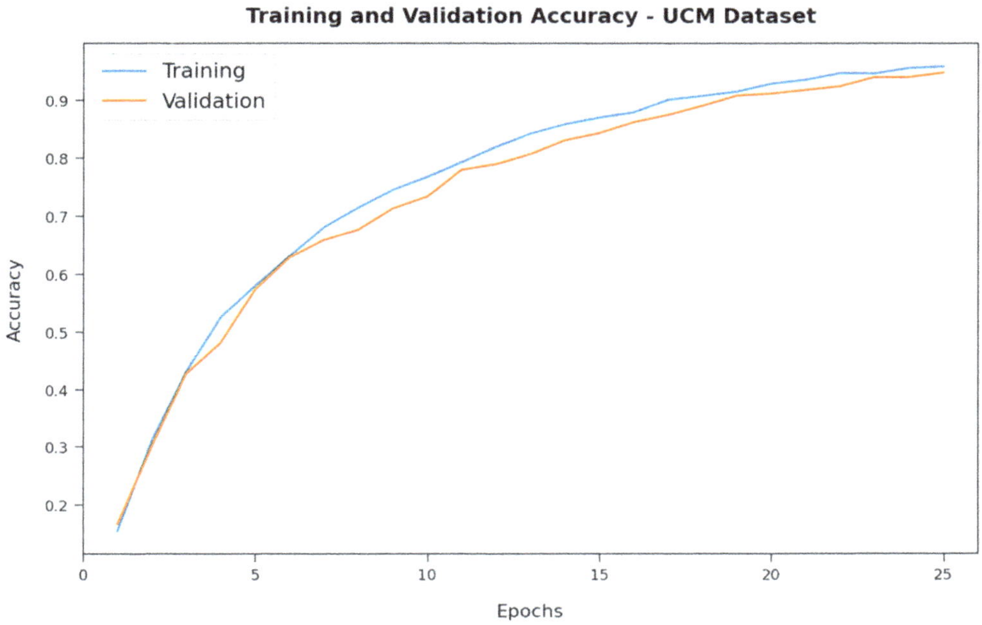

Figure 6. TA and VA analysis of SIDTLD-AIC technique on UCM dataset.

The training loss (TL) and validation loss (VL) achieved by the SIDTLD-AIC model on UCM dataset are established in Figure 7. The experimental outcome inferred that the SIDTLD-AIC model has been able least values of TL and VL. In specific, the VL seemed that lower than TL.

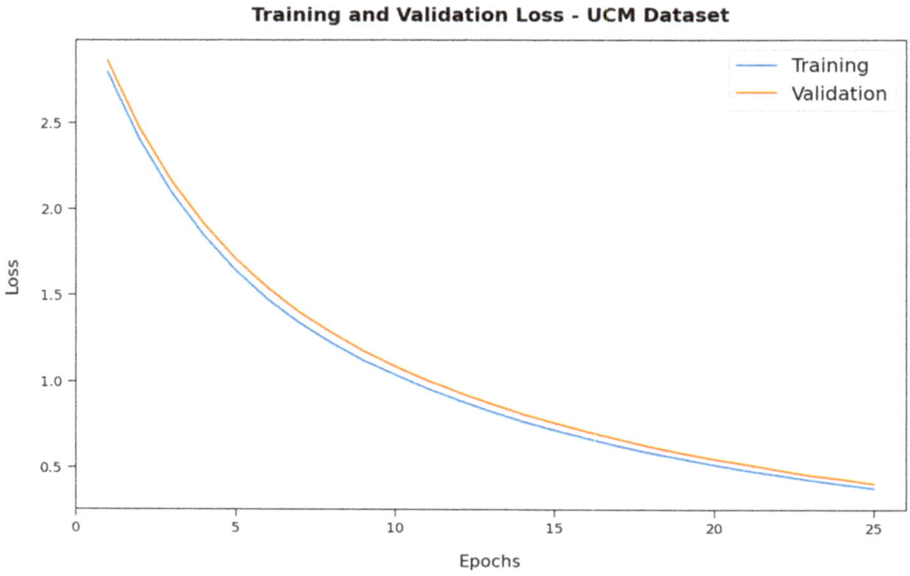

Figure 7. TL and VL analysis of SIDTLD-AIC technique on UCM dataset.

A brief precision-recall examination of the SIDTLD-AIC method on UCM dataset is portrayed in Figure 8. By observing the figure, it can be noticed that the SIDTLD-AIC method has been able maximal precision-recall performance under all classes.

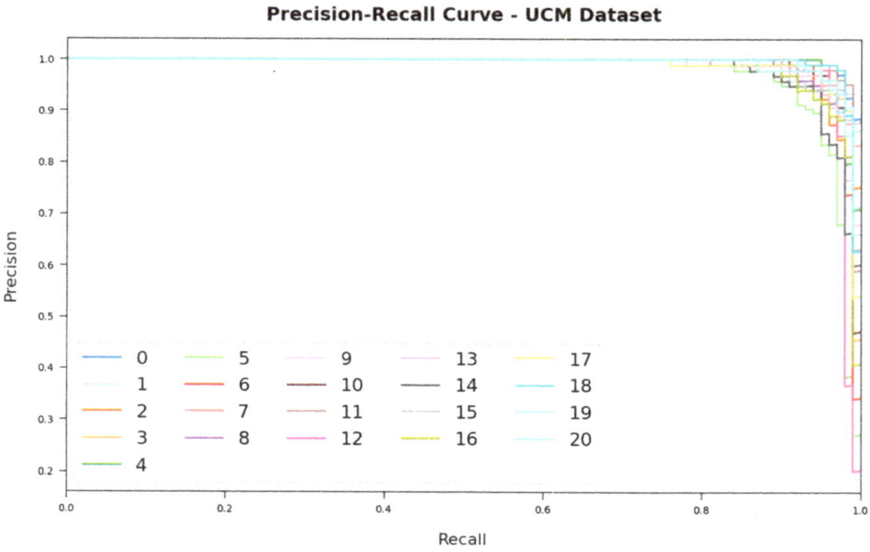

Figure 8. Precision-recall curve analysis of SIDTLD-AIC technique on UCM dataset.

A detailed ROC investigation of the SIDTLD-AIC approach to UCM dataset is represented in Figure 9. The results indicated that the SIDTLD-AIC model has exhibited its ability in categorizing different classes on the UCM dataset.

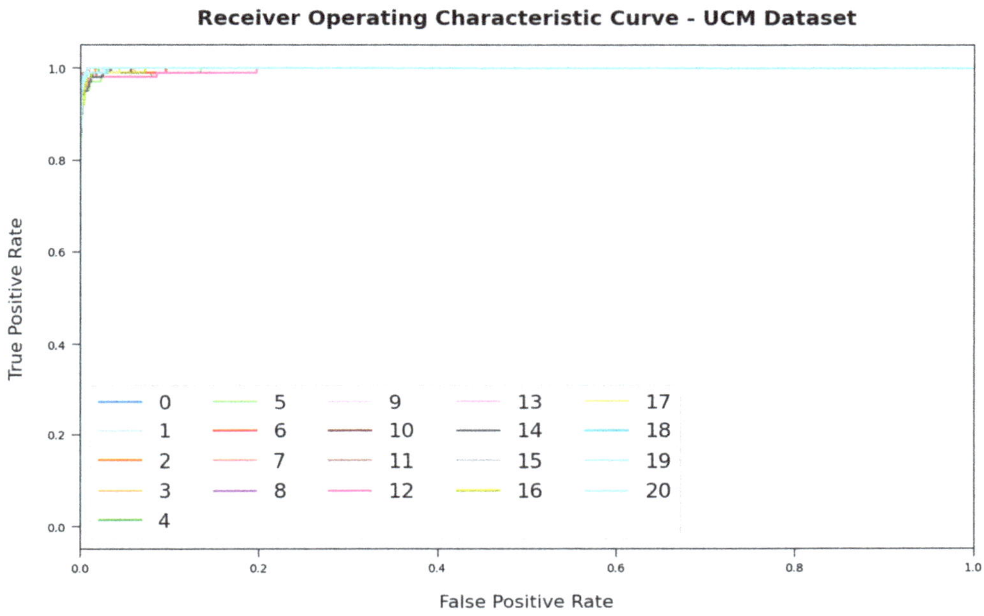

Figure 9. ROC curve analysis of SIDTLD-AIC technique on UCM dataset.

Figure 10 depicts the average image classification results of the SIDTLD-AIC model on 70% of UCM datasets as training datasets and 30% of UCM datasets as testing datasets. The figure shows that the SIDTLD-AIC model has resulted in better classification results under both aspects. On applied 70% of UCM datasets as training datasets, the SIDTLD-AIC model has resulted in average $accu_y$, $prec_n$, $reca_l$, F_{score}, and G_{mean} of 96.60%, 95.89%, 95.86%, 95.85%, and 97.80%, respectively. Likewise, on applied 30% of UCM datasets as testing datasets, the SIDTLD-AIC model has resulted in average $accu_y$, $prec_n$, $reca_l$, F_{score}, and G_{mean} of 99.50%, 94.93%, 94.67%, 94.70%, and 97.15%, respectively.

Figure 11 illustrates a comparative $accu_y$ examination of the SIDTLD-AIC model with recent models. The experimental values implied that the DL-PlacesNet and DL-VGG-VD19 models have shown lower values of $accu_y$. Moreover, the DL-VGG-VD16, DL-VGG-M, DL-VGG-F, DL-CaffeNet, and DL-AlexNet models have resulted to closer $accu_y$ values. Then, the DL-VGG-S and DL based multiobjective PSO (DL-MOPSO) techniques have reached reasonable $accu_y$ values of 95.24% and 95.81%. Though the DL-C-PTRN model has resulted in considerable $accu_y$ of 98.96%, the SIDTLD and SIDTLD+SSA models have accomplished near optimal $accu_y$ of 98.98% and 99.01%. However, the SIDTLD-AIC model has accomplished superior outcome with maximum $accu_y$ of 99.50%.

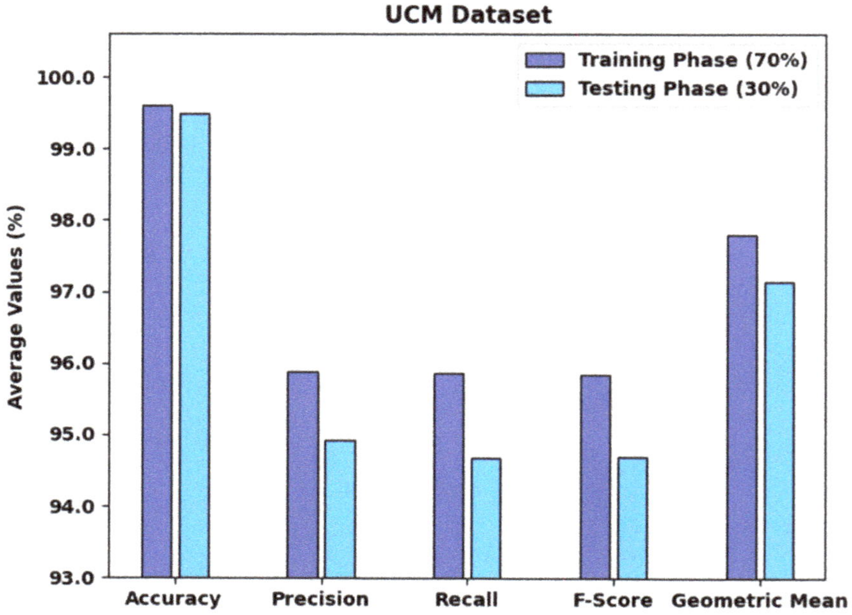

Figure 10. Average analysis of SIDTLD-AIC technique with various measures.

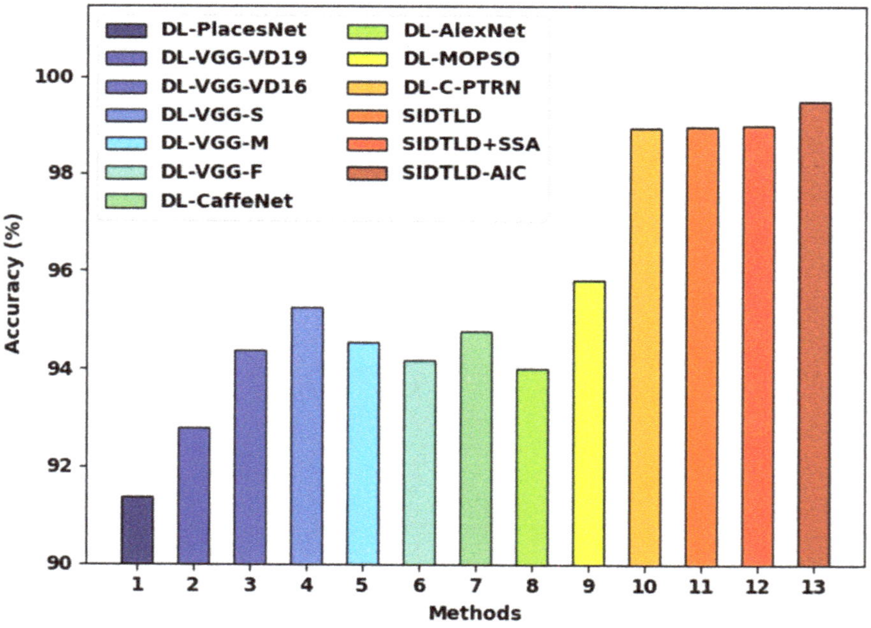

Figure 11. Accuracy analysis of SIDTLD-AIC technique with existing methods.

Finally, a computation time (CT) assessment of the SIDTLD-AIC model with recent models is carried out in Figure 12. The experimental values implied that the DL-PlacesNet, DL-VGG-VD-19, DL-VGG-VD-16, and DL-VGG-S approaches have obtained increased CT values. Followed by, the DL-VGG-F, DL-CaffeNet, and DL-AlexNet models have reached moderately reduced CT values. The DL-MOPSO and DL-C-PTRN models have accomplished reasonable CT of 135s and 95s, respectively. Meanwhile, the SIDTLD and SIDTLD+SSA models have attained CT of 67s and 54s, respectively. Finally, the SIDTLD-AIC model has outperformed other methods with minimal CT of 40s. The results implied that the SIDTLD-AIC model has gained enhanced classification performance due to the inclusion of SSA and SOA based hyperparameter optimizers. From the above results and discussion, it can be stated that the SIDTLD-AIC model has accomplished enhanced image classification results on the UAV networks.

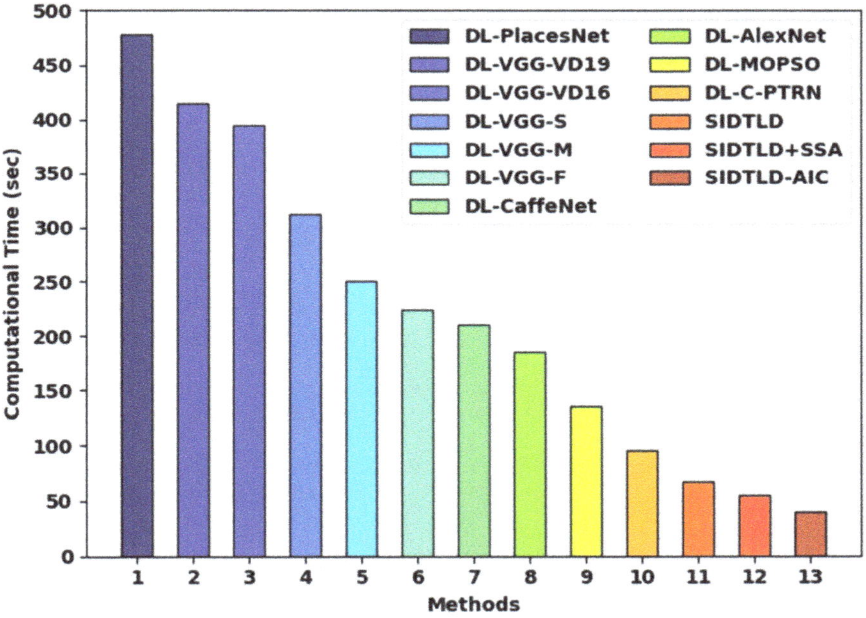

Figure 12. CT analysis of SIDTLD-AIC technique with existing approaches.

5. Conclusions

In this article, an automated SIDTLD-AIC technique was established for the proper identification and classification of images into distinct kinds on UAV networks. The presented SIDTLD-AIC model follows a feature extraction module using RetinaNet model in which the hyperparameter optimization process is performed by the use of SSA. Next, the SOA-CLSTM model is applied to classify the aerial images. For assuring the better performance of the SIDTLD-AIC method, a wide range of simulations are executed and the outcomes are investigated in various aspects. The comparative study reported the better performance of the SIDTLD-AIC model over recent approaches with maximum accuracy of 99.50%. Thus, the presented SIDTLD-AIC model can be exploited for aerial image classification in real time environment such as vegetation mapping, crop classification, disaster management, weather prediction, etc. In future, hybrid metaheuristics should be utilized for improving the overall classification performance. Furthermore, the proposed model can be extended to real-time large-scale databases in future. Moreover, the investigation of the performance using statistical analysis can be done in our future work.

Author Contributions: Conceptualization, S.S.A. and N.N.; methodology, H.A.M.; software, R.M.; validation, A.M.H., R.M. and M.R.; investigation, A.M.H., M.A.S.; resources, M.R.; data curation, R.M.; writing—original draft preparation, S.S.A., H.A.M., N.N., I.Y.; writing—review and editing, A.M., M.A.S., A.S.Z.; visualization, A.S.Z.; supervision, S.S.A.; project administration, R.M.; funding acquisition, H.A.M. All authors have read and agreed to the published version of the manuscript.

Funding: The authors extend their appreciation to the Deanship of Scientific Research at King Khalid University for funding this work through Large Groups Project under grant number (42/43). Princess Nourah bint Abdulrahman University Researchers Supporting Project number (PNURSP2022R114), Princess Nourah bint Abdulrahman University, Riyadh, Saudi Arabia. The authors would like to thank the Deanship of Scientific Research at Umm Al-Qura University for supporting this work by Grant Code: (22UQU4210118DSR21).

Data Availability Statement: Data sharing not applicable to this article as no datasets were generated during the current study.

Conflicts of Interest: The authors declare that they have no conflict of interest. The manuscript was written through contributions of all authors. All authors have given approval to the final version of the manuscript.

References

1. Choi, S.K.; Lee, S.K.; Kang, Y.B.; Seong, S.K.; Choi, D.Y.; Kim, G.H. Applicability of image classification using deep learning in small area: Case of agricultural lands using UAV image. *J. Korean Soc. Surv. Geod. Photogramm. Cartogr.* **2020**, *38*, 23–33.
2. Tetila, E.C.; Machado, B.B.; Astolfi, G.; de Souza Belete, N.A.; Amorim, W.P.; Roel, A.R.; Pistori, H. Detection and classification of soybean pests using deep learning with UAV images. *Comput. Electron. Agric.* **2020**, *179*, 105836. [CrossRef]
3. Öztürk, A.E.; Erçelebi, E. Real UAV-bird image classification using CNN with a synthetic dataset. *Appl. Sci.* **2021**, *11*, 3863. [CrossRef]
4. Ammour, N.; Alhichri, H.; Bazi, Y.; Benjdira, B.; Alajlan, N.; Zuair, M. Deep learning approach for car detection in UAV imagery. *Remote Sens.* **2017**, *9*, 312. [CrossRef]
5. Bashmal, L.; Bazi, Y.; Al Rahhal, M.M.; Alhichri, H.; Al Ajlan, N. UAV image multi-labeling with data-efficient transformers. *Appl. Sci.* **2021**, *11*, 3974. [CrossRef]
6. Anwer, M.H.; Hadeel, A.; Fahd, N.A.-W.; Mohamed, K.N.; Abdelwahed, M.; Anil, K.; Ishfaq, Y.; Abu Sarwar, Z. Fuzzy cognitive maps with bird swarm intelligence optimization-based remote sensing image classification. *Comput. Intell. Neurosci.* **2022**, *2022*, 4063354.
7. Abunadi, I.; Althobaiti, M.M.; Al-Wesabi, F.N.; Hilal, A.M.; Medani, M.; Hamza, M.A.; Rizwanullah, M.; Zamani, A.S. Ederated learning with blockchain assisted image classification for clustered UAV networks. *Comput. Mater. Contin.* **2022**, *72*, 1195–1212.
8. Li, J.; Yan, D.; Luan, K.; Li, Z.; Liang, H. Deep learning-based bird's nest detection on transmission lines using UAV imagery. *Appl. Sci.* **2020**, *10*, 6147. [CrossRef]
9. Youme, O.; Bayet, T.; Dembele, J.M.; Cambier, C. Deep Learning and Remote Sensing: Detection of Dumping Waste Using UAV. *Procedia Comput. Sci.* **2021**, *185*, 361–369. [CrossRef]
10. Mittal, P.; Singh, R.; Sharma, A. Deep learning-based object detection in low-altitude UAV datasets: A survey. *Image Vis. Comput.* **2020**, *104*, 104046. [CrossRef]
11. Bouguettaya, A.; Zarzour, H.; Kechida, A.; Taberkit, A.M. Vehicle detection from UAV imagery with deep learning: A review. *IEEE Trans. Neural Netw. Learn. Syst.* **2021**, 1–21, *in press*. [CrossRef] [PubMed]
12. Huang, H.; Deng, J.; Lan, Y.; Yang, A.; Deng, X.; Zhang, L.; Wen, S.; Jiang, Y.; Suo, G.; Chen, P. A two-stage classification approach for the detection of spider mite-infested cotton using UAV multispectral imagery. *Remote Sens. Lett.* **2018**, *9*, 933–941. [CrossRef]
13. Haq, M.A.; Rahaman, G.; Baral, P.; Ghosh, A. Deep learning based supervised image classification using UAV images for forest areas classification. *J. Indian Soc. Remote Sens.* **2021**, *49*, 601–606. [CrossRef]
14. Zeggada, A.; Melgani, F.; Bazi, Y. A deep learning approach to UAV image multilabeling. *IEEE Geosci. Remote Sens. Lett.* **2017**, *14*, 694–698. [CrossRef]
15. Bashmal, L.; Bazi, Y. Learning robust deep features for efficient classification of UAV imagery. In Proceedings of the 2018 1st International Conference on Computer Applications & Information Security (ICCAIS), Riyadh, Saudi Arabia, 4–6 April 2018; pp. 1–4.
16. Rajagopal, A.; Ramachandran, A.; Shankar, K.; Khari, M.; Jha, S.; Lee, Y.; Joshi, G.P. Fine-tuned residual network-based features with latent variable support vector machine-based optimal scene classification model for unmanned aerial vehicles. *IEEE Access* **2020**, *8*, 118396–118404. [CrossRef]
17. Rajagopal, A.; Joshi, G.P.; Ramachandran, A.; Subhalakshmi, R.T.; Khari, M.; Jha, S.; Shankar, K.; You, J. A deep learning model based on multi-objective particle swarm optimization for scene classification in unmanned aerial vehicles. *IEEE Access* **2020**, *8*, 135383–135393. [CrossRef]

18. Pustokhina, I.V.; Pustokhin, D.A.; Kumar Pareek, P.; Gupta, D.; Khanna, A.; Shankar, K. Energy-efficient cluster-based unmanned aerial vehicle networks with deep learning-based scene classification model. *Int. J. Commun. Syst.* **2021**, *34*, e4786. [CrossRef]
19. Outay, F.; Mengash, H.A.; Adnan, M. Applications of unmanned aerial vehicle (UAV) in road safety, traffic and highway infrastructure management: Recent advances and challenges. *Transp. Res. Part A Policy Pract.* **2020**, *141*, 116–129. [CrossRef]
20. Wang, Y.; Wang, C.; Zhang, H.; Dong, Y.; Wei, S. Automatic ship detection based on RetinaNet using multi-resolution Gaofen-3 imagery. *Remote Sens.* **2019**, *11*, 531. [CrossRef]
21. Mirjalili, S.; Gandomi, A.H.; Mirjalili, S.Z.; Saremi, S.; Faris, H.; Mirjalili, S.M. Salp Swarm Algorithm: A bio-inspired optimizer for engineering design problems. *Adv. Eng. Softw.* **2017**, *114*, 163–191. [CrossRef]
22. Yadav, R.K.; Bhattarai, B.; Jiao, L.; Goodwin, M.; Granmo, O.C. Indoor Space Classification Using Cascaded LSTM. In Proceedings of the 2020 15th IEEE Conference on Industrial Electronics and Applications (ICIEA), Kristiansand, Norway, 9–13 November 2020; pp. 1110–1114.
23. Shafik, M.B.; Chen, H.; Rashed, G.I.; El-Sehiemy, R.A. Adaptive multi objective parallel seeker optimization algorithm for incorporating TCSC devices into optimal power flow framework. *IEEE Access* **2019**, *7*, 36934–36947. [CrossRef]
24. Yang, Y.; Newsam, S. Bag-Of-Visual-Words and Spatial Extensions for Land-Use Classification. In Proceedings of the ACM SIGSPATIAL International Conference on Advances in Geographic Information Systems (ACM GIS), San Jose, CA, USA, 2–5 November 2010.

applied sciences

MDPI

Article

Evaluation of Reinforcement and Deep Learning Algorithms in Controlling Unmanned Aerial Vehicles

Yalew Zelalem Jembre [1], Yuniarto Wimbo Nugroho [1], Muhammad Toaha Raza Khan [2], Muhammad Attique [3], Rajib Paul [4], Syed Hassan Ahmed Shah [5] and Beomjoon Kim [1,*]

1 Department of Electronic Engineering, Keimyung University, Daegu 42601, Korea; zizutg@kmu.ac.kr (Y.Z.J.); wimboyt@kmu.ac.kr (Y.W.N.)
2 School of Computer Science & Engineering, Kyungpook National University, Daegu 41566, Korea; toaha@knu.ac.kr
3 Department of Software, Sejong University, Seoul 05006, Korea; attique@sejong.ac.kr
4 Department of Software and Computer Engineering, Ajou University, Suwon 16499, Korea; rajib@ajou.ac.kr
5 JMA Wireless, Corona, CA 92881, USA; sh.ahmed@ieee.org
* Correspondence: bkim@kmu.ac.kr

Abstract: Unmanned Aerial Vehicles (UAVs) are abundantly becoming a part of society, which is a trend that is expected to grow even further. The quadrotor is one of the drone technologies that is applicable in many sectors and in both military and civilian activities, with some applications requiring autonomous flight. However, stability, path planning, and control remain significant challenges in autonomous quadrotor flights. Traditional control algorithms, such as proportional-integral-derivative (PID), have deficiencies, especially in tuning. Recently, machine learning has received great attention in flying UAVs to desired positions autonomously. In this work, we configure the quadrotor to fly autonomously by using agents (the machine learning schemes being used to fly the quadrotor autonomously) to learn about the virtual physical environment. The quadrotor will fly from an initial to a desired position. When the agent brings the quadrotor closer to the desired position, it is rewarded; otherwise, it is punished. Two reinforcement learning models, Q-learning and SARSA, and a deep learning deep Q-network network are used as agents. The simulation is conducted by integrating the robot operating system (ROS) and Gazebo, which allowed for the implementation of the learning algorithms and the physical environment, respectively. The result has shown that the Deep Q-network network with Adadelta optimizer is the best setting to fly the quadrotor from the initial to desired position.

Keywords: reinforcement learning; UAV; quadrotor; flight control; intelligent control

Citation: Jembre, Y.Z.; Nugroho, Y.W.; Khan, M.T.R.; Attique, M.; Paul, R.; Ahmed, S.H.S.; Kim, B. Evaluation of Reinforcement and Deep Learning Algorithms in Controlling Unmanned Aerial Vehicles. *Appl. Sci.* **2021**, *11*, 7240. https://doi.org/10.3390/app11167240

Academic Editor: Yosoon Choi

Received: 15 July 2021
Accepted: 3 August 2021
Published: 6 August 2021

1. Introduction

In recent times, drone or unmanned aerial vehicle (UAV) technology has advanced significantly, and it can be applied not only in the military sector but also in civilian areas, such as in search and rescue (SAR) and package shipment, due to its high mobility and large overload maneuver [1]. Many researchers worldwide are now working to address issues related to UAVs. Herein, we focus on the application, performance, and implementation of machine learning algorithms for controlling UAVs. Even though there are several types of UAVs, such as fixed wings, quadrotors, blimps, helicopters, and ducted fan [2], due to its small size, low inertia, maneuverability, and cheap price, the quadrotor had become an industry favorite [3]. There are several applications of the quadrotor in industries such as film, agriculture, delivery, infrastructure inspection, etc. [3,4]. A quadrotor or quadcopter (henceforth, the terms UAV, drone, quadcopter, and quadrotor are used interchangeably) is a type of UAV with four rotors designed in a cross configuration with two pairs of opposite rotors rotating in the clockwise direction, whereas the other rotor pair rotates in a counter-clockwise direction to balance the torque [5]. Figure 1 shows the famous arrangements of the rotors of a quadcopter for flight mode.

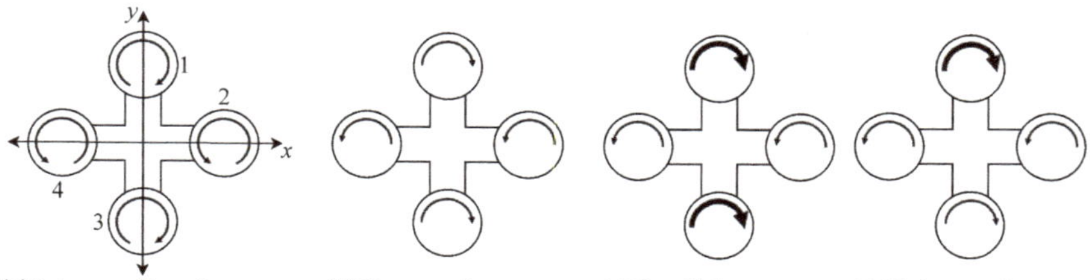

Figure 1. Configuration of rotors in quadcopter.

Configuration "A" shows how the rotors of the drone work when rotors 1 and 3 rotate in the clockwise direction, and rotors 2 and 4 rotate in the same counter-clockwise direction. Configuration "B" depicts the drone in hovering mode. In this case, all rotors have the same torques. Configuration "C" shows how the drone performs a yaw flight, where the strength torques of rotors 1 and 3 exceed those of rotors 2 and 4, or the strength torques of rotors 2 and 4 exceed those of rotors 1 and 3. Both scenarios are dependent on the direction of the drone flight with reference to the Z-axis. Configuration "D" represents the pitch up and pitch down. Here, one rotor has greater torque strength than the rest of the rotors. The rotor that can produce greater torque strength depends on the movements, which provides better flexibility.

Traditionally, a closed-loop proportional integral derivative (PID) controller is used to fly the quadrotor from the initial to the desired position. The PID controller flies the UAV by tuning values such as K_p, K_i, K_d. However, tuning these values is a challenging and cumbersome task. In contrast, recent reinforcement learning-based controllers have shown a more promising way than the conventional method to fly a quadrotor accurately [6,7].

In this paper, to study this phenomenon further, we integrated learning algorithms with a simulation environment and tested their performance under different conditions, optimization, and reward functions. We use the robot operating system (ROS) framework [8], together with Gazebo [9], for simulation, and OpenAI Gym to load the agents (machine learning algorithms) [10]. The drones will be flown from the initial to the desired position autonomously via one of the agents. The agent flying the quadrotor will take one step from the initial to the desired position. Then, for the remaining distance, if the current distance is closer than the previous, the agent will be rewarded; otherwise, it will be punished. This process allows the agent to learn its physical environment.

Q-learning and State–action–reward–state–action (SARSA) reinforcement learning as well as Deep Q-network (DQN) deep learning agents are selected as agents. Since the DQN algorithm can be optimized to improve the performance, Adadelta [11], RMSProp [12], Stochastic Gradient Descent (SGD) [13], and ADAM [14] optimizers are used in this work. On the other hand, two reward functions are used to evaluate the actions taken by an agent, Euclidean distance and mean square error (MSE) of distance. For agents, handling even the simplest task is difficult. Hence, we choose to conduct our simulation using data-position (X, Y, and Z) to specify initial and desired positions. Our work shows that autonomous flight, without the involvement of other additional sensors such as light detection and ranging (LiDAR) or vision, is possible, which saves the power of the UAV and reduces the cost. In addition, we have shown that the learning process is highly dependent on the optimizer and reward function, rather than the learning steps.

The rest of this paper is organized as follows: in Section 2, issues related to this work are discussed. In Section 3, preliminary concepts are presented, while in Section 4, the agents used in this work are explained. The simulation environment and performance evaluation are in Sections 5 and 6, respectively. Finally, a conclusion to our work is given in Section 7.

2. Related Work

Although machine learning-based autonomous flights have been the main focus of current researchers, conventional UAV control algorithms are still in play. In this section, a literature review from both directions is presented. Thus far, the most widely used algorithms for UAV control are traditional control concepts, which do not involve any type of intelligence. Numerous techniques and algorithms can be used in UAV control systems. Among them are the proportional integral derivative (PID), linear quadratic regulators (LQR), sliding mode control (SMC), model predictive control (MPC), integrator backstepping, adaptive control, robust control, optimal control, and feedback control [15]. PID is the most widely used controller with a feedback mechanism and is an industry favorite [16]. The PID controller has achieved better performances for controlling pitch angles, etc. [17]. Generally, the PID controller has been successfully applied in the quadcopter, although with several limitations.

Batikan et al. [18] proposed a technique with the application of a self-tuning fuzzy PID in real-time trajectory tracking of UAVs. The work focused on stabilizing the altitude and trajectory. Meanwhile, Eresen et al. [19] presented the vision for the detection of obstacles. The work demonstrated flying autonomously in urban areas while avoiding obstacles. Goodarzi et al. [20] proposed a full six degree of freedom dynamic model of a quadrotor. The controller was developed to avoid singularities of the minimal altitude representation. Lwin et al. [21] proposed a method that combines a Kalman filter for separating true signal from noise and a PID controller to calculate the error. For the stability of the flight control system, the UAV was adjusted by the PID parameters. Salih et al. [22] presented a new method for autonomous flight control for a quadrotor with a model vertical take-off and landing (VTOL). The work by Zang et al. [23] focused on controlling the UAV height during drone operation. The algorithm in this work uses active disturbance rejection control (ADRC) and Kalman filtering to process controlling the height as well as to enable autonomous flight of the UAV. The authors of [24], Siti et al., use a hierarchical strategy to improve the PID controller to dive the UAV in a predetermined trajectory using only system orientation. First, a reference model (RM) is used to synthesize the PID in the inner loop, and then genetic algorithm is applied to optimize the outer loop. In [25], Hermans et al. proposed a solution to control the UAV in a geofencing application, which is a virtual boundary of a specific geographical area. An explicit reference governor (ERG) that first stabilizes the UAV and then uses the Lyapunov theory to control the UAV is presented in this paper.

Nevertheless, the PID has several limitations, such as complicated and challenging tuning. Furthermore, the PID or classic controller still lacks complete handling and solving the control problem of an autonomous flight of the quadrotor. Due to the strides made in artificial intelligence, specifically machine learning, researchers both in academia and industry are now turning their attention to this matter to solve autonomous flight control in UAVs. Supervised learning is one of the most used methods in attempting UAV control, but the training dataset has been problematic in this regard. Hence, the focus has now been shifted to reinforcement learning (RL), which is also the case in our work. Reinforcement learning entails learning what to do and how the agent resolves some challenges by taking actions in the environment such that the agent maximizes reward [26]. The reinforcement learning algorithm can reduce learning times and increase stability [27]. Currently, deep RL is a powerful approach for controlling complex systems and situations [28].

W Koch et al. [5] used reinforcement learning to improve the accuracy and precision of altitude control of UAVs to replace classic control algorithms, such as the PID. Zhao et al. [29] presented their research on the use of RL to learn a path while avoiding obstacles. At first, the Q-learning algorithm was used to allow UAVs to learn the environment, and then the adaptive and random exploration (ARE) approach was used to accomplish both task navigation and obstacle avoidance. Kim et al. [30,31] proposed a path planning and obstacle avoiding strategy for UAVs through RL. The Q-learning and deep double dueling deep Q-network (DD-DQN) [32] learning algorithms are used to navigate the simulation environment. On the other hand, Cheng et al. [33] presented a method focused on enemy avoidance based on an RL, where a UAV is expected to avoid another UAV coming its way. The authors have shown that the learned policy achieved a higher possibility of reaching the goal compared with the random and fixed-rule policies. Kahn et al. [34] argued that even RL can be unsafe for the robot during training. The aim of the research was to develop an algorithm that takes uncertainty into consideration. On the other hand, Hwangbo et al. [35] proposed a method to control the quadrotor actuators via the RL technique. The drone was able to produce accurate responses, achieving high stability even under poor conditions.

The impact of reinforcement learning on UAV control and path planning has been demonstrated in several dimensions. However, more research output is expected to further verify and solidify the usage of RL in UAV operation than that of conventional PID techniques. Furthermore, researchers focus on the single machine learning technique with a single reward function for improving and testing autonomous flight in UAVs. Our goal is to demonstrate the difference between the widely used machine learning agents for autonomous UAV flight under multiple reward conditions and optimization functions.

3. Preliminaries

Three popular toolkits have been used in our research, namely (i) ROS, (ii) Gazebo simulator, and OpenAI Gym. First, ROS is used to determine the speed, direction, and destination of the drone. The parameters are then used to produce actions that are sent to the Gazebo simulator to visualize the movement of the drone. Following this, ROS uses the reinforcement learning algorithms available in OpenAI Gym to determine the next action. In this section, a brief introduction to these three toolkits is provided.

3.1. ROS

ROS is an open-source and flexible middleware framework for writing robot software [8]. Despite including an operating system in its name, this toolkit is not one. It is more similar to a motherboard where chips and other modules are mounted on, to create a computer. The ROS is a framework that allows developers to collaborate on developing software and firmware for robots. However, it provides services such as a hardware abstraction layer, low-level device control, sending a message in-process, and packet management, which are characteristics of a typical operating system.

Enabling robots to complete a simple task, which could easily be handled by humans, involves several components and complex systems. However, several components written by different people can be assembled using ROS bottom-up architecture, so as to contribute to the collaborative development of the robotic software.

3.2. Gazebo Simulator

Unlike typical software, which is limited to the virtual world, the software for robots and UAVs takes action in the physical world. Hence, visualizing the steps taken and the decisions made by the robots is part of the experiment. For this, we employed the Gazebo Simulator, which is an open-source 3D robotics simulator, to see the simulation and action [9].

In addition to being open-source, there are several advantages to using Gazebo. It has a range of robots that are highly accurate, efficient, and have great visualization. Furthermore, the integration with ROS is simple, and testing as well as training it with AI algorithms in realistic scenarios is possible. A comparison of robotic simulation tools is available in [36].

Drone model: In this work, a drone model, which has been developed by the Technical University of Munich (TUM), is used, which represents most off-the-shelf quadrotors on the market. This package is based on TU Darmstadt ROS PKG and the simulator Ardrone. The simulator can simulate AR.Drone 1.0 and 2.0. This simulator can connect to sticks and other devices, and Figure 2 from [37] shows how a joystick or a mouse can be used to control the drone in the simulator.

The TUM drone simulator has been forked by Shanghai Jiao Tong University for development to test the SLAM algorithm with different sensors such as inertial measurement unit (IMU), range finder, and laser range. This simulator will work on Ubuntu 16.04 and 18.4 and Gazebo 7.

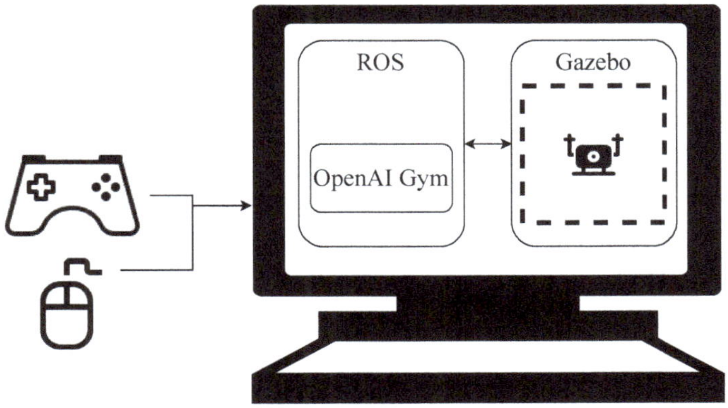

Figure 2. Simulation configuration.

3.3. *OpenAI Gym*

To develop and compare multiple reinforcement learning algorithms, we use the Python-based gym toolkit [10]. All three learning agents used here are from this toolkit. The goal of this work is to enhance productivity in the AI sector by providing an easy-to-set and flexible environment.

4. Agents

In this section, we discuss three popular reinforcement learning algorithms, which are to be used by the agent/learner, the quadrotor in our case. Whenever the agent chooses the best action or policy, it will receive a reward or point. However, the agent will be left with its current state and reward instead of using the information from the environment for future feedback. Usually, to optimize its reward, the agent is forced to decide between choosing a new action to enter a new state or an old action to be in a known state, which is referred to as *"exploration versus exploitation trade-off"*. The agent then considers whether the environment is known or unknown and takes the next action [38]. Table 1 summarizes all three agents used in this study.

Table 1. Type of agent algorithm.

Algorithm	Description	Model	Policy	Action Space	State Space	Operator
Q-Learning	State-action-reward-state	Model-Free	Off-Policy	Discrete	Discrete	Q-Value
SARSA	State-action-reward-state	Model-Free	On-Policy	Discrete	Discrete	Q-Value
DQN	State-action-reward-state	Model-Free	Off-Policy	Discrete	Continuous	Q-Value

4.1. *Q-Learning*

Q-learning is a special case of a temporal difference (TD) learning process, where an agent iteratively takes all actions in all states to obtain the optimal reward. In Q-learning, the next action is taken such that the state will maximize the reward. The goal of this learning process is to find the optimal estimation of the optimal state-action value function Q^* in the case of an unknown model [38]. The Q-learning algorithm samples a new state

s' and takes a new action a', which are used to update the policy value according to the following equation:

$$Q^*(s,a) \leftarrow Q(s,a) + (1-a)[r(s,a) + \gamma^{max}_{a' \epsilon A} Q, (s',a') \quad (1)$$

The aim is to find the optimal policy Q*, which can be represented as follows:

$$\pi^*(s) = argmax\ Q^*(s,a) : a \in A \quad (2)$$

4.2. SARSA

SARSA is another reinforcement learning algorithm used to train an agent in an unknown environment. The name is derived from the quintuples s, a, r', s', a' that are used to update the Q function, which is given as follows [39]:

$$Q(s,a) \leftarrow Q(s,a) + \alpha_t(s,a)\left[r' + \gamma \times Q(s',a') - Q(s,a)\right] \quad (3)$$

SARSA not only depends on the reward to be obtained from the current state and action, but it also takes the state and action it will be in.

4.3. DQN

The Deep Q network (DQN) is one of the popular algorithms in reinforcement learning, also called deep reinforcement learning (DRL). As the name suggests, DQN is a combination of Q-learning with the neural network (NN) and many-layered or deep NN specialization for a spatial processing array of data [26]. This means that the DQN is a multi-layered neural network for a given state 's' that outputs a vector of actions value $Q(s,a;\theta)$, where θ is the trainable weights of the network parameter. Since $Q(s,a;\theta)$ is approximately $Q(s,a)$ [40], the target function used in the DQN is written as:

$$Q(s,a) \leftarrow r + \gamma^{\max}_{a'} Q(s',a') \quad (4)$$

The DQN model was coded by using the Keras and Tensorflow backend framework. We used three hidden layers. The layer of the parameters is shown in Table 2.

Table 2. Layer purpose.

Layer	Output X	Activation
Dense 1	None, 300	RELU
Dense 2	None, 300	RELU
Dense 3	None, 300	RELU

4.4. Optimizers

All the algorithms examined in this work are variants of first-order optimization. However, it is impossible to pick the best one among them provided that the performance depends on the problem environment and dataset.

4.4.1. SGD

One of the most usual methods is SGD, which is used to train a neural network. SGD uses a small collection of data (mini-batch) in comparison to BGD, which uses the entire set of data (batch) [13].

4.4.2. RMSProp

SGD requires many steps, which makes it slower in comparison. Interestingly, RMSProp targets resolving the diminishing learning rate of Adagrad. In RMSProp, the learning is adjusted automatically by using a moving average of the squared gradient [12].

4.4.3. Adadelta

Similarly, an extension of Adagrad is Adadelta, and it accumulates a fixed size past gradient rather than all past squared gradients. At any given time t, the running average depends only on the previous average and the current gradient [11].

4.4.4. ADAM

Instead of a single gradient, Adam adapts multiple gradients, along with an adaptive learning rate according to the magnitude of the gradient [14].

4.5. Reward Computation

In this study, agents are rewarded based on the distance measures between the initial and desired position. Two reward functions are used in our work. First, a simple Euclidean distance is used as a reward function to compare the three agents. Then the agent that showed a better performance is examined with a mean square error computed using training and predicted distance data.

The Euclidean distance is the ordinary straight-line distance between two points in Euclidean space [41]. In this case, we use three-dimensional Euclidean space as shown in Equation (5):

$$d^i_{(p,q)} = \sqrt{\left(q^i_1 - p^i_1\right)^2 + \left(q^i_2 - p^i_2\right)^2 + \left(q^i_3 - p^i_3\right)^2} \quad (5)$$

where $i \in \{0, 1, 2, ..., N\}$, p is the position the UAV is at the ith step, q is the desired position, and N is the number of steps taken from the initial to desired position. We assume that at least one step is taken from the initial point towards the desired position. The reward points are given in Table 3.

To compute the MSE [42] using Equation (6), we used the data obtained during a training session that resulted in the fastest path from the initial to desired position. After each step, the MSE is calculated and compared to the MSE computed from the previous step to generate the reward. The MSE-based reward points are also shown in Table 3.

$$MSE^i = \frac{1}{n}\sum_{i=1}^{n}\left(d^i_{(p,q)} - \overline{d^i_{(p,q)}}\right)^2 \quad (6)$$

where $\overline{d^i_{(p,q)}}$ is the Euclidean distance at step i, whereas $d^i_{(p,q)}$ is the distance after step i during training for the fastest path. The agent uses the Euclidean function to shorten the distance between the UAV and destination, whereas the MSE is used to find the fastest path.

Table 3. Rewards condition.

Condition	Rewards
$d^i_{(p,q)} = 0$	+100, desired position
$d^{i-1}_{(p,q)} - d^i_{(p,q)} > 0$	+100, getting closer
$d^{i-1}_{(p,q)} - d^i_{(p,q)} < 0$	−100, moving away
$d^{i-1}_{(p,q)} - d^i_{(p,q)} = 0$	0, no movement
$MSE_i < MSE_{i-1}$	+100, shortens path
$MSE_i > MSE_{i-1}$	−100, elongate path
$MSE_i = MSE_{i-1}$	0, no movement
Current Altitude > Z position	−100
Pitch Bad	−100
Roll Bad	−100

5. Environment

Here, the simulation environment is discussed. The first subsection entails the experiment setup and general overview of the system, while the next subsection presents the building of the drone environment in the experiment, including an action command for the drone, collection of the data sensor, and reward function.

5.1. Experimental Setup

For visualization, interface, and its highly dynamic physics engine, Gazebo is chosen. The first step is to start it. Then the ROS is used to control the drone. Here, OpenAI Gym, which provides the learning agents, is implemented inside the ROS to control the drone. The drone simulator in Gazebo creates the environment and sends several data sensors to give feedback to the agent, and the agent must send actions (Figure 3).

First, the drone is trained with a certain number of episodes. The training is expected to move the drone from one location to a predetermined desired destination. Data from the drone simulator contain the positions X, Y, and Z that will be sent to the ROS, and the reinforcement learning algorithm was trained in controlling the drone to fly directly to the desired position. The agent will send one of the ten commands (actions) in Table 4.

The final goal is to fly the drone autonomously. In the training process of the drone, reinforcement learning algorithms are used, which involves several parts, agents, and environments.

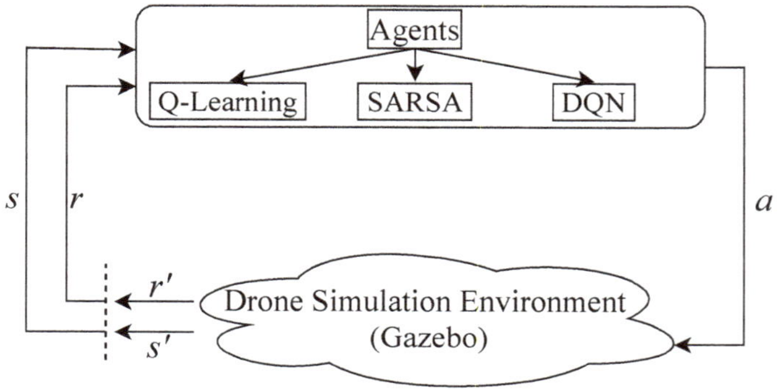

Figure 3. State, reward, agent, and environment interaction.

Table 4. Velocity commands.

Action	Velocity Linear X	Velocity Linear Y	Velocity Linear Z	Velocity Angular Z
0 = Forward	= Speed Value	-	-	0.0
1 = Turn Left	0.05	-	-	= Speed Value
2 = Turn Right	0.05	-	-	= Speed Value
3 = Up	0.05	-	-	= − Speed Value
4 = Down	-	-	= − Speed Value	0.0
5 = Backward	= − Speed Value	-	-	0.0
6 = Fly To Left	-	= Speed Value	-	0.0
7 = Fly To Right	-	= − Speed Value	-	0.0

The first condition from the drone is on the floor, Initial Drone Position (IDP), where the coordinated ground truth is X = 0.0, Y = 0.0, Z = 0.0, as shown in Figure 4. The goal is then to move it to the Desired Drone Position (DDP), such as X = 9.0, Y = 0.0, Z = 1.0 (Figure 4).

The training starts with the agent sending the take-off command. The drone environment must take off with an altitude of 1 m from the floor and send a message to the agent, "take-off success." Then every 0.3 s, the agent sends action and receives a reward feed point as feedback from the environment.

The training session is divided into episodes, each containing 100 steps to arrive from the initial to the desired location. The agent accumulates each reward it receives per step and calculates the average reward at the end of the episode, which it then uses to learn and adopt.

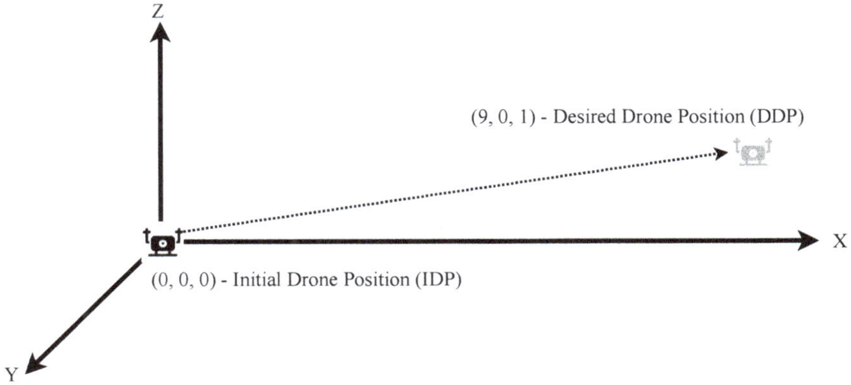

Figure 4. Drone movement from initial to the desired position.

5.2. Drone Environment

One of the main parts of reinforcement learning is a physical environment, and this environment has been developed for the AR drone. Following the OpenAI decisions, the environment only provides the abstraction, not the agent. This means that the environment is independent of the agent, which resides in the ROS.

Directed by the OpenAI Gym rule, the environment must contain registration, steps, and rewards. This will make sure that the interaction between the simulation and OpenAI is smooth. The following tasks are executed to achieve this [10].

1. Registration: registers the training environment in the gym as an available one.
2. Init: in this stage, several parameters such as take-off and landing commands, as well as training parameters such as the value of the speed, desired position, running steps (new command sending time, 0.3 s), the maximum inclination of the drone, and the maximum altitude of the drone, are set. Simulation stabilization is also done at this stage.
3. Reset: this task allows one to reset simulation, pause/resume simulation, reset the robot to initial conditions, and take observation of the state of the drone.
4. Step: for a given action selected by the reinforcement learning algorithm, the quadrotor performs corresponding movements after determining the velocity values as shown in Table 4. The speed value in our simulation is 1 m/s.
5. Observe data: is a function block to obtain data from the drone sensor and also data about position and IMU.
6. Process data: based on the data from the environment and IMU, a reward function is used to compute the progress of the quadrotor to the desired position, such that the quadrotor is given a reward or penalized. Then, the next action will be sent to the drone. Roll, pitch, and altitude movements are also penalized or rewarded.

6. Evaluation

In this section, we first explain the parameters used for the simulation and follow with the result obtained from the experiment.

Since our goal is to understand the behavior of the learning algorithms in flying the drones autonomously, most of the ROS and Gazebo parameters are kept to default settings. The three most influential parameters for our simulation are:

1. Learning Rate (α): when set to 0, robots will not learn; the ideal value is always greater than 0.

2. Discount Factor (γ): setting it to 0 means that agents consider only the current reward; the best value is to arrange it in such a way that the rewards will increase to a higher value for the long-term.
3. Exploration Constant (ϵ): is used to randomize decisions; setting it to a number approaching 1 (such as 0.9) will make 90% of the actions stochastic.

A summary of these and other parameters is shown in Table 5. The simulation is run from 500 to 1000 episodes, each of which is 100 steps. At the end of each episode, the drone will start again in the initial condition and receive feedback and the called observation. In every episode, the drone tries to take a maximum number of steps, learning every step to obtain a high reward point. There are one initial and three desired positions (Table 6).

Table 5. Simulation hyperparameters.

Parameter	Q-Learning and SARSA	DQN
Episode	500–1000	500–1000
Steps	50–100	50–100
α	0.1	0.00025
γ	0.9	0.99
ϵ	0.9	1
Memory Size	-	1,000,000
Network Input	-	3
Network Output	-	8
Network Structure	-	300,300
Update Target Network	-	10,000
Mini Batch Size	-	128
Learn Start	-	128

Table 6. Initial and desired positions.

Coordinates			Description
X	Y	Z	
0.0	0.0	0.0	IDP
9.0	0.0	1.0	X-DDP
0.0	9.0	1.0	Y-DDP
0.0	0.0	10.0	Z-DDP

6.1. Result and Discussion

Here, we discuss the results obtained from all three learning algorithms discussed above. For DQN, we use the RMSProp optimizer [12]. Figure 5 shows the moving average of the rewards that the agent received after completing an episode. The Euclidean distance reward function is used during this run. All three learning algorithms gained rewards as the number of episodes increased, in all directions.

However, DQN showed significant improvement and had no negative moving average reward in any direction. Compared to SARSA, Q-learning has better performance in X-DDP and Y-DDP (Figure 5a,b, respectively). Nevertheless, both algorithms have a negative reward in Z-DDP (Figure 5c). This indicates that DQN has no problem in flying horizontally or vertically, while SARSA and Q-learning are able to fly horizontally but not vertically. As can be seen from the results, the change in reward after the 500th episode is small. Hence, the rest of the evaluations are tested for 500 episodes only, whereas the reward function remains the Euclidean distance.

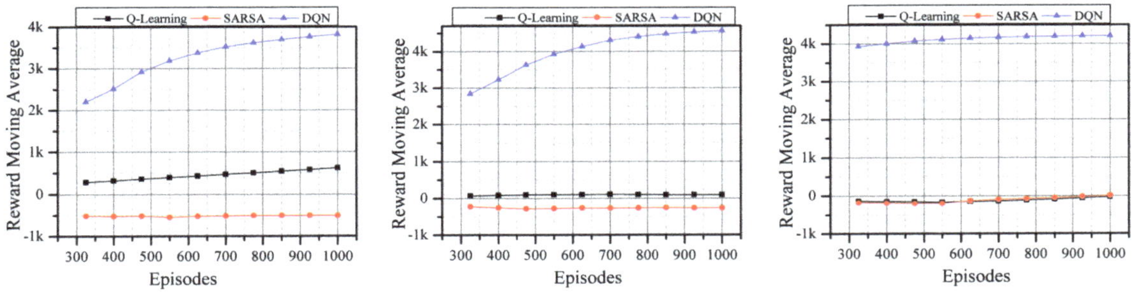

(**a**) X-DDP (**b**) Z-DDP (**c**) Z-DDP

Figure 5. Learning algorithms with 100 steps/episode for a total of 1000 episodes using the Euclidean distance reward function.

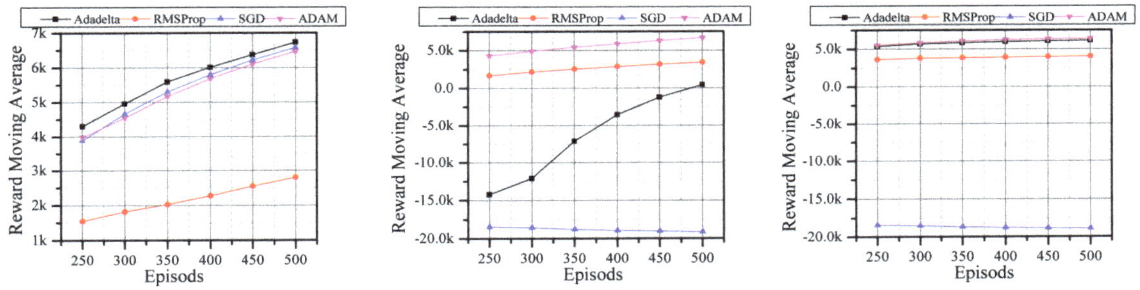

(**a**) X-DDP (**b**) Y-DDP (**c**) Z-DDP

Figure 6. DQN and different optimizers with 100 steps/episode for a total of 500 episodes using the Euclidean distance reward function.

Since DQN with RMSProp optimizer has better rewards, we analyzed the performance of DQN under other optimizers such as Adadelta [11], SGD [13], and ADAM [14]. Here, the agent is expected to reach the desired position in only 500 steps. In X-DDP and Z-DDP, the Adadelta optimizer has a better reward (Figure 6a,c). Even though the SGD optimizer showed a good reward in X-DDP, the result in Figure 6b,c shows that SGD has a negative reward. When the agent uses the RMSProp optimizer, it never outperforms other optimizers in any of the directions, whereas the agent shows good performance with the ADAM optimizer in all directions (Figure 6). In Figure 5, there are six results that obtained negative rewards; by dropping Q-learning and SARSA and adopting other optimizers, we reduced that number to three (Figure 6). Then, we replaced the reward function with MSE, which further reduced the negative reward results to two (Figure 7). Although all optimizers show a good sign in flying the quadrotor autonomously, Adam and Adadelta are the best optimizers in both horizontal and vertical desired positions. In addition, we can see from Figures 6 and 7 that the agent is always improving towards a positive reward when with the Adadelta optimizer. This shows the significant role that the optimizers and the reward functions play in flying the drone autonomously.

Therefore, using the Adadelta optimizer and MSE reward function, we evaluated the performance of the DQN agent under different steps, that is, the maximum number of steps an agent can take between the initial and desired position. The results in Figure 8 indicate that by limiting the number of steps, the reward increases, which means that the agent performs better. This is due to the fact that the agent is not taking unnecessary actions that might lead to negative rewards. However, reducing the number of steps just to improve reward does not result in better performance, as there are small differences between 50 and 75 steps. In addition, realistic scenarios are not as simple as the simulation cases which can be reached in a few steps.

The results obtained in this evaluation showed that learning algorithms can be used to fly drones autonomously. In addition to the learning algorithms, the choice of reward function and optimizer also impacts the performance of autonomous drone flight. Overall, the DQN agent using either the Adadelta or ADAM optimizer and applying the MSE reward function with the number of steps set to 50 shows the best performance in our assessment. In the future, we plan to add more obstacles, use multiple reward functions, and select different learning schemes depending on the next step, such as up, down, or horizontal. We hope that this will reveal more interesting characteristics of the learning schemes.

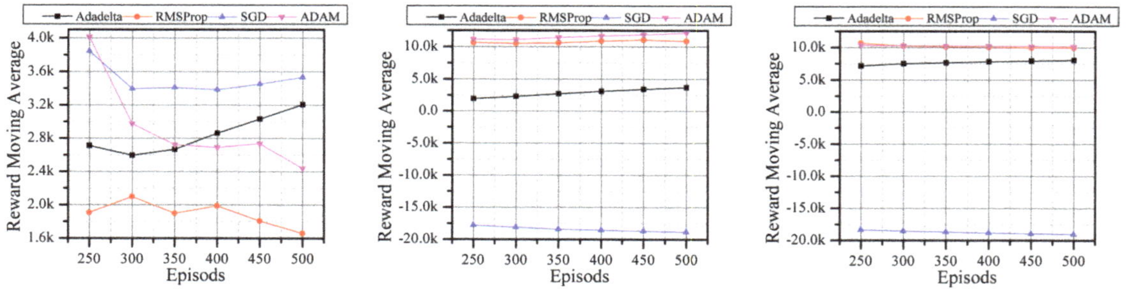

(a) X-DDP (b) Y-DDP (c) Z-DDP

Figure 7. DQN and different optimizers with 100 steps/episode for a total of 500 episodes using MSE reward function.

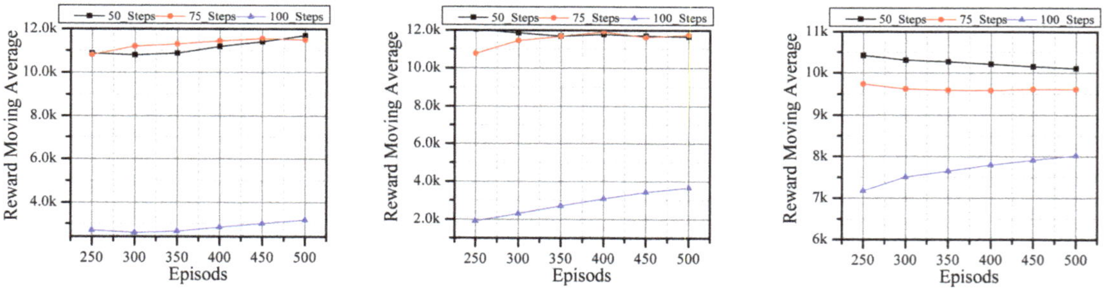

(a) x_steps (b) y_steps (c) z_steps

Figure 8. DQN and Adadelta optimizer with steps/episode varying for a total of 500 episodes using MSE reward function.

7. Conclusions

Autonomously flying UAVs can no longer continue to use traditional controllers such as PID due to tuning, stability, and flexibility issues. However, new reinforcement and deep learning methods are currently showing better control and movement strategies in autonomous UAV flights.

In this work, the simulation and performance evaluation of learning algorithms such as Q-learning, SARSA, and DQN was presented. These algorithms have been evaluated under a combination of positions (X, Y, and Z direction desired positions), optimizers (RMSProp, Adadelta, SGD, and ADAM), and reward functions (Euclidean distance and its MSE). From the evaluation, DQN with the Adadelta optimizer using MSE has shown the best performance in flying drones from the initial to the desired position.

In the future, we plan to investigate the performance of other deep network and neural network learning algorithms under environments that involve obstacles and complicated destinations and to introduce a complex reward function that is more suitable for the autonomous UAV flight.

Author Contributions: Conceptualization, methodology, validation, and writing—original draft preparation, Y.Z.J. and Y.W.N.; supervision, project administration, funding acquisition, B.K. and M.A.; software, resources, visualization, M.T.R.K. and R.P.; writing—review and editing, validation, S.H.A.S. All authors have read and agreed to the published version of the manuscript.

Funding: This research was supported by the Bisa Research Grant of Keimyung University in 2020 (No. 20200195). Muhammad Attique was supported by the National Research Foundation of Korea Grant funded by the Korean Government (2020R1G1A1013221).

Institutional Review Board Statement: Not Applicable.

Informed Consent Statement: Not Applicable.

Data Availability Statement: The data used in the experimental evaluation of this study are available within this article.

Conflicts of Interest: The authors declare no conflict of interest regarding the publication of this manuscript.

References

1. Zhang, Y.; Zu, W.; Gao, Y.; Chang, H. Research on autonomous maneuvering decision of UCAV based on deep reinforcement learning. In Proceedings of the 2018 Chinese Control and Decision Conference (CCDC), Shenyang, China, 9–11 June 2018; pp. 230–235.
2. Valavanis, K.P.; Vachtsevanos, G.J. *Handbook of Unmanned Aerial Vehicles*; Springer: Berlin/Heidelberg, Germany, 2015; Volume 1.
3. Lippitt, C.D.; Zhang, S. The impact of small unmanned airborne platforms on passive optical remote sensing: A conceptual perspective. *Int. J. Remote Sens.* **2018**, *39*, 4852–4868. [CrossRef]
4. Alwateer, M.; Loke, S.W.; Zuchowicz, A. Drone services: Issues in drones for location-based services from human-drone interaction to information processing. *J. Locat. Based Serv.* **2019**, *13*, 94–127. [CrossRef]
5. Koch, W.; Mancuso, R.; West, R.; Bestavros, A. Reinforcement Learning for UAV Attitude Control. *ACM Trans. Cyber-Phys. Syst.* **2019**, *3*, 1–21. [CrossRef]
6. Pham, H.X.; La, H.M.; Feil-Seifer, D.; Nguyen, L.V. Autonomous UAV Navigation Using Reinforcement Learning. *arXiv* **2018**, arXiv:1801.05086.
7. Bou-Ammar, H.; Voos, H.; Ertel, W. Controller design for quadrotor UAVs using reinforcement learning. In Proceedings of the 2010 IEEE International Conference on Control Applications, Yokohama, Japan, 8–10 September 2010; pp. 2130–2135. [CrossRef]
8. Quigley, M.; Conley, K.; Gerkey, B.; Faust, J.; Foote, T.; Leibs; J.; Wheeler, R.; Ng, A.Y. ROS: An open-source Robot Operating System. In Proceedings of the ICRA Workshop on Open Source Software, Kobe, Japan, 12–17 May 2009; Volume 3, p. 5.
9. Koenig, N.; Howard, A. Design and use paradigms for gazebo, an open-source multi-robot simulator. In Proceedings of the 2004 IEEE/RSJ International Conference on Intelligent Robots and Systems (IROS) (IEEE Cat. No. 04CH37566), Sendai, Japan, 28 September–2 October 2004; Volume 3, pp. 2149–2154.
10. Brockman, G.; Cheung, V.; Pettersson, L.; Schneider, J.; Schulman, J.; Tang, J.; Zaremba, W. Openai gym. *arXiv* **2016**, arXiv:1606.01540.
11. Zeiler, M.D. Adadelta: An adaptive learning rate method. *arXiv* **2012**, arXiv:1212.5701.
12. Tieleman, T.; Hinton, G. Divide the Gradient by a Running Average of Its Recent Magnitude. Coursera: Neural Networks for Machine Learning. Technical Report. 2017. Available online: https://www.scirp.org/(S(czeh2tfqyw2orz553k1w0r45))/reference/ReferencesPapers.aspx?ReferenceID=1911091 (accessed on 5 August 2021).
13. Zhang, C.; Liao, Q.; Rakhlin, A.; Miranda, B.; Golowich, N.; Poggio, T. Theory of deep learning IIb: Optimization properties of SGD. *arXiv* **2018**, arXiv:1801.02254.
14. Kingma, D.P.; Ba, J. Adam: A method for stochastic optimization. *arXiv* **2014**, arXiv:1412.6980.
15. Kim, J.; Gadsden, S.A.; Wilkerson, S.A. A comprehensive survey of control strategies for autonomous quadrotors. *Can. J. Electr. Comput. Eng.* **2019**, *43*, 3–16.
16. Lee, K.; Kim, H.; Park, J.; Choi, Y. Hovering control of a quadrotor. In Proceedings of the ICCAS 2012—2012 12th International Conference on Control, Automation and Systems, Jeju Island, Korea, 17–21 October 2012; pp. 162–167.
17. Zulu, A.; John, S. A Review of Control Algorithms for Autonomous Quadrotors. *Open J. Appl. Sci.* **2014**, *04*, 547–556. [CrossRef]
18. Demir, B.E.; Bayir, R.; Duran, F. Real-time trajectory tracking of an unmanned aerial vehicle using a self-tuning fuzzy proportional integral derivative controller. *Int. J. Micro Air Veh.* **2016**, *8*, 252–268. [CrossRef]
19. Eresen, A.; İmamoğlu, N.; Önder Efe, M. Autonomous quadrotor flight with vision-based obstacle avoidance in virtual environment. *Expert Syst. Appl.* **2012**, *39*, 894–905. [CrossRef]
20. Goodarzi, F.; Lee, D.; Lee, T. Geometric nonlinear PID control of a quadrotor UAV on SE(3). In Proceedings of the 2013 European Control Conference (ECC), Zurich, Switzerland, 17–19 July 2013; pp. 3845–3850. [CrossRef]
21. Lwin, N.; Tun, H.M. Implementation Of Flight Control System Based On Kalman And PID Controller For UAV. *Int. J. Sci. Technol. Res.* **2014**, *3*, 309–312.

22. Salih, A.L.; Moghavvemi, M.; Mohamed, H.A.; Gaeid, K.S. Flight PID controller design for a UAV quadrotor. *Sci. Res. Essays* **2010**, *5*, 3660–3667.
23. Zang, Z.; Ma, Z.; Wang, Y.; Ji, F.; Nie, H.; Li, W. The Design of Height Control System of Fully Autonomous UAV Based on ADRC-PID Algorithm. *J. Phys. Conf. Ser.* **2020**, *1650*, 032136. [CrossRef]
24. Siti, I.; Mjahed, M.; Ayad, H.; El Kari, A. New trajectory tracking approach for a quadcopter using genetic algorithm and reference model methods. *Appl. Sci.* **2019**, *9*, 1780. [CrossRef]
25. Hermand, E.; Nguyen, T.W.; Hosseinzadeh, M.; Garone, E. Constrained control of UAVs in geofencing applications. In Proceedings of the 2018 26th Mediterranean Conference on Control and Automation (MED), Zadar, Croatia, 19–22 June 2018; pp. 217–222.
26. Sutton, R.S.; Barto, A.G. *Reinforcement Learning: An Introduction*, 2nd ed.; The MIT Press: Cambridge, MA, USA, 2018.
27. Xia, W.; Li, H.; Li, B. A Control Strategy of Autonomous Vehicles Based on Deep Reinforcement Learning. In Proceedings of the 2016 9th International Symposium on Computational Intelligence and Design (ISCID), Hangzhou, China, 10–11 December 2016; Volume 2, pp. 198–201. [CrossRef]
28. Tuyen, L.P.; Layek, A.; Vien, N.A.; Chung, T. Deep reinforcement learning algorithms for steering an underactuated ship. In Proceedings of the 2017 IEEE International Conference on Multisensor Fusion and Integration for Intelligent Systems (MFI), Daegu, Korea, 16–18 November 2017; pp. 602–607. [CrossRef]
29. Yijing, Z.; Zheng, Z.; Xiaoyi, Z.; Yang, L. Q learning algorithm based UAV path learning and obstacle avoidence approach. In Proceedings of the 2017 36th Chinese Control Conference (CCC), Dalian, China, 26–28 July 2017; pp. 3397–3402. [CrossRef]
30. Kim, J.; Shin, S.; Wu, J.; Kim, S.D.; Kim, C.G. Obstacle Avoidance Path Planning for UAV Using Reinforcement Learning Under Simulated Environment. In Proceedings of the IASER 3rd International Conference on Electronics, Electrical Engineering, Computer Science, Okinawa, Japan, 14–18 July 2017; pp. 34–36.
31. Shin, S.Y.; Kang, Y.W.; Kim, Y.G. Obstacle avoidance drone by deep reinforcement learning and its racing with human pilot. *Appl. Sci.* **2019**, *9*, 5571. [CrossRef]
32. Van Hasselt, H.; Guez, A.; Silver, D. Deep reinforcement learning with double q-learning. In Proceedings of the AAAI Conference on Artificial Intelligence, Phoenix, AZ, USA, 12–17 February 2016; Volume 30.
33. Cheng, Q.; Wang, X.; Yang, J.; Shen, L. Automated Enemy Avoidance of Unmanned Aerial Vehicles Based on Reinforcement Learning. *Appl. Sci.* **2019**, *9*, 669. [CrossRef]
34. Kahn, G.; Villaflor, A.; Pong, V.; Abbeel, P.; Levine, S. Uncertainty-Aware Reinforcement Learning for Collision Avoidance. *arXiv* **2017**, arXiv:cs.LG/1702.01182
35. Hwangbo, J.; Sa, I.; Siegwart, R.; Hutter, M. Control of a Quadrotor With Reinforcement Learning. *IEEE Robot. Autom. Lett.* **2017**, *2*, 2096–2103. [CrossRef]
36. Giuliani, M.; Assaf, T.; Giannaccini, M.E. Towards Autonomous Robotic Systems. In Proceedings of the 19th Annual Conference, TAROS 2018, Bristol, UK, 25–27 July 2018; Springer: Berlin/Heidelberg, Germany, 2018; Volume 10965.
37. Hongrong, H.; Jürgen, S. Tum_Simulator-ROS Wiki. Available online: http://wiki.ros.org/tum_simulator (accessed on 5 August 2021).
38. Mohri, M.; Rostamizadeh, A.; Talwalkar, A. *Foundations of Machine Learning*; The MIT Press: Cambridge, MA, USA, 2012.
39. Suh, J.; Tanaka, T. SARSA (0) reinforcement learning over fully homomorphic encryption. *arXiv* **2020**, arXiv:2002.00506.
40. Nair, A.; Srinivasan, P.; Blackwell, S.; Alcicek, C.; Fearon, R.; De Maria, A.; Panneershelvam, V.; Suleyman, M.; Beattie, C.; Petersen, S.; et al. Massively parallel methods for deep reinforcement learning. *arXiv* **2015**, arXiv:1507.04296.
41. Danielsson, P.E. Euclidean distance mapping. *Comput. Graph. Image Process.* **1980**, *14*, 227–248. [CrossRef]
42. Willmott, C.J.; Matsuura, K. Advantages of the mean absolute error (MAE) over the root mean square error (RMSE) in assessing average model performance. *Clim. Res.* **2005**, *30*, 79–82. [CrossRef]

applied sciences

MDPI

Article

Robust Quadrotor Control through Reinforcement Learning with Disturbance Compensation

Chen-Huan Pi, Wei-Yuan Ye and Stone Cheng *

Department of Mechanical Engineering, National Yang Ming Chiao Tung University, Hsinchu City 30010, Taiwan; john40532.me00@g2.nctu.edu.tw (C.-H.P.); s10030789.me03@g2.nctu.edu.tw (W.-Y.Y.)
* Correspondence: stonecheng@mail.nctu.edu.tw

Abstract: In this paper, a novel control strategy is presented for reinforcement learning with disturbance compensation to solve the problem of quadrotor positioning under external disturbance. The proposed control scheme applies a trained neural-network-based reinforcement learning agent to control the quadrotor, and its output is directly mapped to four actuators in an end-to-end manner. The proposed control scheme constructs a disturbance observer to estimate the external forces exerted on the three axes of the quadrotor, such as wind gusts in an outdoor environment. By introducing an interference compensator into the neural network control agent, the tracking accuracy and robustness were significantly increased in indoor and outdoor experiments. The experimental results indicate that the proposed control strategy is highly robust to external disturbances. In the experiments, compensation improved control accuracy and reduced positioning error by 75%. To the best of our knowledge, this study is the first to achieve quadrotor positioning control through low-level reinforcement learning by using a global positioning system in an outdoor environment.

Keywords: external disturbance; quadrotor; reinforcement learning

Citation: Pi, C.-H.; Ye, W.-Y.; Cheng, S. Robust Quadrotor Control through Reinforcement Learning with Disturbance Compensation. *Appl. Sci.* **2021**, *11*, 3257. https://doi.org/10.3390/app11073257

Academic Editors: Sylvain Bertrand and Hyo-sang Shin

Received: 9 March 2021
Accepted: 3 April 2021
Published: 5 April 2021

1. Introduction

A quadrotor is an underactuated, nonlinear coupled system. Because quadrotors have various applications, researchers have long been focusing on the problems of attitude stabilization and trajectory tracking in quadrotors. Many control methods are used for quadrotors. Proportional–integral–derivative (PID) controllers are widely used in consumer quadrotor products and is often treated as a baseline controller for comparison with other controllers [1]. In practice, the tuning of the PID controller's gain often requires expertise, and the gain is selected intuitively by trial and error. Advanced control strategies using model-based methods have been applied to improve the flight performance of quadrotors. Methods such as feedback linearization [2], model predictive control (MPC) [3,4], robust control [5], sliding mode control (SMC) [6–8], and adaptive control [9,10] have been applied to optimize the flight performance of quadrotors. However, the performance and the robustness of the aforementioned strategies are highly related to the accuracy of the manually developed dynamic model.

During outdoor flight, quadrotors are susceptible to wind gust, which affects the flight performance or even leads to system instability [3]. Although quadrotors are sensitive to external disturbances [11], designers of most controllers have not accounted for this problem. Some active disturbance rejection methods have been proposed to estimate disturbances, and these methods perform well in cases of a sustained disturbance. Chovancova et al. [12] designed proportional–derivative (PD), linear quadratic regulator (LQR), and backstepping controllers for position tracking and compared their performance in a simulation. A disturbance observer with a position estimator was designed to improve controller positioning performance, which was evaluated when external disturbance was applied in simulations. The active disturbance rejection control (ADRC) algorithm treats the total disturbance as a new state variable and estimates it through an extended state observer (ESO). Moreover, the ADRC algorithm does not require the exact mathematical model of the overall system to be known. Therefore, this algorithm has become an attractive technique for the flight control of quadrotor unmanned aerial vehicles (UAVs) [13,14]. Yang et al. proposed the use of ADRC and PD control in a dual closed-loop control framework [15]. An ESO was used to estimate the perturbations of gust wind as dynamic disturbances in the inner loop control. A quadrotor flight controller with a sliding mode disturbance observer (SMC-SMDO) was used in [16]. The SMC-SMDO is robust to external disturbances and model uncertainties without the use of high control gain. Chen et al. [17] constructed a nonlinear disturbance observer that considers external disturbances from wind model uncertainties separately from the controller and compensates for the negative effects of the disturbances. In [18], a nonlinear observer based on an unscented Kalman filter was developed for estimating the external force and torque. This estimator reacted to a wide variety of disturbances in the experiment conducted in [18].

Reinforcement learning (RL) has solved many complicated quadrotor control problems in many studies. RL outperforms other optimization approaches and does not require a predefined controller structure, which limits the performance of an agent. In [19], a quadrotor with a deep neural network (DNN)-based controller was proposed for following trails in an unstructured outdoor environment. In [20], RL and MPC were used to enable a quadrotor to navigate unknown environments. MPC enables vehicle control, whereas RL is used to guide a quadrotor through complex environments. In addition to high-level planning and navigation problems, RL control has been used for achieving robust attitude and position control [21,22]. The control policy generated through the RL training of a neural network achieves low-level stabilization and position control and the policy can control the quadrotor directly from the quadrotor state inputs to four motor outputs. The aforementioned studies have implemented their proposed control strategies in simulations and real environments. Although quadrotors with RL controller exhibit stability under disturbance, the control policy cannot eliminate the steady-state error caused by wind or modeling error and the performance of the controller can be improved. In [23], an integral compensator was used to enhance tracking accuracy. The effect of this compensator on the tracking accuracy of the controller was verified by introducing a constant horizontal wind that flowed parallel to the ground in a simulation. Although the aforementioned integral compensator can eliminate the steady-state error, it slows down the controller response and has a large overshoot.

This paper presents a unique disturbance compensation RL (DCRL) framework that includes a disturbance compensator and an RL controller. The external disturbance observer in this framework is based on the work of [24]. The rest of this paper is organized as follows. Section 2 introduces the dynamic model of a quadrotor and the basics of RL. Section 3 describes the proposed DCRL control strategy. Section 4 describes the training and implementation of the proposed DCRL strategy in a quadrotor experiment in indoor and outdoor environments. Finally, Section 5 concludes the paper.

2. Preliminary Information

This section briefly introduces the dynamic model of a quadrotor, the basics of RL and the use of RL in solving the quadrotor control problem.

2.1. Quadrotor Dynamic Model

In this paper, we assume that a quadrotor is a rigid and symmetrical body whose center of gravity coincides with its geometric center.

The vector $\mathbf{x} = [x, y, z]^T$ denotes the position of the quadrotor in an inertial frame. The translation dynamics of the quadrotor can be expressed as follows:

$$\begin{aligned} \dot{\mathbf{x}} &= \mathbf{v} \\ \dot{\mathbf{v}} &= m^{-1}\left(\sum_{i=1}^{4} T_i \mathbf{b_z} + \mathbf{F}_{ext}\right) - g\mathbf{i_z}, \end{aligned} \tag{1}$$

where m and g are the mass of the quadrotor and the acceleration due to gravity, respectively; $\mathbf{R} = [\mathbf{b_x}\ \mathbf{b_y}\ \mathbf{b_z}] \in SO(3)$ is the rotation matrix, which is used to transform a coordinate from the body-fixed reference frame to the inertial reference frame; and T_i is the thrust generated from motors and applied on the z-axis of the body frame $\mathbf{b_z}$. Figure 1 displays the order of motors placement. Finally, the vector $\mathbf{F}_{ext}$ represents the external disturbance force accounting for all other forces acting on the quadrotor.

For the rotation dynamics of system, we use a quaternion representation of quadrotor attitude to avoid gimbal lock and ensure better computational efficiency.

$$\begin{aligned} \dot{\mathbf{q}} &= 0.5 \cdot \mathbf{q} \otimes \begin{bmatrix} 0 \\ \mathbf{\Omega} \end{bmatrix}^T \\ \dot{\mathbf{\Omega}} &= J^{-1}(\mu - \mathbf{\Omega} \times J\mathbf{\Omega}), \end{aligned} \tag{2}$$

where $\dot{\mathbf{q}} = [q_w, q_x, q_y, q_z]^T$ is the normed quaternion attitude vector, $\otimes$ is the quaternion multiplication. Ω is the angular velocity of body-frame, μ is the control moment vector and J is the matrix of vehicle moment of inertia tensor. The rotation transformation between the quaternion q to the rotation matrix $\mathbf{R}$ can be expressed as follows:

$$\mathbf{R} = \begin{bmatrix} 1-2(q_y^2+q_z^2) & 2(q_xq_y - q_zq_w) & 2(q_xq_z + q_yq_w) \\ 2(q_xq_y + q_zq_w) & 1-2(q_x^2+q_k^2) & 2(q_yq_z - q_xq_w) \\ 2(q_xq_z - q_yq_w) & 2(q_yq_z + q_xq_w) & 1-2(q_x^2+q_y^2) \end{bmatrix} \tag{3}$$

Each thrust from the propeller axis is assumed to be aligned perfectly with the z-axis. The force T_i and motor moment μ produced at a motor spinning speed of ω_i can be expressed as follows:

$$\begin{aligned} T_i &= c_f {\omega_i}^2 \\ \mu &= \frac{1}{\sqrt{2}} \begin{bmatrix} -lc_f & lc_f & lc_f & -lc_f \\ lc_f & -lc_f & lc_f & -lc_f \\ c_d & c_d & -c_d & -c_d \end{bmatrix} \begin{bmatrix} \omega_1 \\ \omega_2 \\ \omega_3 \\ \omega_4, \end{bmatrix}^2 \end{aligned} \tag{4}$$

where $i = 1, 2, 3, 4$. Ω_i is the speed of motors; l is the arm length of the quadrotor; and c_f, c_d are the coefficients of the generated force and z-axis moment, respectively.

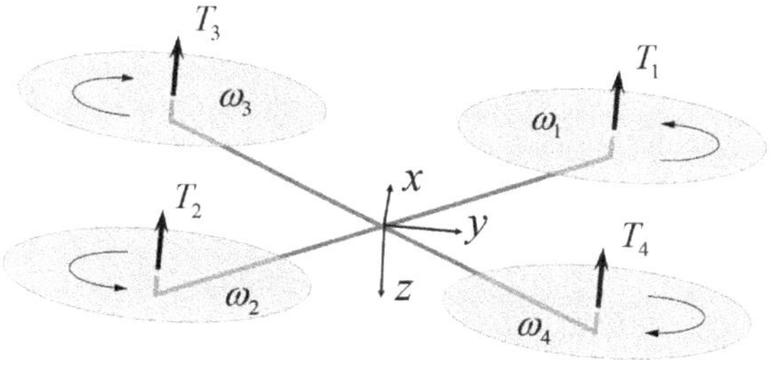

Figure 1. Body-fixed frame of the quadrotor and motor placement.

The developed dynamic model is based on the following assumptions: (a) the quadrotor structure is rigid, (b) the center of mass of the quadrotor and the rotor thrusts are in the same plane, and (c) blade flapping and aerodynamics can be ignored.

2.2. Reinforcement Learning

The standard RL framework comprises a learning agent interacting with the environment according to a Markov decision process (MDP). A state transition has a Markov property if the probability of this transition is independent of its history. The MDP involves solving decision problems with the Markov property, and RL theories are based on the MDP. The standard MDP is defined by the tuple $(\mathcal{S}, \mathcal{A}, P^a_{ss'}, r, \rho_0, \gamma)$, where $\mathcal{S}$ is the state space, $\mathcal{A}$ is the action space, $P^a_{ss'} : \mathcal{S} \times \mathcal{A} \times \mathcal{S} \to \mathbb{R}+$ is the transition probability density of the environment, $r : \mathcal{S} \times \mathcal{A} \to \mathbb{R}$ is the reward function, $\rho_0 : \mathcal{S} \to \mathbb{R}^+$ is the distribution of the initial state s_0, and $\gamma \in (0, 1)$ is the discount factor. In modern deep RL conducted using neural networks, the agent selects an action at each time step according to the policy $\pi(a|s;\theta) = Pr(a|s;\theta)$, where $\theta \in \mathbb{R}^{N_\theta}$ is the weight of the neural network.

The goal of the MDP is to find a policy $\pi(a \in \mathcal{A}|s)$ that can maximize the cumulative discounted reward.

$$\sum_{t=0}^{\infty} \gamma^t r(s_t, a_t). \tag{5}$$

A state-dependent value function V^π that measures the expected discounted reward with respect to π can be defined as follows:

$$V^\pi(s) = \mathbb{E}\left[\sum_{l \geq 0} \gamma^l r_{t+l} \middle| s_t = s, \pi\right]. \tag{6}$$

The state-action-dependent value function can be defined as follows:

$$Q^\pi(s,a) = \mathbb{E}\left[\sum_{l \geq 0} \gamma^l r_{t+l} \middle| s_t = s, a_t = a, \pi\right]. \tag{7}$$

The advantage function can be defined as follows:

$$A^\pi(s,a) = Q^\pi(s,a) - V^\pi(s), \tag{8}$$

where A^π is the difference between the expected value when selecting some specific action a. The advantage function can be used to determine whether the selected action is suitable with respect to policy π. Many basic RL algorithms, such as the policy gradient method [25], off-policy actor–critic algorithm [26], and trust region policy optimization [27] can be used to optimize a policy. To maximize the expected reward function $V^\pi(s)$, the neural network parameterized policy $\pi = \pi(a|s;\theta)$ is adjusted as follows:

$$\theta \leftarrow \theta + \alpha \sum_{s \in \mathcal{S}, a \in \mathcal{A}} \rho^\pi(s) Q^\pi(s,a) \frac{\partial \pi(a|s)}{\partial \theta}, \tag{9}$$

where

$$\rho^\pi(s) = \sum_{t \geq 0} \gamma^t Pr(s_t = s | t, a \sim \pi) \tag{10}$$

is the state occurrence probability and $\alpha > 0$ is the size of the learning step. Equation (9) is an expression for the policy gradient [28]. By using the state distribution ρ and state-action value function Q, a policy can be improved without any environmental information. The state distribution $\rho^\pi(s)$ depends on the policy π, which indicates that it must be re-estimated when the policy is changed. In [26], the policy gradient was analyzed by replacing the original policy π with another policy μ; therefore, (9) had the following form in [26].

$$\sum_{s \in \mathcal{S}, a \in \mathcal{A}} \rho^\mu(s) Q^\pi(s,a) \frac{\partial \pi(a|s)}{\partial \theta}. \tag{11}$$

Equation (11) can still maximize V^π with distinct policy gradient strategies.

To solve the nonlinear dynamic control problem for a quadrotor, we used the proximal policy optimization algorithm (PPO) [29] and off-policy training method [22] to train the actor and critic functions. The following inequalities are valid for the actor function:

$$V^{\pi}(s) - V^{\mu}(s) \geq 0, \forall (s,a) \in \mathcal{S} \times \mathcal{A}, \tag{12}$$

if

$$[\pi(a|s) - \mu(a|s)] A^{\mu}(s,a) > 0 \tag{13a}$$

or

$$[\pi(a|s) - \mu(a|s)] A^{\pi}(s,a) > 0. \tag{13b}$$

Therefore, for a policy search iteration, (13) provides improvement criteria for the action policy under a certain state.

3. Disturbance Observation and Control Strategy

Figure 2 depicts a block diagram of the quadrotor control with the DCRL framework. The proposed DCRL strategy enhances the RL control policy with external disturbance observer and compensator to strengthen the system robustness. The compensation algorithm estimates the external forces and adjusts the input command for the RL controller. The RL controller then changes the motor thrusts accordingly. In Figure 2, the observer takes the attitude $\mathbf{q}_f$ and acceleration $\mathbf{a}_f$ as input, and outputs the estimated external disturbance. $\mathbf{F}_{ext}, \hat{\mathbf{F}}_{ext}$ are the external disturbance and the external disturbance estimated by the observer. The disturbance compensator calculates the $\mathbf{q}_{comp}$ from quadrotor attitude $\mathbf{q}_f$ and $\hat{\mathbf{F}}_{ext}$. The original RL actor was trained to hover at the original point by receiving the state s which contains the position, velocity, attitude and angular velocity of the quadrotor, and output four motors thrust follows the policy $\alpha(s)$. In DCRL, to make the quadrotor hover at the reference position $\mathbf{x}_{ref}$, the original point can be shifted with an off-set of reference command and as the input of position $\mathbf{x}_{dev}$ to the RL actor. The DCRL generates thrust command with the sum of RL actor output and compensation force.

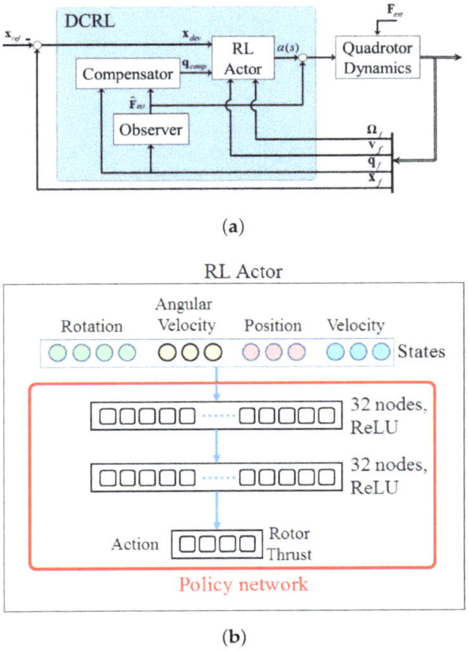

Figure 2. The structures of quadrotor control. (**a**) Block diagram of the quadrotor control structure with reinforcement learning control and external disturbance compensator. (**b**) The neural network in the reinforcement learning (RL) actor.

The RL controller in the DCRL was trained to recover and hover at the original point under an ideal simulation environment, while the external disturbances were not considered in RL control policy training. Several reasons exist for not adding external disturbances in RL training. First, the RL controller sometimes has superior performance to traditional methods if the simulation environment is highly similar to a real-world controller model. However, such a model has numerous uncertainties and is highly difficult to reproduce in a simulator. In this study, the sensor noise was one of the uncertain factors because each inertial measurement unit (IMU) sensor on the flight computer had different physical characteristics. Second, the sensor noise does not follow the assumption of the MDP in RL theory; thus, the final performance of the trained policy cannot be guaranteed to be suitable. Finally, the aforementioned traditional external disturbance estimation methods have been demonstrated to be effective. Therefore, we focused on eliminating known disturbances by using an RL controller with a traditional observation method for achieving a superior positioning performance in this study.

3.1. Disturbance Observer

In general, by rearranging the terms in (1), the external force may be calculated directly from the acceleration information as follows:

$$\hat{\mathbf{F}}_{ext} = m\mathbf{R}^T a^b - \sum_{i=1}^{4} T_i \mathbf{b_z} + mg\mathbf{i_z}, \tag{14}$$

where the thrust forces are only applied on the z-axis of the quadrotor in the body frame and a^b is the acceleration measured by the onboard IMU sensor. The parameter a^b includes the gravitational acceleration.

A low-pass filter (LPF) with cut-off frequency at 30 Hz is used to reduce the effects of noise caused by rotor spinning vibrations or the IMU. The thrust and acceleration are transformed to the inertial reference frame prior to filtering. The reason of this pre-processing is that the external force in the inertial reference frame $\hat{\mathbf{F}}_{ext}$ is assumed to have a lower rate of change than do be slow-changing relative to the LPF dynamics.

3.2. Disturbance Compensator

When an external disturbance $\mathbf{F}_{ext}$ is acting on the quadrotor (Figure 3), this disturbance generates a translational acceleration vector $\mathbf{a}_{ext}$. For disturbance compensation, a new compensation thrust vector $\mathbf{g}_{fc}$ is defined. This vector combines $\mathbf{a}_{ext}$ and the gravitational acceleration vector $g\mathbf{i_z}$ and can be expressed as follows:

$$\mathbf{i_{zc}} = \frac{\mathbf{a}_{ext} + g\mathbf{i_z}}{|\mathbf{a}_{ext} + g\mathbf{i_z}|}, \tag{15}$$

which only considers the hovering situation without an acceleration command from the trajectory tracking reference. The normalized vector $\mathbf{g}_{fc}$ is then used to formulate a new coordinate frame (force compensation frame) with a rotation matrix $\mathbf{R}_{ci}$ relative to the inertial frame.

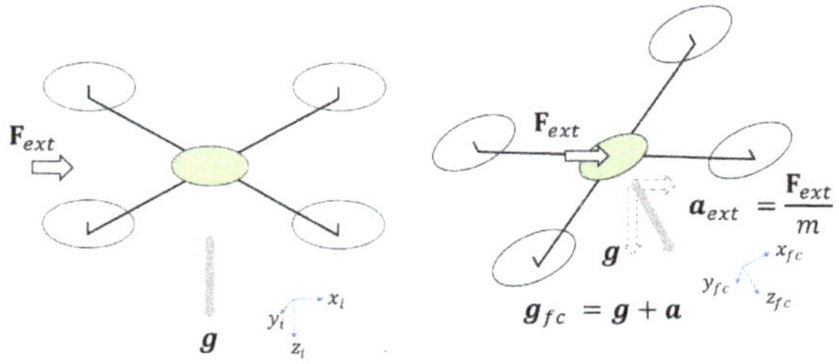

Figure 3. Force compensation frame.

In three-dimensional space, any rotation coordinate system about a fixed point is equivalent to a single rotation by an angle θ about a fixed axis (called the Euler axis) that passes through the fixed point.

To obtain the rotation matrix $\mathbf{R}_{ci}$ from $\mathbf{i_z}$ to $\mathbf{g}_{fc}$ represent in quaternion with the following equations can be used:

$$\begin{aligned} \mathbf{v} &= [0, \mathbf{g}_{fc}] \\ \mathbf{q} &= [\cos(\frac{\theta}{2}), \sin(\frac{\theta}{2})\mathbf{u}] \\ \mathbf{v}' &= \mathbf{q}\mathbf{v}\mathbf{q}^* = \mathbf{q}\mathbf{v}\mathbf{q}^{-1} = \mathbf{i_z}, \end{aligned} \tag{16}$$

where $\mathbf{v}$ is the original quaternion vector, $\mathbf{u}$ is the unit vector of the rotation axis, $\mathbf{v}'$ is the rotated quaternion vector, $\mathbf{q}$ is the rotation vector between $\mathbf{v}$ and $\mathbf{v}'$, and θ is the rotation angle.

By substituting $\mathbf{q}_{ci}$ into (3), $\mathbf{R}_{ci}$ can be determined as follows:

$$\begin{aligned} \mathbf{v}' = \mathbf{q}_{ci}\mathbf{v}\mathbf{q}_{ci}^* &= \mathbf{q}_{ci}\mathbf{v}\mathbf{q}_{ci}^{-1} \\ &= \mathbf{R}_{ci}\mathbf{v}. \end{aligned} \tag{17}$$

The aforementioned equation is equivalent to Rodrigues' rotation formula. After obtaining the rotation matrix $\mathbf{R}_{ci}$, the quadrotor attitude in the force compensation frame $\mathbf{R}_{cb}$ can be calculated using the following equation:

$$\mathbf{R}_{cb} = \mathbf{R}_{ci}\mathbf{R}_{ib}. \tag{18}$$

After obtained the corrected coordinate for compensation, the magnitude of thrust of motors with compensation is

$$T_{ci} = \frac{\hat{\mathbf{F}}_{ext} \cdot \mathbf{b_z}}{4} + \alpha_i(s), \tag{19}$$

where $\alpha_i(s)$ is the i-th motor action output of neural network which would be specified in following section. By modifying the quadrotor attitude in the compensation frame $\mathbf{R}_{cb}$ and using this attitude as the input state of the RL controller, the controller can generate corresponding motor thrust to maintain the target attitude and therefore eliminate the disturbance acting on the quadrotor.

4. Experiments

In this section, we introduce our training method for a low-level quadrotor control policy. The RL controller receives the information on the quadrotor state (position, velocity, attitude, and angular velocity) from sensors and directly outputs the control commands of four rotors. The training was first performed and tested in a simulator. The quadrotor simulator was established using Python according to the dynamic model described in Section 2.1 for training and verifying the flight performance.

After verifying that the RL control policy was trained successfully, we transported the controller into our DCRL structure and performed a real flight with the quadrotor. To implement the proposed DCRL control algorithm in this study, PixRacer flight controller hardware was developed and implemented using Simulink. The DCRL control strategy was examined in an indoor environment by performing fixed-point hovering under an external wind disturbance. Then, the quadrotor was set to track a square trajectory in an outdoor experiment. The position and velocity of the quadrotor were obtained using an OptiTrack motion capture system on the ground station computer. These data were transmitted through Wi-Fi to the onboard PixRacer flight controller within 10 m range in the indoor experiment. The physical parameters of the quadrotor platform are presented in Table 1. For outdoor trajectory tracking, position information was only obtained from an onboard global positioning system (GPS) sensor.

Table 1. Physical parameters of the quadrotor.

Weight (g)	I_{xx}, I_{yy}, I_{zz} (kgm^2)
665	0.0023, 0.0025, 0.0037

4.1. RL Controller Training

In the RL training process, we followed the dynamic equations in Section 2.1 and constructed a simulation environment in Python to generate training data. In the simulation environment, the state space of the MDP comprised the position, velocity, attitude, and angular velocity of the quadrotor. Moreover, the four motors thrust outputs were chosen as the action space. The training process follows the work in [22] using two processes, one for data collection and another for value and policy network update. The update is based on off-policy training, and the main difference between on-policy is the on-policy only uses the collected data once and be cleaned up after each time neural network updates. On the contrary in the off-policy training, the collection thread keeps generating the trajectory data and neural network updating thread can reuse the data from collection which accelerates the learning process.

In data collection process, the quadrotor was randomly launched in a 2 m cubic space with random states. The training data, which comprised the quadrotor states, action, probability, and reward, were recorded in each episode which contained 200 steps in two seconds flight, and then saved as a single data trajectory in a memory buffer. The normalized reward function for evaluating the current state of the quadrotor is as follows:

$$r = -(0.002\|e_q\| + 0.002\|e_p\| + 0.002\|a\|), \tag{20}$$

where e_q is the vehicle angle error, e_p is the vehicle position error, and a is the motor thrust command for constraining the energy cost.

When the number of data trajectories in memory buffer exceeded 10, the training process starts. In training process, trajectories were randomly sampled from the memory buffer. The advantage and value functions were defined recursively and calculated in reverse direction which depend on the future time $t+1$. The functions were estimated according to

$$A_t^{trace} = A_t + \gamma \min(1, \frac{\pi_{t+1}}{\mu_{t+1}}) A_{t+1}^{trace}, \tag{21}$$

and

$$V_t^{trace} = V_t + \min(1, \frac{\pi_t}{\mu_t}) A_t^{trace}. \tag{22}$$

With the two equations above, we use stochastic gradient descent to optimize the objectives as follows:

$$\begin{aligned} maximize\ L_{policy} &= \sum_{(s,a)\in T} \min\left[\left(\frac{\pi(a|s)}{\mu(a|s)} - 1\right) A^{trace}, \epsilon |A^{trace}|\right] \\ minimize\ L_{value} &= \frac{1}{|T|} \sum_{(s,a)\in T} \left(V(s) - V^{trace}\right)^2. \end{aligned} \tag{23}$$

To approximate the function $\pi(a|s)$ for proposing actions and $V(s)$ for predicting the state value, two neural network were formulate as following equations, where stochastic Gaussian policy was used for the actor network:

$$\theta(s) = \tilde{h}_1^{32} \circ \tilde{h}_0^{32}(s) \tag{24}$$

$$\alpha(s) = \sin \circ \tilde{y}_2^4 \circ \theta(s) \tag{25}$$

$$\sigma(s) = \frac{1}{2} + \frac{1}{2} \cos \circ \hat{y}_2^4 \circ \theta(s) \tag{26}$$

$$\pi(a|s) = \frac{1}{\sqrt{2\pi\sigma^2(s)}} e^{-\frac{(a-\alpha(s))^2}{2\sigma^2(s)}}. \tag{27}$$

The state value function can be approximated as follows:

$$V(s) = y_2^1 \circ h_1^{128} \circ h_0^{128}(s), \tag{28}$$

where h_i^j and y_i^j are the ith hidden layer and output layer of neural networks with width j, $\circ$ is the fully connected activation function. In both the actor and critic networks, the input state s is the quadrotor's position, velocity, attitude and angular velocity. The sin and cos functions are used to constrain the output range. A rectified linear unit (ReLU) is used as the activation function due to its characteristic of fast calculation and easy implementation in a microcontroller unit. When implementing the RL controller in a quadrotor flight computer system, only (25) is used to control the quadrotor.

To apply the developed RL controller in an outdoor environment, we extracted the parameters of a trained neural network and loaded them into a Simulink model. The input state of position was limited to the same finite range as that adopted in the training environment to prevent an untrained condition from occurring when using the developed RL controller with a GPS in outdoor environments. With the successful training of the external disturbance observer and RL neural network controller, the DCRL control policy was transferred to the PixRacer flight computer in real quadrotors to replace the original PID controller.

4.2. Results of the Indoor Experiment

In the indoor experiment, we put an electric fan to simulate a constant wind disturbance (Figure 4). We used a self-made quadrotor with a flight control board and GPS mounted in the plane. An OptiTrack motion capture system provides reliable state information, and a multisensor fusion framework in the flight computer fuses the measurement from the onboard IMU and the motion capture data to compensate for the time delay and low update frequency of the OptiTrack system. We compared the position errors between RL with and without compensation under wind disturbance. The measured wind speed was 3.6 m/s at the center of the x-axis of the quadrotor. Figure 5 displays the position error histogram. The mean errors of the original RL and DCRL controllers were 8.4 and 2 cm, respectively, which reduced the hovering error by 75%. Figure 6 presents the estimated position error and external disturbance force for a 30-s flight. Video clips of the indoor experiment can be found at https://youtu.be/RtAoiljZTSI (accessed on 5 April 2021).

4.3. Results of the Outdoor Experiment

After verifying the DCRL control algorithm under motion capture accurate position and velocity measurement and relative steady wind perturbation in a laboratory environment, the quadrotor was moved outdoors and a GPS was used to obtain position feedback. The maximum wind speed was measured to be 4.2 m/s by an anemometer. The position trajectory was a 10-m^2 square with a constant height and a velocity of 1 m/s. The results of the outdoor experiment are shown in Figure 7, and the estimated external forces acting on the x-axis and y-axis are presented in Figure 8. Table 2 summarizes the position errors in the indoor and outdoor experiments. In the outdoor experiment, a position waypoint was used as a reference for trajectory tracking without a velocity command. Thus, the quadrotor had to maintain a certain position error to obtain the moving velocity for following the waypoint. The tracking errors may have also been caused by the 2.5-m horizontal position accuracy and 10-Hz update rate of the adopted GPS sensor. However, the experimental results still indicate that the DCRL structure can reduce the quadrotor positioning error.

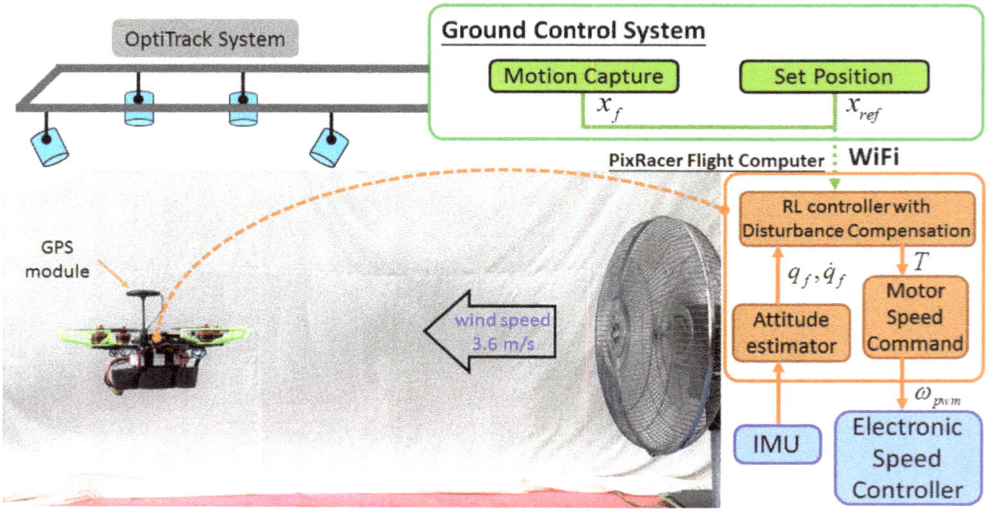

Figure 4. Experimental setup for indoor fixed-position hovering. An electric fan was used for simulating a constant wind disturbance with a speed of 3.6 m/s which was applied along the x-axis of the quadrotor.

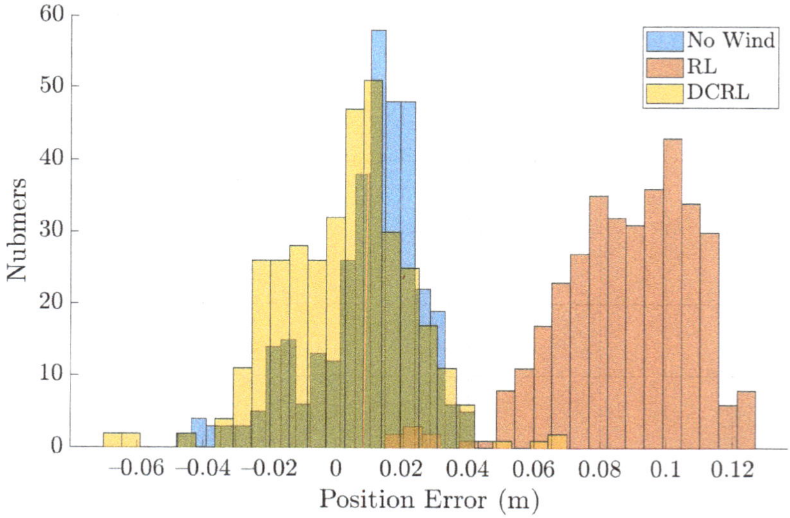

Figure 5. The position error histogram distribution of indoor fix-position hover experiment. The mean error of the original RL controller and the disturbance compensation RL (DCRL) is 8.4 cm and is 2 cm respectively.

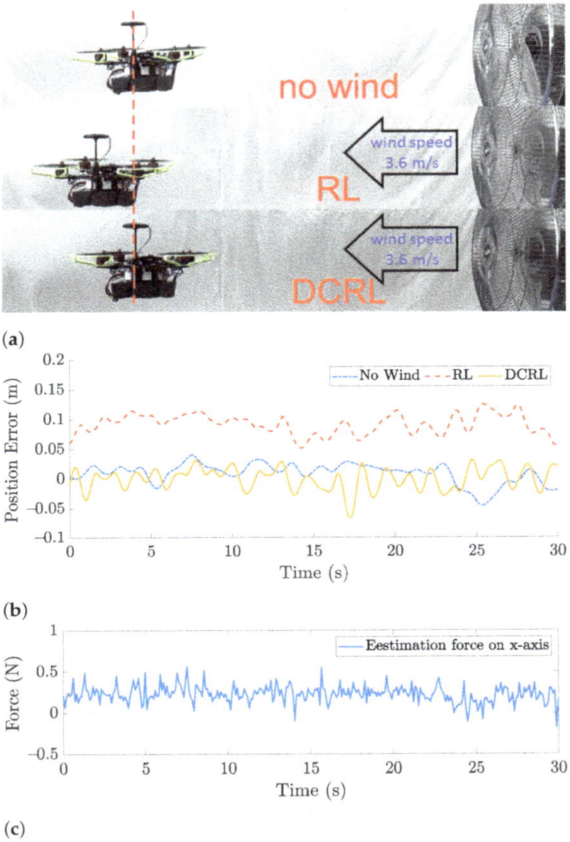

Figure 6. (**a**) The indoor fix-position hover experiment setup. An electric fan was used for simulating a constant wind disturbance with wind speed 3.6 m/s applied onto quadrotor x-axis. (**b**) The position error along x-axis. (**c**) The estimation of external force on x-axis.

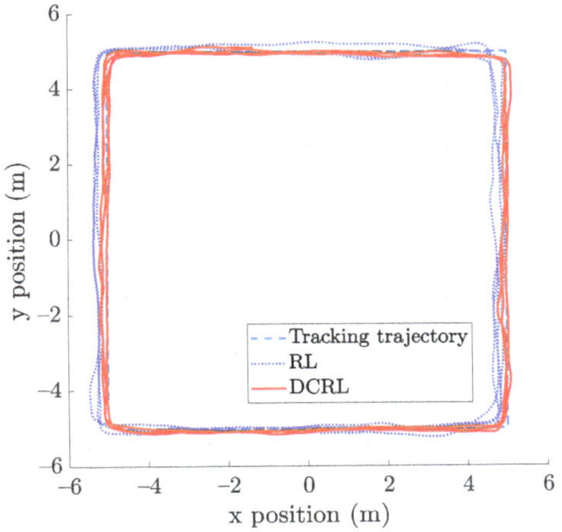

Figure 7. The position of tracking a 10 m square trajectory in an outdoor experiment. The GPS sensor was used for position information measurement.

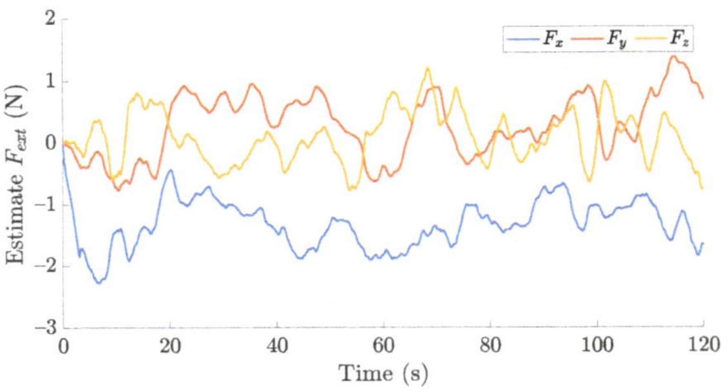

Figure 8. The estimation of external forces on x and y-axis in outdoor tracking experiment.

Table 2. Root mean square errors along the three quadrotor axes in the indoor and outdoor experiments.

		e_x **RMSE (m)**	e_y **RMSE (m)**	e_z **RMSE (m)**
Indoor	RL	0.08	0.04	0.02
	DCRL	0.02	0.06	0.02
Outdoor	RL	0.80	0.46	0.17
	DCRL	0.45	0.41	0.07

5. Conclusions

In this paper, an RL control structure with external force compensation and an external force disturbance observer is proposed for quadrotors. The DCRL controller can reduce the effects of wind gust on quadrotors in fixed-position hovering and trajectory tracking tasks and improve their flight performance. In the outdoor experiments, compared with the original RL control algorithm, the proposed control strategy reduced the fixed-position hovering error by 75%. To the best of our knowledge, this study is the first to use a low-level RL controller with a GPS in an outdoor environment to eliminate the external disturbance acting on flying quadrotors.

Author Contributions: Conceptualization, C.-H.P.; Methodology, C.-H.P.; Project administration, S.C.; Software, C.-H.P. and W.-Y.Y.; Supervision, S.C.; Validation, C.-H.P. and W.-Y.Y.; Writing—original draft, C.-H.P. and W.-Y.Y.; Writing—review & editing, S.C. All authors have read and agreed to the published version of the manuscript.

Funding: This research received no external funding.

Institutional Review Board Statement: Not applicable.

Informed Consent Statement: Not applicable.

Data Availability Statement: The data presented in this study are available on request from the corresponding author.

Conflicts of Interest: The authors declare no conflict of interest.

References

1. Bouabdallah, S.; Noth, A.; Siegwart, R. PID vs. LQ control techniques applied to an indoor micro quadrotor. In Proceedings of the 2004 IEEE/RSJ International Conference on Intelligent Robots and Systems (IROS) (IEEE Cat. No.04CH37566), Sendai, Japan, 28 September–2 October 2004; Volume 3, pp. 2451–2456.
2. Lee, D.; Kim, H.J.; Sastry, S.S. Feedback linearization vs. adaptive sliding mode control for a quadrotor helicopter. *Int. J. Control. Autom. Syst.* **2009**, *7*, 419–428. [CrossRef]
3. Alexis, K.; Nikolakopoulos, G.; Tzes, A. Switching model predictive attitude control for a quadrotor helicopter subject to atmospheric disturbances. *Control Eng. Pract.* **2011**, *19*, 1195–1207. [CrossRef]
4. Alexis, K.; Nikolakopoulos, G.; Tzes, A. Model predictive quadrotor control: Attitude, altitude and position experimental studies. *IET Control Theory Appl.* **2012**, *6*, 1812–1827. [CrossRef]

5. Lee, T. Robust Adaptive Attitude Tracking on ${SO}(3)$ With an Application to a Quadrotor UAV. *IEEE Trans. Control Syst. Technol.* **2013**, *21*, 1924–1930.
6. Wang, H.; Ye, X.; Tian, Y.; Zheng, G.; Christov, N. Model-free–based terminal SMC of quadrotor attitude and position. *IEEE Trans. Aerosp. Electron. Syst.* **2016**, *52*, 2519–2528. [CrossRef]
7. Xu, B. Composite Learning Finite-Time Control With Application to Quadrotors. *IEEE Trans. Syst. Man Cybern. Syst.* **2018**, *48*, 1806–1815. [CrossRef]
8. Xu, R.; Özgüner, Ü. Sliding Mode Control of a Quadrotor Helicopter. In Proceedings of the 45th IEEE Conference on Decision and Control, San Diego, CA, USA, 13–15 December 2006; pp. 4957–4962.
9. Dydek, Z.T.; Annaswamy, A.M.; Lavretsky, E. Adaptive Control of Quadrotor UAVs: A Design Trade Study with Flight Evaluations. *IEEE Trans. Control Syst. Technol.* **2013**, *21*, 1400–1406. [CrossRef]
10. Zou, Y.; Meng, Z. Immersion and Invariance-Based Adaptive Controller for Quadrotor Systems. *IEEE Trans. Syst. Man Cybern. Syst.* **2019**, *49*, 2288–2297. [CrossRef]
11. Liu, H.; Zhao, W.; Zuo, Z.; Zhong, Y. Robust control for quadrotors with multiple time-varying uncertainties and delays. *IEEE Trans. Ind. Electron.* **2016**, *64*, 1303–1312. [CrossRef]
12. Chovancova, A.; Fico, T.; Hubinský, P.; Duchon, F. Comparison of various quaternion-based control methods applied to quadrotor with disturbance observer and position estimator. *Robot. Auton. Syst.* **2016**, *79*, 87–98. [CrossRef]
13. Zhang, Y.; Chen, Z.; Zhang, X.; Sun, Q.; Sun, M. A novel control scheme for quadrotor UAV based upon active disturbance rejection control. *Aerosp. Sci. Technol.* **2018**, *79*, 601–609. [CrossRef]
14. Xu, L.X.; Ma, H.J.; Guo, D.; Xie, A.H.; Song, D.L. Backstepping sliding-mode and cascade active disturbance rejection control for a quadrotor UAV. *IEEE/ASME Trans. Mechatronics* **2020**, *25*, 2743–2753. [CrossRef]
15. Yang, H.; Cheng, L.; Xia, Y.; Yuan, Y. Active Disturbance Rejection Attitude Control for a Dual Closed-Loop Quadrotor Under Gust Wind. *IEEE Trans. Control Syst. Technol.* **2018**, *26*, 1400–1405. [CrossRef]
16. Besnard, L.; Shtessel, Y.; Landrum, B. Quadrotor vehicle control via sliding mode controller driven by sliding mode disturbance observer. *J. Frankl. Inst.* **2012**, *349*, 658–684. [CrossRef]
17. Chen, F.; Lei, W.; Zhang, K.; Tao, G.; Jiang, B. A novel nonlinear resilient control for a quadrotor UAV via backstepping control and nonlinear disturbance observer. *Nonlinear Dyn.* **2016**, *85*, 1281–1295. [CrossRef]
18. McKinnon, C.D.; Schoellig, A.P. Unscented external force and torque estimation for quadrotors. In Proceedings of the 2016 IEEE/RSJ International Conference on Intelligent Robots and Systems (IROS), Daejeon, Korea, 9–14 October 2016; pp. 5651–5657.
19. Smolyanskiy, N.; Kamenev, A.; Smith, J.; Birchfield, S. Toward low-flying autonomous MAV trail navigation using deep neural networks for environmental awareness. In Proceedings of the 2017 IEEE/RSJ International Conference on Intelligent Robots and Systems (IROS), Vancouver, BC, Canada, 24–28 September 2017; pp. 4241–4247.
20. Greatwood, C.; Richards, A. Reinforcement learning and model predictive control for robust embedded quadrotor guidance and control. *Auton. Robot.* **2019**, 1–13. [CrossRef]
21. Hwangbo, J.; Sa, I.; Siegwart, R.; Hutter, M. Control of a quadrotor with reinforcement learning. *IEEE Robot. Autom. Lett.* **2017**, *2*, 2096–2103. [CrossRef]
22. Pi, C.H.; Hu, K.C.; Cheng, S.; Wu, I.C. Low-level autonomous control and tracking of quadrotor using reinforcement learning. *Control Eng. Pract.* **2020**, *95*, 104222. [CrossRef]
23. Wang, Y.; Sun, J.; He, H.; Sun, C. Deterministic Policy Gradient With Integral Compensator for Robust Quadrotor Control. *IEEE Trans. Syst. Man Cybern. Syst.* **2020**, *50*, 3713–3725. [CrossRef]
24. Tomić, T.; Ott, C.; Haddadin, S. External wrench estimation, collision detection, and reflex reaction for flying robots. *IEEE Trans. Robot.* **2017**, *33*, 1467–1482. [CrossRef]
25. Sutton, R.S.; McAllester, D.A.; Singh, S.P.; Mansour, Y. Policy gradient methods for reinforcement learning with function approximation. In *Advances in Neural Information Processing Systems*; ACM: New York, NY, USA, 2000; pp. 1057–1063.
26. Degris, T.; White, M.; Sutton, R.S. Off-Policy Actor-Critic. *arXiv* **2012**, arXiv:1205.4839.
27. Schulman, J.; Levine, S.; Abbeel, P.; Jordan, M.; Moritz, P. Trust region policy optimization. In Proceedings of the International Conference on Machine Learning, Lille, France, 6–11 July 2015; pp. 1889–1897.
28. Sutton, R.S.; Barto, A.G. Reinforcement Learning: An Introduction. *IEEE Trans. Neural Netw.* **1988**, *16*, 285–286. [CrossRef]
29. Schulman, J.; Wolski, F.; Dhariwal, P.; Radford, A.; Klimov, O. Proximal policy optimization algorithms. *arXiv* **2017**, arXiv:1707.06347.

applied sciences

MDPI

Article

Terminal Distributed Cooperative Guidance Law for Multiple UAVs Based on Consistency Theory

Zhanyuan Jiang [1], Jianquan Ge [1,*], Qiangqiang Xu [2] and Tao Yang [1]

1 College of Aerospace Science and Engineering, National University of Defense Technology, Changsha 410073, China; jiangzhanyuan@nudt.edu.cn (Z.J.); yt_yangtao@nudt.edu.cn (T.Y.)
2 College of Space Command, Space Engineering University, Beijing 101416, China; 2003010218@st.btbu.edu.cn
* Correspondence: nudt_gejq@nudt.edu.cn; Tel.: +86-15290198937

Abstract: In order to realize a saturation attack of multiple unmanned aerial vehicles (UAVs) on the same target, the problem is transformed into one of multiple UAVs hitting the same target simultaneously, and a terminal distributed cooperative guidance law for multiple UAVs based on consistency theory is proposed. First, a new time-to-go estimation method is proposed, which is more accurate than the existing methods when the leading angle is large. Second, a non-singular sliding mode guidance law (NSMG) of impact time control with equivalent control term and switching control term is designed, which still appears to have excellent performance even if the initial leading angle is zero. Then, based on the predicted crack point strategy, the NSMG law is extended to attack maneuvering targets. Finally, adopting hierarchical cooperative guidance architecture, a terminal distributed cooperative guidance law based on consistency theory is designed. Numerical simulation results verify that the terminal distributed cooperative guidance law is not only applicable to different forms of communication topology, but also effective in the case of communication topology switching.

Keywords: UAVs; impact time control; sliding mode control; cooperative guidance law; consistency theory

Citation: Jiang, Z.; Ge, J.; Xu, Q.; Yang, T. Terminal Distributed Cooperative Guidance Law for Multiple UAVs Based on Consistency Theory. *Appl. Sci.* **2021**, *11*, 8326. https://doi.org/10.3390/app11188326

Academic Editor: Juan-Carlos Cano

Received: 6 August 2021
Accepted: 6 September 2021
Published: 8 September 2021

1. Introduction

With the rapid iterative update of the air and antimissile defense system equipped around an enemy's high-value targets, it becomes increasingly difficult for a single unmanned aerial vehicle (UAV) to attack high-value targets [1]. In order to solve this problem, there are usually two solutions: one is to adopt a cluster cooperative attack to break through with intelligent cooperation and quantitative advantage; the second is to break through with a speed advantage [2]. For the first, an effective method to achieve multiple UAVs cooperative attack is to control the impact time, which will realize the simultaneous attack of multiple UAVs on targets, thereby improving the impact effect [3].

The design of the impact time control guidance law is actually a tracking problem in which the final impact time error is the tracking error. After defining the impact time error, many system control theories, such as bias proportional guidance with error feedback, sliding mode control theory, Lyapunov function, etc., can be used to make the tracking error zero [4–6]. In [7], a guidance law with impact time control was proposed for the first time in 2006, which consists two parts: one is the classic proportional navigation guidance (PNG), and the other is the feedback item of impact time error.

In [8,9], considering the impact angle constraint, the fast terminal sliding mode algorithm is applied to meet the requirements of guidance accuracy and landing angle by adjusting the line of sight angular velocity. In [10], the second-order sliding mode control theory was used to make the time-to-go estimation curve converge to the desired time-to-go curve in finite time. On this basis, the desired time-to-go was planned by using a double-layer cooperative guidance structure, so as to meet the impact time cooperative guidance of multiple aircraft. In [11], the space is expanded from two-dimensional to three-dimensional, and a three-dimensional impact time control cooperative guidance law satisfying the line-of-sight constraint was proposed. Refs. [12,13] proposed a guidance law training framework based on reinforcement learning theory, which was robust to uncertainties and different parameters.

The above research mainly focuses on the cooperative guidance laws for stationary targets, and there is relatively little research on the cooperative guidance laws for maneuvering targets. References [14–21] study the problem of cooperative guidance for maneuvering targets, but reference [14] needs to assume that the tracking equation can meet the linearization condition of small disturbance. References [14–19] need to assume that the direct measurement information of target acceleration can be obtained, which is usually difficult to achieve in engineering practice; although reference [20] studies the cooperative guidance of maneuvering targets, its method is centralized. The method adopted in reference [21] requires that the communication topology is undirected, which usually leads to more traffic and energy consumption.

In recent years, when the terminal impact angle constraint has been considered at the same time, the impact time and angle control guidance law has gradually developed. Based on the non-singular terminal sliding mode control theory (NTSMC), a guidance law satisfying both impact time and impact angle constraints was designed in [22]. Compared with the traditional sliding mode guidance law, the proposed guidance law did not need to design the line of sight curve offline, nor did it need to switch between the impact time control guidance law and the impact angle control guidance law. In [23], an impact angle control guidance law was designed based on backstepping control method, and an impact time control guidance law was designed based on proportional guidance. The constraints of impact time and impact angle were finally satisfied by using segments. In [24], a conversion scheme was designed. When the impact time error was greater than a certain specified value, the impact time control guidance law based on sliding mode theory was adopted. When the impact time error was less than a certain fixed value, the optimal guidance law designed in [25] satisfying the impact angle constraint was adopted in order to finally realize the cooperative control of impact time and angle. In [26], the trajectory optimization problem with impact time and impact angle constraints was firstly transformed into a nonlinear trajectory planning problem, and then the Gauss pseudo-spectral method was adopted to solve the problem with the optimization objective of minimizing the total control energy. Reference [27] proposed a two-stage guidance law with auxiliary stage. By appropriately modifying the switching conditions of the two-stage guidance law with auxiliary stage, the impact time and angle can be controlled at the same time.

Considering the mutual communication among aircraft, in [28] the average estimated value of the time-to-go of each member was taken as the coordination variable to design the variation curve of range, and the control quantity was designed to track the nominal trajectory, so as to realize the impact time cooperative guidance. Based on the principle of distributed communication and network synchronization, a distributed time cooperative guidance law was designed by taking the "lead-followers" mode to realize the simultaneous convergence of multiple aircrafts to the target position in [29]. In [30], the desired time-to-go was directly set as the average of each member's time-to-go, so as to design a hybrid guidance law satisfying both impact time and impact angle. The research in [31] designed a guidance and control integrated guidance law satisfying the impact time constraint, in which not only the time-varying velocity, but also the constraints such as uncertainty and line-of-sight were considered.

The above cooperative guidance laws based on communication adopted a centralized coordination strategy, and the coordination variable existed only in one member of the formation, which was easy to implement. However, this strategy required the information of the whole formation, and when the members were attacked and failed, the coordinated control of the entire formation would fail, which would reduce the robustness and reliability of the system. Therefore, a distributed cooperative guidance law design based on consistency theory is proposed in this paper. The main contributions of this paper are as follows:

(1) A new time-to-go estimation method is proposed, which is more accurate than the existing method in [32] when the leading angle is large.
(2) A non-singular sliding mode guidance law (NSMG) of impact time control with equivalent control term and switching control term is designed, which still appears to have excellent performance even if the initial leading angle is zero, while some existing impact time control laws in [4,8,33] are invalid. Then the guidance law is extended to attack maneuvering targets.
(3) Adopting hierarchical cooperative guidance architecture, a terminal distributed cooperative guidance law based on consistency theory is designed, which is not only applicable to different forms of communication topology, but also effective in the case of communication topology switching.

The other parts of this paper are arranged as follows: In Section 2, the problem statement and motion models are given. The new time-to-go estimation method and the bottom layer guidance law based on sliding mode control theory are proposed in Section 3. The upper-level distributed coordination strategy based on the consistency theory is given in Section 4. Several numerical simulation examples are provided and compared in Section 5. The conclusions are given in the final section.

2. Problem Statement

Two points are explained before establishing the cooperative guidance model. First, during the flight, the thrust of the UAV is adjusted in a small range, which can keep the velocity of the UAV basically unchanged, and the terminal attack distance is short, usually only a few kilometers to more than 10 kilometers. Therefore, it can be assumed that the velocity of each UAV is a constant. Second, in the process of designing the guidance law, the guidance law can be designed separately in the longitudinal plane and the horizontal plane. The guidance law designed in the longitudinal plane is to keep the UAV flying at a fixed height, and the cooperative guidance law designed in the horizontal plane to is meet the relevant cooperative strike requirements. Therefore, it can be assumed that the UAV and the target are in the same horizontal plane. Therefore, four assumptions can be made, as below:

Assumption 1: *The UAV velocity can be considered as a constant value.*

Assumption 2: *The UAV and target are considered as ideal point-mass models.*

Assumption 3: *The target is stationary.*

Assumption 4: *Compared with the guidance loop dynamics of UAV, the dynamic response speed of the UAV detection device and autopilot is fast enough, so it can be ignored.*

Based on the above assumptions, it is assumed that the UAV in the two-dimensional plane attacks the stationary target at a constant speed, and the relative motion relationship is shown in Figure 1.

The UAV and the target are denoted by M and T, respectively. The equations describing the motions between the UAV and the target can be expressed as follows:

$$\dot{r} = -V_M \cos \sigma_M \tag{1}$$

$$\dot{\lambda} = -\frac{V_M \sin \sigma_M}{r} \tag{2}$$

$$\dot{\gamma}_M = a_M / V_M \tag{3}$$

$$\sigma_M = \gamma_M - \lambda \tag{4}$$

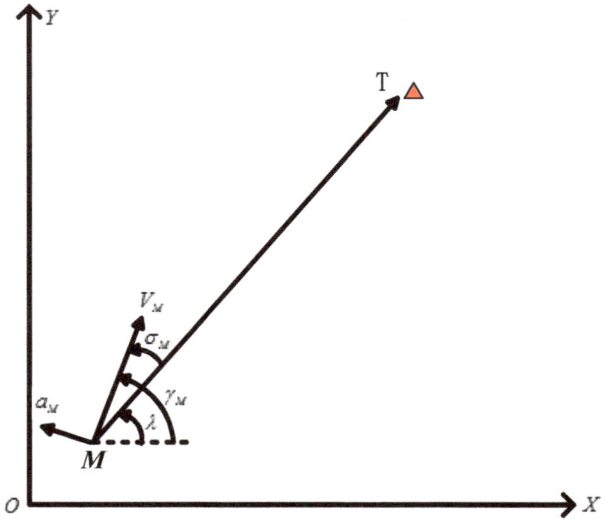

Figure 1. Relative motion relationship between unmanned aerial vehicle (UAV) and target.

In the above equations, V_M is the UAV velocity, the symbol r is the relative distance between the UAV and target, namely the range-to-go. Symbols γ_M, λ and σ_M represent the flight path angle, the line of sight (LOS) angle and the leading angle, respectively. Symbol a_M is the acceleration command.

3. Design of Bottom Guidance Law Based on Sliding Mode Control Theory

In this section, a new time-to-go estimation method is first proposed and compared with the method in [32]. Then, the NSMG law for impact time control based on sliding mode control theory is designed.

3.1. Time-to-Go Estimation of PNG Law

Assuming that the UAV is guided by the PNG law, the acceleration is expressed as follows:

$$a_M = NV_M\dot{\lambda} \tag{5}$$

where, N is the navigation gain and $\dot{\lambda}$ is the rate of the LOS angle.

Substituting Equation (5) into Equation (3), yields

$$\dot{\gamma}_M = N\dot{\lambda} \tag{6}$$

Differentiating Equation (4) and substituting Equation (6), yields:

$$\dot{\sigma}_M = \dot{\gamma}_M - \dot{\lambda} = (N-1)\dot{\lambda} \tag{7}$$

Substituting Equation (2) into Equation (7), yields:

$$\dot{\sigma}_M = -\frac{(N-1)V_M \sin\sigma_M}{r} \tag{8}$$

It can be obtained from Equations (1) and (8) that:

$$\frac{d\sigma_M}{dr} = \frac{\dot{\sigma}_M}{\dot{r}} = \frac{(N-1)\tan\sigma_M}{r} \tag{9}$$

Integrating Equation (9) and its solution can be obtained as follows:

$$r = r_0 \left(\frac{\sin \sigma_M}{\sin \sigma_{M0}} \right)^{\frac{1}{N-1}} \tag{10}$$

where r_0 is the initial relative distance and σ_{M0} is the initial leading angle.

Substituting Equation (10) into Equation (8), yields

$$\dot{\sigma}_M = -(N-1) V_M \sin \sigma_M / r_0 \left(\frac{\sin \sigma_M}{\sin \sigma_{M0}} \right)^{\frac{1}{N-1}} = K (\sin \sigma_M)^{\frac{N-2}{N-1}} \tag{11}$$

where $K = -\frac{(N-1)V_M}{r_0} (\sin \sigma_{M0})^{\frac{1}{N-1}}$.

It can be obtained from Equation (11) that:

$$dt = \frac{1}{K} (\sin \sigma_M)^{\frac{2-N}{N-1}} d\sigma_M \tag{12}$$

Integrating Equation (12) and using Taylor series expansion, ignore advanced items, $\sin x = x - \frac{1}{6}x^3$ and $(1+x)^\alpha = 1 + \alpha x$, yields:

$$\begin{aligned} t - t_0 &= \tfrac{1}{K} \int_{\sigma_{M0}}^{\sigma_M} (\sin \sigma_M)^{\frac{2-N}{N-1}} d\sigma_M \\ &\approx \tfrac{1}{K} \int_{\sigma_{M0}}^{\sigma_M} \left(\sigma_M - \tfrac{\sigma_M^3}{6} \right)^{\frac{2-N}{N-1}} d\sigma_M \\ &= \tfrac{1}{K} \int_{\sigma_{M0}}^{\sigma_M} \sigma_M^{\frac{2-N}{N-1}} \left(1 - \tfrac{\sigma_M^2}{6} \right)^{\frac{2-N}{N-1}} d\sigma_M \\ &\approx \tfrac{1}{K} \int_{\sigma_{M0}}^{\sigma_M} \sigma_M^{\frac{2-N}{N-1}} \left(1 - \tfrac{2-N}{N-1} \tfrac{\sigma_M^2}{6} \right) d\sigma_M \\ &= \tfrac{1}{K} \int_{\sigma_{M0}}^{\sigma_M} \left(\sigma_M^{\frac{2-N}{N-1}} + \tfrac{2-N}{N-1} \tfrac{\sigma_M^{\frac{N}{N-1}}}{6} \right) d\sigma_M \end{aligned} \tag{13}$$

Equation (13) can be further simplified as follows:

$$t = t_0 + \frac{r_0}{V_M} \left(1 + \frac{2-N}{6(N-1)(2N-1)} \sigma_{M0}^2 \right) \left(\frac{\sigma_{M0}}{\sin \sigma_{M0}} \right)^{\frac{1}{N-1}} - \frac{r_0}{V_M} \left(1 + \frac{2-N}{6(N-1)(2N-1)} \sigma_M^2 \right) \left(\frac{\sigma_M}{\sin \sigma_{M0}} \right)^{\frac{1}{N-1}} \tag{14}$$

When the UAV attacks the target, the leading angle σ_M equals zero. Therefore, the time-to-go t_{go} at the moment t can be expressed as follows:

$$t_{go} = \frac{r}{V_M} \left(1 + \frac{2-N}{6(N-1)(2N-1)} \sigma_M^2 \right) \left(\frac{\sigma_M}{\sin \sigma_M} \right)^{\frac{1}{N-1}} \tag{15}$$

Defining,

$$N\prime = \frac{2-N}{6(N-1)(2N-1)} \tag{16}$$

and the new time-to-go estimation method proposed in Equation (15) can be rewritten as follows:

$$t_{go} = \frac{r}{V_M} \left(1 + N\prime \sigma_M^2 \right) \left(\frac{\sigma_M}{\sin \sigma_M} \right)^{\frac{1}{N-1}} \tag{17}$$

Here, another time-to-go estimation method proposed in [32] is also given as below:

$$t_{go} = \frac{r}{V_M} \left(1 + \frac{\sigma_M^2}{2(2N-1)} \right) \tag{18}$$

3.2. Design of the Impact Time Control Guidance Law

3.2.1. Design of the Guidance Law for Stationary Target

For stationary targets, considering the impact time control, the sliding mode surface is designed as below:

$$s = t_f - t_f^d = t + t_{go} - t_f^d = t_{go} - t_{go}^d \tag{19}$$

where, t_f^d and t_{go}^d are the desired impact time and the desired time-to-go respectively. t_{go} is the time-to-go under proportional navigation law and the expression is shown in Equation (18).

The time derivative of Equation (19) is expressed as follows:

$$\begin{aligned}\dot{s} &= \dot{t}_{go} - \dot{t}_{go}^d \\ &= (1+K_1) + (K_2+K_3)\dot{\sigma}_M \\ &= (1+K_1) + (K_2+K_3)\left(\tfrac{a_M}{V_M} + \tfrac{V_M \sin\sigma_M}{r}\right) \\ &= (1+K_1) + (K_2+K_3)\tfrac{V_M \sin\sigma_M}{r} + (K_2+K_3)\tfrac{a_M}{V_M}\end{aligned} \tag{20}$$

where, K_1, K_2 and K_3 are the corresponding coefficients, and the specific expressions can be expressed as:

$$K_1 = -\cos\sigma_M\left(1 + N\prime\sigma_M^2\right)\left(\frac{\sigma_M}{\sin\sigma_M}\right)^{\frac{1}{N-1}} \tag{21}$$

$$K_2 = \frac{r}{V_M}\frac{1}{N-1}\left(\frac{\sigma_M}{\sin\sigma_M}\right)^{\frac{1}{N-1}-1}\frac{\sin\sigma_M - \sigma_M\cos\sigma_M}{\sin^2\sigma_M}\left(1 + N\prime\sigma_M^2\right) \tag{22}$$

$$K_3 = \frac{r}{V_M}(2N\prime\sigma_M)\left(\frac{\sigma_M}{\sin\sigma_M}\right)^{\frac{1}{N-1}} \tag{23}$$

According to the sliding surface designed by Equation (19), the impact time control guidance law based on Lyapunov non-linear control theory is designed as follows:

$$a_M = a_M^{eq} + a_M^{sw} \tag{24}$$

$$a_M^{eq} = -\frac{V_M}{K_2+K_3}\left((1+K_1) + (K_2+K_3)\frac{V_M\sin\sigma_M}{r}\right) \tag{25}$$

$$a_M^{sw} = -ks\sin\sigma_M - M(p\,\mathrm{sign}(K_2+K_3)+1)\mathrm{sign}(s) \tag{26}$$

where, a_M^{eq} and a_M^{sw} are the equivalent part and switching part of the guidance law, respectively, and the parameters $k > 0, M > 0, p > 1$. The equivalent control item is used to control the line-of-sight angular velocity to ensure that the UAV can impact the target, and to maintain the sliding mode surface reaching law $\dot{s} = 0$. The switching control term is to satisfy the impact time constraint, while ensuring that the designed sliding mode guidance law Equation (24) satisfies the Lyapunov stability condition as well as being non-singular, that is, not containing singular points.

3.2.2. Proof of Stability

Choose $V = (1/2)s^2$ as the Lyapunov function, then,

$$\dot{V} = s\dot{s} = -\frac{K_2+K_3}{V_M}ks^2\sin\sigma_M - \frac{M}{V_M}(p|K_2+K_3| + (K_2+K_3))|s| \tag{27}$$

It can be seen from $(K_2+K_3)\sin\sigma_M \geq 0$ and $p|K_2+K_3| + (K_2+K_3) \geq 0$ that $\dot{V}$ is negative semidefinite, which means that when the leading angle, the sliding surface $s = 0$ may not be satisfied. In order to satisfy the attack time control constraints and make the sliding surface $s = 0$, it is necessary to explain that the leading angle $\sigma_M = 0$ is not an attractor.

It can be seen from Equation (4) that:

$$\dot{\sigma}_M = \dot{\gamma}_M - \dot{\lambda} \tag{28}$$

When the leading angle $\sigma_M = 0$, it can be seen from Equation (2) that the rate of change of line of sight angle is as follows:

$$\dot{\lambda} = -\frac{V_M \sin \sigma_M}{r} = 0 \tag{29}$$

Equivalent control term of the guidance law:

$$a_M^{eq} = -\frac{V_M}{K_2 + K_3}\left((1 + K_1) + (K_2 + K_3)\frac{V_M \sin \sigma_M}{r}\right) = 0 \tag{30}$$

Switching term of the guidance law:

$$a_M^{sw} = -ks \sin \sigma_M - M(p\mathrm{sign}(K_2 + K_3) + 1)\mathrm{sign}(s) = -M\mathrm{sign}(s) \tag{31}$$

It can be obtained from Equation (3) that:

$$\dot{\gamma}_M = \frac{a_M}{V_M} = \frac{a_M^{eq} + a_M^{sw}}{V_M} = -\frac{M}{V_M}\mathrm{sign}(s) \tag{32}$$

Then the change rate of the leading angle satisfies,

$$\dot{\sigma}_M = \dot{\gamma}_M - \dot{\lambda} = -\frac{M}{V_M}\mathrm{sign}(s) \tag{33}$$

Therefore, when the leading angle $\sigma_M = 0$ but $s \neq 0$, $\dot{\sigma}_M \neq 0$, it means that the leading angle is not an attractor. At the same time, it can be seen from Equation (33) that when the sliding surface $s > 0$, the change rate of the leading angle $\dot{\sigma}_M < 0$, the leading angle decreases; when the sliding surface $s < 0$, the change rate of the leading angle $\dot{\sigma}_M > 0$, the leading angle increases. This means that only when the sliding surface $s = 0$ and the leading angle $\sigma_M = 0$, the leading angle $\sigma_M = 0$ is an attractor of the system. For the leading angle $\sigma_M \neq 0$, the stability of Lyapunov function has been proved by Equation (27).

At the same time, it should be noted that when the leading angle $\sigma_M = 0$, this can be known according to the law of Robida:

$$\lim_{\sigma_M \to 0} \frac{\sin \sigma_M - \sigma_M \cos \sigma_M}{\sin^2 \sigma_M} = \lim_{\sigma_M \to 0} \frac{\sigma_M}{2 \cos \sigma_M} = 0 \tag{34}$$

$$\lim_{\sigma_M \to 0} \frac{\sigma_M}{\sin \sigma_M} = \lim_{\sigma_M \to 0} \frac{1}{\cos \sigma_M} = 1 \tag{35}$$

Therefore, the coefficients K_2 and K_3 are not singular. When the UAV's leading angle is zero, the guidance law can also be activated. According to the above analysis, the guidance law, Equation (24), is a non-singular sliding mode guidance law with impact time control, which is recorded as NSMG.

3.2.3. The Extension of the Guidance Law under the Maneuvering Target

For maneuvering target, in order to achieve the effective attack on the target under the designated time, the strategy of predicting the collision point is adopted. The target prediction point (x_{TP}, y_{TP}) is shown in Figure 2.

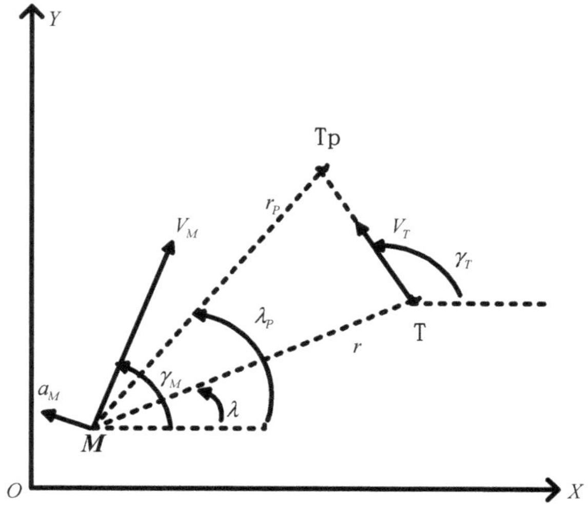

Figure 2. Relative motion relationship based on predicted collision point.

The coordinates of the target prediction collision point can be expressed as:

$$\begin{aligned} x_{TP} &= x_T + (V_T \cos \gamma_T)t_{go} \\ y_{TP} &= y_T + (V_T \sin \gamma_T)t_{go} \end{aligned} \tag{36}$$

where, (x_T, y_T) is the target coordinates at the current time, V_T is the target velocity, γ_T is the target flight path angle. r_P is the relative distance between the UAV and predicted collision point, λ_P is the corresponding leading angle. By replacing r and λ with r_P and λ_P respectively, and bringing them into the guidance law Equation (24), it can attack the maneuvering target in the designated time.

4. The Upper-Level Distributed Coordination Strategy Based on Consistency

When the upper layer of the two-layer guidance architecture adopts the centralized coordination strategy, the coordination variable only exists in one member of the formation, which is easier to realize. However, this strategy requires the information of the global system, and when the centralized cooperative member is attacked and fails, the cooperative control of the whole formation will fail, which reduces the robustness and reliability of the system. Therefore, a distributed cooperative guidance law based on consistency theory is designed in this paper.

Let us assume that n UAVs launched simultaneously are required to carry out a saturation attack on a fixed target at the same time. The formation structure composed of multiple UAVs is regarded as the network communication topology structure, and each UAV is regarded as the network communication node. The acceleration command of the *ith* UAV is expressed as:

$$a_{M,i} = a_{M\_1,i} + a_{M\_2,i}(i = 1, \ldots, n) \tag{37}$$

where, $a_{M\_1,i}$ is the local control term of the *ith* UAV for zero miss distance, and $a_{M\_2,i}$ is the cooperative control item for realizing cooperative attack. The local control item selected in this paper is:

$$a_{M\_1,i} = a_{M,i}^{eq} = -\frac{V_{M,i}}{K_{2,i} + K_{3,i}}\left((1 + K_{1,i}) + (K_{2,i} + K_{3,i})\frac{V_{M,i} \sin \sigma_{M,i}}{r_i}\right) \tag{38}$$

The collaborative control item can be designed as:

$$a_{M\_2,i} = f\left(s_{i1}(t)t_{go,1}, \ldots, s_{iq}(t)t_{go,q}, \ldots, s_{in}(t)t_{go,n}\right) \tag{39}$$

where, $s_{ij}(t)$ is the function of time, and f is the network communication connection. At time t, when the jth UAV can receive the information transmitted by the ith UAV, $s_{ij}(t) = 1$, otherwise, $s_{ij}(t) = 0$, and $s_{ii}(t) = 1$. Then the instantaneous communication matrix describing the information exchange between UAVs in formation can be defined as:

$$S(t) = \begin{bmatrix} s_{11}(t) & s_{12}(t) & \ldots & s_{1n}(t) \\ s_{21}(t) & s_{22}(t) & \ldots & s_{2n}(t) \\ \ldots & \ldots & \ldots & \ldots \\ s_{n1}(t) & s_{n2}(t) & \ldots & s_{nn}(t) \end{bmatrix} \tag{40}$$

The following formula can be obtained by deriving the time-to-go:

$$\begin{aligned} \dot{t}_{go,i} &= K_{1,i} + (K_{2,i} + K_{3,i})\dot{\sigma}_{M,i} \\ &= K_{1,i} + (K_{2,i} + K_{3,i})\left(\frac{a_{M,i}}{V_{M,i}} + \frac{V_{M,i}\sin\sigma_{M,i}}{r_i}\right) \\ &= K_{1,i} + (K_{2,i} + K_{3,i})\frac{V_{M,i}\sin\sigma_{M,i}}{r_i} + (K_{2,i} + K_{3,i})\frac{a_{M\_1,i}}{V_{M,i}} + (K_{2,i} + K_{3,i})\frac{a_{M\_2,i}}{V_{M,i}} \\ &= f_{1,i}(r_i, V_{M,i}, \sigma_{M,i}) + f_{2,i}(r_i, V_{M,i}, \sigma_{M,i})a_{M\_2,i} \end{aligned} \tag{41}$$

where, $f_{1,i}(r_i, V_{M,i}, \sigma_{M,i}) = K_{1,i} + (K_{2,i} + K_{3,i})\frac{V_{M,i}\sin\sigma_{M,i}}{r_i} + (K_{2,i} + K_{3,i})\frac{a_{M\_1,i}}{V_{M,i}}$, $f_{2,i}(r_i, V_{M,i}, \sigma_{M,i}) = (K_{2,i} + K_{3,i})/V_{M,i}$.

When $a_{M\_2,i} = 0$, it means that there is no need to adjust the impact time of the UAV, so the impact time of the UAV satisfies $\dot{t}_{f,i} = 0$. From $t_{go,i} = t_{f,i} - t_i$, it can be known that $\dot{t}_{go,i} = -1$. Considering that $f_{1,i}(r_i, V_{M,i}, \sigma_{M,i})$ does not explicitly contain $a_{M\_2,i}$, then $f_{1,i}(r_i, V_{M,i}, \sigma_{M,i}) = -1$. Therefore, whether $a_{M\_2,i}$ is zero or not, the derivative of the time-to-go can be expressed as:

$$\dot{t}_{go,i} = -1 + f_{2,i}(r_i, V_{M,i}, \sigma_{M,i})a_{M\_2,i} \tag{42}$$

The dynamic change of the UAV's impact time can be expressed as:

$$\dot{t}_{f,i} = f_{2,i}(r_i, V_{M,i}, \sigma_{M,i})a_{M\_2,i} \tag{43}$$

For the dynamic system described in Equation (43), according to the cooperative control theory, the following cooperative control algorithm is designed:

$$a_{M\_2,i} = f_{2,i}^{-1}(r_i, V_{M,i}, \sigma_{M,i})\left(\sum_{j=1}^{n}\frac{s_{ij}t_{f,j}}{\sum\limits_{j=1}^{n} s_{ij}} - t_{f,i}\right) = f_{2,i}^{-1}(r_i, V_{M,i}, \sigma_{M,i})\sum_{j=1}^{n}\frac{s_{ij}}{\sum\limits_{j=1}^{n} s_{ij}}\left(t_{f,j} - t_{f,i}\right) \tag{44}$$

By using this algorithm, the operational requirement of impact time cooperative guidance can be satisfied. Substitute Equation (44) into Equation (43) to obtain,

$$\dot{t}_{f,i} = \sum_{j=1}^{n}\frac{s_{ij}}{\sum\limits_{j=1}^{n} s_{ij}}\left(t_{f,j} - t_{f,i}\right) = \sum_{j=1}^{n} d_{ij}\left(t_{f,j} - t_{f,i}\right) \tag{45}$$

where, $d_{ij} = s_{ij}/\sum\limits_{j=1}^{n} s_{ij}$

For the first-order closed-loop cooperative control system described in Equation (45), the research results of the consistent cooperative control theory show that the necessary and sufficient conditions for the communication network topological system to converge to consistency are as follows: if and only if the communication network topology of the system is strongly connected, that is, there is connectivity between any two nodes in the communication network structure. Therefore, for the cooperative guidance system composed of multiple UAVs, the ultimate goal of the *ith* UAV can be expressed as:

$$t_{f,i} \to \sum_{j=1}^{n} s_{ij} t_{f,j} / \sum_{j=1}^{n} s_{ij} \tag{46}$$

A non-singular sliding mode guidance law with impact time constraint is designed in Section 3. When the desired impact time is designated in advance, the guidance law can be used to attack the target at the designated time. Based on this, the collaborative control is designed:

$$a_{M\_2,i} = k\varepsilon_i \sin\sigma_{M,i} + M(p\text{sign}(K_{2,i} + K_{3,i}) + 1)\text{sign}(\varepsilon_i) \tag{47}$$

where, $\varepsilon_i = \sum_{j=1}^{n} d_{ij}\left(t_{f,j} - t_{f,i}\right) = \sum_{j=1}^{n} d_{ij}\left(t_{go,j} - t_{go,i}\right)$

To sum up, the distributed time cooperative guidance law with time constraint designed in this paper can be expressed as:

$$\begin{aligned} a_{M,i} &= a_{M\_1,i} + a_{M\_2,i} \\ &= -\frac{V_{M,i}}{K_{2,i}+K_{3,i}}\left((1+K_{1,i}) + (K_{2,i}+K_{3,i})\frac{V_{M,i}\sin\sigma_{M,i}}{r_i}\right) + \\ & k\varepsilon_i \sin\sigma_{M,i} + M(p\text{sign}(K_{2,i}+K_{3,i}) + 1)\text{sign}(\varepsilon_i) \end{aligned} \tag{48}$$

It can be seen from Equation (48) that the architecture of the distributed time cooperative guidance law is a two-layer cooperative guidance architecture. The bottom layer is the guidance law based on sliding mode control theory, and the upper layer is the distributed cooperative strategy based on consistency.

5. Numerical Simulation

5.1. Comparison of Methods for Time-to-Go Estimation

The time-to-go estimation methods proposed in this paper and in [32] can be expressed by Equations (17) and (18), respectively. In order to compare the accuracy of the two methods, the navigation gain is set to $N = 3$; the initial range is set to $r_0 = 10,000$ m; the constant velocity of the UAV is set to $V_M = 330$ m/s. The time-to-go calculated by the two methods is compared with the actual time-to-go, which can be shown in Figure 3.

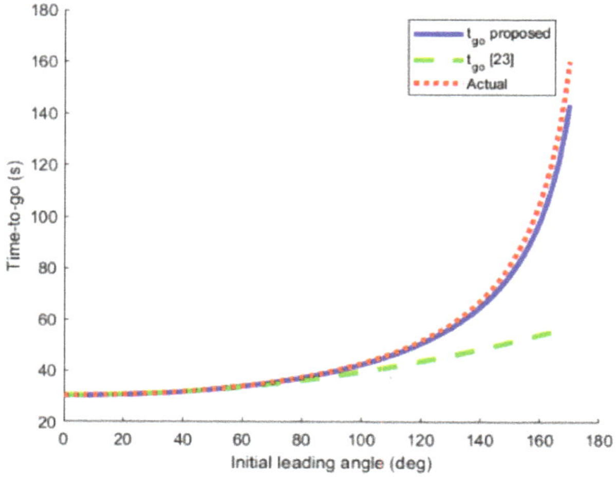

Figure 3. Time-to-go with different initial leading angles.

It can be seen from the figure that when the leading angle is less than 60 degrees (deg.), the time-to-go calculated by the two methods is close to the actual time-to-go; however, when the leading angle is greater than 60 deg., the time-to-go calculated by the method proposed in this paper is almost the same as the actual time-to-go, while the time-to-go calculated by the method in [32] is quite different from the actual time-to-go, so the time-to-go estimation method proposed in this paper is more accurate.

5.2. Verification of the Bottom Non-Singular Sliding Mode Guidance Law

In order to fully verify the non-singular sliding mode guidance law with impact time control designed in this paper, the following simulation examples are designed for simulation verification.

5.2.1. Comparison of Non-Singular Sliding Mode Guidance Law (NSMG) Law and Sliding Mode Control (SMC) Law

In this case, the performance of the NSMG law and the SMC law are compared. The velocity of the UAV is 330 m/s, the initial position is $(0,0)$ m, the initial flight path angle is 0 deg. The navigation gain is set to $N = 3$. The maximum acceleration of the UAV is 5 g and $g = 9.8\ \mathrm{m/s^2}$. The target position is $(10,0)$ km and the designated impact time is set to 45 s.

$$\begin{aligned} a_M &= a_M^{eq} + a_M^{dis} \\ &= \left[\left\{1 + \frac{\dot{r}}{V_M}\left[1 + \frac{\sigma_M^2}{2(2N-1)}\right] + \frac{-r\dot{\lambda}\sigma_M}{(2N-1)V_M}\right\} \times C\mathrm{sign}\left(\dot{\lambda}\right) - \frac{2\dot{r}\dot{\lambda}}{r}\right] \Big/ \left[\frac{\cos\sigma_M}{r} - \frac{Cr\sigma_M \mathrm{sign}\left(\dot{\lambda}\right)}{(2N-1)V_M^2}\right] + K_M^{dis}\mathrm{sign}(S) \end{aligned} \quad (49)$$

where

$$K_M^{dis} = M/\mathrm{sign}\left[\frac{\cos\sigma_M}{r} - \frac{Cr\sigma_M \mathrm{sign}\left(\dot{\lambda}\right)}{(2N-1)V_M^2}\right] \quad (50)$$

C and M are positive constants, and the form of the guidance law is denoted as SMC. The non-singular sliding mode guidance law denoted NSMG proposed in this paper is compared with the guidance law denoted SMC proposed in [33] for simulation, and the variation curves of the UAV's flight trajectory, leading angle, impact time error and acceleration command over time are obtained, as shown in Figure 4.

It can be seen from Figure 4c that the SMC law cannot make the UAV attack the target at the designated time. At this time, the acceleration of the UAV is 0, and the UAV directly flies to the target at a constant speed with a flight time of 30.3 s, which shows that when the leading angle is 0, the SMC law cannot be started, while the NSMG law can attack the target at the designated time. Therefore, when the initial leading angle is 0, the performance of the non-singular sliding mode guidance law with impact time control proposed in this paper is better than the SMC law.

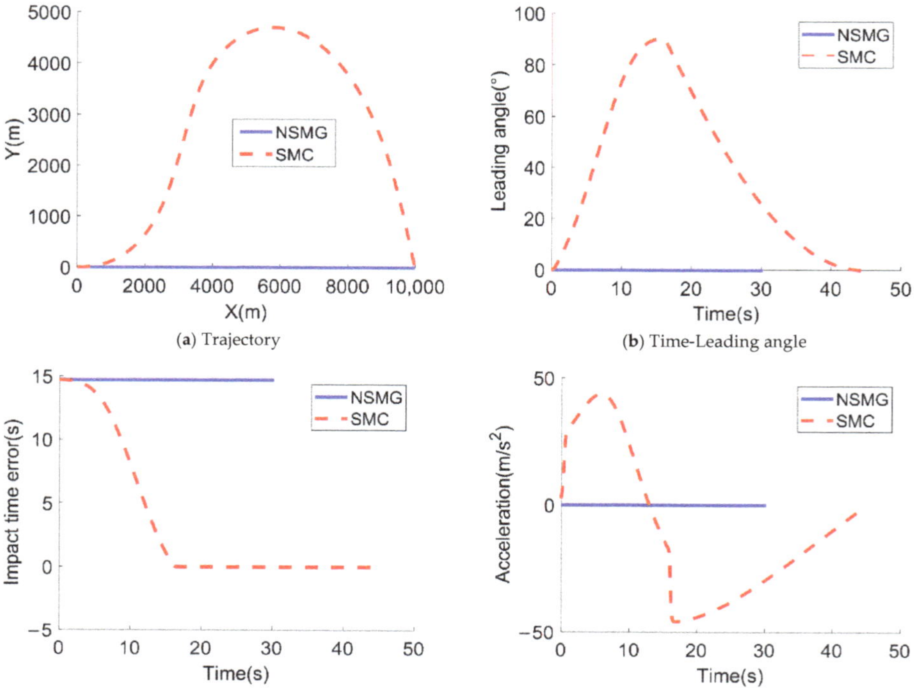

(**a**) Trajectory (**b**) Time-Leading angle (**c**) Time-Impact time error (**d**) Time-Acceleration

Figure 4. Simulation results of the non-singular sliding mode guidance (NSMG) law and the sliding mode control (SMC) law.

5.2.2. Performance of the NSMG Law with Different Impact Time

To evaluate the performance of the NSMG law under different impact time, the designated impact time is set to 45 s, 65 s, 85 s and 105 s, respectively. Other parameters are the same as the parameters used in Section 5.2.1. Simulation results are shown in Figure 5.

The legends represent different simulation situations. For example, "t_d = 45 s" represents the simulation results obtained by using the NSMG law when the designated impact time is 45 s.

As can be seen from Figure 5a, the UAV can attack the target at a designated time. The longer the designated impact time is, the more obvious the lateral maneuverability of the UAV will be. As can be seen from Figure 5b,c, in the initial stage, since the estimated value of the UAV's time-to-go is less than the desired time-to-go, the leading angle increases and then gradually converges to zero. Therefore, the corresponding acceleration command increases in the initial stage, and then converges to zero with the decrease of the leading angle, as shown in Figure 5d. At the same time, when the designated time is small, that is, the error of the initial impact time is small, although the acceleration command of the UAV increases in the initial stage, it does not exceed the boundary of the acceleration command. However, when the designated time is large, that is, the error of initial impact time is large, the acceleration command of the UAV will reach the specified boundary in the initial stage, and the larger the error of initial impact time is, the longer the duration will be.

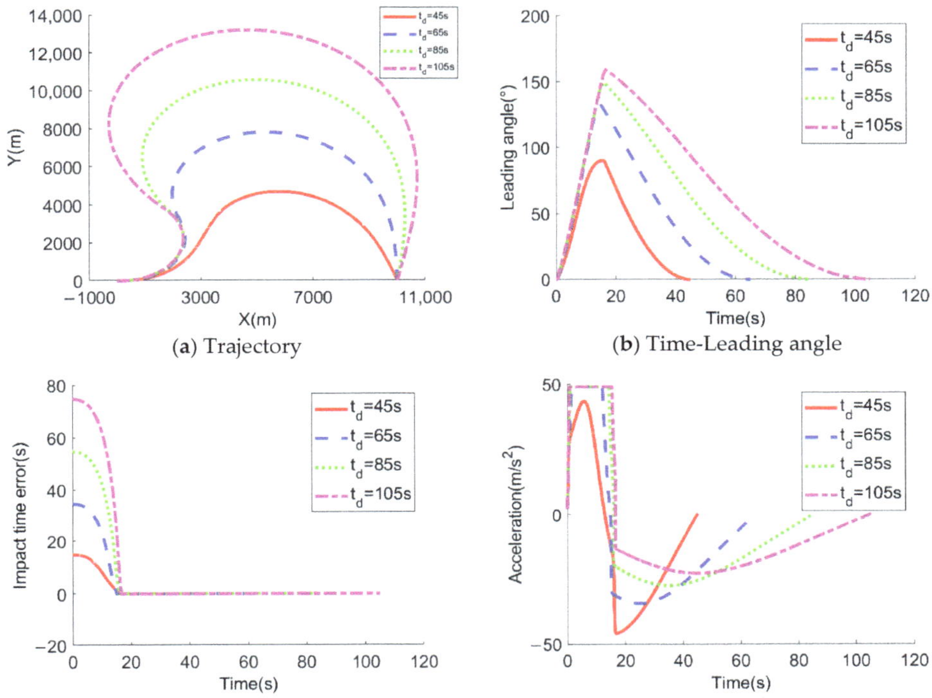

(a) Trajectory
(b) Time-Leading angle
(c) Time-Impact time error
(d) Time-Acceleration

Figure 5. Simulation results under different designated impact time.

5.2.3. Performance of the NSMG Law with Different Initial Leading Angles

The initial leading angles are set to 20, 40, 60 and 80 deg., respectively. The designated impact time is set to 45 s. Other parameters are the same as the parameters used in Section 5.2.1. Simulation results are shown in Figure 6.

The legends represent different simulation situations. For example, "" represents the simulation results obtained by using the NSMG law when the initial leading angle is 20 deg.

It can be seen from Figure 6a that for different initial leading angles, including the case of large initial leading angle, the UAV can reach the target in a designated time. It can be seen from Figure 6b,c that the leading angle increases in the initial stage to extend the flight time and reduce the impact time error. When the designated impact time is fixed, the larger the initial leading angle is, the smaller the initial impact time error will be, the faster the convergence speed of the impact time error will be, and the smaller the corresponding acceleration command will be. When the impact time error converges to 0, the UAV will fly with pure proportional guidance. When it reaches the target, the relative distance between the UAV and the target is 0, the leading angle is 0, and the acceleration also converges to 0.

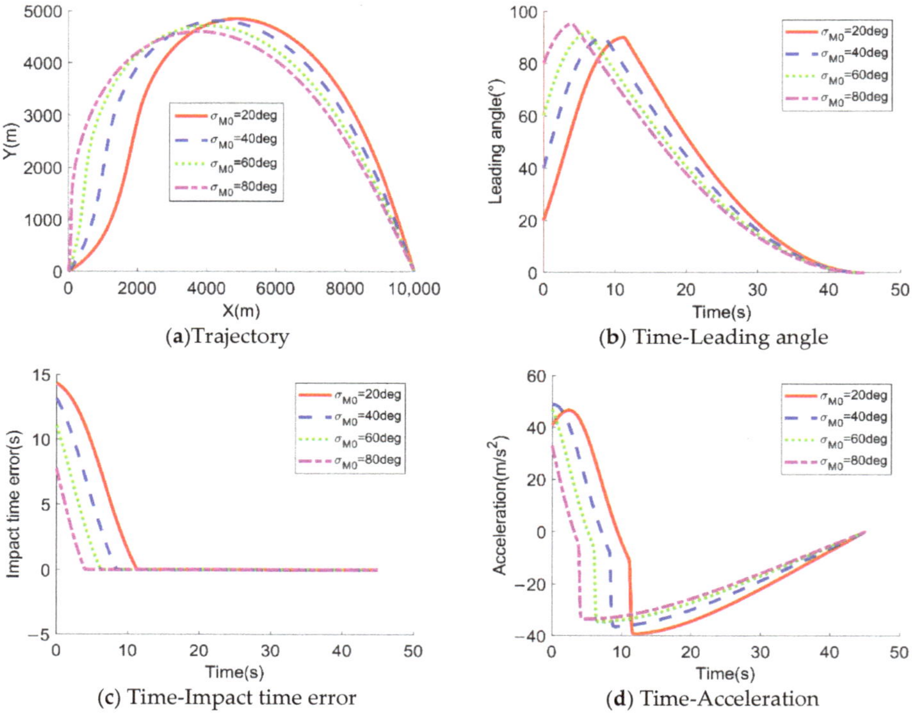

Figure 6. Simulation results under different the initial leading angles.

5.2.4. Salvo Attack on Maneuvering Target with the NSMG Law

The above simulation examples show that the non-singular sliding mode guidance law NSMG can be applied to strike missions under different initial conditions and different designated impact time. Therefore, the NSMG law can be applied to cooperative combat scenarios. At the same time, in order to verify the effectiveness of the extended guidance law for a maneuvering target, it is assumed that four UAVs with different initial conditions attack the same uniformly moving target. The proportional guidance coefficients are all 3, and the initial launch time is consistent. The other simulation parameters are shown in Table 1.

Table 1. Simulation parameters of multiple UAVs' cooperative strike against maneuvering target.

UAVs/Target	Initial Position (km)	Velocity (m/s)	Initial Flight Path Angle (deg.)	Designated Time (s)
M1	(0,0)	330	0	
M2	(5,8)	320	30	
M3	(15,5)	310	5	45
M4	(5,−8)	300	45	
Target	(10,0)	50	30	

When all four UAVs adopt the extended form of guidance law under maneuvering target in Section 3.2, the simulation results are shown in Figure 7.

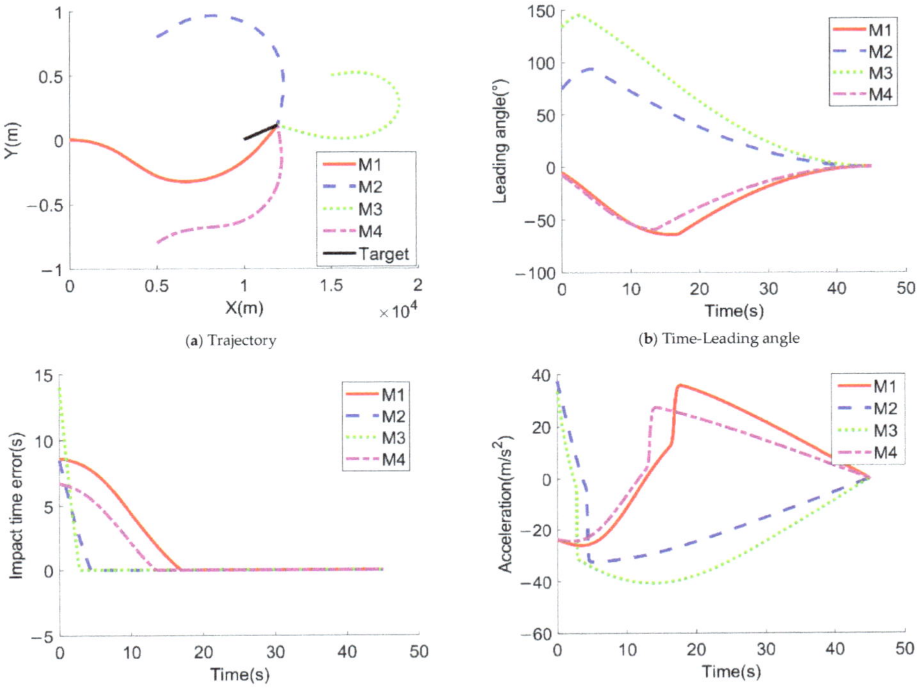

(**a**) Trajectory (**b**) Time-Leading angle

(**c**) Time-Impact time error (**d**) Time-Acceleration

Figure 7. The simulation results of cooperative attack on maneuvering targets with NSMG.

Figure 7a–d show the variation curves of the flight trajectory, leading angle, impact time error and acceleration command of four UAVs over time when the NSMG law is applied to a cooperative attack scenario. It can be seen from the figures that the four UAVs under different initial conditions can strike the target at the same designated time, meeting the demand of time coordination. The terminal leading angle of each UAV is 0, and the corresponding acceleration command is also 0. At the initial moment, the estimated value of the time-to-go of each UAV is less than the desired time-to-go. Under the action of the acceleration command, the amplitude of the leading angle increases to extend the flight time, and finally the impact time error gradually converges to 0.

In conclusion, the impact time control cooperative guidance law based on sliding mode control theory has been fully verified, and the guidance law is suitable for strike missions under different initial conditions and different designated impact times. For maneuvering targets, the predictive collision point strategy is used to extend the form of the guidance law, which can realize an accurate attack.

5.3. Verification of Upper Level Distributed Coordination Strategy

Let us assume that three UAVs form a network formation to attack the same fixed target, and all UAVs are required to attack the target at the same time. The proportional guidance coefficient of each UAV is 3, and the maximum acceleration is no more than 5 g. The other simulation parameters are shown in Table 2.

Table 2. Simulation parameters of upper distributed coordination strategy.

UAVs	Initial Position (km)	Velocity (m/s)	Initial Flight Path Angle (deg.)	Target Position (km)
M1	(0,0)	330	0	
M2	(5,8)	320	30	(10,0)
M3	(15,5)	310	−120	

Suppose that the network communication matrix of the UAV formation has the following three forms.

$$S_1 = \begin{bmatrix} 1 & 1 & 1 \\ 1 & 1 & 1 \\ 1 & 1 & 1 \end{bmatrix} \quad S_2 = \begin{bmatrix} 1 & 1 & 0 \\ 1 & 1 & 1 \\ 0 & 1 & 1 \end{bmatrix} \quad S_3 = \begin{bmatrix} 1 & 1 & 1 \\ 1 & 1 & 0 \\ 1 & 0 & 1 \end{bmatrix} \tag{51}$$

The corresponding network topologies of the three communication matrices are shown in Figure 8.

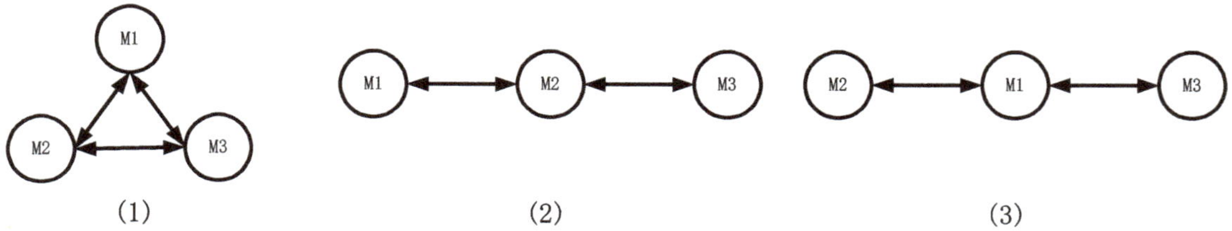

Figure 8. Different network topologies.

As can be seen from the figure, in the first network topology, the three UAVs can exchange information with each other, which can be called the ring network topology. However, in the second and third network topologies, all interconnections cannot be realized. These two forms can be called chained network topologies. The following is the simulation verification for different network topologies.

5.3.1. Ring Network Topology

Based on the double-layer cooperative guidance architecture, when the network topology of the formation is a loop, the variation curves of the flight trajectory, the leading angle, time-to-go and acceleration command of each member of the formation obtained by simulation over time are shown in Figure 9.

As can be seen from Figure 9, the initial time-to-go of the three UAVs is different, respectively, 30.29 s, 37.97 s and 22.96 s. However, after mutual coordination, they gradually become consistent in about 12 s. In the later stage, the cooperative guidance law degenerates into the UAV's own control item, and finally the saturation attack on the target is carried out simultaneously in 33.86 s. The leading angle and acceleration command converge to 0 at the terminal moment.

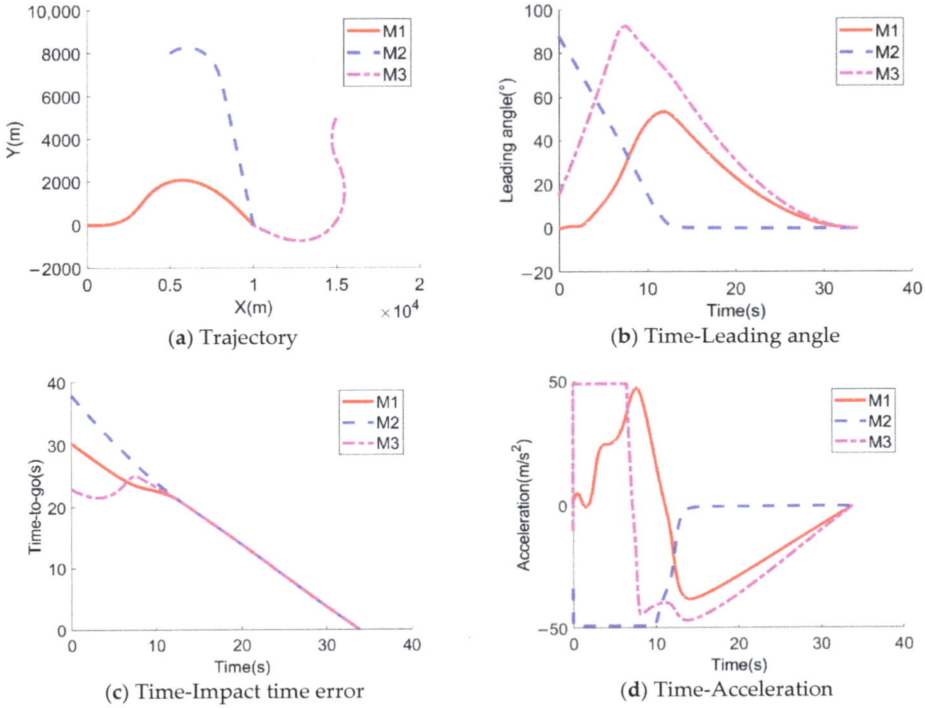

Figure 9. Simulation results of cooperative attack under ring network topology.

5.3.2. Chain Network Topology

Taking the chain network topology as an example, the simulation curves of the flight trajectory, the leading angle, the time-to-go and the acceleration command of each member of the formation are shown in Figure 10.

As can be seen from Figure 10, the initial time-to-go of the three UAVs is different, namely 30.29 s, 37.97 s, and 22.96 s. However, after mutual coordination, they gradually become consistent in about 8 s. In the later stage, the cooperative guidance law degenerates into the UAV's own control item, and finally the saturation attack on the target is carried out simultaneously in 34.44 s. The leading angle and acceleration command converge to 0 at the terminal moment.

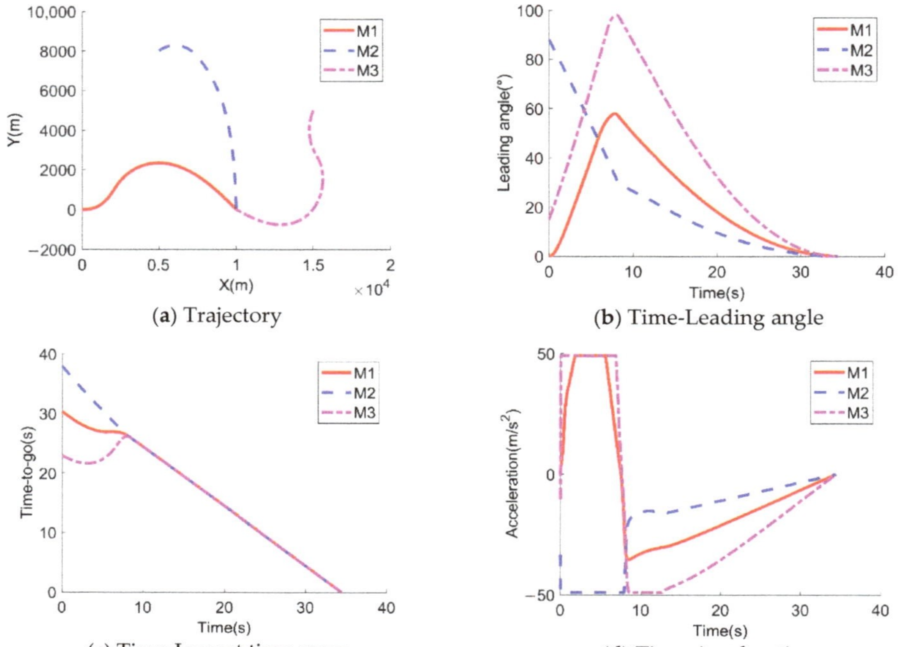

(**a**) Trajectory (**b**) Time-Leading angle

(**c**) Time-Impact time error (**d**) Time-Acceleration

Figure 10. Simulation results of cooperative attack under chain network topology.

5.3.3. The Situation of Network Topology Switching

In order to verify the time characteristics of cooperative attack in the case of the switching network communication topology of a multiple UAV formation, it is assumed that there is a switching network topology among the above three structures with a switching period of 5 s and a switching sequence of (1) $\rightarrow$ (2) $\rightarrow$ (3) $\rightarrow$ (1). The simulation curves of the flight trajectory, the leading angle, the time-to-go and the acceleration command over time of each member of the formation are shown in Figure 11.

As can be seen from Figure 11, the initial time-to-go of the three UAVs is different at, respectively, 30.29 s, 37.97 s and 22.96 s. However, after mutual coordination, they gradually become consistent in about 11 s. In the later stage, the cooperative guidance law degenerates into the UAV's own control item, and finally the saturation attack on the target is carried out simultaneously in 34.21 s. The leading angle and acceleration command converge to 0 at the terminal moment. It is worth noting that from Figure 11d, it can be seen that the acceleration commands of M1 at 10 s, M2 at 10 s and 15 s have obvious jumps. This is mainly because the coordination information obtained by the UAV has obvious changes when the network topology is switched.

Based on the simulation results of the above three different situations, it can be seen that the upper-layer distributed coordination strategy and the lower-layer non-singular sliding mode guidance law designed in this paper can realize the impact time cooperative guidance under the fixed or switching network topology of the UAV formation through the information exchange between them.

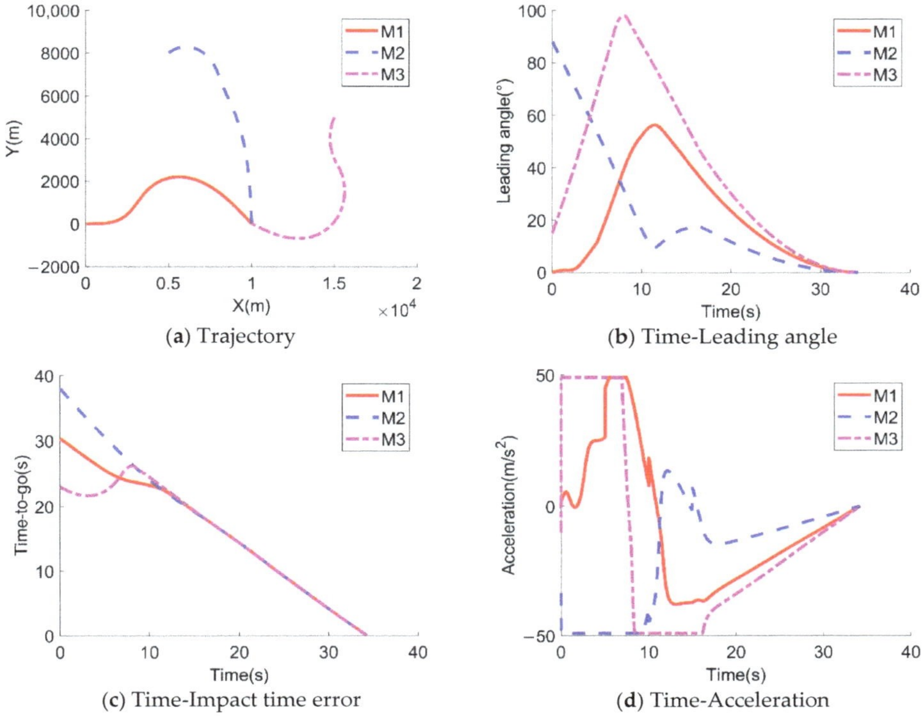

Figure 11. Simulation results of switching the network topology.

6. Conclusions

In order to solve the problem of system instability when the centralized cooperative strategy is adopted, a distributed cooperative guidance law is designed based on the consistency theory. The structure of the guidance law consists of two parts: the bottom non-singular sliding mode guidance law and the upper distributed coordination strategy. First, a new time-to-go estimation method is proposed, which is more accurate than the existing methods when the leading angle is large. Second, a non-singular sliding mode guidance law (NSMG) of impact time control with equivalent control term and switching control term is designed, which still appears to have excellent performance even if the initial leading angle is zero. Then, based on the predicted crack point strategy, the NSMG law is extended to attack maneuvering targets. Finally, adopting hierarchical cooperative guidance architecture, a terminal distributed cooperative guidance law based on consistency theory is designed. The simulation results show that:

(1) The time-to-go estimation method proposed in this paper is more accurate than [27] under large leading angles.
(2) The non-singular sliding mode guidance law with impact time constraint at the bottom layer can be applied to strike missions under different initial conditions and designated impact time. For maneuvering targets, the predictive collision point strategy is used to extend the form of the guidance law, which can still achieve precise strike.
(3) In this paper, the upper-layer distributed coordination strategy and the lower-layer non-singular sliding mode guidance law are combined to make the formation members exchange information with each other, so as to realize the time cooperative online closed-loop guidance under the condition of fixed or switching network topology of the UAV formation.

Author Contributions: Conceptualization, Z.J. and J.G.; Formal analysis, Q.X. and Z.J.; Methodology, Z.J., T.Y. and J.G.; Software, Z.J.; Validation, Z.J. and J.G.; Writing—original draft, Z.J.; Writing-review and editing, T.Y. and Q.X. All authors have read and agreed to the published version of the manuscript.

Funding: This research received no external funding.

Institutional Review Board Statement: Not applicable.

Informed Consent Statement: Not applicable.

Data Availability Statement: Not applicable.

Conflicts of Interest: The authors declare no conflict of interest.

Nomenclature

Acronyms	
UAV	Unmanned Aerial Vehicle
NSMG	nonsingular sliding mode guidance
PNG	Proportional Navigation Guidance
NTSMC	nonsingular terminal sliding mode control theory
LOS	line of sight
SMC	Sliding mode control

References

1. Lyu, T.; Guo, Y.; Li, C.; Ma, G.; Zhang, H. Multiple missiles cooperative guidance with simultaneous attack requirement under directed topologies. *Aerosp. Sci. Technol.* **2019**, *89*, 100–110. [CrossRef]
2. Jiang, H.; An, Z.; Yu, Y.; Chen, S.; Xiong, F. Cooperative guidance with multiple constraints using convex optimization. *Aerosp. Sci. Technol.* **2018**, *79*, 426–440. [CrossRef]
3. Chen, Y.; Xiang, S.; Jiang, D.; Li, J.; Chen, S. Research on Modified Proportional Guidance Method with Attack Time Control. *Acta Sichuan Armory Eng.* **2018**, *39*, 88–92.
4. Kumar, S.R.; Ghose, D. Impact time guidance for large heading errors using sliding mode control. *Aerosp. Electron. Syst. IEEE Trans.* **2015**, *51*, 3123–3138. [CrossRef]
5. Liu, Y.; He, G.; Qiao, Z.; Guo, Z.; Wang, Z. Measurement Compensation for Time-Delay Seeker and Three-Dimensional Adaptive Guidance Law Design. *Sensors* **2021**, *21*, 3977. [CrossRef] [PubMed]
6. Che, F.; Niu, Y.; Li, J.; Wu, L. Cooperative Standoff Tracking of Moving Targets Using Modified Lyapunov Vector Field Guidance. *Appl. Sci.* **2020**, *10*, 3709. [CrossRef]
7. Jeon, I.S.; Lee, J.I.; Tank, M.J. Impact-Time-Control Guidance Law for Anti-Ship Missiles. *IEEE Trans. Control. Syst. Technol.* **2006**, *14*, 260–266. [CrossRef]
8. Zhang, S.; Zang, X.; Liang, S. Research on the Terminal Guidance Law for TV-Guided UCAV Ground Attacking with high Impact angle. *Aircr. Des.* **2016**, *36*, 34–37.
9. Wang, Y. Sliding-Mode Guidance Law on UAV with LOS Controllable. *Mod. Navig.* **2018**, *9*, 65–69.
10. Cui, N.; Wang, B.; Ji, Y. A Cooperative Guidance Law Based on Second Order Sliding Mode Control Theory. *Aerosp. Control.* **2018**, *36*, 40–46.
11. Liu, X.; Han, Y.; Li, P.; Guo, H.; Wu, W. Target Tracking Enhancement by Three-Dimensional Cooperative Guidance Law Imposing Relative Interception Geometry. *Aerospace* **2021**, *8*, 6. [CrossRef]
12. Kim, M.; Hong, D.; Park, S. Deep Neural Network-Based Guidance Law Using Supervised Learning. *Appl. Sci.* **2020**, *10*, 7865. [CrossRef]
13. Hong, D.; Kim, M.; Park, S. Study on Reinforcement Learning-Based Missile Guidance Law. *Appl. Sci.* **2020**, *10*, 6567. [CrossRef]
14. Nikusokhan, M.; Nobahari, H. Closed-form optimal cooperative guidance law against random step maneuver. *IEEE Trans. Aerosp. Electron. Syst.* **2016**, *52*, 319–336. [CrossRef]
15. Zhao, J.; Zhou, R.; Dong, Z. Three-dimensional cooperative guidance laws against stationary and maneuvering targets. *Chin. J. Aeronaut.* **2015**, *28*, 1104–1120. [CrossRef]
16. Zhao, J.; Zhou, R. Unified approach to cooperative guidance laws against stationary and maneuvering targets. *Nonlinear Dyn.* **2015**, *81*, 1635–1647. [CrossRef]
17. Zhao, Q.; Dong, X.; Song, X.; Ren, Z. Cooperative time-varying formation guidance for leader-following missiles to intercept a maneuvering target with switching topologies. *Nonlinear Dyn.* **2018**, *95*, 129–141. [CrossRef]
18. Zhou, J.; Lyu, Y.; Li, Z.; Yang, J. Cooperative Guidance Law Design for Simultaneous Attack with Multiple Missiles Against a Maneuvering Target. *J. Syst. Sci. Complex.* **2018**, *31*, 287–301. [CrossRef]

19. Shaferman, V.; Shima, T. Cooperative Optimal Guidance Laws for Imposing a Relative Intercept Angle. *J. Guid. Control. Dyn.* **2015**, *38*, 1395–1408. [CrossRef]
20. Shaferman, V.; Shima, T. Cooperative Differential Games Guidance Laws for Imposing a Relative Intercept Angle. *J. Guid. Control. Dyn.* **2017**, *40*, 2465–2480. [CrossRef]
21. Wang, X.; Tan, C. 3-D Impact angle constrained distributed cooperative guidance for maneuvering targets without angular-rate measurements. *Control. Eng. Pract.* **2018**, *78*, 142–159. [CrossRef]
22. Hou, Z.; Yang, Y.; Liu, L.; Wang, Y. Terminal sliding mode control based impact time and angle constrained guidance. *Aerosp. Sci. Technol.* **2019**, *93*, 105142. [CrossRef]
23. Jung, B.; Kim, Y. Guidance Laws for Anti-Ship missiles Using Impact Angle and Impact Time. In Proceedings of the AIAA Guidance, Navigation, and Control Conference and Exhibit, Keystone, CO, USA, 21–24 August 2006.
24. Kumar, S.; Ghose, D. Impact Time and Angle Control Guidance. In Proceedings of the AIAA Guidance, Navigation, and Control Conference, Kissimmee, FL, USA, 5–9 January 2015.
25. Ryoo, C.; Cho, H.; Tahk, M. Optimal Guidance Laws with Terminal Impact Angle Constraint. *J. Guid. Control. Dyn.* **2005**, *28*, 724–732. [CrossRef]
26. Zhang, L.; Sun, M.; Chen, Z.; Wang, Z.; Wang, Y. Receding Horizon Trajectory Optimization with Terminal Impact Specifications. *Math. Probl. Eng.* **2014**, *2014*, 693–697. [CrossRef]
27. Tang, Y.; Zhu, X.; Zhou, Z.; Yan, F. Two-phase guidance law for impact time control under physical constraints. *Chin. J. Aeronaut.* **2020**, *33*, 126–138. [CrossRef]
28. Zhang, B.; Song, J.; Song, S. Study on Coordinated Guidance of Multiple Missiles with Angle Constraints. *J. Proj. Rocket. Rocket. Guid.* **2014**, *34*, 13–15.
29. Wu, Z.; Guan, Z.; Yang, C.; Li, J. Terminal Guidance Law for UAV Based on Receding Horizon Control Strategy. *Complexity* **2017**, *2017*, 1–19. [CrossRef]
30. Harrison, G.A. Hybrid Guidance Law for Approach Angle and Time-of-Arrival Control. *J. Guid. Control. Dyn.* **1971**, *35*, 1104–1114. [CrossRef]
31. Wang, X.; Zheng, Y.; Lin, H. Integrated guidance and control law for cooperative attack of multiple missiles. *Aerosp. Sci. Technol.* **2015**, *42*, 1–11. [CrossRef]
32. Jeon, I.; Lee, J.; Tahk, M. Homing guidance law for cooperative attack of multiple missiles. *J. Guid. Control. Dyn.* **2010**, *33*, 275–280. [CrossRef]
33. Kumar, S.R.; Ghose, D. Sliding mode control based guidance law with impact time constraints. In Proceedings of the American Control Conference IEEE, Washington, DC, USA, 17–19 June 2013.

applied sciences

MDPI

Article

Simulation and Analysis of Grid Formation Method for UAV Clusters Based on the 3 × 3 Magic Square and the Chain Rules of Visual Reference

Rui Qiao, Guili Xu *, Yuehua Cheng, Zhengyu Ye and Jinlong Huang

College of Automation, Nanjing University of Aeronautics and Astronautics, Nanjing 211100, China; qiaorui@nuaa.edu.cn (R.Q.); chengyuehua@nuaa.edu.cn (Y.C.); kasoll076@outlook.com (Z.Y.); huangjinlong@nuaa.edu.cn (J.H.)
* Correspondence: guilixu@nuaa.edu.cn

Abstract: Large-scale unmanned aerial vehicle (UAV) formations are vulnerable to disintegration under electromagnetic interference and fire attacks. To address this issue, this work proposed a distributed formation method of UAVs based on the 3 × 3 magic square and the chain rules of visual reference. Enlightened by the biomimetic idea of the plane formation of starling flocks, this method adopts the technical means of airborne vision and a cooperative target. The topological structure of the formation's visual reference network showed high static stability under the measurement of the network connectivity index. In addition, the dynamic self-healing ability of this network was analyzed. Finally, a simulation of a battlefield using matlab showed that, when the loss of UAVs reaches 85% for formations with different scales, the UAVs breaking formation account for 5.1–6% of the total in the corresponding scale, and those keeping formation account for 54.4–65.7% of the total undestroyed fleets. The formation method designed in this paper can maintain the maximum number of UAVs in formation on the battlefield.

Keywords: large-scale unmanned aerial vehicle formations; electromagnetic interference; 3 × 3 magic square; chain rules of visual reference; network connectivity; dynamic self-healing capacity

Citation: Rui, Q.; Xu, G.; Cheng, Y.; Ye, Z.; Huang, J. Simulation and Analysis of Grid Formation Method for UAV Clusters Based on the 3 × 3 Magic Square and the Chain Rules of Visual Reference. *Appl. Sci.* **2021**, *11*, 11560. https://doi.org/10.3390/app112311560

Academic Editors: Sylvain Bertrand and Hyo-sang Shin

Received: 18 October 2021
Accepted: 29 November 2021
Published: 6 December 2021

1. Introduction

In August 2018, the U.S. Department of Defense released *the Unmanned Systems Integrated Roadmap 2017–2042*, which reemphasized that the development of autonomous technology is of great importance for improving the efficiency and performance of unmanned systems as well as soldiers [1]. The development of UAVs is an essential part of studying unmanned military systems [2], of which UAV autonomous clusters have become an important direction for the future [3]. Moreover, UAV clusters have begun to play a key role in targeted attacking in the future battlefield with advantages including "defeating the most enemy with the least resources", a flexible and straightforward delivery mode, and ease of avoiding enemy's Air Defense Radar System (ADRS). With this attacking strategy, the successful attack rate can be improved because attacking UAVs require expensive and high precision strike weapons; furthermore, it is difficult for the enemy to find, defend against, and destroy UAVs quickly. Therefore, studying the stable formation method of large-scale UAV clusters has practical implications for military operation.

At present, there are five commonly used plane formation methods: the leader-follower method [4–10], the behavior-based method [11–16], the virtual structure method [17–26], the graph theory method [27–33], and the consistency method [34–45]. However, these methods do not consider the stability of UAV formation planes when they are destroyed or decoyed by the enemy on the battlefield. If the "Leading goose" UAV in the formation or a UAV on a certain critical node faces such a situation, formations using the methods above will be disrupted.

The UAV cluster formations can be disrupted by strong electromagnetic communication or enemy fire attacks. To address these issues with ideal and mature formation methods, this paper studied the bionic mechanism of the maturely evolved flocks and compared the characteristics of the classical models proposed by scholars worldwide. For example, Vicsek established an essential but straightforward cluster model—the Vicsek model (VM) [46,47]—based on the assumption that the individual field of view (FOV) is 360°, which is not realistic given that this range for most creatures is limited.

Considering the limited FOV, Tian et al. [48] established the RFVN model by upgrading the VM. The Couzin model also considered the FOV issue in studying cluster motion modeling [49]. However, the RFVN model assumes that the direction of FOV is consistent with the individual's moving direction, which is inconsistent with the actual biological perception mode. Therefore, based on the RFVN model, Calvao et al. [49] introduced the limited FOV and the strategy of random line-of-sight (LOS) to establish the Random LOSVM (RLosVM). Furthermore, based on the above models [3], Duan Haibin and Qiu Xinhua et al. proposed a fixed neighborhood region (FNR) model and a fixed number of neighbors (FNN) model according to the topological distance interaction rules of the starling movement.

In the FNN model, when one individual refers to the motion state of another in the perception range, its sight may be blocked by others in the formation, making it unable to obtain information about its neighbors effectively. After improving the FNN model, the MFNN model was built, with which individuals can dynamically sense the motion of the nearest "neighbor" in all directions. In addition, Duan Haibin and Qiu Xinhua et al. believed that the VM only considers the information of the previous moment when updating, but the individuals in the actual cluster motion have "memories". This means that the individual decision-making considers not only neighbors' information at the current time, but also previous ones. Therefore, they introduced the fractional calculus idea to the VM and established the fractional order VM (FOVM). The simulation contrast experiments on the above models found that a higher number of neighbors is not necessarily better for the interactions between individuals within a biological cluster. If there are redundancies in the perception information among individuals, the cluster motion cannot achieve faster synchronization, and the synchronised movement of the system will also be interfered with. Therefore, the reasonable distribution of neighboring individuals in space is helpful to reduce redundancies' interactions and improve the information utilization rate [3]. Furthermore, historical information also enhances the efficiency of instant decision making for individuals. However, the above ideas about biomimetic cluster formation models have not been applied to large-scale UAV formations.

In order to integrate the advantages of the VM and its improved models into a large-scale UAV formation method, this paper summarized the advantages in each model and proposed the 3×3 magic square formation method that is capable of anti-jamming and anti-deception visually. This biomimetic formation method is enlightened by the plane formation of starling flocks and is based on the chain rules for visual reference. It adopts the technical means of airborne vision and cooperative targets and possesses strong anti-electromagnetic interference and anti-deception capabilities. In addition, this formation has strong network resilience and regeneration capabilities concerning its network topological structure. With this method, the maximum number of UAVs can be kept in form on the battlefield. The main contributions of this paper are as follows:

(1) A distributed formation method for UAVs based on the 3×3 magic square and the chain rules of visual reference are proposed in this work;
(2) The biomimetic method is enlightened by the formation of starling flocks, and draws on the strengths of the Vicsek model and its refinements [3,46–49], overcoming the disadvantages of poor resilience and regeneration capabilities of the existing formation methods [4–45];
(3) Matlab simulations and the network connectivity test revealed the strong network resilience and topological regeneration capabilities of this proposed method;

(4) This proposed method will significantly improve the ability of formations to resist electromagnetic interference and destruction in the battlefield environment.

The following sections are arranged as follows: Section 2 describes the relevant formation work, such as the formation mechanism of the starling flocks, how a single UAV simulates the distribution of starling's visual sensors, and the cooperative targets' division in the fuselage. Section 3 details the proposed 3 × 3 magic square formation method and describes the matlab simulation of the 11 × 11-scale UAV grid formations. Section 4 analyzes the topological structure stability of the visual reference network based on nested loop nine-grids. Section 5 conducts the matlab simulation experiments and results analyses on different scale UAV formations on the battlefield. Section 6 is the conclusion.

2. Relevant Formation Work

Before describing the specific formation methods, we need to explain various issues, including the formation mechanism of starling flocks, the distributions of visual sensors, and cooperative targets in the UAVs, etc. These explanations will specify the pre-conditions of the proposed formation methods.

2.1. Characteristics of the Formation Mechanism of Starling Flocks

As the most widely distributed birds in the world, starlings are gregarious birds with strong imitation abilities. Thousands of starlings often fly together with a small distance between individuals, and their formations are complex and change frequently with frequent splitting and merging, enabling them to evade predators. Biologists and physicists found that, when a starling flock flies [50–52], there is a mutual reference between neighboring individuals, and each starling only interacts with the surrounding 6–7 individuals, as shown in Figure 1. In addition, scholars verified that the choice of reference neighbors is based on the topological model rather than the Euclidean geometric model, as shown in Figures 2 and 3. The position of each bird, i, and its velocity were represented by p_i and V_i, repetitively, and the dynamics model is

$$\vec{p_i}(t+1) = \vec{p_i}(t) + \vec{V_i}(t+1) \tag{1}$$

$$\vec{V_i}(t+1) = \frac{\left[\theta_i(t) + \sum_j \theta_j(t)\right]}{N_i + 1} \tag{2}$$

where N_i is the the total number of individuals that bird i can interact with.

In the Euclidean geometric model, bird i interacts with all neighboring individuals within a fixed distance $\bar{r}$, while in the topological model, bird i interacts with its n_c neighboring individual, i.e., $N_i = n_c$. The specific mathematical model is as follows:

Let $A = [a_{ij}]$ be the adjacency matrix among individuals; then, the Euclidean model is:

$$a_{ij}(t) = \begin{cases} 1 & \text{if } ||r_{ij}(t)|| \leq \bar{r} \\ 0 & \text{if } ||r_{ij}(t)|| > \bar{r} \end{cases} \tag{3}$$

where $r_{ij}(t)$ is the distance from individual i to j, and $\bar{r}$ is the distance range established for communication.

Additionally, the topological mode is:

$$a_{ij}(t) = a_{ij}(t_0)\ \forall t \geq 0,\ ||r_{ij}(t)|| \in \mathbf{R}^+ \tag{4}$$

where $a_{ij}(t_0)$ is the flag bit of the communication at the initialization time $a_{ij}.\ (t_0) = 0$ indicates no communication connection, and $a_{ij}(t_0) \neq 0$ means such a connection exists.

Second, when the predator is moving in the opposite direction to the flock and there is a vertical offset d, the predator exerts a repulsive force on each bird, which attenuates as the bird moves further away from the predator. As shown by a large number of simulated numerical experiments, under different initial conditions, the clusters of two models present different grouping probabilities after being attacked by predators. Specifically, under the Euclidean model, the flock is usually dispersed into five groups, indicating low restoration capacity of the model. In contrast, it is highly possible for flocks to maintain a complete group under the topological model, and the original group is not easily dispersed, showing strong cohesion. Therefore, it is concluded that when flocks of starlings fly in nature, the choice of reference neighbors is not based on the Euclidean geometric model, but on the topological model [50].

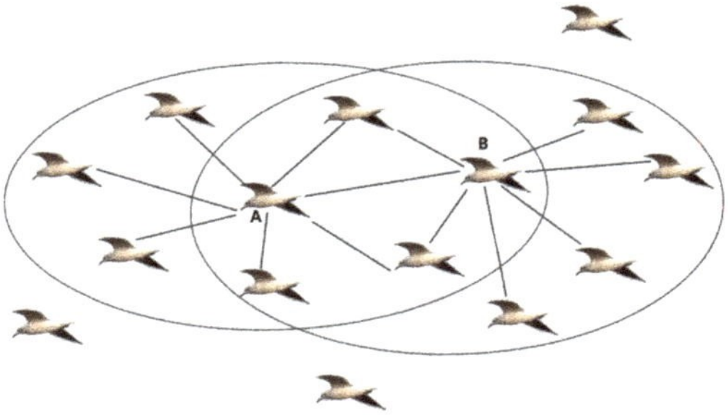

Figure 1. Visual reference diagram of starlings A and B in a formation.

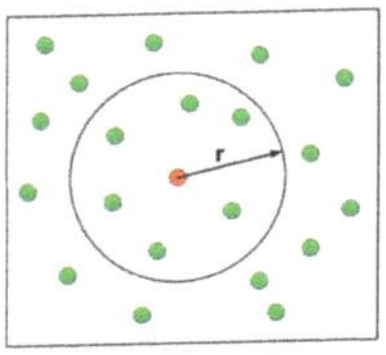

Figure 2. Euclidean model.

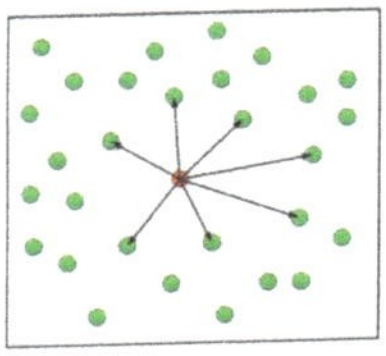

Figure 3. Topological model.

When starlings fly in flocks, the plane direction of the entire formation is integrated. Specifically, the direction and speed of individual movements are initially haphazard, but through continued local interactions between individuals, they eventually fly in the same direction and speed as the movement of the entire flock. The Φ-order parameter is generally used to characterize the synchronization index for the motion direction of all individuals in the starling cluster system. The formula is as follows:

$$\Phi = \| \frac{1}{N} \sum_{i=1}^{n} \frac{\overrightarrow{V_i}}{\|V_i\|} \| \quad (5)$$

where V_i represents the speed of the ith starling, and N denotes the total number of the entire flock. The value of Φ will be zero if each starling flies in a different direction and speed; conversely, it will be close to one if most starlings fly in the same direction. Scholars analyzed 24 starling flocks and found that their flight direction has global orderliness [51]. When the perception is uncertain, interacting with the neighboring 6–7 starlings is an optimal choice to balance the cohesion of the flock and individual cost. The plane status of starling flocks can change correlatively: the plane state change of a single starling will affect all other individuals in the entire flock, regardless of the flock size.

2.2. The Distribution of Visual Sensors and Cooperative Targets in UAVs Based on the Bionics of Starlings

As the whole plane formation system is based on the formation principle of starling flocks, each UAV in the fleet shall have a similar visual function as a single starling. The compared architecture between starling flocks and UAV fleets is shown in Figure 4:

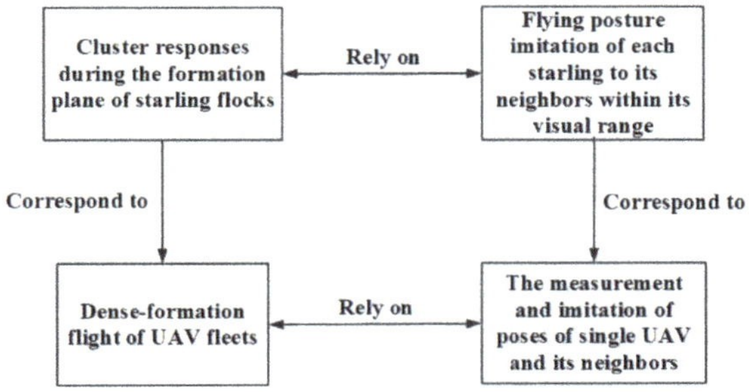

Figure 4. Compared architecture between starling flocks and UAV fleets.

To enable the UAV to observe the flying posture of its surrounding UAVs as starlings do, each UAV was equipped with visual sensors and high-precision ranging sensors on the left side, right side, directly behind and in front (these items of equipment are not necessarily on the directly above and below orientations because the plane formation was conducted on a single plane). For a more visual indication of the orientation, we give a top view of the FOV distribution of a 3 × 3 size UAV formation in Figure 5. As can be seen, there are eight basic directional positions (see details in Figure 6a) determined by the inertial navigation equipment. The flying postures on these positions can be observed by the two sensors equipped. For example, the UAVs numbered 1, 6, 7, 2, 9, 4, 3, and 8 locate the 8th, 1st, 2nd, 3rd, 4th, 5th, 6th, and 7th directions of the No.5 UAV, respectively.

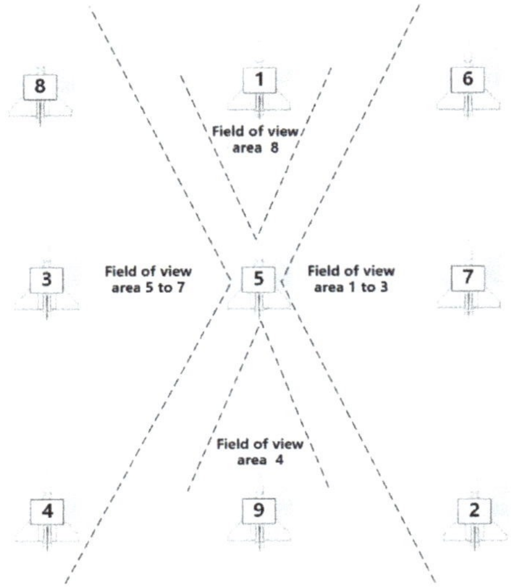

Figure 5. Corresponding directions diagram for the visual range of a single UAV in a 3 × 3 magic square.

These directions were fixed after the UAV joined the formation. No matter how the UAV turned during flying, the eight directions would always remain the initial state (as shown in Figure 6b), so that each UAV can obtain a fixed reference versus the surrounding UAVs. At the same time, the vision system of the UAV can collect the signal conditions of cooperative signal lamps located on the UAV surface in different directions (as shown in Figure 7), thus determining the flying posture of a referenced UAV in each direction. Each UAV can also collect the real-time flying distance between the referenced UAVs and itself together with the high-precision ranging sensors.

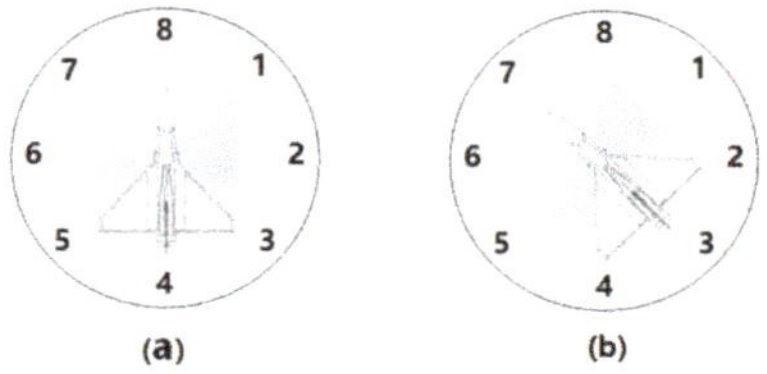

Figure 6. 8 Basic directions: (**a**) Schematic diagram of the eight directional positions of the UAV in initial formation; (**b**) Schematic diagram of the eight directional positions of the UAV after turning.

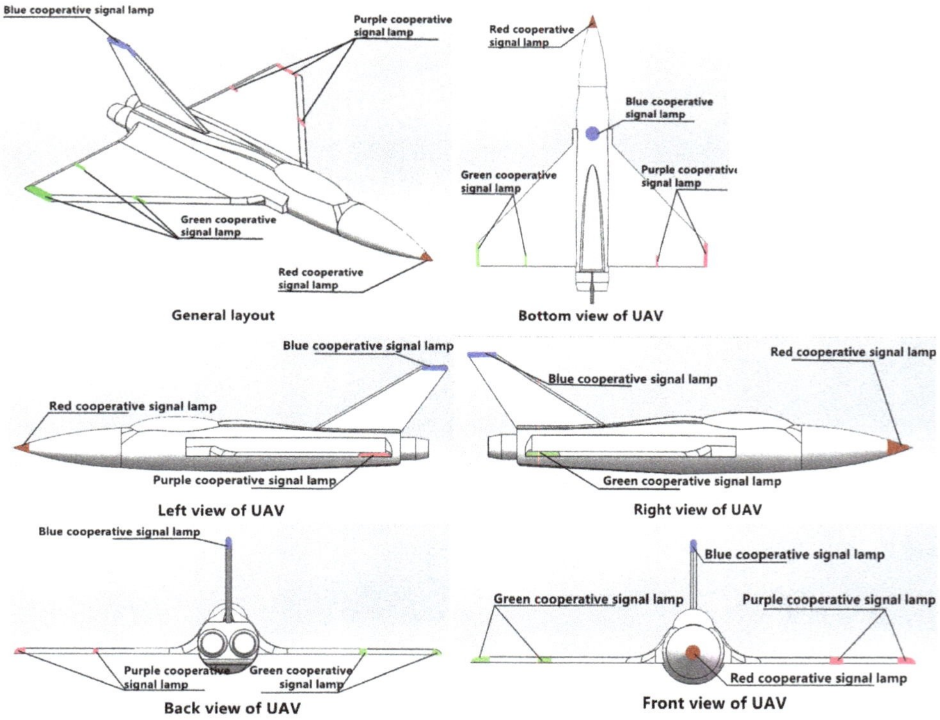

Figure 7. Distribution of the cooperative signal lamps located on the UAV surface in different directions.

3. Formation Methods and Simulation

Based on the work above, this chapter elaborates the distributed formation method based on the 3 × 3 magic square and the chain rules of visual reference. Using the method, the advantages of the VM and its improved models are integrated into the large-scale UAV cluster formation, so that the VM's redundant neighborhood information can be avoided in its formation. Notably, this method is characterized by a more stable neighborhood information collection than the RLosVM model and the memory function of the FOVM model. In addition, the dynamic visual reference in the FNN model has been improved to enhance the formation's anti-jamming and anti-deception capacity.

First, the formation was divided into two areas, kept at a certain distance to be anti-jamming. One was the unformatted UAV area, and the other was the formatted area. The involved UAVs could fly freely in the first area and at a random position outside the formatted area. When entering the formatted area, UAVs have their designated routes until arriving at the terminal. However, the routes of all UAVs were constrained by the grid formation, in which each UAV in flight maintained a certain distance, the same altitude and the same speed between them, using airborne distance sensors and their vision system. Based on the 3×3 magic square and the chain rules of visual reference, the vision system determines which drones in which directional positions can be referenced to guide the formation.

3.1. Distributed Formation Method Based on the 3×3 Magic Square and the Chain Rules of Visual Reference

For the formatted areas, a suppositional 3×3 magic square grid was set. The size of the square varied according to the scale of formation. Each square was marked with a number to show its position. For instance, Figure 8 is a typical 3×3 magic square diagram.

8	1	6
3	5	7
4	9	2

Figure 8. 3×3 magic square formation code.

When the first UAV entered the formatted area, the very place it arrived was the square numbered 5, as shown in Figure 9. Afterward, the second UAV flew from the unformatted area to the square numbered 1.

	1	
	5	

Figure 9. 3×3 magic square formation.

As mentioned above, the visual sensor of each UAV could sense 8 basic directions in the same plane (Figure 6). Thus, the eight directions of UAVs in grids 5 and 1 are shown in Figure 10.

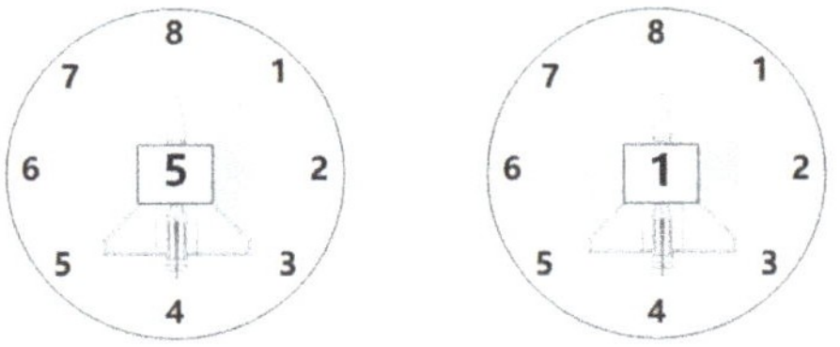

Figure 10. Eight directions of UAVs in square 5 and square 1.

According to Figures 9 and 10, the UAV in square 1 was in direction 8 of the UAV in square 5, whose airborne visual sensor identified the cooperation signal of the UAV in square 1. Thus, the poses of the UAV in square 1 could be obtained. The UAV in square 1 could offer reference to that in square 5 in direction 8. Similarly, the UAV in square 5 was in direction 4 of the UAV in square 1, whose airborne visual sensor identified the cooperation signal of the UAV in square 5. Therefore, the poses of the UAV in square 5 could be obtained. The UAV in square 5 was set as the reference for the UAV in square 1 in its direction 4. Similarly, through this visual cross-reference, UAVs could be formatted in other parts of the 3 × 3 magic square.

After the formation, a visual reference topological structure diagram of the 3 × 3 magic square was formed, as shown in Figure 11, where node numbers of the square referred to individual UAVs, and the lines between nodes showed the visual reference among UAVs.

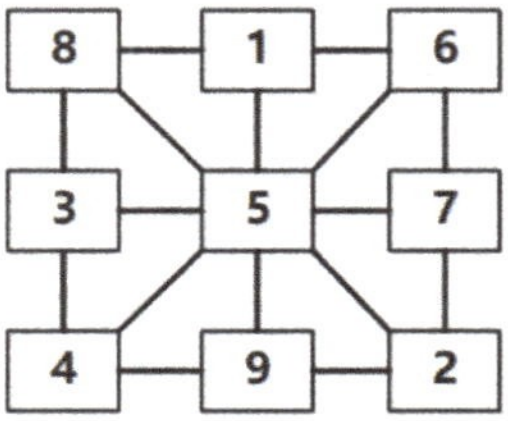

Figure 11. A visual reference topological structure diagram of the 3 × 3 magic square.

According to the 3 × 3 magic square agreement and chain rules of visual reference, UAVs to be referred must meet two prerequisites. First, the numbers of UAVs and their reference must be in the same line in the topological structure diagram. Second, in the same line, there must be three nodes in that direction, with each of their numbers adding up to be 15. With these two prerequisites, UAVs at the nodes could be viewed as references. For instance, in Figure 11, UAVs at square 8 would refer to UAVs in square 1 and square 6 in direction 2, UAVs in square 5 and square 2 in direction 3, and UAVs in square 3 and square 4 in direction 4. In these three reference directions (2, 3, and 4), the sum of numbers in the three nodes was 15, satisfying the 3 × 3 magic square agreement and chain rules of visual reference. Thus, UAVs at square 8 could refer to squares 1, 6, 5, 2, 3, and 4. UAVs at square 3 could refer to squares 8, 4, 5, and 7. Similarly, we could get the reference for UAVs at other squares based on this principle. For example, 6 UAVs could be the reference for UAVs at squares 2, 4, 6, and 8, 8 for UAVs at square 5, and 4 for UAVs at square 1, 3, 7, and 9.

3.2. Visual Reference Topological Structure Diagram of the Nesting 3 × 3 Magic Squares

To expand the scale of the UAV formation, we expand the magic square by nesting under the exact mechanism of the first 3 × 3 magic square (circling the black dotted bordered rectangle in Figure 12). 3 × 3 magic squares were nested, forming a 7 × 7 magic square formation.

2	9	4	9	2	9	4
7	5	3	5	7	5	3
6	1	8	1	6	1	8
7	5	3	5	7	5	3
2	9	4	9	2	9	4
7	5	3	5	7	5	3
6	1	8	1	6	1	8

Figure 12. 7 × 7-scale nested magic square formation.

For the convenience of studying the formation of UAV clusters, the formation structure after each expansion should be in line with magic squares. For different scale square arrays, the grid numerical codes can be described by the following Equations (6) and (7):

$$n_5 = (2n+1)^2, (n = 0, 1, 2, \ldots) \tag{6}$$

$$M = [3 + 2(\sqrt{n_5} - 1)] \tag{7}$$

where M refers to the number of clusters and n_5 refers to the number of 3×3 magic squares.

Based on the above formation mechanism and the above equations, we could achieve $11 \times 11, 15 \times 15, \ldots$ expanded UAV formations. The expanded versions were more complex than the topologies of 3×3 magic squares, whose nesting structures made UAV formation more closely related, enhancing the formation stability. For instance, in the 7×7 visual reference topological structure diagram of UAV formation, UAVs at square 4 in the red dotted bordered rectangle (Figure 13) satisfied the 3×3 magic square agreement and the chain rules of visual reference, as shown in Figure 14. According to the 3×3 magic square agreement and chain rules of visual reference, the UAV at square 4 in the red-dotted bordered rectangle could refer to UAVs at squares 5, 9, 5, 3, 5, 9, 5, and 3 (as marked by the blue dashed box in Figure 13) in direction 1–8 as well as squares 6, 2, 6, 8, 6, 2, 6, and 8 (as marked by the green dashed box in Figure 13) in direction 1–8 of the extended nodes. In total, there are 16 UAVs in line with the prerequisites of UAVs for reference, as shown in Figure 14. If they were destroyed, the UAV at square 4 in the red dotted bordered rectangle would be out of the formation.

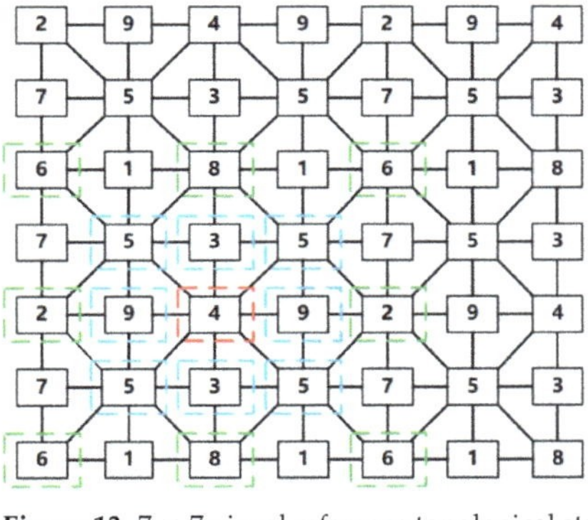

Figure 13. 7×7 visual reference topological structure diagram of UAV formation.

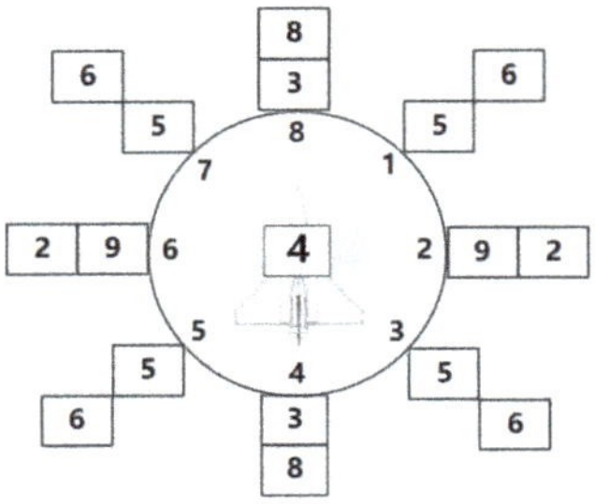

Figure 14. Reference for UAV at square 4.

Similarly, for the UAV at square 1 (as marked by the red dashed box in Figure 15), 4 UVAs meeting the 3×3 magic square agreement and the chain rules of visual reference, as shown in Figure 16, respectively, were at neighboring squares 8, 6, and 5 (as marked by the blue dashed box in Figure 15) in direction 2, 6, and 8, as well as square 9 (as marked by the green dashed box in Figure 15) in direction 8 of the extended node. Without these 4 UAVs for reference, the UAV at square 1 will be out of formation. It could be seen that nodes with fewer reference UAVs were located at the margin of the formation. Such is the case of Figure 15, where the UAV at square 1 in the dotted bordered rectangle was in an individual 3 × 3 magic square without nested relation with others.

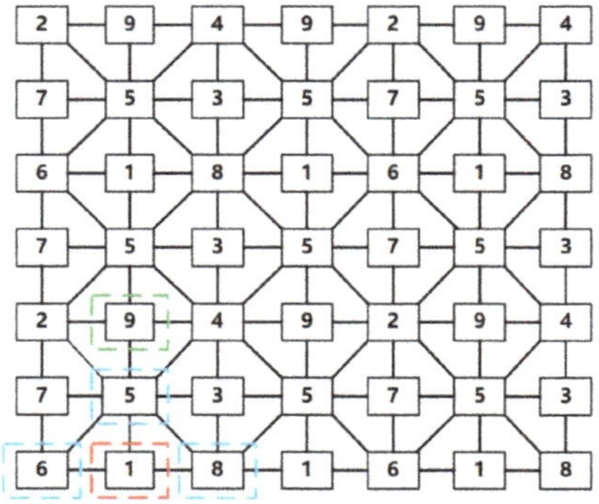

Figure 15. A visual reference topological structure diagram of a 7 × 7 nested magic square UAV formation.

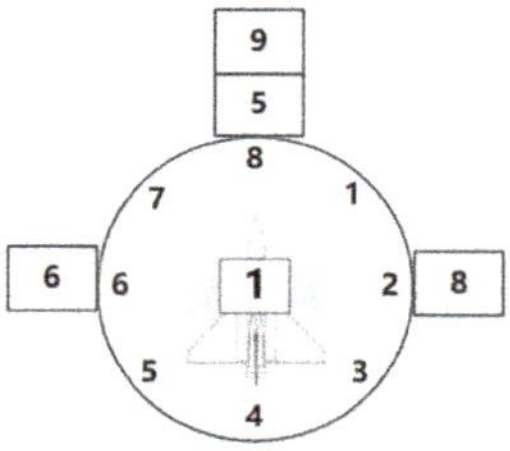

Figure 16. Reference for the UAV at square 1.

3.3. 11 × 11 Matlab Simulations of UAVs Magic Square Formation

3.3.1. UAV Model

In real UAVs with different model parameters, there are multiple aerodynamic configurations, causing the variance of mathematical modeling. To simplify the algorithm of upper control, we suppose that the UAV internal-loop is controlled by autopilot. Thus, the model could be built with the UAV position and velocity external-loop model as the upper control algorithm, as shown in Figure 17.

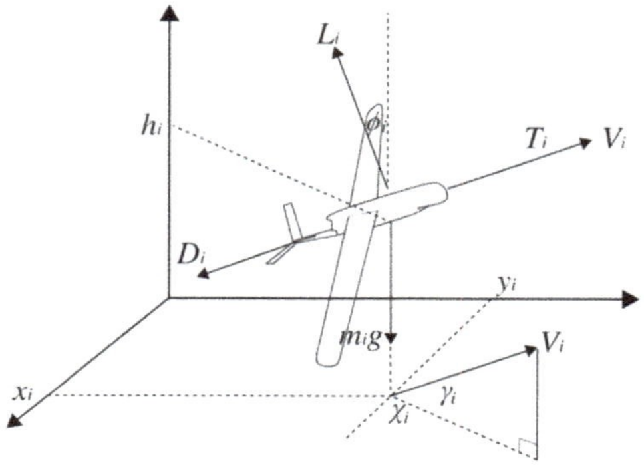

Figure 17. UAV model.

The mathematical model of the UAV_i is expressed as Equation (8)

$$\begin{cases} \dot{x}_i = V_i \cos\gamma_i \cos\chi_i \\ \dot{y}_i = V_i \cos\gamma_i \sin\chi_i \\ \dot{h}_i = V \sin\gamma_i \\ \dot{V}_i = \dfrac{T_i - D_i}{m_i} - g\sin\gamma_i \\ \dot{\gamma}_i = \dfrac{L\cos\Phi_i - m_i g\cos\gamma_i}{m_i V_i} \\ \dot{\chi}_i = \dfrac{L_i \sin\Phi_i}{m_i V_i \cos\gamma_i} \end{cases} \tag{8}$$

where $i = 1, \cdots, N$. x_i, y_i, and h_i correspond to the down-range, cross-range, and altitude displacement. V_i refers to the airspeed of UAV_i, γ_i is the plane path angle, and χ_i represents the heading angle. T_i is the engine thrust, D_i refers to drag, m_i is the quality of UAV_i, and g represents the gravity acceleration. Furthermore, L_i refers to lift, and Φ_i is the bank angle.

Equation (9) can be achieved with the transformation of the mathematical model.

$$\begin{cases} \ddot{x}_i = u_{xi} \\ \ddot{y}_i = u_{yi} \\ \ddot{z}_i = u_{hi} \end{cases} \tag{9}$$

u_{xi}, u_{yi}, and u_{hi} are the subjunctive control input, and the transformation relationship between the executive order and subjunctive control input can be expressed as Equation (10),

$$\begin{cases} \Phi_i = \arctan\left(\dfrac{u_{yi}\cos\chi_i - u_{xi}\sin\chi_i}{(u_{hi}+g)\cos\gamma_i - (u_{xi}\cos\chi_i + u_{yi}\sin\chi_i)\sin\gamma_i}\right) \\ L_i = m_i \dfrac{(u_{hi}+g)\cos\gamma_i - (u_{xi}\cos\chi_i + u_{yi}\sin\chi_i)\sin\gamma_i}{\cos\Phi_i} \\ T_i = m_i[(u_{hi}+g)\sin\gamma_i + (u_{xi}\cos\chi_i + u_{yi}\sin\chi_i)\cos\gamma_i] + D_i \end{cases} \tag{10}$$

where $\tan(\chi_i) = \dot{y}_i/\dot{x}_i$, and $\sin(\gamma_i) = \dot{h}_i/V_i$. Therefore, the subjunctive control input is designed as Equation (9), and the real input of the UVA could be calculated through Equation (10), which can be expressed as the state place:

$$\begin{cases} \dot{z}_i = Az_i + Bu_i \\ p_i = C_p z_i \\ v_i = C_v z_i \end{cases} \tag{11}$$

where $z_i = [{p_i}^T, v_i]^T$, p_i refers to the position vector, v_i is the speed vector, and $u_i = [u_{xi}^T, u_{yi}^T, u_{hi}^T]^T$ shows the subjunctive control input.

$$A_i = \begin{bmatrix} 0 & 1 \\ 0 & 0 \end{bmatrix} \otimes I_3, B_i = \begin{bmatrix} 0 \\ 1 \end{bmatrix} \otimes I_3, C_p = \begin{bmatrix} 1 & 0 \end{bmatrix} \otimes I_3, C_v = \begin{bmatrix} 0 & 1 \end{bmatrix} \otimes I_3 \tag{12}$$

$I_3 \in R^{3\times3}$ refers to the identity matrix, and $\otimes$ is the Kronecker product.

In Equation (10), the air resistance D_i can be expressed as Equation (13).

$$D_i = 0.5\rho(V_i - V_{wi})^2 SC_{D0} + \frac{2k_d k_n^2 L^2}{\rho(V_i - V_{wi})^2 Sg^2} \tag{13}$$

where ρ refers to the air density, C_{D0} represents the zero-lift drag coefficient, V_{wi} refers to gust, S is the wing area, k_d is the induced drag, and k_n refers to the load-factor effectiveness.

The mathematical modeling of gust can be expressed as Equation (14).

$$\begin{cases} V_{wi} = \overline{V}_{wi} + \delta V_{wi} \\ \overline{V}_{wi} = 0.215V_m \log_{10}(h_i) + 0.285V_m \end{cases} \quad (14)$$

where $\overline{V}_{wi}$ is normal wind shear, V_m refers to the mean wind speed and δV_{wi} is the wind gust turbulence. The zero mean equals 0, and the standard deviation was $0.9V_m$ for this Gaussian random variable.

3.3.2. Design of UAV Controller

Through an algorithm based on the 3 × 3 magic square grid, which was illustrated in Section 3, the expected position p_{di} and expected speed of every UAV_i could be calculated. Thus, the controller form of individual UAVs can be expressed as Equation (15).

$$u_i = k_p(p_{di} - p_i) + k_d(q_{di} - q_i) \quad (15)$$

where $k_p > 0$ and $k_d > 0$ are parameters of UAV PID controllers.

The values of each item in simulations are as shown in Table 1.

Table 1. Settings of UAV Parameters.

Symbol	Value	Unit
m_i	20	kg
g	9.81	kg/m^2
ρ	1.225	kg/m^3
S	1.37	m^2
C_{D0}	0.02	Non-dimensional
k_d	0.1	Non-dimensional
k_n	1	Non-dimensional
V_m	4	m/s (at $h_i = 80$ m)
T_i	$[0, 125)$	N
L_i	$(-294.3, 392.4)$	N
Φ_i	$[-80, 80]$	N
χ_i	$[-180, 180]$	deg
γ_i	$[-90, 90]$	deg

3.3.3. Simulations of Scale UAV Grid Formations

Considering different scales of nested magic squares, this study will not illustrate them one by one. However, they share the same formation rule and topological structure, so the 11 × 11-scale UAV grid formation (121 UAVs) was used as an example. Its simulation results are as shown in Figures 18–20.

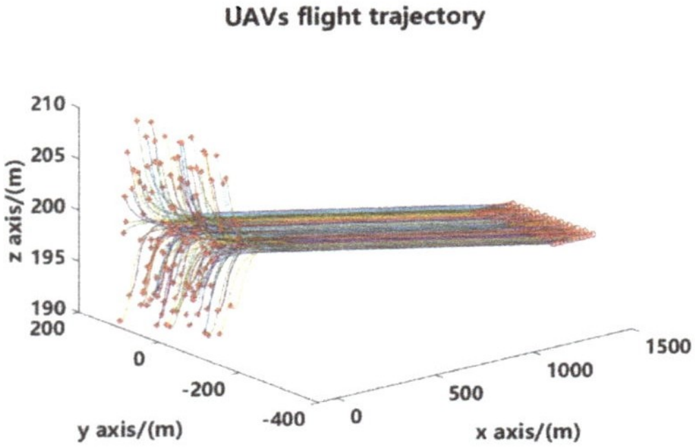

Figure 18. 121 UAVs' flight trajectories.

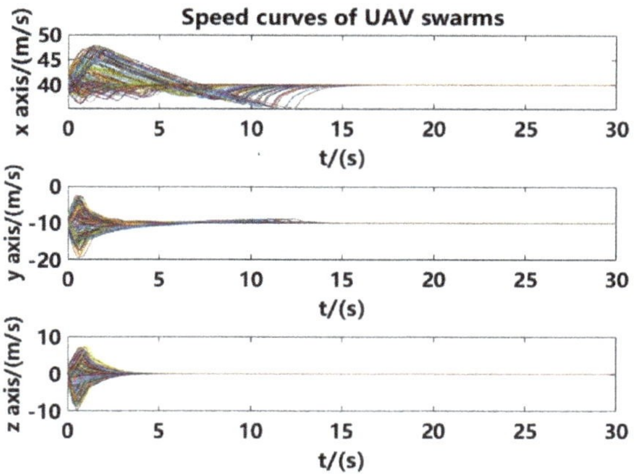

Figure 19. Speed curves of UAV swarms.

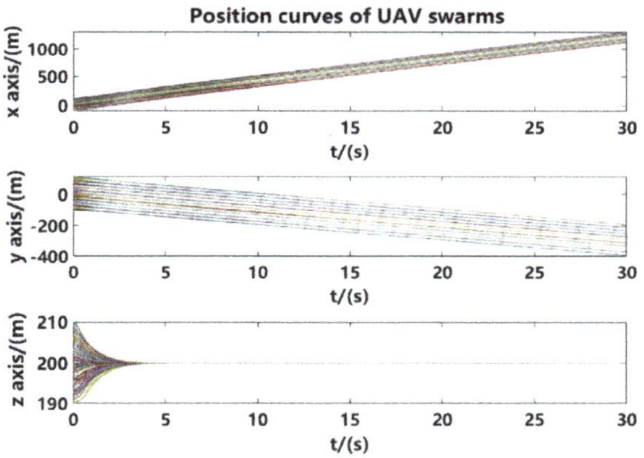

Figure 20. Position curves of UAV swarms.

According to the UAV flying trajectory in the simulation results, it could be concluded that the UAV cluster initially moved from the unformatted sector to the formatted one. In addition, the initial flying orientation was along the x axis. From the curve graph, the cluster converged to 200 m in height within 5 s and soon entered the formatted sector. Based on the speed graph, all UAVs achieved uniform convergence in axis x, y, and z at 15 s, when the curve graph of controller output, controller input, and executive output achieved convergence. Thus, it can be seen that the 121 UAVs in that formation generally realized convergence in speed and completed the formation in 15 s. This formation is large in scale, stable in plane, and swift in convergence compared with other formations.

4. An Analysis on the Stability of the Visual Reference Topological Structure

In this chapter, the network connectivity index of graph theory was introduced to analyze the static stability of the visual reference topological structure of the nested magic squares. Meanwhile, a detailed description of the self-healing dynamic visual reference grid of UAV formations will be given based on the principle and argument mentioned in this study.

In this analysis, only nodes with close relations would be taken into account. For instance, in Figure 21, the UAV at square 3 in the red-dotted bordered rectangle could only refer to the UAVs at squares 4, 5, 8, and 5 in directions 8, 2, 4, and 6, respectively. If these nodes were destroyed, the UAV at square 3 would be out of the topology.

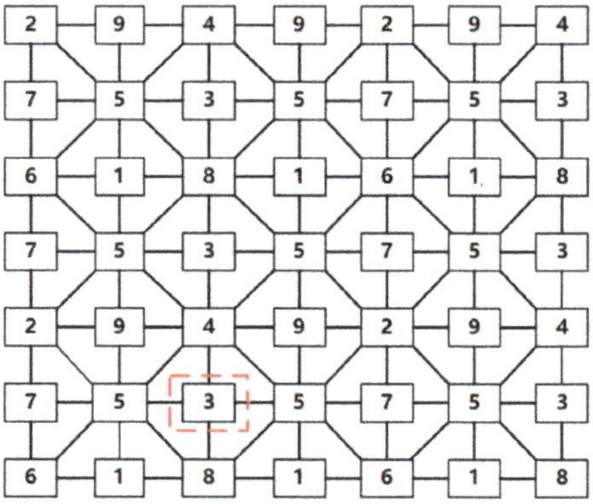

Figure 21. A visual reference topology diagram of nested 7 × 7 magic squares.

In this analysis, its basic concepts include network connectiveness, network resistance to destruction, network cutpoint, network vertex cutpoint, minimum vertex cutpoint, vertex impact, network impact, and network connectivity. Their specific definitions are given as follows:

Definition 1 (network connectiveness). *In the network G(V,E), if there is a path from vertex v to v′, the two vertexes are connected. If for every pair of vertexes (v_i,v_j ∈ V) in the network G(V,E), v_i and v_j are connected, then G is connected.*

Definition 2 (network resistance to destruction). *Several vertexes or chains should be destroyed to impede the connectivity of certain vertexes. The cohesion strength and connectivity degree are often used to show the resistance to destruction.*

Definition 3 (network cutpoint). *In the network G(V,E), if, for vertex v, its connected lines are deleted, the connected component of the network will be divided into two or more connected components. The vertex v will be called a cutpoint of G.*

Definition 4 (network vertex cutpoint). *In the network G(V,E), suppose V′⊆V; if G-V′ are disconnected, V′ will be called G's cutpoint or vertex cutpoint. The vertex cutpoint with k vertexes will be called the k vertex cutpoint.*

Definition 5 (minimum vertex cutpoint). *In the network G(V,E), the vertex with the least points is called G's minimum vertex cutpoint.*

Definition 6 (vertex impact). *In the network G(V,E), suppose that d_i (i = 1,2,· · · , n). For the degrees of vertex v_i, the vector $L = (\frac{1}{d_1}, \frac{1}{d_2}$; then, $\cdots, \frac{1}{d_n})$ is called the vertex impact, showing the influence of vertexes on adjacent ones.*

Definition 7 (network impact). *In the network G(V,E), suppose A is the adjacent matrix of network G, and D is the vector showing the impact degree between adjacent vertexes. The network impact can be expressed as P = D·A, which indicates the influence of other vertexes on the network G.*

Definition 8 (network connectivity). *G(V,E) is an n-order connected network. If vertex cutpoints exist at G, the point of G's minimum vertex cutpoint is called its connectivity. Otherwise, $n - 1$ will be its connectivity. In other words, the sub-graph is still connected after $k - 1$ vertexes are eliminated in a network with n vertexes ($1 \leq k \leq n - 1$). However, when k vertexes are removed, the graph will be disconnected or become a trivial graph. In this way, k refers to the connectivity of G, expressed as k(G) = k.*

4.1. Calculation of Network Connectivity in the Undirected Topological Diagram

To calculate the network connectivity of the visual topological diagram for different scale UAV clusters, we adopted the algorithm mentioned in the Reference [53], which is more straightforward than the traditional algorithm. The flow chart of the algorithm is as shown in Figure 22. Condition: Suppose that G has n vertexes v_i ($i = 1, 2, \cdots, n$); then, the adjacent matrix is $C = (c_{ij})_{n \times n}$. If v_i and v_j are adjacent, $c_{ij} = 1$. Otherwise, $c_{ij} = 0$. Here, d_i refers to the degree of the vertex v_i.

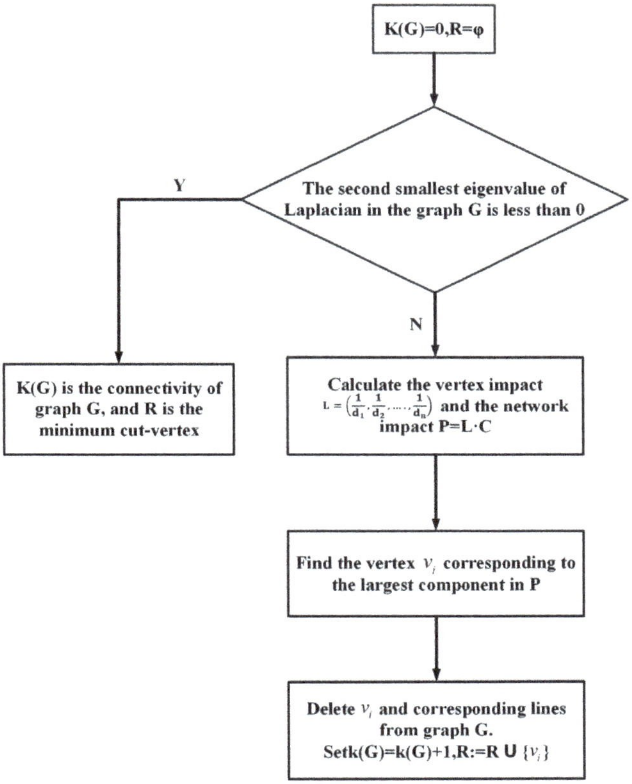

Figure 22. The algorithm flow chart of the undirected network connectivity.

4.2. Matlab Simulations of Network Connectivity of Nested Magic Squares' Topological Structure under Different-Scale UAV Formations

This chapter employed the matlab simulation of the network connectivity of the topological structure from 3 × 3 to 83 × 83 nested magic squares formation according to the connectivity algorithm. The regression curve equation of the connectivity was concluded, as shown in Table 2, and Figure 23.

Table 2. Network connectivity values of different-scale UAV formations.

UAV Cluster Number	Network Connectivity Value
9	3
49	11
121	20
225	33
361	37
529	46
729	51
961	61
1225	70
1521	75
1849	85
2209	96
2601	100
3025	109
3481	115
3969	124
4489	135
5041	140
5625	150
6241	156
6889	166

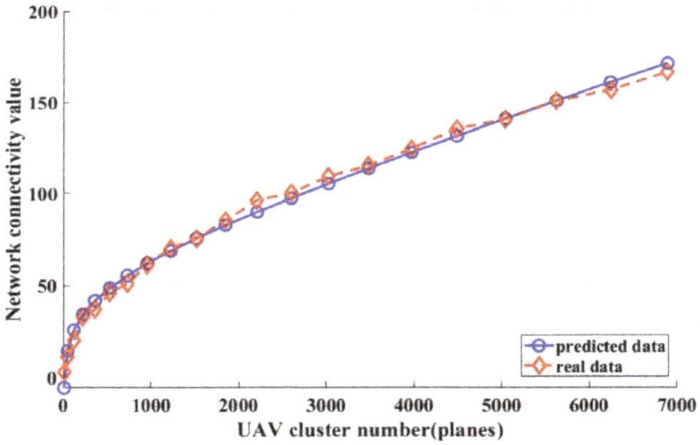

Figure 23. Connectivity regression curve of different-scale UAV clusters.

This study used the least square method to establish the regression model curve via the network values of different-scale network topological diagrams, as shown in Figure 23. Thus, the network connectivity values of topological structures of any scale nested magic squares can be calculated.The regression model curve equation is:

$$K(G) = -30.7292 - 0.0146(M_n) + 26.3306\log_{10}(M_n) \quad (16)$$

where $K(G)$ represents the network connectivity index and M_n refers to the UAV cluster number.

According to the simulation results, the 95% confidence intervals of the gradients were [0.0131, 0.0161] and [21.9585, 30.7027], and the 95% confidence interval of the intercepts was [−41.4515, −20.0070]. The intercepts and gradients of the regression model curve equation satisfied the requirement. The network connectivity index of the visual reference topological diagram had an R2 variance-explained rate of 0.9937, proving the significance test of the regression equation with excellent fitting.

According to the fitted curve equation, the cluster accelerated in expanding, but the network connectivity increased rather slowly. However, the network connectivity is an index to evaluate the trivial graph formed after deleting k nodes in the network topological diagram. Thus, applying nested magic squares' network topological diagrams into large-scale formations could help to greatly enhance the stability of UAV formations. The simulation results show that, at a formation size of 961 UAVs, the resulting visual reference network topology subgraph is still connected after the loss of a random 60 UAVs.

4.3. Dynamic Self-Healing of Grid Formation Based on the 3 × 3 Magic Square and the Chain Rules of Visual Reference

We calculated the network connectivity and concluded that the topological structure of nested magic squares has relatively high static stability. Still, the formation based on the 3 × 3 magic square and the chain rules of visual reference could lead to better stability. For instance, the UAV at square 4 in the dotted rectangle in Figure 13 has 16 planes that satisfy the reference principle, as shown in Figure 14. If the adjacent UAVs at squares 5, 9, 5, 3, 5, 9, 5, and 3 (UAVs in the blue dashed box) in directions 1–8 were destroyed due to fire attacks, the UAV at square 4 could seek reference from 8 UAVs (UAVs in the green dashed box) in its periphery. In this way, the formation could be maintained, and the regenerated topological structure diagram is shown in Figure 24. The general visual reference topology graph changes, but the UAV at square 4 in the dotted rectangle will be kept in the formation. Therefore, the formation based on the 3 × 3 magic square and the chain rules of visual reference not only has great stability but enjoys dynamic self-healing ability.

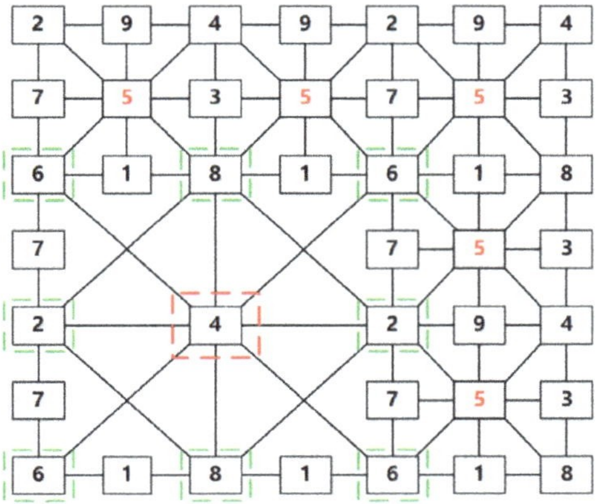

Figure 24. Regenerated topological structure diagram of 7 × 7 nested magic squares formation.

5. Simulations and Analysis in Battlefields

5.1. The Procedure of Matlab Simulations of UAV Formations in Battlefields

To evaluate the survival rate of a formation based on the 3 × 3 magic square and the chain rules of visual reference in battlefields, we used matlab to simulate the attacks on UAV formation in battlefields. There are six premises of the simulation experiments. First, different-scale UAV clusters will enter the enemy region and will be attacked after the formation. Second, once the grid formation is completed, all UAVs' plane height, speed, and relative distance will remain unchanged until they reach the destination. Third, each fire attack on UAVs has a random aim and is completed once it is exerted. The number of UAVs to be destroyed can be set before simulation. Fourth, UAVs out of the formation are those which lose all reference planes in the grid formation. Fifth, surviving UAVs are those which are not destroyed and for which there is at least one reference UAV. Sixth, the UAV clusters will not defend or dodge, so the stability of the formation in worst-case scenarios can be obtained. Figure 25 is the flow chart of the detailed simulation.

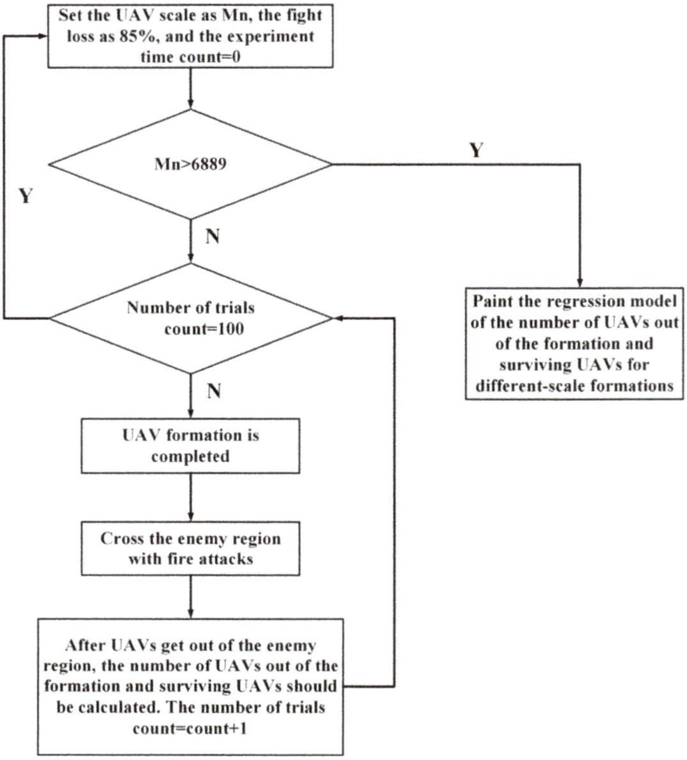

Figure 25. Flow chart of simulation experiments of attacking the UAV formation in battlefields.

Although the number of drones set to be destroyed is the same, there will be some variation in the number of drones out of formation as the aimed destructed areas were randomly set. For this reason, the simulation experiments were conducted 100 times with the same fight loss for the same-scale formation to obtain the average values of the UAVs which were out of formation and those which survived. Next, this study simulated the 3×3 to 83×83 grid formations and calculated the number of UAVs out of formation and surviving UAVs at 85% fight loss.

5.2. The Procedure of Matlab Simulations of UAV Formations in Battlefields

To test and verify the survival rate of formations with nested magic square topological structures based on the 3×3 magic square and the chain rules of visual reference in battlefields, we adopted matlab simulations to obtain the regression curve of UAVs out of the formation and surviving UAVs in different-scale formations with the fight loss set at 85%. These values can be expressed in the following equation:

$$H_n = M_n - D_n - Is_o \tag{17}$$

where H_n is the remaining UAVs, M_n refers to the UAVs before entering the battlefield, D_n represents the total destructed UAVs, and Is_o stands for the undestroyed UAVs that get out of formation.

The simulation results are as shown in Table 3 and Figure 26.

Table 3. Simulation results of UAV formations with 85% fight loss.

UAV Formation Scale (Planes)	UAVs Out of Clusters (Planes)	Surviving UAV Clusters (Planes)
49	3	4
121	7	10
225	12	21
361	20	34
529	29	50
729	39	71
961	51	94
1225	62	122
1521	79	150
1849	96	182
2209	116	216
2601	136	255
3025	156	298
3481	179	344
3969	204	392
4489	230	444
5041	258	499
5625	291	553
6241	320	617
6889	355	679

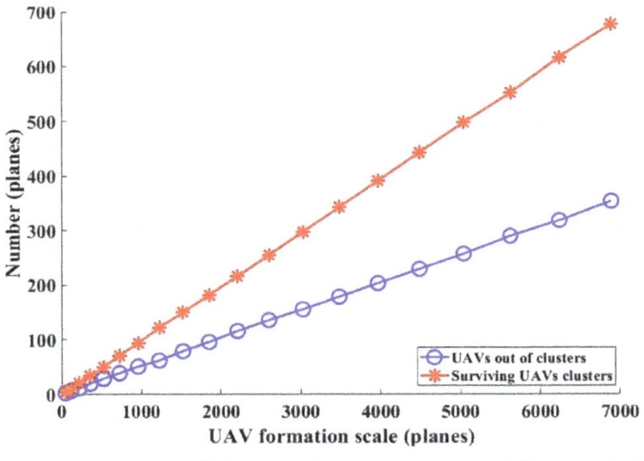

Figure 26. UAVs out of clusters and surviving UAVs in different-scale formations with the fight loss set at 85%.

The least square method was adopted to make the curve fitting simulations of UAVs out of clusters and surviving UAVs clusters with 85% fight loss, and the results are as shown in Figures 27 and 28.

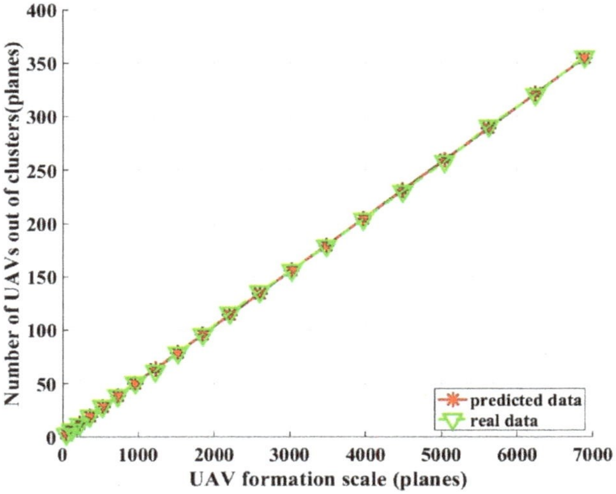

Figure 27. The regression model of the number of UAVs out of clusters under 85% fight loss in different-scale formations.

The regression model equation of UAVs out of clusters can be expressed as:

$$R_a = 0.0512M_n + 1.1267 \tag{18}$$

where R_a is the number of UAVs out of clusters, and M_n refers to the number of clusters.

According to the simulation results, the 95% confidence interval of gradients in the curve model was [0.0510, 0.0515], and that of intercepts was [0.3620, 1.8914], so the intercepts and gradients of the regression model curve equation satisfied the requirement. The R2 variance explained rate was 0.9999, proving the significance test of the regression equation with excellent fitting.

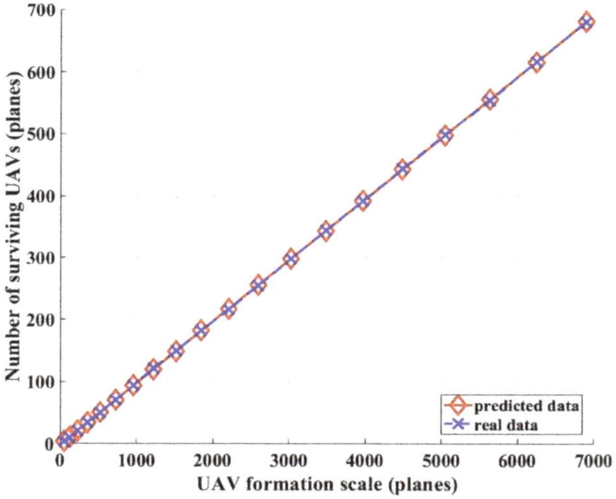

Figure 28. The regression model of the number of surviving UAVs under 85% fight loss in different-scale formations.

The regression model equation of UAVs out of clusters can be expressed as:

$$R_b = 0.0989M_n - 1.1704 \tag{19}$$

R_b is the number of surviving UAVs, and M_n refers to the number of clusters.

According to the simulation results, the 95% confidence interval of gradients in the curve model was [0.0987, 0.0992], and the confidence interval of intercepts was [−1.9811, −0.3596], so the intercepts and gradients of the regression model curve equation satisfied the requirement. The R2 variance explained rate was 0.9999, proving the significance test of the regression equation with excellent fitting.

Based on the simulation results, in the 20 different-scale formation clusters based on the mentioned method, even when the fight loss accounts for 85% in each formation, only 5.1–6% UAVs would be out of the formation. In the remaining 15% undestroyed clusters, 54.4–65.7% of the surviving UAVs could continue fighting.

6. Conclusions

This study proposed a UAV formation method based on a 3 × 3 magic square and the chain rules of visual reference. The formation mainly adopted visual references in diverse directions, which greatly enhanced its anti-electromagnetic interference ability and the regeneration capacity of topological structures. Matlab simulations of real fights showed that when the fight loss of different-scale formations reached 85%, 5.1–6% of UAVs would be out of the formation. More importantly, in the remaining 15% undestroyed clusters, 54.4–65.7% of the surviving UAVs could continue fighting. The simulation results verified that the formation of this study has faster convergence and a larger scale in formation. Moreover, with the expansion of formation scales, the network resistance to destruction increases, leading to a higher survival rate of UAVs to maintain the formation.

Moreover, the simulation experiments were conducted without defensive measures. Otherwise, combat losses would be significantly reduced if the UAV clusters fire weapons at the enemy or have interception or attack capabilities. The formation approach in this study can provide some insight into future large-scale UAV formations for military use.

Author Contributions: Conceptualization, R.Q., G.X. and Y.C.; Methodology, R.Q.; Resources, G.X. and Y.C.; Software, R.Q., Z.Y. and J.H.; formal analysis, R.Q.; Writing—original draft, R.Q.; Writing—review and editing, G.X. and R.Q. All authors have read and agreed to the published version of the manuscript.

Funding: This work was supported in part by the National Key Research and Development Plan under Grant 2018YFB2003803, and in part by the National Natural Science Foundation of China under Grant 62073161, 61905112 and U1804157.

Institutional Review Board Statement: Not applicable.

Informed Consent Statement: Not applicable.

Data Availability Statement: Not applicable.

Acknowledgments: This work was supported in part by National Key Research and Development Plan under Grant 2018YFB2003803, and in part by the National Natural Science Foundation of China under Grant 62073161, 61905112 and U1804157. The authors would like to thank Cheng Yuehua's team, for useful discussions.

Conflicts of Interest: The authors declare no conflict of interest.

References

1. Fahey, K.; Miller, M. *Unmanned Systems Integrated Roadmap 2017–2042*; Office of the Secretary of Defense: Washington, DC, USA, 2018.
2. Lu, J.W.; Wang, Q.-W. Review on Evolution and development of UAV. *Aerodyn. Missile J.* **2017**, *11*, 45–48. 68. [CrossRef]
3. Duan, H.; Qiu, H. *Unmanned Aerial Vehicle Swarm Autonomous Control Based on Swarm Intelligence*, 1st ed.; Science Press: Beijing, China, 2018; p. 12.

4. Desai, J.P.; Ostrowski, J.; Kumar, V. Controlling formations of multiple mobile robots. In Proceedings of the 1998 IEEE International Conference on Robotics and Automation (Cat. No. 98CH36146), Leuven, Belgium, 20–20 May 1998; Volume 4, pp. 2864–2869. [CrossRef]
5. Turpin, M.; Michael, N.; Kumar, V. Trajectory design and control for aggressive formation flight with quadrotors. *Auton. Robot.* **2012**, *33*, 143–156. [CrossRef]
6. Saska, M.; Baca, T.; Thomas, J.; Chudoba, J.; Preucil, L.; Krajnik, T.; Faigl, J.; Loianno, G.; Kumar, V. System for deployment of groups of unmanned micro aerial vehicles in GPS-denied environments using onboard visual relative localization. *Auton. Robot.* **2017**, *41*, 919–944. [CrossRef]
7. Nägeli, T.; Conte, C.; Domahidi, A.; Morari, M.; Hilliges, O. Environment-independent formation flight for micro aerial vehicles. In Proceedings of the 2014 IEEE/RSJ International Conference on Intelligent Robots and Systems, Chicago, IL, USA, 14–18 September 2014; pp. 1141–1146. [CrossRef]
8. Ghamry, K.A.; Dong, Y.; Kamel, M.A.; Zhang, Y. Real-time autonomous take-off, tracking and landing of UAV on a moving UGV platform. In Proceedings of the 2016 24th Mediterranean conference on control and automation (MED), Athens, Greece, 21–24 June 2016; pp. 1236–1241. [CrossRef]
9. Aghdam, A.S.; Menhaj, M.B.; Barazandeh, F.; Abdollahi, F. Cooperative load transport with movable load center of mass using multiple quadrotor UAVs. In Proceedings of the 2016 4th International Conference on Control, Instrumentation, and Automation (ICCIA), Qazvin, Iran, 27–28 January 2016; pp. 23–27. [CrossRef]
10. Liu, H.; Wang, X.; Zhu, H. A novel backstepping method for the three-dimensional multi-UAVs formation control. In Proceedings of the 2015 IEEE International Conference on Mechatronics and Automation (ICMA), Beijing, China, 2–5 August 2015; pp. 923–928. [CrossRef]
11. Reif, J.H.; Wang, H. Social potential fields: A distributed behavioral control for autonomous robots. *Robot. Auton. Syst.* **1999**, *27*, 171–194. [CrossRef]
12. Jadbabaie, A.; Lin, J.; Morse, A.S. Coordination of groups of mobile autonomous agents using nearest neighbor rules. *IEEE Trans. Autom. Control* **2003**, *48*, 988–1001. [CrossRef]
13. Kim, S.; Kim, Y.; Tsourdos, A. Optimized behavioural UAV formation flight controller design. In Proceedings of the 2009 European Control Conference (ECC), Budapest, Hungary, 23–26 Auguat 2009; pp. 4973–4978. [CrossRef]
14. Song, Y.Z.; Yang, F.F. On Formation Control Based on Behavior For Second-order Multi-agent System. *Control Eng. China* **2012**, *19*, 687–690. [CrossRef]
15. Shin, J.; Kim, S.; Suk, J. Development of robust flocking control law for multiple UAVs using behavioral decentralized method. *J. Korean Soc. Aeronaut. Space Sci.* **2015**, *43*, 859–867. [CrossRef]
16. Qiu, H.-X.; Duan, H.-B.; Fan, Y.-M. Multiple unmanned aerial vehicle autonomous formation based on the behavior mechanism in pigeon flocks. *Control Theory Appl.* **2015**, *32*, 1298–1304. [CrossRef]
17. Lewis, M.A.; Tan, K.H. High precision formation control of mobile robots using virtual structures. *Auton. Robot.* **1997**, *4*, 387–403. [CrossRef]
18. Beard, R.W.; Lawton, J.; Hadaegh, F.Y. A coordination architecture for spacecraft formation control. *IEEE Trans. Control Syst. Technol.* **2001**, *9*, 777–790. [CrossRef]
19. Olfati-Saber, R.; Murray, R.M. Distributed structural stabilization and tracking for formations of dynamic multi-agents. In Proceedings of the 2002 41st IEEE Conference on Decision and Control, Las Vegas, NV, USA, 10–13 December 2002; Volume 1, pp. 209–215. [CrossRef]
20. Ren, W.; Beard, R.W. Formation feedback control for multiple spacecraft via virtual structures. *IEE Proc.-Control Theory Appl.* **2004**, *151*, 357–368. [CrossRef]
21. Ren, W.; Beard, R.W. Decentralized scheme for spacecraft formation flying via the virtual structure approach. *J. Guid. Control Dyn.* **2004**, *27*, 73–82. [CrossRef]
22. Yang, E.; Masuko, Y.; Mita, T. Dual controller approach to three-dimensional autonomous formation control. *J. Guid. Control Dyn.* **2004**, *27*, 336–346. [CrossRef]
23. Lalish, E.; Morgansen, K.A.; Tsukamaki, T. Formation tracking control using virtual structures and deconfliction. In Proceedings of the 45th IEEE Conference on Decision and Control, San Diego, CA, USA, 13–15 December 2006; pp. 5699–5705. [CrossRef]
24. Li, N.H.; Liu, H.H. Formation UAV flight control using virtual structure and motion synchronization. In Proceedings of the 2008 American Control Conference, Seattle, WA, USA, 11–13 June 2008; pp. 1782–1787. [CrossRef]
25. Cai, D.; Sun, J.; Wu, S. *AsiaSim 2012: UAVs Formation Flight Control Based on Behavior and Virtual Structure*; Xiao, T., Zhang, L., Fei, M., Eds.; Springer: Berlin/Heidelberg, Germany, 2012; pp. 429–438. [CrossRef]
26. Askari, A.; Mortazavi, M.; Talebi, H. UAV formation control via the virtual structure approach. *J. Aerosp. Eng.* **2015**, *28*, 04014047. [CrossRef]
27. Laman, G. On graphs and rigidity of plane skeletal structures. *J. Eng. Math.* **1970**, *4*, 331–340. [CrossRef]
28. Hendrickx, J.M.; Anderson, B.D.; Delvenne, J.C.; Blondel, V.D. Directed graphs for the analysis of rigidity and persistence in autonomous agent systems. *Int. J. Robust Nonlinear Control IFAC-Affil. J.* **2007**, *17*, 960–981. [CrossRef]
29. Barca, J.C.; Sekercioglu, A.; Ford, A. Controlling Formations of Robots with Graph Theory. In *Intelligent Autonomous Systems 12*; Lee, S., Cho, H., Yoon, K.J., Lee, J., Eds.; Springer: Berlin/Heidelberg, Germany, 2013; pp. 563–574. [CrossRef]

30. Zhang, P.; de Queiroz, M.; Cai, X. Three-Dimensional Dynamic Formation Control of Multi-Agent Systems Using Rigid Graphs. *J. Dyn. Syst. Meas. Control* **2015**, *137*, 111006. [CrossRef]
31. Ramazani, S.; Selmic, R.; De Queiroz, M. Multiagent layered formation control based on rigid graph theory. In *Control of Complex Systems*; Elsevier: Amsterdam, The Netherlands, 2016; pp. 397–419. [CrossRef]
32. Luo, X.Y.; Shao, S.K.; Zhang, Y.Y.; Li, S.B.; Guan, X.P.; Liu, Z.X. Generation of minimally persistent circle formation for a multi-agent system. *Chin. Phys. B* **2014**, *23*, 614–622. [CrossRef]
33. Li, S.; Shao, S.-K.; Guan, X.-P.; Zhao, Y.-J. Dynamic generation and control of optimally persistent formation for multi-agent systems. *Acta Autom. Sin.* **2013**, *39*, 1431–1438. [CrossRef]
34. Murray, R.M.; Olfati-Saber, R. Consensus problems in networks of agents with switching topology and time-delays. *IEEE Trans. Autom. Control* **2004**, *49*, 1520–1533. [CrossRef]
35. Ren, W.; Beard, R.W.; McLain, T.W. Coordination Variables and Consensus Building in Multiple Vehicle Systems. In *Cooperative Control: A Post-Workshop Volume 2003 Block Island Workshop on Cooperative Control*; Kumar, V., Leonard, N., Morse, A.S., Eds.; Springer: Berlin/Heidelberg, Germany, 2005; pp. 171–188. [CrossRef]
36. Ren, W.; Beard, R.W. Consensus of information under dynamically changing interaction topologies. In Proceedings of the 2004 American Control Conference, Boston, MA, USA, 30 June–2 July 2004; Volume 6, pp. 4939–4944. [CrossRef]
37. Ren, W.; Beard, R.W.; Atkins, E.M. Information consensus in multivehicle cooperative control. *IEEE Control Syst. Mag.* **2007**, *27*, 71–82. [CrossRef]
38. Seo, J.; Ahn, C.; Kim, Y. Controller Design for UAV Formation Flight Using Consensus Based Decentralized Approach. In Proceedings of the AIAA Infotech@Aerospace Conference, Seattle, WA, USA, 6–9 April 2009; pp. 1–11. [CrossRef]
39. Jamshidi, M.; Gomez, J.; Jaimes, B.; Aldo, S. Intelligent control of UAVs for consensus-based and network controlled applications. *Appl. Comput. Math.* **2011**, *10*, 35–64.
40. Kuriki, Y.; Namerikawa, T. Consensus-based cooperative formation control with collision avoidance for a multi-UAV system. In Proceedings of the 2014 American Control Conference, Portland, OR, USA, 4–6 June 2014; pp. 2077–2082. [CrossRef]
41. Kuriki, Y.; Namerikawa, T. Formation control with collision avoidance for a multi-UAV system using decentralized MPC and consensus-based control. *SICE J. Control Meas. Syst. Integr.* **2015**, *8*, 285–294. [CrossRef]
42. Li, S.; Wang, X. Finite-time consensus and collision avoidance control algorithms for multiple AUVs. *Automatica* **2013**, *49*, 3359–3367. [CrossRef]
43. Xing, G.-S.; Du, C.-Y.; Zong, Q.; Chen, H.-Y. Consensus-based distributed motion planning for autonomous formation of miniature quadrotor groups. *Control Decis.* **2014**, *29*, 2081–2084. [CrossRef]
44. Zong, Q.; Shao, S. Decentralized finite-time attitude synchronization for multiple rigid spacecraft via a novel disturbance observer. *ISA Trans.* **2016**, *65*, 150–163. [CrossRef]
45. Kownacki, C. Multi-UAV Flight using Virtual Structure Combined with Behavioral Approach. *Acta Mech. Autom.* **2016**, *10*, 92–99. [CrossRef]
46. Vicsek, T.; Czirók, A.; Ben-Jacob, E.; Cohen, I.; Shochet, O. Novel type of phase transition in a system of self-driven particles. *Phys. Rev. Lett.* **1995**, *75*, 1226–1229. [CrossRef]
47. Czirók, A.; Vicsek, M.; Vicsek, T. Collective motion of organisms in three dimensions. *Phys. A Stat. Mech. Its Appl.* **1999**, *264*, 299–304. [CrossRef]
48. Tian, B.M.; Yang, H.X.; Li, W.; Wang, W.X.; Wang, B.H.; Zhou, T. Optimal view angle in collective dynamics of self-propelled agents. *Phys. Rev. E* **2009**, *79*, 052102. [CrossRef]
49. Calvao, A.M.; Brigatti, E. The role of neighbours selection on cohesion and order of swarms. *PLoS ONE* **2014**, *9*, e94221. [CrossRef]
50. Ballerini, M.; Cabibbo, N.; Candelier, R.; Cavagna, A.; Cisbani, E.; Giardina, I.; Lecomte, V.; Orlandi, A.; Parisi, G.; Procaccini, A.; et al. Interaction ruling animal collective behavior depends on topological rather than metric distance: Evidence from a field study. *Proc. Natl. Acad. Sci. USA* **2008**, *105*, 1232–1237. [CrossRef] [PubMed]
51. Cavagna, A.; Cimarelli, A.; Giardina, I.; Parisi, G.; Santagati, R.; Stefanini, F.; Viale, M. Scale-free correlations in starling flocks. *Proc. Natl. Acad. Sci. USA* **2010**, *107*, 11865–11870. [CrossRef]
52. Young, G.F.; Scardovi, L.; Cavagna, A.; Giardina, I.; Leonard, N.E. Starling flock networks manage uncertainty in consensus at low cost. *PLoS Comput. Biol.* **2013**, *9*, e1002894. [CrossRef] [PubMed]
53. Sun, X.-J.; Liu, S.-Y.; Wang, Z.-Q. New algorithm for solving connectivity of networks. *Comput. Eng. Appl.* **2009**, *45*, 82–84. [CrossRef]

applied sciences

MDPI

Article

Machine Learning Approach to Real-Time 3D Path Planning for Autonomous Navigation of Unmanned Aerial Vehicle

Abera Tullu [1], Bedada Endale [2], Assefinew Wondosen [2] and Ho-Yon Hwang [3,*]

1 Department of Aerospace Engineering, Sejong University, 209, Neungdong-Ro, Gwangjin-Gu, Seoul 05006, Korea; tuab@sejong.ac.kr
2 Department of Aerospace Engineering, Pusan National University, Busan 46241, Korea; endale@pusan.ac.kr (B.E.); wondebly@pusan.ac.kr (A.W.)
3 Department of Aerospace Engineering, and Convergence Engineering for Intelligence Drone, Sejong University, Seoul 05006, Korea
* Correspondence: hyhwang@sejong.edu; Tel.: +82-10-6575-2282

Abstract: The need for civilian use of Unmanned Aerial Vehicles (UAVs) has drastically increased in recent years. Their potential applications for civilian use include door-to-door package delivery, law enforcement, first aid, and emergency services in urban areas, which put the UAVs into obstacle collision risk. Therefore, UAVs are required to be equipped with sensors so as to acquire Artificial Intelligence (AI) to avoid potential risks during mission execution. The AI comes with intensive training of an on-board machine that is responsible to autonomously navigate the UAV. The training enables the UAV to develop humanoid perception of the environment it is to be navigating in. During the mission, this perception detects and localizes objects in the environment. It is based on this AI that this work proposes a real-time three-dimensional (3D) path planner that maneuvers the UAV towards destination through obstacle-free path. The proposed path planner has a heuristic sense of A^* algorithm, but requires no frontier nodes to be stored in a memory unlike A^*. The planner relies on relative locations of detected objects (obstacles) and determines collision-free paths. This path planner is light-weight and hence a fast guidance method for real-time purposes. Its performance efficiency is proved through rigorous Software-In-The-Loop (SITL) simulations in constrained-environment and preliminary real flight tests.

Keywords: vision-based navigation; cluttered environment; three-dimensional path planner; obstacle avoidance; machine learning

Citation: Tullu, A.; Endale, B.; Wondosen, A.; Hwang, H.-Y. Machine Learning Approach to Real-Time 3D Path Planning for Autonomous Navigation of Unmanned Aerial Vehicle. *Appl. Sci.* **2021**, *11*, 4706. https://doi.org/10.3390/app11104706

Academic Editors: Hyo-sang Shin and Sylvain Bertrand

Received: 10 February 2021
Accepted: 1 May 2021
Published: 20 May 2021

1. Introduction

The cost-effectiveness, ease of access, and mission versatility are the primary compelling qualities of UAVs that attract many aerospace and related sectors. Hence, UAVs are being integrated into tasks such as package delivery, first aid, law enforcement, disaster management, infrastructure inspection, agriculture mechanization, rescue, military intelligence, and many more. As low-altitude aerial vehicles, however, UAVs often encounter obstacles such as trees, mountains, high storey buildings, electric poles, and so on during their missions. Therefore, these aerial vehicles should be equipped with sensors to perceive the environment around them and avoid potential dangers.

To leverage the use of UAVs in cluttered environments, studies have been conducted on the types and ways of integrating various sensors for autonomous navigation. Vehicle localization is one of the pillars of autonomous navigation. In an open-air space, Global Positioning System (GPS) is often used for UAV localization. However, GPS-based UAV localization in cluttered environment is unreliable. In such environment, sensors onboard the UAV are used for localization as well as collision avoidance. Ivan Konovalenko et al. [1] fused inputs from visual camera and Inertial Navigation System (INS) to localize a UAV. Based on computer simulation, the team analyzed various approaches to vision-based UAV position estimation. Jinling Wang et al. [2] combined inputs from GPS, INS, and vision sensors to autonomously navigate UAVs. In their report, the inclusion of GPS input reduces vision-based UAV localization errors and hence enhances the accuracy of navigation. Jesus Garcia et al. [3] presented a methodology of assessing the performance of sensors fusion for autonomous flight of UAVs. Their methodology systematically analyzes the efficiency of input data for accurate navigation of UAVs.

Computer vision technology has evolved over the years to the stage that enables not only UAV localization but also obstacle detection and avoidance. This is realized through the advent of high-performance computers with the ability to process data and perform complex calculations at high speeds. With the promising progress in computer vision technology, many vision-based navigation algorithms have been developing. A comprehensive review of computer vision algorithms and their implementations for UAVs' autonomous navigation was presented by Abdulla Al-Kaff [4]. Lidia et al. [5] provided a detailed analysis on the implementation of computer vision technologies for navigation, control, tracking, and obstacle avoidance of UAVs. Wagoner et al. [6] also explored various computer vision algorithms and their capabilities to detect and track a moving object such as a UAV in flight.

Alongside computer vision technology, Artificial Intelligence (AI) is being implemented into UAVs navigation system to enable them to acquire humanoid perception. The idea is to train the computer that is either onboard a UAV or integrated with ground-based command system so that it takes control of UAV navigation with little to no human intervention. Su Yeon Choi and Dowan Cha [7] reviewed the historical development of AI and its implementation to UAVs with a particular focus on UAVs control strategies and object recognition for autonomous flight of UAVs. They also considered machine-learning-based UAV path planning and navigation methods.

The integration of AI and computer vision technology brings a remarkable importance in civilian application of UAVs. Many challenging tasks such as wildlife monitoring, disaster managment, and search and rescue are being addressed by UAVs equipped with AI and computer technology. Luis F. Gonzalez et al. [8] reported how AI- and computer-vision-enabled UAVs have solved the challenges of wildlife monitoring. The study reported by Christos and Theocharis [9] reflects the importance of UAVs equipped with AI and computer vision for autonomous monitoring of disaster-stricken areas. Eleftherios et al. [10] combined AI with a computer vision system onboard a UAV to enable real-time human detection during search and rescue operations.

The integration of the two aforementioned key technologies—AI and computer vision—provides environment acquaintance to UAVs. This helps the UAVs to plan their collision-free paths. For autonomous navigation, a UAV has to have either a predetermined path or a capacity to plan a path in real-time. A mission with predetermined route requires less number of sensors as compared to a mission with real-time path planning. The challenges with real-time path planning are the complexity of multiple sensors integration, input data synchronization, and computational burdens thereof. Valenti et al. [11] developed techniques to enrich a UAV with capabilities of localizing itself and autonomously navigate in a GPS-denied environment. In their report, stereo cameras on-board the UAV-based vision data were used for UAV localization and to build a 3D map of the surroundings. Based on this information, an improved A^* path-planning algorithm was implemented for autonomous navigation of the UAV collision-free along the shortest path to the goal.

System-resource-intensive computational burdens on the companion computer onboard a UAV is always a setback to real-time path planning for the UAV. The companion computer has to deal with visual data processing for UAV localization, obstacle detection, and path planning. A comprehensive literature review on vision-based UAV localization, obstacle avoidance, and path planning was reported by Yuncheng et al. [12]. In their study, the challenges of acquiring real-time data processing for safe navigation of the UAV are reflected. They also reported the challenges of autonomous navigation of a UAV due to intensive computation and high storage consumption of 3D map of the surroundings. Yan et al. [13] developed a computer-simulation-based deep reinforcement learning technique towards real-time path planning for UAV in dynamic environments. Although this is a promising step towards real-time path planning in dynamic environments, the assumption of predetermined global situational data and the absence of real flight test that verifies the efficiency the technique may degrade its attention.

To ease the computational burden on a companion computer dedicated to UAV localization, obstacle detection, and 3D path planning, we propose the integration of the fastest object detection algorithm with a light-weight 3D path planner that relies on few obstacle-free points to generate a 3D path. The proposed 3D path planner is based on AI acquired through YOLO (You Only Look Once), which is the fastest object detection algorithm.

The study presented in this report is organized into sections. In Section 2, the problem to be addressed in this study is stated and the implemented methodology is explained. In Section 3, the overall descriptions of the implemented hardware and software components and their configurations are given. The machine learning approach for object detection is explained in Section 4. Then, the commonly known 3D path planning algorithms are discussed with their advantages and disadvantages in Section 5. The developed real-time 3D path planner is detailed in this section, followed by its performance tests in Section 6. Results and discussion are given in the final Section 7.

2. Problem Statement

The challenge in autonomous navigation of a UAV in urban environment is recognizing and localizing obstacles at the right time and continuously adjusting the path of the UAV in such a way that it can avoid the obstacles and navigate to the destination safely. To this end, it requires integrating effective object detection and path planning algorithms that run on a companion computer onboard the UAV.

Most of the widely used object detection algorithms are based on scanning the entire environment and discretizing the scanned region to create a dense mesh of grid points from which objects are detected. This process requires a companion computer with high storage capacity and intensive computational power. Moreover, the well-known path-planning algorithms either randomly sample or exhaustively explore the entire consecutive obstacle-free grid points to generate optimal path towards destination. This incurs additional computational burden on companion computer and compromises the real-timeness of the navigation commands . Liang et al. [14] conducted a comprehensive review on the most popular 3D path planing algorithms. In their review, a detailed analysis of the advantages and disadvantages of these commonly used algorithms is given. They reported that despite the intensive applications of these algorithms, the problem of real-time path planning in a cluttered environment remains unsolved.

Dai et al. [15] proposed light-weight CNN-based network structure for both object detection and safe autonomous navigation of a UAV in indoor/outdoor environments. However, the whole process of object detection and UAV path planning was performed on a ground-based computer and communication with the UAV was through a Wifi connection. This had a catastrophic drawback on the safe navigation of the UAV in indoor environment where Wifi connection failure is likely. Moreover, the Wifi data transfer rate may create a delay in navigation commands to be sent to the UAV. In an attempt to remove the dependency of the UAV on ground-based commands, Juan et al. [16] proposed a UAV framework for autonomous navigation in a cluttered indoor environment based on companion computer on-board the UAV. The performance of this framework was validated through hardware-in-the-loop simulation, and it appears to be promising to put an end to ground-based navigation command. However, an occupancy map of the cluttered environment in which the UAV navigated was pre-loaded on the companion computer. This undermines the applicability of the framework in dynamic environment.

To avoid computational burden on the companion computer, Antonio et al. [17] applied a data-driven approach, where data about the cluttered environment must be collected prior to the UAV mission. As proposed in their work, DroNet makes use of the collected data and safely navigates a UAV in the streets of a city. However, this approach, again, has limitations when it comes to dynamic or unknown environments.

This study, therefore, tends to address the challenges of computational burden subjected to companion computer onboard a UAV by integrating the available fastest object detection algorithm and the proposed light-weight real-time 3D path planner. Such an approach by-passes the challenges of dynamic or unknown environments. In the preliminary performance test, we assumed limited number of objects: pedestrian, window, electric poles, tunnel, trees, and barely visible nets as plausible obstacles that the UAV may encounter in a disaster monitoring scenario. Once the proposed 3D path planner is validated in a complete real-flight tests, further objects will be included in the machine learning process.

Methodology

To enable a companion computer onboard a UAV for simultaneous object detection and 3d path planning in real-time, it is essential to integrate the fastest object detection algorithm and 3D path planner that requires less computational burden. YOLO, as explained in Section 4.1, is selected as the fastest object detection algorithm. In addition to object detection, this algorithm also localizes the object(s). The proposed 3D path planner relies on the relative locations of the detected objects to calculate a collision-free path for the UAV. Although the proposed 3D path planner resembles A^* path planning algorithm in implementing heuristic function for cost minimization, it avoids an exhaustive search for consecutive collision-free nodes and storage method of A^*. Unlike A^*, the proposed 3D path planner maps the current location of the UAV to a few nodes between consecutive obstacles. These few nodes are determined based on the size of the UAV and the gap between consecutive obstacles, as explained in Sections 5.1.1 and 5.1.2. A Euclidean function is used as a heuristic function in this 3D path planner.

Prior to a real flight test, the performance of the proposed 3D path planner must be checked in a simulated environment. For this performance test, software tools are essential components. One of the software tools specifically designed for such task is Gazebo 3D dynamic environment simulator. This software was primarily designed to evaluate algorithms for robots [18] and provides realistic rendering of the environment in which the robot navigates. Moreover, it is enriched by various types of simulated sensors. We designed a simulated cluttered 3D environment in Gazebo and used it to test performance of the proposed 3D path planner during its successive development.

3. Utilized Tools and Their Integration

Various open-source software was implemented in both the gazebo-based simulation and real flight tests for the development and validation of the 3D path planner. The type and implementations of this software are explained in the following two subsections.

3.1. Setup for Software-In-The-Loop Simulation

It is very common that Software-In-The-Loop (SITL) simulation is often used for testing the performance of an algorithm under development. This utility saves time and cost of repair of probable crashes in real flight test scenarios.

Open-source software such as px4 flight control firmware, Gazebo simulator, and Robot Operating System (ROS) were integrated and used for the development and performance testing of 3D path planner. Gazebo is a dynamic 3d model simulation environment particularly suitable for obstacle avoidance and computer vision. This simulation environment is enriched with simulated sensors that mimic the real sensors on-board the UAVs. YOLO object detector, with its Darknet architecture wrapped with ROS, was also used to publish information about obstacles in the UAV's navigation environment. For training and validation of YOLO, images of the 3D models of the objects simulated in the gazebo simulation environment were taken. The Images were taken under various backgrounds and lighting conditions.

The 3D path planner algorithm that prompts px4 flight controller to send actuator commands to quadcopter model in gazebo simulator was developed as an ROS node. Hardware models implemented in this SITL Gazebo simulation are iris quadcopter, depth stereo camera, three ultrasonic sensors, and LiDAR as shown in Figure 1.

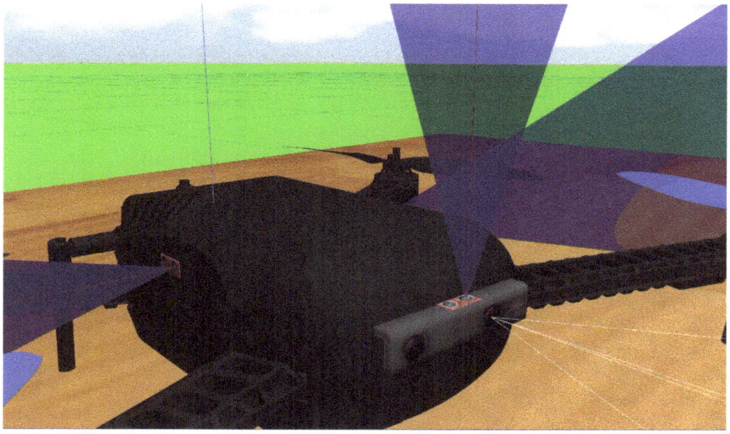

Figure 1. SITL: quadcopter equipped with on-board components.

The camera is for the frontal environment's image input, LiDAR is for quadcopter's altitude estimation in combination with GPS, and the ultrasonic sensors are used to detect lateral obstacles that may be encountered during takeoff and rolling. The 3D path planner acquires information from the aforementioned sensors in Gazebo simulator using Gazebo_ros packages that enables sensors to publish their information. The whole process runs on the desktop computer, whose software specifications are given in Table 1.

Table 1. Desktop Computer specification and software used for simulation.

Type	Specification
Operating System	Ubuntu 16.04
Memory	2 GB
Processor	intel i7 CPU 972@2.67 GHz x 8
Graphics	NV92
Gazebo	version 7 with its dependencies
ROS	Kinetic with dependencies
PX4 firmware	version 1.9.2

3.2. *Setup for Real-Flight Based Performance Test*

Following SITL simulation-based performance validation, the 3D path planner was uploaded onto NVIDIA Xavier companion computer. The computer was integrated with Pixhawk 4 autopilot on-board Tarot 650 quadcopter platform. The platform components and their specifications are given in Table 2.

Table 2. Tarot quadcopter components specifications.

Parameter/Item	Specification
Frame weight	750 g
Motor to motor length	600 mm
Payload weight	1665.5 g
4 motors	MN4006-23 KV: 380 T-motor
4 propellers	13 × 5.5 Carbon Prop
Battery	Poly-Tronics 14.8 V, 10,000 mAh
Electronic Speed Controller	Arris Simonk 30 A

Hardware components used for real autonomous navigation are shown in Figure 2a. The Tarot quadcopter was equipped with a forward-looking ZED mini stereo camera, downward-looking LiDAR sensor, and upward-, right-, and left-looking ultrasonic sensors. The tasks of these hardware are as mentioned in the SITL simulation counterpart. The integration of the quadcopter, mounted sensors, and companion computer is shown in Figure 3.

The autopilot board is Pixhawk 4, which is mounted on the quadcopter underneath the companion computer: NVIDIA Xavier. The companion computer and LiDAR are connected to the Telem 2 and I2C ports, respectively, of autopilot board. ZED mini stereo camera and three ultrasonic sensors are connected to the companion computer.

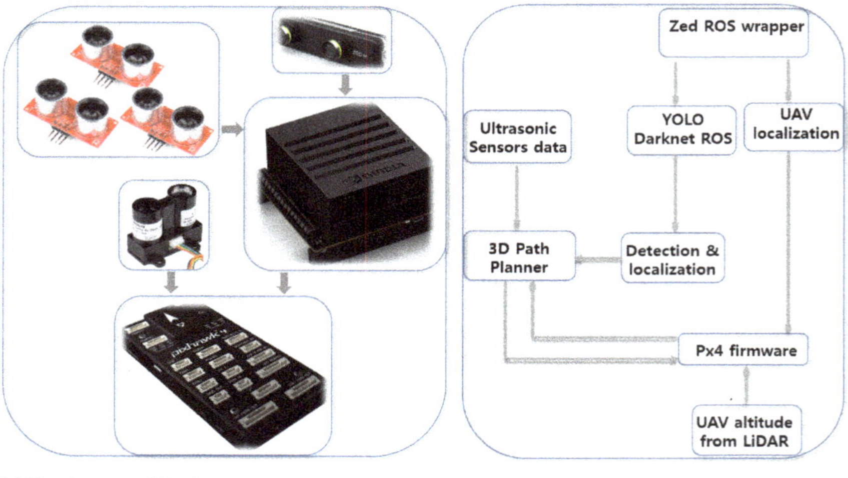

(**a**) Hardware architecture (**b**) Software architecture

Figure 2. Architectures of hardware and software.

Figure 3. Quadcopter equipped with on-board components.

System Calibration and Configuration

PX4 firmware version 1.9.2 was installed on Pixhawk 4, and x-configuration type quadcopter airframe was selected. All the necessary sensors calibrations were done and parameters were set in such a away that the autopilot could communicate with external hardware. Quadcopter localization was enabled by GPS, LiDAR, and ZED mini stereo camera fusion. Pixhawk autopilot supports a Micro Aerial Vehicle Link (MAVLink) protocol that serializes messages. Telem 2 serial port of Pixhawk 4 was set to convey messages to-and-from Pixhawk through MAVLink protocol.

Robot Operating Software (ROS) Kinetic version was installed on the companion computer. ROS provides software tools that enable communication among hardware. Communication between the autopilot and companion computer was enabled by MAVROS: a ROS package that bridges ROS topics (message buses) with MAVLink messages. To extract information from the ZED mini stereo camera and publish in the form of specific message types through ROS topics, an open source, named ZED_ROS wrapper node, was installed on the companion computer. YOLO version 3 (YOLOv3) object detection algorithm and its framework Darknet_ROS were installed on the companion computer for obstacle detection and localization. The 3D path planner module runs on companion computer and communicates with autopilot through MAVROS. The configuration of software components is shown in Figure 2b.

4. Machine Learning for Object Detection

In the machine learning process, a companion computer onboard a quadcopter was trained to identify assumed obstacles that it may encounter during a disaster monitoring mission. The assumed obstacles are pedestrians, windows, electric poles, tunnels, trees, and barely visible nets. The companion computer can be trained to identify a large number of objects once the performance of the proposed path planner is validated on the assumed ones.

4.1. YOLO Object Detection and Localization

Object detection is a task in computer vision that involves identifying the presence and type of one or more objects in a given image. There are various types of object detection algorithms [19–25], and YOLO is one of them with its fastest detection and localization mechanisms. Matija Radovic et al. [26] reported the preference of YOLO over the other detection algorithms that runs on CNN. The key features underpinning YOLO as the fastest detection means are applying a single neural network on the entire image and considering detected object localization as a regression problem. The architecture of this neural network is called Darknet: a type of CNN. It has 24 convolutional layers working as feature extractors and 2 dense layers for doing the predictions. A detailed discussion on the neural network and its architecture is given by Joseph Redmon et al. [27]. There is a configuration file with a given architecture. This file contains information about:

- layers and activations of the architecture
- anchor boxes
- number of classes
- learning rates
- optimization techniques
- input size
- probability score threshold
- batch size

Each configuration file has corresponding pre-trained weights. For training, YOLO requires two files: a file with list of names of objects and a file with a list of training images that contain desired objects with their corresponding labels. The labels are relative centers and dimensions of objects in the image. The configuration file can be modified as per the need of a user. For instance, increasing the batch value improves and speeds up the training but at the cost of demanding more memory. Two of the most important parameters in the configuration file that need to be checked are classes and final layer filters. The values of these parameters should match with the total number of objects in the training.

Once the training is over, the configuration and corresponding weight files are integrated with YOLO Darknet ROS module for object detection and localization during autonomous navigation of UAVs. Along with the detection of each object, there is a bounding box, which is characterized by the following parameters.

- confidence score that the object is detected
- center of the bounding box (U_c,V_c)
- dimension of the bounding box (w, h)

where U and V are coordinate axes of an image frame in which U increases from left to right and V increases from top to bottom. Both the center and dimensions of the box are normalized to fall between 0 and 1. Based on these parameters, the sides of the bounding box can be calculated as:

$$U_{min} = U_c - \frac{w}{2} \quad \text{and} \quad U_{max} = U_c + \frac{w}{2} \tag{1}$$

and

$$V_{min} = V_c - \frac{h}{2} \quad \text{and} \quad V_{max} = V_c + \frac{h}{2} \tag{2}$$

where U_{min} and U_{max} are the locations of left and right sides of the bounding box along the U-axis. Similarly, V_{min} and V_{max} are the locations of upper and lower sides of the bounding box along the V-axis. The coordinate transformation from the image frame to camera frame follows the procedure shown in [28]. Since the path planner was written as the ROS node that follows a reference frame FLU (Forward (x), left (y), and upward (z)), coordinate transformation from camera frame to ROS frame (FLU) was done. Moreover, PX4 uses FRD (Forward (x), right (y), and Down (z)). The ROS package MAVROS handles coordinate frame transformation from ROS frame to PX4 frame.

5. Three-Dimensional Path Planning Algorithms

The top challenge in autonomous navigation of UAVs is planning an obstacle-free route from the start to the destination. Encountering obstacles is possible, especially for missions like law enforcement, package delivery, and first aid in urban areas. Most of the path planning algorithms for UAVs are derived from pre-existing algorithms designed for ground robots. These algorithms are often 2D and need to be modified into 3D for aerial vehicles. The complexity to design and the demand for high performance computers on-board the UAVs are challenges that incurred by the 3D path planners. The obstacle-free 3D path planning process demands an intensive computational burden that often limits the maximum cruising capability of the UAV. The effect of this computational burden is true for both free and cluttered environments as long as image processing has to occur.

Commonly known 3D path planning algorithms are $A^\star$ with its variants, Rapidly–Exploring Random Tree (RRT) with its variants, Probabilistic RoadMaps (PRM), Artificial Potential Field (APF), and Genetic or Evolutionary algorithms. These algorithms can be categorized into two: sampling-based and node/grid-base algorithms. Sampling-based algorithms connect randomly sampled points (subset of all points) all the way from start to the goal points thereby creating random graphs from which a graph with shortest path-length is selected. The algorithms include RRT, PRM, and APF.

Node/grid-based algorithms, unlike sampling-based algorithms, exhaustively explore throughout consecutive nodes. These algorithms include $A^\star$ and its variants. In search for an obstacle-free path, the algorithm takes in an image of the environment and discretizse it into grid cells that includes the current (start) location of the UAV and the goal location. The $A^\star$ algorithm has two functions to prioritize the cells to be visited. These two functions are the cost function, which calculates the distance from the current cell to the next cell, and the heuristic function, which calculates the distance from the next cell to the cell that contains the goal. With the objective of minimizing the sum of these two functions, the cells to be visited are heuristically prioritized. In the case of 3D search, the cost function calculates distances from the current cell to all 26 neighboring cells, and the heuristic function calculates distance from the 26 cells to the cell that contains the goal. In a cluttered environment with complex occlusion, highly dense grid cells are required, which in turn increase the computational burden, and thus the selected path may not be optimal.

5.1. Machine Learning-Based 3D Path Planner

Training an on-board computer to quickly identify objects and avoid collision with them in an environment in which UAV is set to navigate can be taken as a paradigm shift as it inherits the mechanism that a human being takes to avoid collision. The computational intelligence of a human brain is the degree that it is trained to, as is the artificial intelligence of the computer onboard a UAV. This is why intensive training of on-board computer is compulsory.

Apart from the capabilities of ensuring the presence of objects and their relative locations from the UAV, the companion computer may be required to know the type of objects it detected. The YOLO object detection algorithm installed on the companion computer has such a capability. Strategies to avoid collision with an object may depend on the type of the object. For instance, the avoidance mechanism for a window (open obstacle) is different from the mechanism for a tree (closed obstacle). Our 3D path planner includes those capabilities, as explained below.

5.1.1. Open Obstacles

In this type of obstacle, there is a possibility in which the UAV has no other option but to pass through the opening, such as in the case when the mission is to enter or exit a closed room through open window. Missions like in-house first aid or disaster monitoring may encounter such a scenario. In this case, the algorithm determines the relative position of the UAV with respect to the center of the bounding box around the obstacle. The center of the bounding box, as shown in Figure 4, has coordinate axes (x_c, y_c, z_c) with respect to the ZED mini stereo camera frame, whose origin is located at the center of left camera.

The x, y, and z axes of this frame point forward, right-to-left, and upward, respectively. Therefore, x represents the depth of the detected object (e.g., x_c depth of the window). The depth information is directly extracted from the ZED mini camera, whereas y and z are derived from the (U,V) coordinate values through coordinate transformation. Information obtained with respect to image frame, including Equations (1) and (2), are transformed to the camera frame.

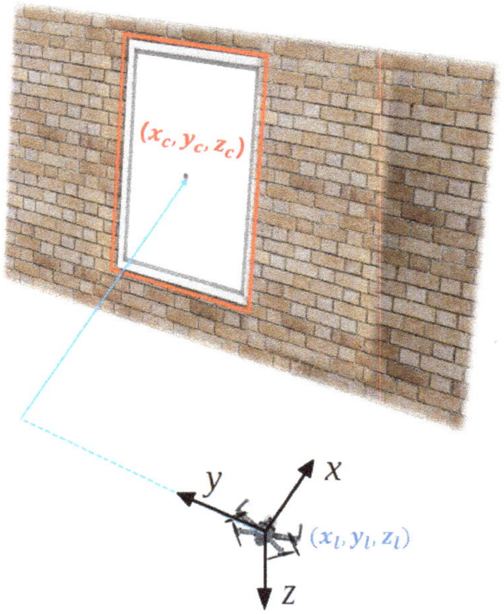

Figure 4. Open obstacle passing strategy.

The UAV's local position (x_l, y_l, z_l) is acquired from GPS embedded in the Pixhawk 4 autopilot, LiDAR and ZED mini stereo camera. Before the UAV tries to pass through the window, it has to align itself with a vector normal to the plane of the window through appropriate attitude and altitude changes. In the figure, the setpoint (x_l, y_s, z_s) is sent by the 3D path planner to the autopilot to command the UAV to adjust itself before advancing forward. The variables y_s and z_s are the y and z axes' setpoint values, respectively, obtained as follows:

$$y_s = y_l - y_c \qquad \text{and} \qquad z_s = z_l - z_c \tag{3}$$

While the UAV is responding to the command, the ultrasonic sensors mounted on the sides of the UAV check whether there are objects or not in the way. Once alignment is done, the UAV advances through the window with the setpoint (x_s, y_s, z_s), where x_s is the relative depth of the bounding box with clearance.

$$x_s = \mid x_l - x_c \mid + \quad obj_{clr} \tag{4}$$

The variable obj_{clr} is a minimum object clearance or distance of the UAV behind the window that ensures the UAV has completely passed through the window with clearance. Moreover, to confirm the passage of the UAV through the window, the readings from the ultrasonic sensors mounted on the left and right sides of the UAV are considered. This method is implemented in cases like passing through tunnels or holes alike.

5.1.2. Closed Obstacles

If the obstacle is closed, our path planner considers the pass-by option with a minimum side clearance from the obstacle. The 3D path planning algorithm takes in bounding boxes information of all detected objects and assigns an identity index to each of them based on the locations of their centers along the y-axis. All information about the bounding box are with respect to the camera frame onboard the UAV. As shown in Figure 5a, the index value increases towards the increasing y-axis of the ZED min stereo camera (in this case, from right to left).

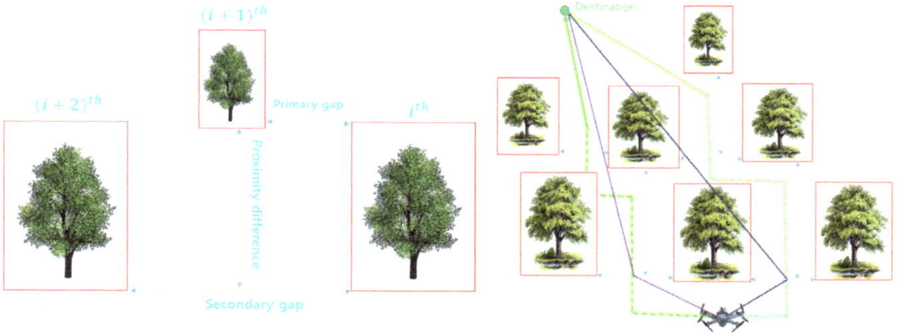

(**a**) Avoidance strategy (**b**) Optimal path selection.

Figure 5. Obstacle-free optimal path selection

There are three conditions to be considered to determine next setpoint for the UAV. These are searching for

- Wide primary gaps: gaps between consecutive obstacles;
- Narrow primary gaps but with proximity difference: depth difference between consecutive obstacles; and
- Narrow primary gaps with small or no proximity difference.

Based on Figure 5a, the algorithm calculates the primary gap (between the ith and $(i+1)^{th}$) and secondary gap (between i^{th} and $(i+2)^{th}$). The importance of calculating the secondary gap is that if the primary gap is narrow (less than twice UAV width) but with proximity difference more than twice the UAV length, there is the possibility that the UAV can advance forward but should check whether the secondary gap is wide enough or not to let the UAV pass in between. The gaps and proximity differences are calculated as follows:

$$\begin{aligned} Y_{min}^{i+1} - Y_{max}^{i} &\quad \text{primary gap} \\ Y_{min}^{i+2} - Y_{max}^{i} &\quad \text{secondary gap} \\ \mid X^{i+1} - X^{i} \mid &\quad \text{proximity difference} \end{aligned} \tag{5}$$

The pseudo-algorithm of our 3D path planner in the presence of multiple detected obstacles, as shown in Figure 5b, is given below.

- index the bounding boxes of the obstacles based on y-axis values of their centers. The box with the smallest y-axis value is indexed as the ith box;
- calculate Y_{min} and Y_{max} for each bounding box;
- calculate the primary gap between the ith and $(i+1)^{th}$
- if the gap is greater than or equal to twice UAV width;
  - calculate the midpoint of the gap;
  - calculate distances from the current location of the UAV to the midpoint and from the midpoint to the goal point. Save the sum of these two distances as path-length;
- else if the primary gap is smaller, calculate the proximity difference of the two consecutive bounding boxes i^{th} and $(i+1)^{th}$;
  - if proximity difference is greater than or equal to twice UAV length, calculate the secondary gap;
  - if secondary gap is greater than or equal to twice the UAV width, check the following conditions:
    * if the ith obstacle is closer than the $(i+1)^{th}$, then set $(X^i, Y_{max}^i + obj_{clr}, Z^i)$ as a potential setpoint;
    * else, set $(X^{i+2}, Y_{min}^{i+2} - obj_{clr}, Z^{i+2})$ as a potential setpoint;

  - calculate distances from the current location of the UAV to the potential setpoint and from potential setpoint to the goal point. Save the sum of these two distances as path-length;
- apply the above steps for the remaining bounding boxes;
- compare the path-lengths and set the setpoint that leads to a minimum path length as the next setpoint for the UAV;
- else if the secondary gap is less than twice the UAV width, hover at a current altitude and yaw to search for any possible path applying the above procedure;
- if no path is discovered, land the UAV.

6. Path Planner Performance Tests

Performance tests were carried out during the developmental stage of the the path planner. Prior to real flight performance tests, rigorous computer-simulation-based tests were conducted. The implemented software tools and their integration as well as real flight test procedures are described in the following subsections.

6.1. SITL Test

Gazebo simulation environments shown in Figure 6 (front view) and Figure 7 (top views) were built, in which the path planner was to be tested.

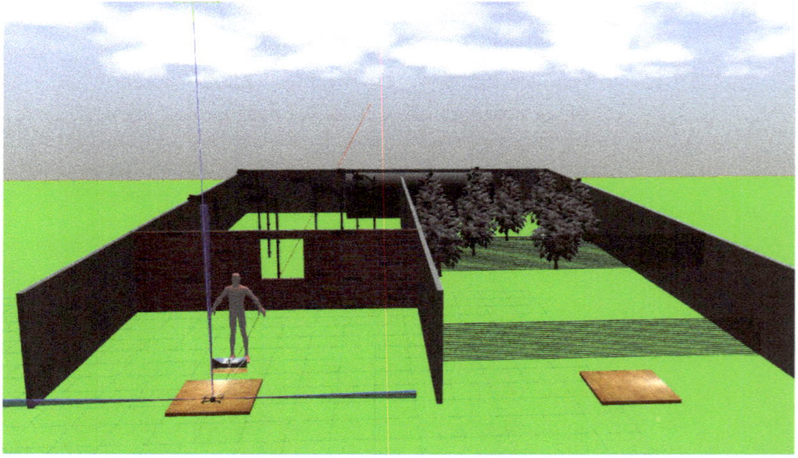

Figure 6. Front view of gazebo environment

The gazebo world has left and right sections. Each section has a width (y-axis) of 10 m and a length (x-axis) of 30 m. The UAV located in the left section has to avoid the obstacles on its mission to arrive at landing pad, which is located in the right section. During path planner performance tests, the poles and trees were randomly re-located in the simulation environment. Every time the arrangements of these obstacles are changed, the path followed by the UAV changes. Figure 7c,d shows two traced trajectories for the obstacles' arrangements shown in Figure 7a,b, respectively.

The 3D models of the obstacles imported to gazebo world were pedestrian, open window, poles, tunnel, trees, and two consecutive nets. The obstacles were designed in consideration of UAV mission for in-house first aid, law enforcement during suspect monitoring and door-to-door package delivery services in urban areas where the aforementioned obstacles are assumed to be potential threats in such missions. The UAV is supposed to pass by or through these obstacles on its way to a targeted location, in this case, the landing pad.

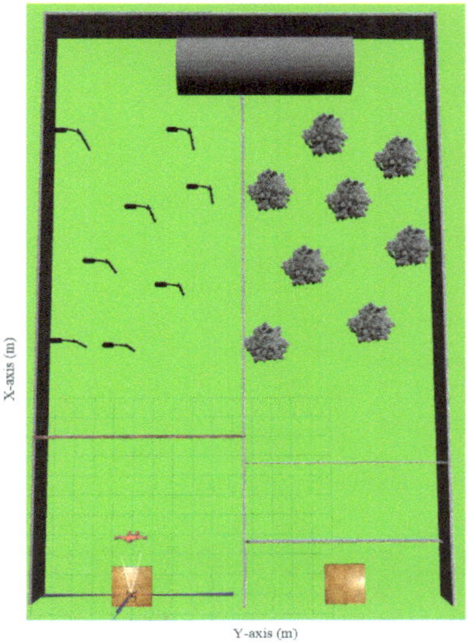

(**a**) Top view of environment 1

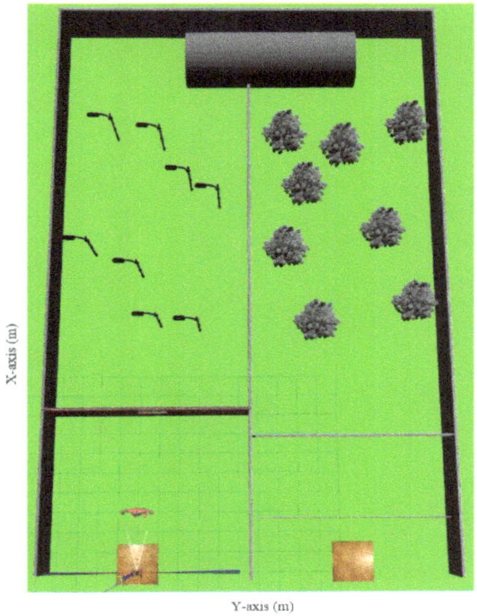

(**b**) Top view of environment 2

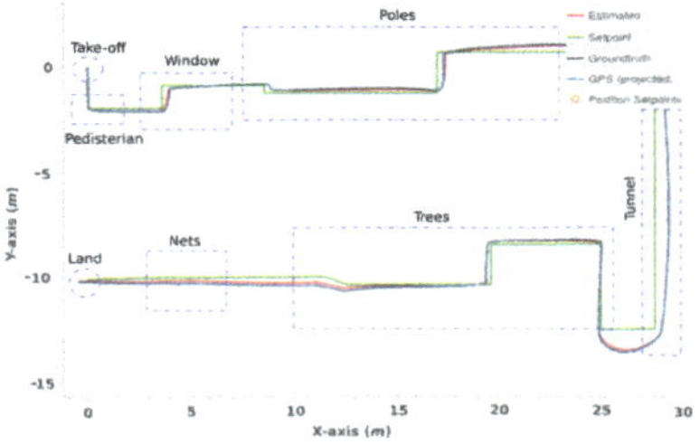

(**c**) Traced trajectory for environment 1

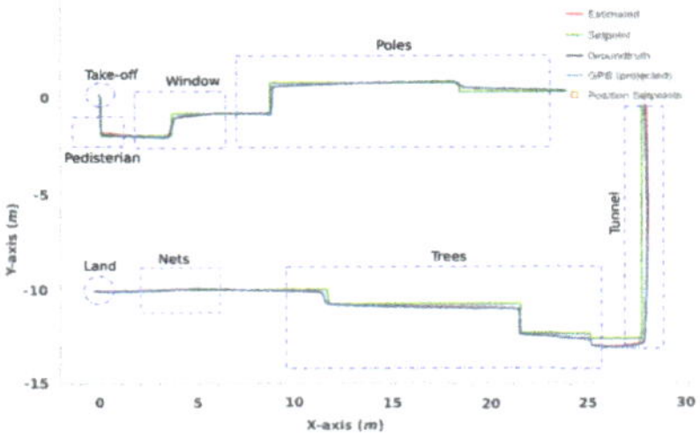

(**d**) Traced trajectory for environment 2

Figure 7. Top views of two simulation environments and traced trajectories.

The overall simulation infrastructure is shown in Figure 8. The 3D path planner written as ROS node communicates with the PX4 module named Mavlink_main. MAVROS bridges the ROS topics of the path planner with MAVLink messages of PX4 firmware. In addition to bridging ROS topics with MAVLink messages, MAVROS has extra-advantage in taking care of coordinate transformation between the ROS frame and PX4 Flight Control Unit (FCU) frame. ROS works with the East–North–Up (ENU) frame, and FCU works with the North–East–Down (NED) frame. PX4 firmware has a module called simulator_mavlink that lets the firmware interact with the 3D model of the UAV in the Gazebo world. The message exchanges between the PX4 firmware and gazebo simulator are handled by simulator MAVLink protocol.

As part of the 3D path planner's efficiency verification tests, video (named as Video S1) is submitted with this manuscript. The livestreamed videos on qgroundcontrol (a ground control station for UAVs) were recorded, and the snapshots of a video at the instants of attitude or altitude changes, to avoid obstacles, are displayed in Figure 9.

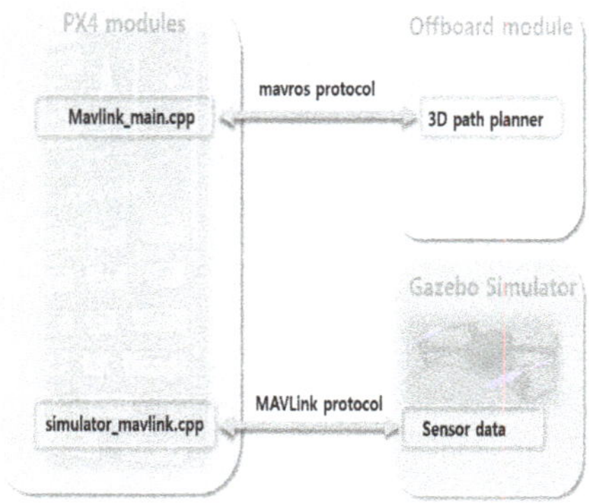

Figure 8. Software_In_The_Loop infrastructure.

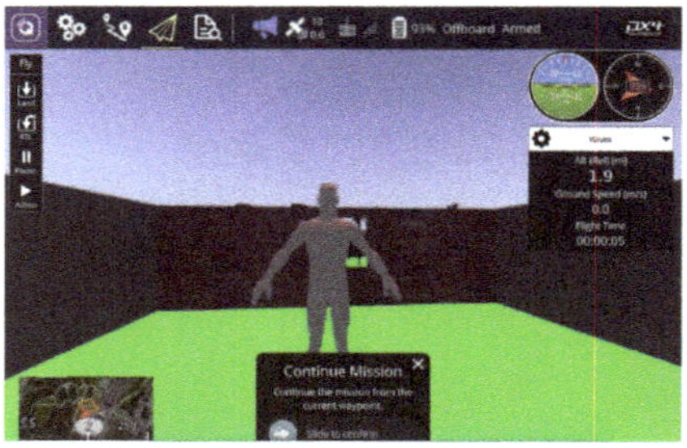

(**a**) Pass-by pedestrian

(**b**) Pass-through window

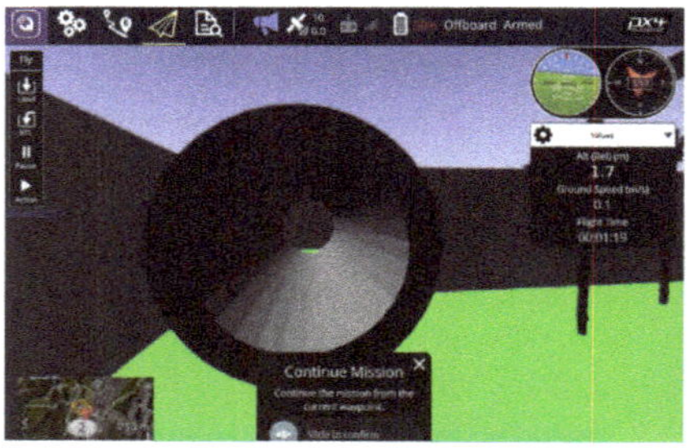

(**c**) Pass-through tunnel

(**d**) Pass-by poles

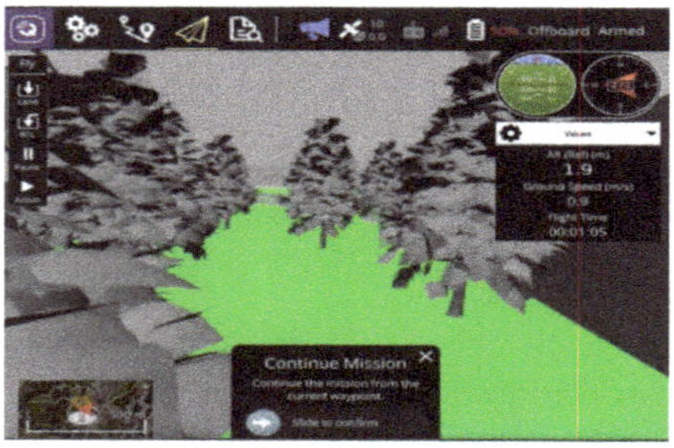

(**e**) Pass-through trees

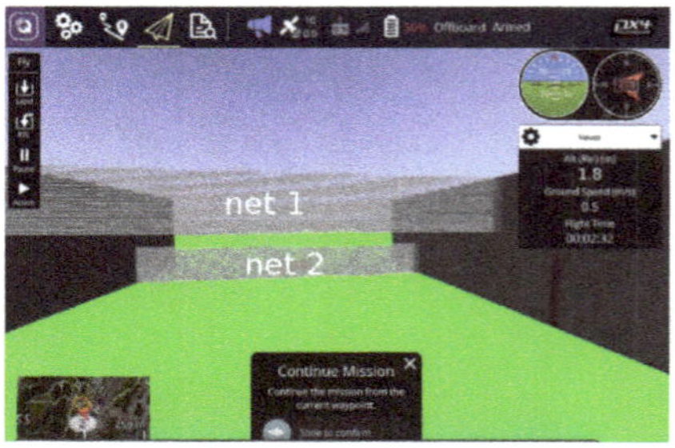

(**f**) Net under/over pass

Figure 9. Instant snapshots during obstacle avoidance phases.

The position and attitude accuracy for the environments Figure 7c,d are shown in the first and second columns of Figure 10, respectively.

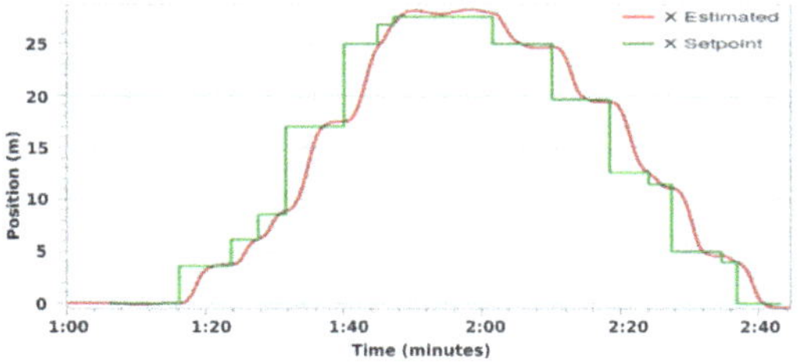

(**a**) Environment 1: position x

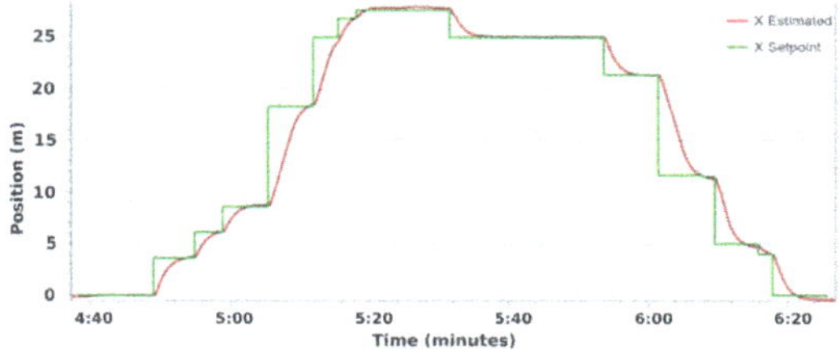

(**b**) Environment 2: position x

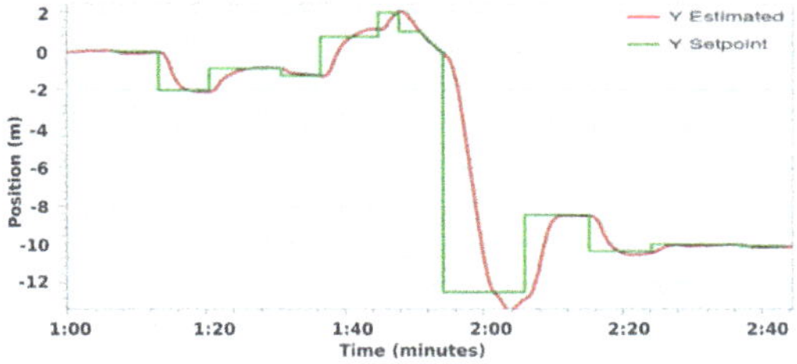

(**c**) Environment 1: position y

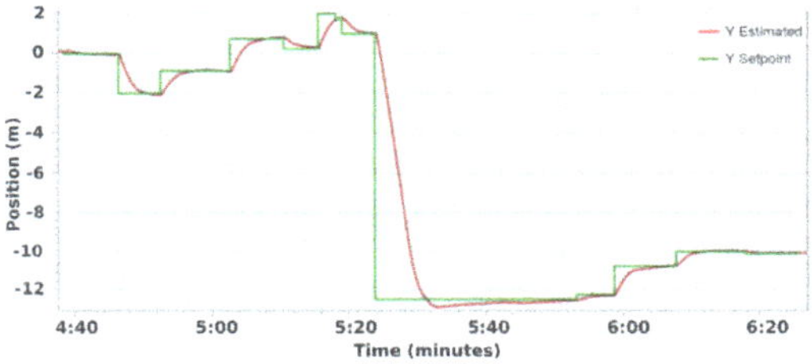

(**d**) Environment 2: position y

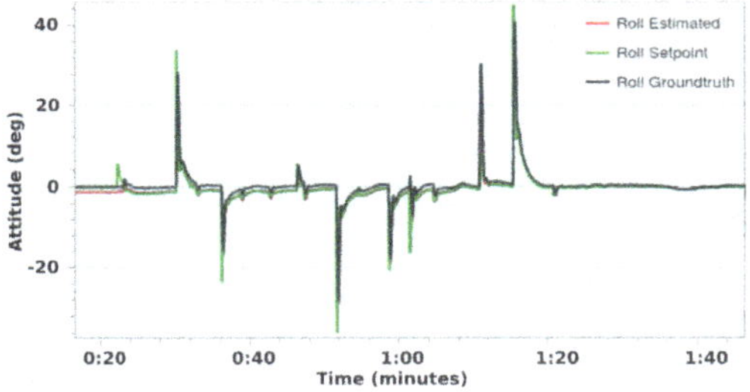

(**e**) Environment 1: roll

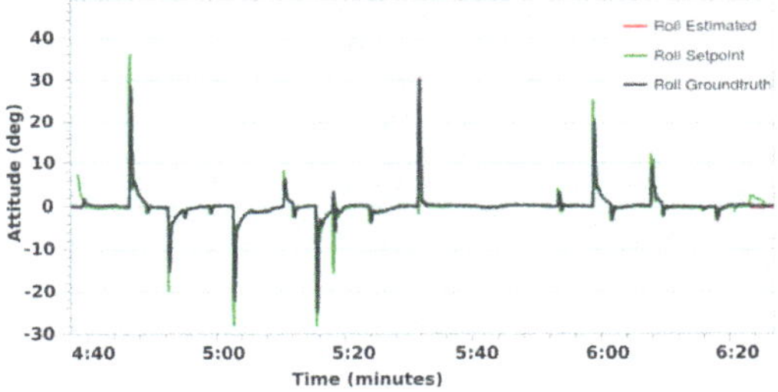

(**f**) Environment 2: roll

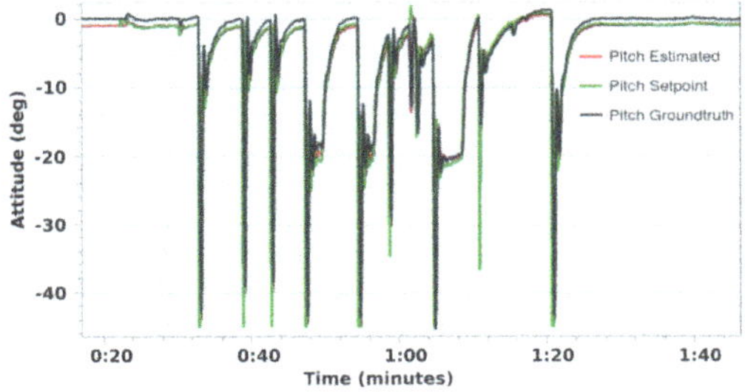

(**g**) Environment 1: pitch

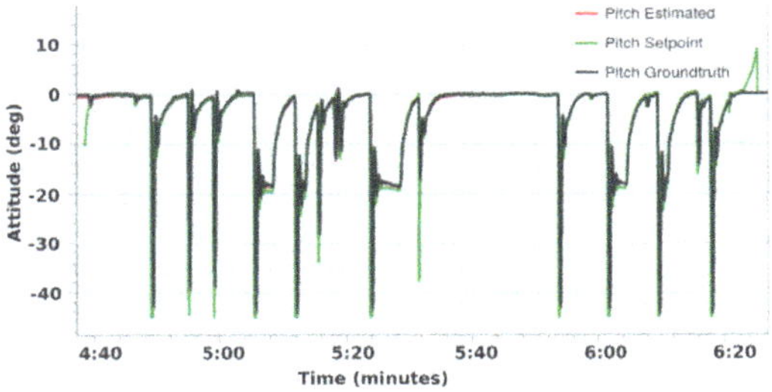

(**h**) Environment 2: pitch

Figure 10. Simulation: position and attitude accuracy tests.

6.2. Real Flight Test

The real flight test requires us to do an intensive machine learning or training the companion computer to identify the obstacles simulated in the gazebo environment. This training process is not over yet: at least up to the report of this work. To get the sense of the efficiency of the path planner, real flight tests were conducted for the first obstacle pass, as shown in Figure 11. As the quadcopter approaches the pedestrian, it has to evaluate the best route based on the conditions given in the pseudo-algorithm Section 5.1.2. For this test phase, short videos (named Videos S2 and S3) accompany this manuscript.

Considering the fact that building a real constrained environment as the simulated one requires time and money, a ROS node that sequentially publishes the simulated locations of obstacles was developed. The node publishes all the information that the 3D path planner requires from ZED mini stereo in a real scenario. Based on this, the quadcopter was deployed to arrive at a given destination, avoiding collisions with the obstacles. The effectiveness of the path planner is validated as shown in Figure 12, where the setpoints sent by the path planner and the estimated positions of the quadcopter throughout the whole mission overlap.

(**a**) Pedestrian pass test 1

(**b**) Pedestrian pass test 2

Figure 11. Pedestrian as obstacle pass tests.

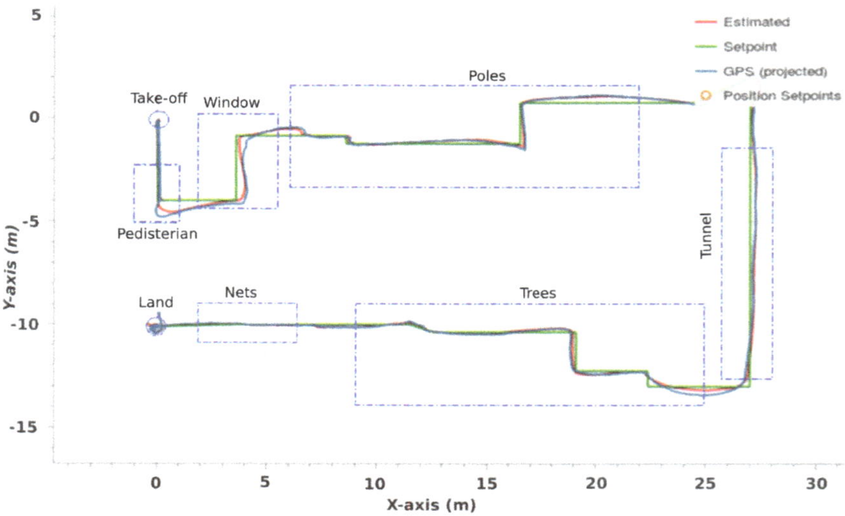

Figure 12. Estimated position and position setpoint comparison in real flight test.

Furthermore, Figure 13 shows component-wise position and attitude accuracy validation.

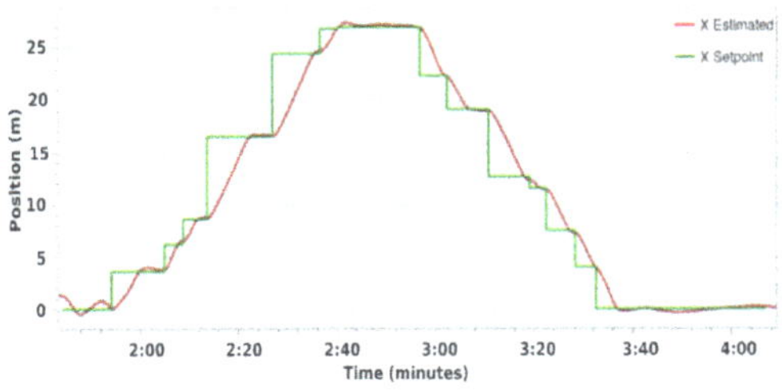

(**a**) position x

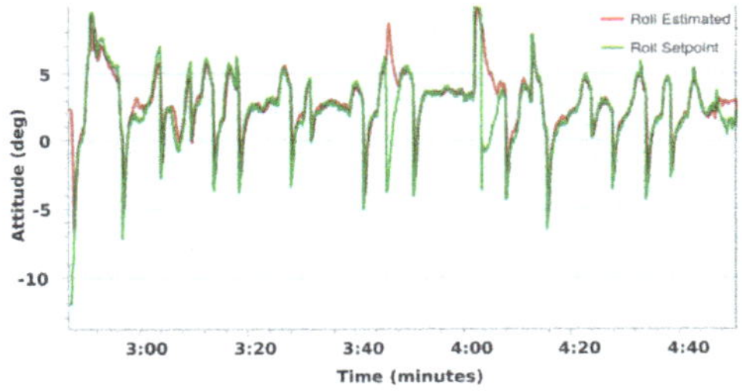

(**b**) attitude: roll

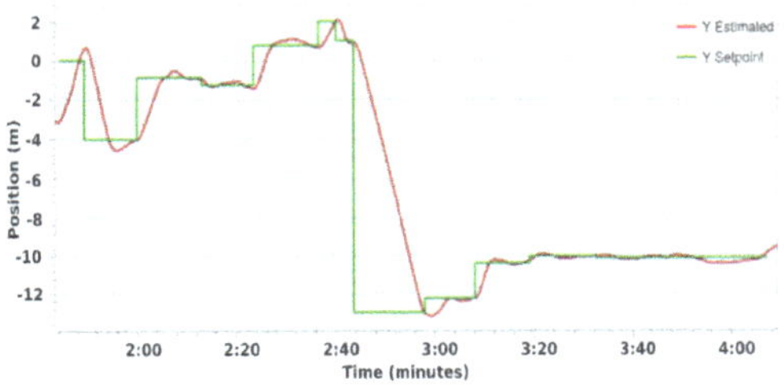

(**c**) position y

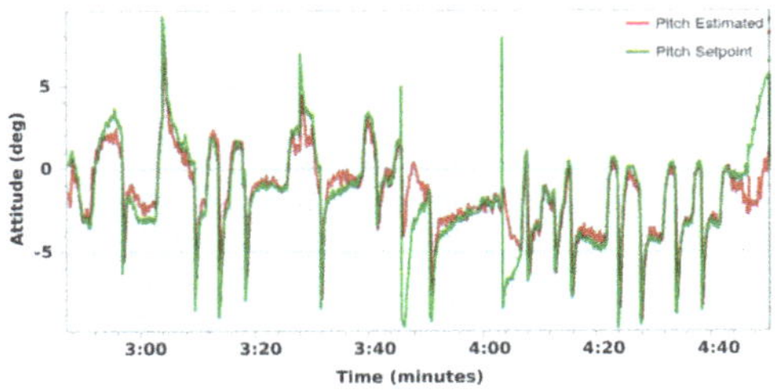

(**d**) attitude: pitch

Figure 13. Real flight test for path planner efficiency validation.

7. Results and Discussion

The validation of the developed 3D path planner was conducted through both SITL and preliminary real flight tests. Gazebo 3D model simulation environment was thoroughly used to develop and validate our 3D path planner prior to its upload into Pixhawk autopilot. The Gazebo environment shown in Figure 6 was set in such a way that it has obstacles like human, window, poles, tunnel, trees, and nets. These obstacles implicate the plausible encounters that the drone may face during missions such as package delivery, disaster monitoring, law enforcement, and first aid. For a complete navigation from the left section to the right section of the environment, the drone has to avoid collision with any of the mentioned obstacles and safely land at the landing pad.

Rigorous simulation tests were done where the two randomly arranged environments shown in Figure 7 are some of the environments in which the tests were done. The path followed by the quadcopter in the environment in Figure 7a is shown in Figure 7c. Similarly, the path followed in the environment in Figure 7b is shown in Figure 7d. As can be seen in these figures, the quadcopter followed two different trajectories in response to the two different arrangements of the obstacles in the environments. Moreover, the setpoints sent by the path planner and the estimated locations of the quadcopter overlap throughout the trajectories. This overlap validates the effectiveness of the path planner to autonomously navigate the quadcopter in a cluttered and GPS-denied environment.

Components of position and attitude responses in the two environments, Figure 7a,b, are shown in Figure 10. The well-traced setpoints of both position and attitude prove the efficiency of the path planner. In the path planner, a setpiont acceptance radius is set to 0.30 m. The differences observed at setpoint nodes are due to this acceptance radius. The quadcopter advances to the next setpoint assuming that the current setpoint is achieved at the moment the quadcopter crosses the acceptance radius, though the quadcopter may not reach the actual setpoint. This causes a gap between the estimated position and position setpoint. The attitude estimates of the quadcopter in both environments conform to the setpoints.

The preliminary real flight tests were conducted for collision avoidance with pedestrians. Machine learning was done for a pedestrian with different posture, clothes, and light exposures. As shown in Figure 9a, the quadcopter attempts to avoid collision with the pedestrian by rolling either right or left, implementing the conditions given in the pseudo-algorithm. For reference, the recorded two short videos on pedestrian collision avoidance are submitted with this manuscript.

For a complete mission test, a real environment, similar to the simulated environment shown in Figure 6, should have been constructed. This would take time and money. For this report, the real environment was modeled by an ROS node that publishes required information to the 3D path planner. This node publishes simulated locations of obstacles, and the 3D path planner takes those locations and calculates an obstacle-free path. With this, the UAV was commanded to autonomously head to the landing pad avoiding all possible obstacles on its way. The path followed by the UAV during this mission is shown in Figure 12. The overlap of the estimated quadcopter positions and intended setpoints shows that the 3D path planner effectively executed the mission.

In the real flight test, which was conducted in an open field, the quadcopter localization was limited to GPS and LiDAR. LiDAR is only for altitude estimation. ZED mini stereo camera, combined with GPS for quadcopter localization, does not provide proper localization of the quadcopter in an open field as it is required to get reflected rays from objects in its operation range. Therefore, for localization, the quadcopter in this circumstance relies on GPS whose accuracy is about 2 m. Depending on the number of satellites accessed and the environment in which the quadcopter is, the accuracy of the GPS drifts. The initial location of the quadcopter before takeoff had high drifts as can be seen in Figure 13c.

The test results obtained so far show that the 3D path planning algorithm is effectively guiding the UAV through collision-free paths. The future work includes the real flight tests in the environment similar to the simulated one as well as in unconstrained environments. Moreover, machine learning for various objects will be conducted based on the mission profile of the UAV.

Supplementary Materials: The following are available online at https://www.mdpi.com/article/10.3390/app11104706/s1, Video S1: Performance of path planner in cluttered environment, Video S2: Path planner in avoiding collision with pedestrian left pass, Video S3: Path planner in avoiding collision with pedestrian right pass.

Author Contributions: In this manuscript, machine learning for the aforementioned obstacles was done by B.E. and hardware integration and sensors fusion were done by A.W. The 3D path planning was done by A.T., while H.-Y.H. was responsible for the overall conceptual design of the methodology followed in this manuscript. All authors have read and agreed to the published version of the manuscript.

Funding: This work is supported by the Korea Agency for Infrastructure Technology Advancement (KAIA) grant funded by the Ministry of Land, Infrastructure and Transport (Grant 21CTAP-C157731-02).

Institutional Review Board Statement: Not applicable.

Informed Consent Statement: Not applicable.

Data Availability Statement: This study did not report any data.

Acknowledgments: This work is supported by the Korea Agency for Infrastructure Technology Advancement (KAIA) grant funded by the Ministry of Land, Infrastructure and Transport (Grant 21CTAP-C157731-02).

Conflicts of Interest: Authors mentioned in this manuscript have strongly involved in this study from the start to the end. The manuscript were thoroughly reviewed by the authors before its submission to journal of Applied Science. This manuscript has not been submitted to another journal for publication.

References

1. Konovalenko, I.; Kuznetsova, E.; Miller, A.; Miller, B.; Popov, A.; Shepelev, D.; Stepanyan, K. New approaches to the integration of navigation systems for autonomous unmanned vehicles (UAV). *Sensors* **2018**, *18*, 3010. [CrossRef] [PubMed]
2. Wang, J.; Garrat, M.; Lambert, A.; Wang, J.J.; Han, S.; Sinclair, D. Integration of GPS/INS/vision sensors to navigate unmanned aerial vehicles. *ISPRS Int. Arch. Photogramm. Remote Sens. Spatial Inform. Sci.* **2008**, *37*, 963–970.
3. García, J.; Molina, J.M.; Trincado, J.; Sánchez, J. Analysis of sensor data and estimation output with configurable UAV platforms. In Proceedings of the 2017 Sensor Data Fusion: Trends, Solutions, Applications (SDF), Bonn, Germany, 10–12 October 2017; pp. 1–6.
4. Abdulla, A.-K.; David, M.; Fernando, G.; Arturodela, E.; José, M.A. Survey of computer vision algorithms and applications for unmanned aerial vehicles. *Expert Syst. Appl.* **2018**, *92*, 447–463.
5. Belmonte, L.M.; Morales, R.; Fernández-Caballero A. Computer Vision in Autonomous Unmanned Aerial Vehicles- A Systematic Mapping Study. *Appl. Sci.* **2019**, *9*, 3196. [CrossRef]
6. Wagoner, A.R.; Schrader, D.K.; Matson, E.T. Survey on Detection and Tracking of UAVs Using Computer Vision. In Proceedings of the First IEEE International Conference on Robotic Computing (IRC), Taichung, Taiwan, 10–12 April 2017; pp. 320–325.
7. Choi, S.Y.; Dowan, C. Unmanned aerial vehicles using machine learning for autonomous flight; state-of-the-art. *Adv. Robot.* **2019**, *33*, 265–277. [CrossRef]
8. Gonzalez, L.F.; Montes, G.A.; Puig, E.; Johnson, S.; Mengersen, K.; Gaston, K.J. Unmanned Aerial Vehicles (UAVs) and Artificial Intelligence Revolutionizing Wildlife Monitoring and Conservation. *Sensors* **2016**, *16*, 97. [CrossRef] [PubMed]
9. Kyrkou, C.; Theocharides, T. Deep-Learning-Based Aerial Image Classification for Emergency Response Applications using Unmanned Aerial Vehicles. In Proceedings of the IEEE Conference on Computer Vision and Pattern Recognition Workshops (CVPRW), Long Beach, CA, USA, 16–17 June 2019.
10. Lygouras, E.; Santavas, N.; Taitzoglou, A.; Tarchanidis, K.; Mitropoulos, A.; Gasteratos, A. Unsupervised Human Detection with an Embedded Vision System on a Fully Autonomous UAV for Search and Rescue Operations. *Sensors* **2019**, *19*, 3542. [CrossRef] [PubMed]
11. Valenti, F.; Giaquinto, D.; Musto, L.; Zinelli, A.; Bertozzi, M.; Broggi, A. Enabling computer vision-based autonomous navigation for unmanned aerial vehicles in cluttered gps-denied environments. In Proceedings of the IEEE International Conference on Intelligent Transportation Systems (ITSC), Maui, HI, USA, 4–7 November 2018.
12. Lu, Y.; Xue, Z.; Xia, G.-S.; Zhang, L. A Survey on vision-based UAV navigation. *Geo-Spat. Inf. Sci.* **2018**, *21*, 21–32. [CrossRef]
13. Yan, C.; Xiang, X.; Wang, C. Towards Real-Time Path Planning through Deep Reinforcement Learning for a UAV in Dynamic Environments. *J. Intell. Robot. Syst.* **2019**. [CrossRef]
14. Yang, L.; Qi, J.; Song, D.; Xiao, J.; Han, J.; Xia, Y. Survey of Robot 3D Path Planning Algorithms. *J. Control Sci. Eng.* **2016**, 1–22. [CrossRef]
15. Dai, X.; Mao, Y.; Huang, T.; Qin, N.; Huang, D.; Li, Y. Automatic obstacle avoidance of quadcopter UAV via CNN-based learning. *Neurocomputing* **2020**, *402*, 346–358. [CrossRef]
16. Sandino, J.; Vanegas, F.; Maire, F.; Caccetta, P.; Sanderson, C.; Gonzalez, F. UAV Framework for Autonomous Onboard Navigation and People/Object Detection in Cluttered Indoor Environment. *Remote Sens.* **2020**, *12*, 3386. [CrossRef]
17. Loquercio, A.; Maqueda, A.I.; Del-Blanco, C.R.; Scaramuzza, D. DroNet: Learning to Fly by Driving. *IEEE Robot. Autom. Lett.* **2018**, *3*, 1088–1095. [CrossRef]
18. Nathan, K.; Andre, H. Design and Use Pradigms for Gazebo, An Open-Source Multi-Robot Simulator. In Proceedings of the 2004 IEEE/RSJ International Conference Intelligent Robot System, Sendai, Japan, 28 September–2 October 2004; pp. 2149–2154.
19. Chahal, K.; Dey, K. A Survey of Modern Object Detection Literature using Deep Learning. *arXiv* **2018**, arXiv:1808.07256v1.
20. Girshick, R.; Donahue, J.; Darrell, T.; Malik, J. Region-Based Convolutional Networks for Accurate Object Detection and Segmentation. *IEEE Trans. Pattern Anal. Mach. Intell.* **2016**, *38*, 142–158. [CrossRef] [PubMed]
21. Nayagam, M.G.; Ramar, D.K. A Survey on Real time Object Detection and Tracking algorithms. *Int. J. Appl. Eng. Res.* **2015**, *10*, 8290–8297.
22. Najibi, M.; Rastegari, M.; Davis, L.S. G-CNN: An Iterative Grid Based Object Detector. In Proceedings of the IEEE Conference Computer Vision and Pattern Recognition, Las Vegas, NV, USA, 27–30 June 2016; pp. 2369–2377.
23. Zhao, Z.-Q.; Zheng, P.; Xu, S.-T.; Wu, X. Object Detection with Deep Learning: A review. *IEEE Trans. Neural Netw. Learn. Syst.* **2019**, *30*, 3212–3232. [CrossRef] [PubMed]
24. Liu, Q.; Xiang, X.; Wang, Y.; Luo, Z.; Fang, F. Aircraft detection in remote sensing image based on corner clustering and deep learning. *Eng. Appl. Artif. Intell.* **2020**, *87*, 103333. [CrossRef]
25. Tan, J. Complex object detection using deep proposal mechanism. *Eng. Appl. Artif. Intell.* **2020**, *87*, 103234. [CrossRef]
26. Radovic, M.; Adarkwa, O.; Wang, Q. Object recognition in aerial images using convolutional neural networks. *J. Imaging* **2017**, *3*, 21. [CrossRef]
27. Redmon, J.; Divvala, S.; Girshick, R. You only look once: Unified, real-time object detection. In Proceedings of the IEEE International Conference Computer Vision and Pattern Recognition (CVPR), Las Vegas, NV, USA, 27–30 June 2016; pp. 779–788.
28. Sheikh, T.S.; Afanasyev, I.M. Stereo Vision-based Optimal Path Planning with Stochastic Maps for Mobile Robot Navigation. Intelligent Autonomous Systems 15. IAS 2018. In *Advances in Intelligent Systems and Computing*; Springer: Cham, Switzerland, 2018; Volume 867.

applied sciences

MDPI

Article

Path Planning Method for UAVs Based on Constrained Polygonal Space and an Extremely Sparse Waypoint Graph

Abdul Majeed * and Seong Oun Hwang *

Department of Computer Engineering, Gachon University, Seongnam 13120, Korea
* Correspondence: ab09@gachon.ac.kr (A.M.); sohwang@gachon.ac.kr (S.O.H.); Tel.: +82-31-750-5327 (S.O.H.)

Abstract: Finding an optimal/quasi-optimal path for Unmanned Aerial Vehicles (UAVs) utilizing full map information yields time performance degradation in large and complex three-dimensional (3D) urban environments populated by various obstacles. A major portion of the computing time is usually wasted on modeling and exploration of spaces that have a very low possibility of providing optimal/sub-optimal paths. However, computing time can be significantly reduced by searching for paths solely in the spaces that have the highest priority of providing an optimal/sub-optimal path. Many Path Planning (PP) techniques have been proposed, but a majority of the existing techniques equally evaluate many spaces of the maps, including unlikely ones, thereby creating time performance issues. Ignoring high-probability spaces and instead exploring too many spaces on maps while searching for a path yields extensive computing-time overhead. This paper presents a new PP method that finds optimal/quasi-optimal and safe (e.g., collision-free) working paths for UAVs in a 3D urban environment encompassing substantial obstacles. By using Constrained Polygonal Space (CPS) and an Extremely Sparse Waypoint Graph (ESWG) while searching for a path, the proposed PP method significantly lowers pathfinding time complexity without degrading the length of the path by much. We suggest an intelligent method exploiting obstacle geometry information to constrain the search space in a 3D polygon form from which a quasi-optimal flyable path can be found quickly. Furthermore, we perform task modeling with an ESWG using as few nodes and edges from the CPS as possible, and we find an abstract path that is subsequently improved. The results achieved from extensive experiments, and comparison with prior methods certify the efficacy of the proposed method and verify the above assertions.

Keywords: constrained polygonal space; path length; path planning; obstacles; maps; unmanned aerial vehicles; urban environments; time complexity; extremely sparse waypoint graph

Citation: Majeed, A.; Hwang, S.O. Path Planning Method for UAVs Based on Constrained Polygonal Space and an Extremely Sparse Waypoint Graph. *Appl. Sci.* **2021**, *11*, 5340. https://doi.org/10.3390/app11125340

Academic Editors: Sylvain Bertrand and Hyo-sang Shin

Received: 22 March 2021
Accepted: 4 June 2021
Published: 8 June 2021

1. Introduction

Unmanned aerial vehicles (UAVs) are highly useful for executing diverse missions not only in urban environments but also in hazardous areas that are not easily reachable, such as forests, deserts, and hilly areas. UAVs (being lightweight, low-cost, and with the abilities to fly at lower altitudes) are now extensively used for a wide range of both military and civilian tasks. Owing to military and civilian investments in UAV technology, this field continuously advances with the passage of time. Based on a forecast by the Teal Group, the market for UAVs is constantly growing globally, and yearly spending on this technology is expected to be higher than US $12 billion by 2024 [1]. Advancements in the technology, such as improved computation capacity, low-cost sensors, artificial intelligence-based algorithms, and fuzzy logic–based decision-making abilities, enable UAVs to easily perform many practical applications in complex environments that otherwise would take a long time and require significantly high costs. The economic and potential applications of UAVs in the real world are most lucrative, including distribution of vaccines [2], tourism security and safety [3], vegetable inspection [4], document delivery for libraries [5], industrial applications [6], forest and urban firefighting [7], sensing of large areas [8], forestry applications [9], aerial forest fire detection [10], estimating forest structure [11], traffic monitoring [12], retrieving tree volumes in forests [13], scientific research data collection [14], optical remote sensing [15], disaster assessment and management [16], mountain anti-terrorism combat [17], crust detection on steel bridges [18], vehicle detection in real-time, tracking and speed estimation [19], and ocean exploration assignments [20], among others. Moreover, the UAVs' next generation will offer more unique advancements that may increase their use in military applications around the globe [21,22].

In the majority of civilian or military applications, a UAV usually needs the ability to search for the target location in a short time while avoiding collisions with obstacles it may face during the mission. However, without human onboard control, UAV use brings many challenges that need robust solutions, and, among those challenges, one is searching for an optimal/quasi-optimal, safe, and time-efficient path between two locations in a 3D map. Due to the large-scale utilization of UAVs in countless sectors, the Path Planning (PP) problem has become a very vibrant research topic. PP is a method of finding a workable path between two locations while safely bypassing obstacles present in the underlying 3D environment map, simultaneously satisfying one or more optimization objectives, such as distance, time, and consumption of energy [23]. PP is regarded as a Non-deterministic Polynomial-time (NP)-hard optimization problem in the robotics field. Generally, there are two types of PP problems: global PP and local PP. In global PP, finding a path that is performed in an environment that is known. However, local PP is relatively complicated because the UAV operating environment can be partially or fully unknown. Taking into account the mission scenarios of UAVs, pathfinding problems can be divided into two categories: single-agent and multi-agent. In the latter scenario, the number of deployed UAVs is more than one, unlike the former in which only one UAV is deployed. The process for finding a path generally begins with searching a waypoint/visibility graph from one location and progressing until the target is found. The quality of a PP method usually relies on choosing low-cost path waypoints from a given graph that contributes to an optimal/quasi-optimal path with the fewest computations. This study focuses on a single-UAV PP problem, and our aim is to lower the time complexity without degrading the path quality.

Many global PP solutions have been designed for augmenting a UAV's autonomy in various practical missions in the airspace [24–28]. The pathfinding procedure mainly encompasses three key steps: (i) modeling the operating environment (e.g., the environment's representation with a graph), (ii) employing a search algorithm on the graph to determine a path, and (iii) applying a heuristic function (e.g., smoothness, energy, distance, or turns) that accompanies the path search. UAV operating-environment depiction with precise geometry is imperative in order to determine a low-cost path. Roadmap [29], cell decomposition [30], and potential field [31] are renowned environment representation approaches for the configuration space. The search algorithm analyzes the graph for low-cost pathfinding. Many algorithms for PP on graphs have been developed since 1959 such as Dijkstra's algorithm [32] and best first search-based greedy algorithm [33]. Both of these algorithms are regarded as pioneer pathfinding algorithms based on a graph search. However, the A* algorithm [34] is known as the benchmark and is extensively used for a low-cost path search. It is more robust than Dijkstra's algorithm and its variants. Aside from these famous algorithms, many improved versions of the A* algorithm such as IDA* [35], Theta* [36], Lazy-theta* [37], LPA* [38], and D*-Lite [39] also have the ability to find a working path.

Most of the prior PP methods for UAVs do not present deep insights into space reduction with a good-quality path guarantee, specifically regarding the effective resolution of the speed-versus-optimality trade-off in complex 3D urban environments. The prior PP solutions mainly focus on constructing better heuristic functions, and, thereby, memory overhead can occur. Most algorithms sacrifice either optimality or speed while finding paths. Meanwhile, in many practical applications for a UAV, the trade-off on any of the given metrics (e.g., speed or optimality) is not tolerable. Hence, it is mandatory to reduce exploration and modeling of low-probability spaces to overcome these computing issues. Various space modification methods have been designed to increase the pathfinding speed, such as abstractions in a hierarchical form [40], symmetry breaking [41], subgoal graphs [42], jump-point searches [43], accurate heuristics [44], compressed path databases [45], pruning dominant states [46], swamp hierarchies [47], influence-aware pathfinding [48], and constraint-aware methods for navigation [49]. Besides the validity of these latest developments, in most cases, either many low-priority locations of a map are searched uselessly or path quality significantly degrades. Recently, a number of studies considered reducing computation times by dealing with pertinent obstacles that are only crossed along a straight axis in the pathfinding process [50,51]. However, these mechanisms have higher computational complexity and yield non-taut paths if obstacle density is high. Hence, these methods are vulnerable to either returning longer paths or demanding more computing power in determining a path. To address the above limitations, this study presents a new PP method that significantly lowers pathfinding computing time without impacting path lengths by leveraging a Constrained Polygonal Space (CPS) and an Extremely Sparse Waypoint Graph (ESWG) while finding a working path from a 3D urban environment.

The rest of this paper is structured as follows: Section 2 presents the background and related work on renowned PP algorithms. Section 3 illustrates the proposed PP method and describes its main steps. Section 4 explains the results obtained from the simulations. Finally, the conclusions and future avenues for research are discussed in Section 5.

2. Background and Related Work

In this section, we briefly discuss the UAV operating environment's modeling techniques, the pathfinding algorithms, and geometric- and sampling-based PP methods. The initial step of the global PP is to model the real environment with correct geometric shapes. It is closely linked to the choice of search algorithm because most search algorithms yield good performance when they are collectively employed with a particular environment's illustration. A comprehensive discussion about the performance impacts of distinct environment modeling techniques collectively tested with their respective search methods was given by Sariff et al. [52]. Many UAV operating environment methods have been discussed in the published studies. These modeling methods are categorized as RoadMap (RM), Cell Decomposition (CD), and Potential Fields (PF). Hyungil et al. [53] presented a comprehensive survey on environment modeling techniques used in PP. Each modeling method differs in terms of the scale of space/time complexity, the modeling method's accuracy, and the path quality. For example, when the cell sizes are relatively small, CD-based methods yield poor path quality. In contrast, if the cells are too wide, they are vulnerable to very high time and space complexity. The PF-based methods are prone to getting trapped in local minima, and, thereby, solution quality can be degraded. After modeling the environment with a visibility/waypoint graph, a search algorithm is utilized for the graph's exploration in order to find a path.

Most of the existing search algorithms explore and model whole maps during the PP that can lead to various overheads, such as resource-hogging, needless exploration of many parts of a map, and latency issues during pathfinding. Generally, they do not take advantage of the available useful knowledge related to obstacles' geometries from underlying environments in order to lower the complications in path computing. While finding a path from a provided graph, they mostly hold all edges that are visible in the memory, thereby memory requirements of these algorithms are high. Current bio-inspired search algorithms are vulnerable to pre-mature convergence by relying solely on the specified parameters that can lead to poor path quality. In addition, they were mostly tested in semi-urban environments, and their completeness property may yield infeasible results in realistic-urban environments. To address these technical problems, we proposed a new PP method for computing low-cost paths in order to facilitate UAV's aerial missions in urban environments.

2.1. Geometric Path Planning Methods

A geometric PP method that assists in determining a good-quality path in a 3D environment of relatively higher complexity was given in [54]. The environment is represented by using a height reduction strategy to solve the trade-off between path-finding efficiency and accuracy in environment modeling. Unfortunately, it does not reduce the searches in the left-over parts of the area, and, thereby, computing time can be higher in most cases. An incremental PP algorithm considering both local and global constraints for good quality pathfinding was designed by Hu et al. [55]. It is fast, and it reduces the set of good-quality path candidates to only four, with minimal computing time. However, the study ignores space reduction to efficiently find a candidate solution. A geometric PP method considering the minimum turn radius of a UAV for optimal paths in a 3D space was designed by Sikha et al. [56]. The suggested concept is reliable and assists in determining a path with the least complexity. An enhanced heuristic-based PP method to find a good-quality path efficiently by considering UAV flight limits was designed by Kun et al. [57]. A new over-segmentation-based method to determine the free-space overlay of a connected region set was suggested by Plaku et al. [58]. This method quickly finds a safe and good-quality path. However, the approach does not take into account information about sharp turns, narrow passages, and other environmental constraints, which may degrade the suggested method's utility. Furthermore, to augment both efficiency and accuracy, it is extremely important to find irrelevant areas that can be discarded if they cannot help to find an optimal/quasi-optimal path in an environment [59]. Several studies have designed closely related PP methods with undoubtedly reduced time cost, such as the Approximation with Visibility Line (ApVL) method [60]. The ApVL PP method [60] is an improvement of the Base Line-Oriented Visibility Line (BLOVL) algorithm [50], and it is regarded as a highly suitable algorithm for finding an approximate shortest path in 3D urban environments. It reduces the obstacle count significantly (e.g., it processes obstacles that are on a straight line only), and constructs visibility graphs from the chosen obstacles' corners only to incrementally find a low-cost path. Meanwhile, it has relatively higher time complexity. In addition, it either yields longer paths or requires more processing to find a working path. In some scenarios, it is even unable to find a flyable path owing to connected obstacles with straight-line obstacles.

2.2. Sampling-Based Path Planning Methods

Sampling-based methods include the Probabilistic Road Map (PRM) [61], Rapidly Exploring Random Trees (RRTs) [62], and their refined versions. These methods have demonstrated effectiveness at quickly generating near-optimal/optimal global solutions. Their algorithmic simplicity makes sampling-based methods applicable to solving both real-time and single-query PP problems. The RRT PP method and its subsequent versions such as informed RRT* [63], Transition-aware RRT (T-RRT) [64], RRT-connect [65], and AnyTime-RRT (AT-RRT) [66] are all complete probabilistically. Most RRT-based methods yield slow convergence rates in complex environments, and they mostly fail to resolve the trade-off between length and time while finding reasonable-quality paths. Sertac et al. [67] designed a better version of the original RRT, named RRT*. This method has a fast rate of convergence compared to RRT, and it has an ability to find a quasi-optimal path with minor post-processing. However, computing issues such as pre-mature convergence, high space complexity, path searching from a whole map, and discarding beneficial samples while converging into a solution pre-maturely make it unreliable for solving practical missions. Jauwairia et al. [68] designed a new variant of RRT* named RRT*-Smart. It has a faster convergence rate, compared to the traditional RRT* algorithm, by using smart-sampling and optimization techniques. However, the main limitations of RRT*-Smart are higher sensitivity to the operating environment, too many iterations, and extensive memory consumption. Yanjie et al. [69] suggested a sampling-based PP method with improved convergence rate. Iram et al. [70] presented a concept relatively closer to our PP method, called RRT*-AB (Adjustable Bounds), to determine low-cost paths. It shows better results than the traditional RRT* method. However, it yields time performance issues due to the near-neighbor search and extensive rewiring operations while optimizing the path lengths. Hence, the shortest paths determined by the existing methods have higher time complexity. Accordingly, a constrained space complexity analysis with in-depth complexity parameters and an obstacle's geometry information has not been simultaneously explored to find a good-quality path with the least time cost.

3. The Proposed Method

A constrained polygonal space and an extremely sparse waypoint graph–based PP method are imperative for addressing the time complexity issues that emerge due to unnecessary path exploration of low-probability spaces on an obstacles-rich map. The proposed PP method limits path exploration to only the constrained spaces that have a higher probability of containing optimal/quasi-optimal paths, and it safely removes the unlikely spaces in order to hasten the pathfinding computations. It removes the only spaces from a map that likely cannot assist in finding a low-cost solution with high probability. It effectively resolves the two competing goals of efficiency and path length while finding paths for UAVs in urban environments. This section provides a brief overview of our proposed PP method and outlines its workings. In Figure 1, we demonstrate the proposed PP method's conceptual overview.

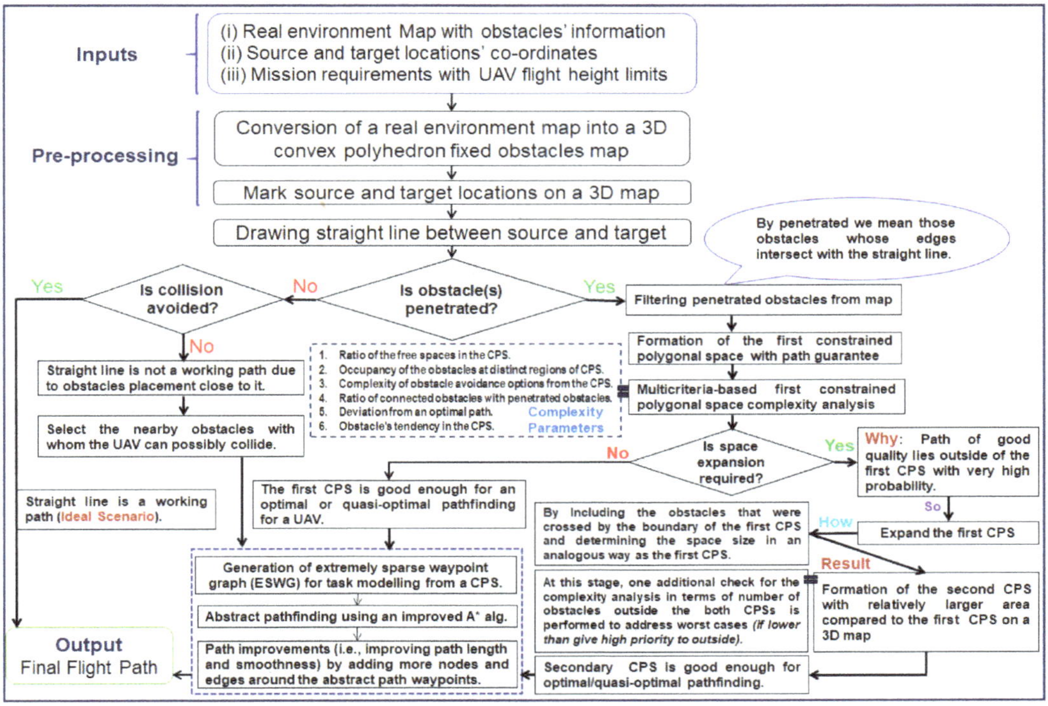

Figure 1. The proposed PP method's conceptual overview.

To find a path, P, between source s and target location t for a UAV, while safely bypassing obstacles in UAV flying environment W, the following six key conceptualizations are introduced: (i) operating-environment modeling by using data from a real environment map; (ii) generation of a constrained polygonal space by exploiting obstacle geometry information; (iii) determining and analyzing the complexity of the constrained polygonal space using a multiple criteria-based method leveraging six different complexity parameters; (iv) providing need-based extension of the constrained polygonal space to the next level; (v) task-modeling with an extremely sparse waypoint graph that has as few nodes and edges as possible by utilizing the concepts of far distance reachability and direction guidance; and (vi) abstract pathfinding with the A* algorithm from the CPS using an ESWG, and enhancing the abstract path quality by generating additional nodes and edges in the vicinity of abstract path nodes. Concise descriptions of each main component, with relevant equations/procedures, are summarized below.

3.1. Representation of the Environment Where the UAV Operates

The initial step in the PP process is to represent the UAV's moving/flying environment from a real environment map with the help of relevant geometrical shapes. Generally, it is a process of dividing W into obstacle-free regions (ξ_{free}) and the obstacle regions ($\xi_{obstacles}$). An example of ξ_{free} and $\xi_{obstacles}$ is presented in Figure 2, in which black and yellow regions represent the ξ_{free} and the $\xi_{obstacles}$ regions, respectively. The obstacles in W can be modeled with sufficient accuracy by leveraging geometric shapes, such as cubes, rectangles, cylinders, circles, polygons, prisms, etc. In our work, obstacle modeling from a raw environment map is carried out with the help of 3D convex polyhedrons having six faces each. The minimum height (e.g., z_{min}) of each obstacle is 0, and the three other dimensions are random. To generate a convex-obstacles set, we extract the digital map's elevation readings in accordance with digital environment elevation data standards, and we find a convex-hull to accomplish the obstacle modeling task. After the real map's processing, we create a map with fixed convex-polyhedron 3D obstacles. The beginning location of the mission is denoted with s (i.e., a 3D point), where $s = (x_s, y_s, z_s)$. The target location is denoted with t (also a 3D point), where $t = (x_t, y_t, z_t)$. Taking into account W's representations, the UAV profile information, and path searching from a 3D environment, the objective of the proposed PP method is to find good-quality paths with the least computing complexity. The proposed PP method fulfills the stated assertions by finding a path from high-probability regions of the map, and the UAV is considered a single 3D point, just like s or t.

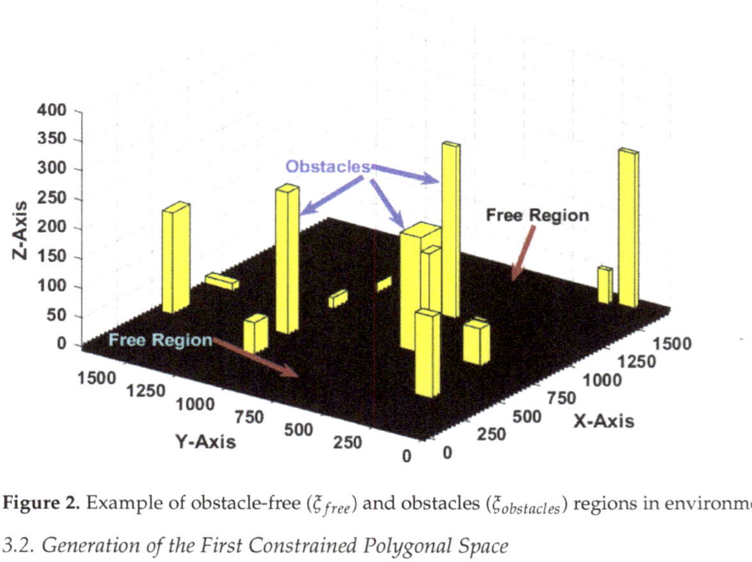

Figure 2. Example of obstacle-free (ξ_{free}) and obstacles ($\xi_{obstacles}$) regions in environment map W.

3.2. Generation of the First Constrained Polygonal Space

Searching for a path by leveraging full map information can be very time consuming, and it may result in serious computing overhead in complex and large urban environments. To address these issues, we convert the full 3D map into a CPS that guarantees P from s to

t, and at the same time makes pathfinding computations faster. The space can be reduced using five main steps: (i) stipulating both s and t for a mission, (ii) sketching a straight-line $\overline{l_o}$ between s and t, (iii) extracting pertinent obstacles (i.e., only those obstacles where edges/vertices cross with the $\overline{l_o}$), (iv) analyzing the geometry of the extracted obstacles, and (v) drawing a minimum-span ζ_{min} polygonal space S_1^3 from the base vertices of pertinent obstacles in such a way that a cross section of the reduced space (e.g., S_1^3) is not completely blocked by all of the pertinent obstacles' cross sections. More specifically, the ζ is selected in such a way that the path can be found in each scenario from the constrained space. Obstacles that are on the $\overline{l_o}$ between s and t are called pertinent obstacles. After receiving the s and t locations for the flight, we sketch a $\overline{l_o}$ from s to t. After sketching the $\overline{l_o}$, four outputs can be derived, which are: (I) neither penetration nor collision, (II) not penetration, but the possibility of a collision exists, (III) penetration, but all obstacles have a lower height, and the UAV can go over them safely, and (IV) both penetration and collision. All four results are demonstrated in Figure 3.

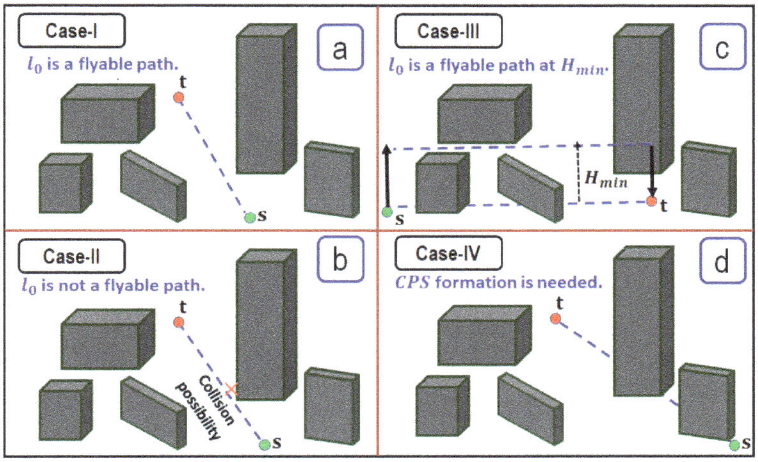

Figure 3. Results of sketching a straight line ($\overline{l_o}$) between s and t locations.

If the $\overline{l_o}$ does not collide with any obstacle, and the collision possibility is zero, then path $P = \overline{l_o}$, which is an optimal path (e.g., a straight line) as given in Figure 3a. Moreover, if the $\overline{l_o}$ does not penetrate any obstacles, but there remain some obstacles in close proximity to the $\overline{l_o}$ with which the UAV could collide with a higher probability, then we deal with these obstacles to generate a safe P, as demonstrated in Figure 3b. In the third case, as given in Figure 3c, few obstacles are penetrated by the $\overline{l_o}$, all obstacles have a lower height, and the UAV can go over them safely. In the fourth case, as given in Figure 3d, some obstacles are penetrated by the $\overline{l_o}$, and we need to bypass them at a lower cost to find a P between s and t while fulfilling the stated objectives. Such obstacles are extracted from the map and utilized for pathfinding from the CPS. The complete pseudo-code utilized for extracting pertinent obstacles is illustrated in Algorithm 1. In Algorithm 1, map W encompassing N distinct obstacles, s, and t, are given as input. The set E, where ($E \subseteq N$) of the pertinent obstacles is retrieved as output. Line 2 can do $\overline{l_o}$ sketching between s and t. Lines 3–7 can do obstacles' extraction that are crossed by the $\overline{l_o}$. At the end, set E of pertinent-obstacles is collected. Moreover, if no intersection occurs between the obstacles and the $\overline{l_o}$, and, then, $E = \varnothing$ will be the output.

After getting set E of pertinent obstacles, we enlarge the obstacles by a safe distance (d_{safe}), and apply the flying minimum and maximum limits, denoted as h_{min} and h_{max}, respectively. Subsequently, we analyze the pertinent obstacle cross sections that are on the $\overline{l_o}$ between s and t. Then, we draw a bottom boundary, Ω_b, around the bottom vertices of the pertinent obstacles, and a top boundary, Ω_t, from minimal height h with path guarantees keeping s and t as two endpoints. This forms a 3D constrained space where the shape can resemble a 3D polygon, and we call this constrained region the CPS, represented with S_1^3. This CPS converts the more difficult problem of UAV pathfinding into the relatively easier problem of pathfinding for a 3D point. A visual overview of S_1^3 is demonstrated in Figure 4. The S_1^3 can be simply defined as full map W partitioned into a small space/region where the outline is the same as a 3D polygon, with s and t as two endpoints. It is obtained by drawing a boundary around the outermost vertices of the pertinent obstacles in such a way that a path can be found from it regardless of its quality. The process of transforming a full map into S_1^3 is given in Figure 4a–c. In Figure 4a, environment map W is shown, which will be converted into S_1^3. Figure 4b shows the $\overline{l_o}$ drawing, and identifies the corresponding pertinent obstacles (e.g., yellow obstacles) that were crossed by it; these obstacles will be used to subsequently generate S_1^3. Figure 4c shows the outline of S_1^3 in 2D form that is obtained by drawing a boundary around obstacles identified in Figure 4b as a consequence of $\overline{l_o}$ penetration.

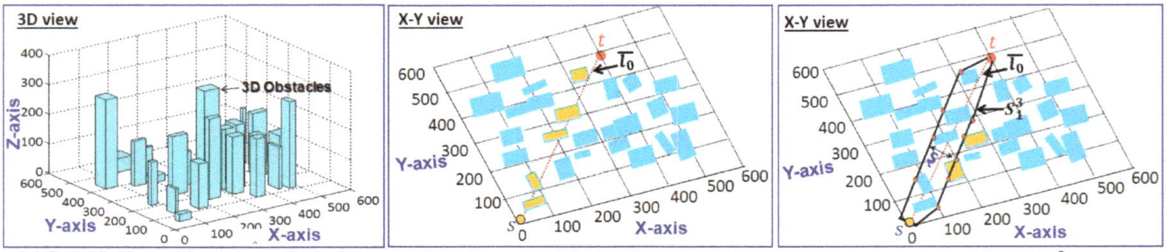

(a) Full 3D Map with various obstacles (b) Drawing a straight line between s & t (c) Formation of the 1st CPS (S_1^3) on a map

Figure 4. Overview of transforming a full 3D map W into a constrained polygonal space (S_1^3).

Algorithm 1: Extracting pertinent obstacles from a 3D obstacle map.

Input : (i) Environment map W with N distinct obstacles
(ii) Starting location (s)
(iii) Ending location (t)
Output : Pertinent obstacles' set E
Procedure:
Initialize, $E = \varnothing$
Sketch straight line $\overline{l_o}$ between (s) and (t) // Assuming case IV given in Figure 3
  for every obstacle O_j, beginning from $O_j = O_1$ to the $O_n \in N$ do
    if CROSSES ($\overline{l_o}$, O_j) then
      $E = E \cup \{O_j\}$
    End if
  End for
return E

The CPS can enclose the pertinent obstacles—just part of them or as a whole. In some cases, due to complex 3D environments, S_1^3 can enclose obstacles that do not belong to set E but that are part of S_1^3, either partially or completely, and we include such obstacles in E and utilize them in pathfinding. The essence of S_1^3 is that it guarantees a flyable path between s and t. However, S_1^3 may or may not be an ideal choice for a good-quality path (e.g., optimal or quasi-optimal) due to several complexity parameters (as illustrated in Figure 1) about obstacles. By considering such potential complexity parameters and a probabilistic analysis of optimal paths, we conducted a CPS complexity analysis leveraging six complexity parameters (also known as complexity constraints), prevailing in S_1^3 that relate to the obstacles' geometries, and a low-cost path tends to lie outside S_1^3, in most cases, with a significantly higher probability.

3.3. Determining and Analyzing the Constrained Polygonal Space Complexity Using a Multicriteria-Based Method

In order to check whether S_1^3 is good enough for optimal/quasi-optimal pathfinding or not, we performed a multi-criteria–based complexity analysis of S_1^3 using multiple complexity parameters before task modeling and pathfinding. We computed the complexity, χ, of S_1^3 by leveraging detailed information regarding the obstacles' geometries. We employed six complexity parameters: the proportion of free spaces, the obstacle occupancy in distinct regions of the CPS, the complexity of obstacle–avoidance options, the proportion of connected obstacles, the length deviations from the optimal path, and obstacles' tendency in the CPS that hinders the solution quality. Through extensive simulations and analysis, it is found that there exists a very firm relationship between the complexity parameters of the CPS and path quality. The total χ is the weighted sum of six parameters cited above. Brief overviews of those six complexity parameters, with their procedures and equations, are described below.

3.3.1. Free Spaces' Ratio

To compute a feasible, safe, and smooth P, it is highly enticing that the amount of free spaces must be high in the CPS. To measure the obstacle-free spaces in the CPS, we first determine the size (ξ) of the S_1^3, blocked-spaces ($\xi_{obstacles}$), and free-spaces (ξ_{free}). The overall size (ξ) of the CPS that is in the form of a polygon can be obtained by the Gauss determinant using Equation (1):

$$\xi = \frac{1}{2}\sum_{i=1}^{|g|}(x_i y_{i+1} - x_{i+1} y_i) \times h \tag{1}$$

where x, y are the coordinate values of the CPS boundary, $|g|$ denotes total vertices of the CPS boundary, and h is the height of the CPS. Out of the ξ-sized CPS, we find the amount occupied by the obstacles ($\xi_{obstacles}$) using Equation (2):

$$\xi_{obstacles} = \sum_{i=1}^{n} \omega O_i \tag{2}$$

where n denotes total obstacles' count present in a S_1^3 and ωO_i denotes the obstacles' occupancy. The occupancy of an O_i obstacles can be determined using Equation (3):

$$\omega O_i = O_H \times O_L \times O_W \tag{3}$$

where O_W, O_H, O_L denote the width, height, and length of an obstacle, respectively. The free space (X_{free}) amount, where UAV can fly safely, can be determined using Equation (4):

$$\xi_{free} = \xi - \xi_{obstacles} \tag{4}$$

The ratio (r_f) of the free spaces can be calculated using Equation (5):

$$r_f = \frac{\xi_{free}}{\xi} \tag{5}$$

The value of r_f ranges between 1 and 0. We represent this ratio as $(1 - r_f)$ to compute the occupied spaces value in overall CPS S_1^3 complexity computation.

3.3.2. Deviation in Length from an Optimal Path

The proposed method is a global PP approach, and all information about the obstacles geometries is known in advance. By utilizing the obstacles geometry information, we can estimate the path length without calculating the actual path. We estimate length of an optimal path L_o, where $L_o = \overline{l_o}$ as an optimal path and estimate the deviation in it that can occur due to obstacles that are on $\overline{l_o}$. We call it deviation $D^{'}$ from the optimal path and formalization used to estimate is explained below.

- Computing the optimal path L_o that is a straight line between s and t using Equation (6).

$$L_o = \sqrt{(x_t - x_s)^2 + (y_t - y_s)^2 + (z_t - z_s)^2} \quad (6)$$

- Calculating the deviation $D^{'}$ in optimal path L_o due to the presence of obstacles on the $\overline{l_o}$ between s and t locations. The D in the paths length due to an obstacle (e.g., O_i) is calculated using Equation (7).

$$D_i = min(\frac{O_W}{2}, O_H) \quad (7)$$

- Estimating the total deviation $D^{'}$ that can likely occur due to the presence of the obstacles between s and t in the CPS using Equation (8).

$$D^{'} = \sum_{i=1}^{n} D_i \quad (8)$$

- Calculating the length of the estimated paths (L_1) avoiding all obstacles that are on the $\overline{l_0}$ using Equation (9).

$$L_1 = L_o + \sum_{i=1}^{n} D_i \quad (9)$$

  where L_1 denotes the lengths of the paths avoiding obstacles in the selected space, L_0 denotes the Euclidean distance between s and t locations, n represents obstacles' strength in the selected space, and D_i denotes the degradation in path length due to obstacles.
- Computing the complexity C_p of the estimated path that can assist in analyzing the S_1^3 complexity using Equation (10):

$$C_p = \frac{L_o}{L_1} \quad (10)$$

  The value of the C_p ranges between 1 and 0. We use this value to represent the estimated path complexity in terms of length in overall CPS S_1^3 complexity evaluation in Equation (20).

3.3.3. Complexity of the Obstacles' Avoidance Options

There are usually four options in total to bypass any obstacle present in a W such as right, left, up, and down (in the case of hanged obstacles or flying obstacles). Meanwhile, after the space reduction, the number of options to avoid obstacle will likely be reduced, and there can be an increase in the complexity of remaining options. Because of this, a path may be taut and path length can be prolonged. Hence, while determining the CPS complexity, we take into account the complexity of options needed to bypass obstacles. To calculate the complexity (C_{AO}) of each option, the entropy concept is employed. Entropy is acknowledged as the most effective and accurate measurement for similar tasks in numerous fields. In this work, we consider the urban environment; therefore, the P cannot go beneath the obstacles since bottom height of each obstacle is zero, and, hence, there are only three options in total to avoid any obstacle. The strategy below is employed to calculate the C_{AO}.

1. Find the proportion (p_i) value of every avoidance option (i.e., AO_l, AO_r, AO_t) category using Equation (11).

$$p_i = \frac{AO_i}{b} \quad (11)$$

   where b denotes the total AO, and its value can be determined using Equation (12).

$$b = \sum_{i=1}^{3} AO_j \tag{12}$$

2. The complexity C_{AO} of avoidance options can be calculated using Equation (13):

$$C_{AO} = -\sum_{i=1}^{b} p_i log_2 p_i \tag{13}$$

The normalized value of the C_{AO} lies between 0 and 1, denoted as $C_{AO} \in [0, 1]$. The C_{AO} with 0 value means that avoidance complexity is low (e.g., all obstacles can be avoided from the same side). In contrast, the C_{AO} value 1 means that enough variations exist in options to avoid all obstacles, and the path can contain many turns. In the CPS analysis, we take into account the C_{AO} values.

3.3.4. Occupancy of Obstacles at Distinct Regions of the CPS

Besides the other complexity parameters described earlier, another important parameter that can lead to genuine performance concerns while pathfinding is the obstacles' occupancy at distinct regions of the CPS. If obstacles in large numbers are clustered at one location (e.g., obstacles' placement in the CPS is uneven), then the path quality likely degrades. The obstacle occupancy at one place introduces cycles/sharp-turns in a P because the P revolves around boundaries of many obstacles before approaching t. To calculate occupancy Π_o of obstacles, we partition S_1^3 into n sub-spaces $\{s_1, s_2, s_3, \ldots, s_n\}$ and find the obstacles occupancy in each subspace s_i. The obstacles' occupancy in a s_i subspace can be determined by taking the ratio of the obstacles' occupancy $\xi_{obstacles}$ in the s_i divided by overall obstacles' occupancy in the CPS. To calculate occupancy, the S_1^3 is partitioned in five equal-size sub-spaces. The occupancy Π_{o_i} of the s_i can be mathematically expressed as

$$\Pi_{o_i} = \frac{s^i_{obstacles}}{\xi_{obstacles}} \tag{14}$$

where $\xi_{obstacles}$ denotes occupancy of all obstacles in total from the CPS as given in Equation (2) and $s^i_{obstacles}$ represent the ith subspace's obstacles occupancy, and its value can be computed using Equation (15):

$$s^i_{obstacles} = \sum_{i=1}^{O'} \omega O'_i \tag{15}$$

where $\omega O'_i$ represents an obstacle's volume, and O' denotes all obstacles count in the CPS. After computing the occupancy of five sub-spaces, we determine the overall occupancy Π_o of the S_1^3 using the following equation:

$$\Pi_o = max\{\Pi_{o_1}, \Pi_{o_2}, \Pi_{o_3}, \ldots, \Pi_{o_n}\} \tag{16}$$

where Π_o is the obstacles' occupancy in the S_1^3. The rationale to choose maximum values is to effectively deal with the worst cases. The occupancy analysis assists with finding the smooth paths by giving considerable attention to the regions of high occupancy.

3.3.5. Ratio of Obstacles' Tendency in the CPS

In some scenarios, the CPS can enclose more obstacles compared to the W (e.g., the tendency of obstacles on $\overline{l_o}$ is high compared to the whole W). Hence, it is viable to assess the impact of obstacles' tendency to yield a good quality path. To analyze the obstacles' tendency T_o, we find the number of obstacles in a CPS, and take a ratio with the obstacles' count present in a W. We denote the number of obstacles present in the CPS with n and number of obstacles present in a full map with N, respectively. We determine the value of T_o using Equation (17):

$$T_o = \frac{n}{N} \tag{17}$$

The value of T_o can lie between 0 and 1, $T_o \in [0, 1]$. The higher value of T_o means that more obstacles are present in the CPS. In our work, we take into account the T_o value while determining and analyzing the complexity of the S_1^3.

3.3.6. Ratio of the Connected Obstacles

In some cases, some obstacles exist that are not directly penetrated with $\overline{l_0}$, but they have connections with the pertinent obstacles (e.g., obstacles directly crossed with $\overline{l_0}$). These obstacles can escalate the time of path computation and yield unnecessary turns in the path. Hence, while analyzing the CPS complexity, it is paramount to take into account the connected obstacles' effect along with other five complexity parameters. The ratio of the connected obstacles can be found by counting the connected obstacles in the CPS divided by the obstacles' count in the S_1^3. The count of connected obstacles can be determined using Equation (18):

$$n' = \sum_{j=1}^{n} (O_j \cup O_{CON}) \tag{18}$$

where O_j is the pertinent obstacle, and O_{CON} denotes the connected obstacle with O_j (e.g., pertinent obstacles). The overall ratio of the connected obstacles (r_{co}) can be computed using Equation (19):

$$r_{co} = \frac{n'}{n} \tag{19}$$

where n' represents connected obstacles' strength, and n shows the number of obstacles in E.

When all six complexity parameters' values have been calculated, the total complexity χ of the S_1^3 can be quantified using Equation (20):

$$\chi(S_1^3) = w_1 \times (1 - r_f) + w_2 \times C_p + w_3 \times C_{AO} + w_4 \times \Pi_o + w_5 \times T_o + w_6 \times r_{co} \tag{20}$$

In Equation (20), $1 - r_f$ denotes the ratio of spaces occupied by obstacles, C_p denotes the deviation in path length from an optimal path, C_{AO} means the complexity of options while avoiding obstacles, Π_o is the occupancy of the obstacles at the distinct region of the CPS, T_o denotes the tendency of obstacles in the CPS in relation to a full map, and r_{co} denotes the ratio of connected obstacles. The drawing of all six complexity parameters described in prior subsections (e.g., Sections 3.3.1–3.3.6) is given in Figure 5.

For calculation simplicity, we used complexity parameters values in normalized form; therefore, the $\chi(S_1^3)$ ranges between 0 and 1. In Equation (20), w_i, where $i = 1, 2, 3, 4, 5, 6$ represents each complexity parameter's weight, and they fulfill two conditions, (i) $w_1 + w_2 + w_3 + w_4 + w_5 + w_6 = 1$, and (ii) $w_i > 0$. We adjust each parameter weight by taking into account the significance and influence of every parameter in the CPS complexity analysis. The probability σ of a path P to be found from the S_1^3 with good quality is given as follows:

$$\sigma(P) = \begin{cases} 1, & \text{if } 0 < \chi(S_1^3) < T. \\ 0, & \text{otherwise.} \end{cases} \tag{21}$$

where $\chi(S_1^3)$ represents the complexity of the S_1^3, and T denotes a threshold. If $\sigma(P) = 1$, no extension of space is required because S_1^3 is appropriate for low-cost pathfinding. Meanwhile, if $\sigma(P) = 0$, then an additional space will be needed to find good quality paths. The threshold T value relies on numerous global factors of the W, and local constraints (e.g., related to the UAV). In simulations, we set the threshold value to 0.7 to make a decision about the space expansion. We did substantial experiments to validate the T value using P's length as a main criteria. However, T's value can be tuned flexibly based on the UAV's workspace and resources.

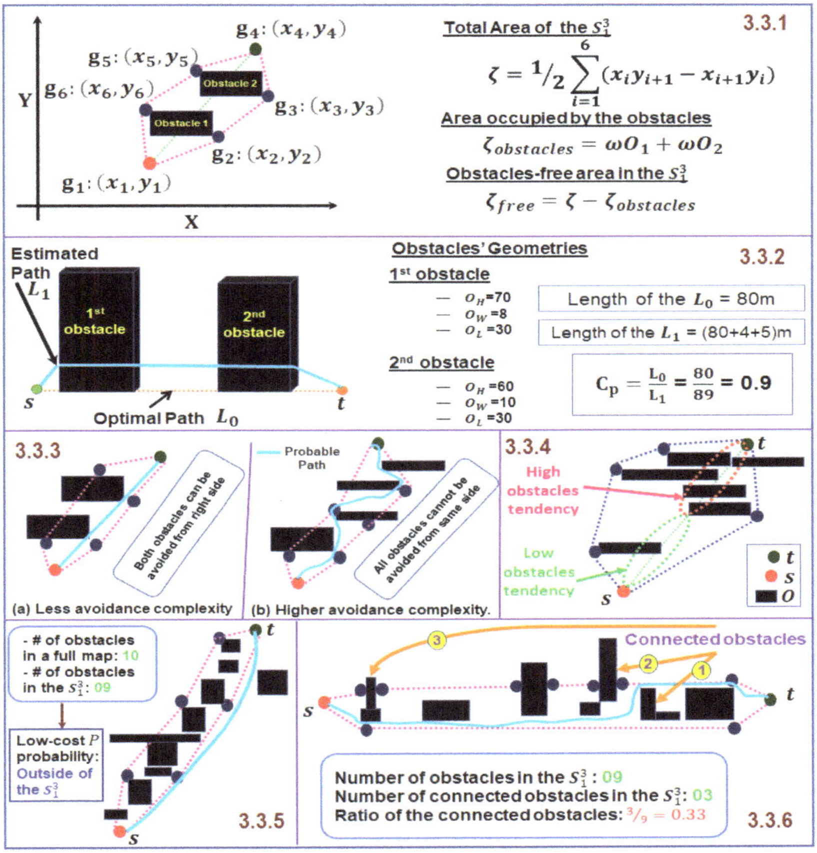

Figure 5. Drawing of the complexity parameters employed to analyze the complexity of the S_1^3.

3.4. Expansion of the First Constrained Polygonal Space

Although S_1^3 always finds path P, it does not guarantee the quality of P in each scenario due to higher complexity in the obstacles. To circumvent this issue and ensure consistent quality for P in each scenario, the scenarios that require a bigger space are identified carefully through the CPS complexity analysis utilizing the six different parameters. With the assistance of the complexity analysis of S_1^3, for a good-quality P, we can accurately identify the cases that require relatively more space than already in S_1^3. Having sufficient information about the obstacles connected with the boundary of S_1^3 enables us to flexibly expand the space to the next level. We adopted this method to expand the space, since it yields less computing overhead and significantly enhances path quality. Hence, by processing obstacles that are penetrated by the boundary of S_1^3, and by marking a polygonal boundary in an analogous way, the first CPS formation emanates into a second CPS of a relatively bigger size, compared to the S_1^3 as visually depicted in Figure 6b.

We denote this expanded space with S_2^3, and it encompasses S_1^3 fully. The S_2^3 includes pertinent obstacles fully, and it provides greater opportunities for P to be determined from S_2^3 solely. The utility of the S_2^3 is that it is highly desirable space for producing P of shortest lengths. The space can be extended to nested levels in identical manner. Meanwhile, we expand the spaces only up to 2-levels because an optimal/quasi-optimal P tends to lie in S_1^3 and S_2^3 with acceptable probability. After the selection of appropriate highest priority space, an ESWG is constructed for pathfinding.

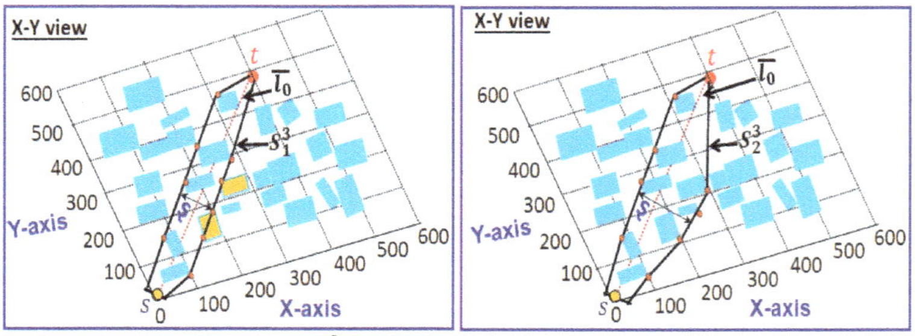

(a) Formation of the 1st CPS (S_1^3) on a map (b) Formation of the secondary CPS (S_2^3)

Figure 6. Overview of constrained polygonal space (S_1^3) and its expansion to the next level (S_2^3).

3.5. Extremely Sparse Waypoint Graph Generation from the Selected CPS

Waypoint graph (WG) is one of the approaches for task modeling and pathfinding, respectively. The WG constructs an indirect and compact graph connecting s with t by catching the connectivity of the ξ_{free} to form a multiple paths' network. However, generating a WG is very expensive in terms of computation. The overall time complexity of generating a WG with n nodes is $O(n^3)$. Many studies that have focused on lowering the complexity of the WG generation have been reported by joining adjacent obstacles, altering obstacles' shape, and ignoring small obstacles. More recent research [60] highlights that WG's time complexity in the 3D environments can be decreased to the $O(n^2)$ by only considering the obstacles crossed by the $\overline{l_o}$. To expedite the time complexity reduction of the WG, this paper suggests a new concept of an ESVG construction method which does not compose a dense WG. This method forms an ESVG from the CPS like a roadmap with connectivity between s and t through intermediate nodes. An ESVG is a double-edge type graph G of reachable and mutually-visible locations, mathematically expressed as: $G = \{X, Y\}$.

To construct a G from the 3D CPS, two steps are generally applied: making a nodes' set X and generating an edge set Y. The initial step is about creating nodes set X. We utilized three vertices of obstacles, bottom, top, and mid to make an ESWG for the first time. Both bottom and top vertices' geometry values are known, and mid vertices can be found leveraging the midpoint formula on top and bottom axis values. Every obstacle has total eight vertices (e.g., four bottom and four top). An ith obstacle's vertices and their respective values can be expressed mathematically in a matrix as demonstrated in Equation (22). The height of bottom four vertices of each obstacles are transformed to the h_{min} and top vertices of the obstacles have the same height as of the CPS height (e.g., h). In below metrics, the value of z_{min} is zero but after adjustment becomes $h_{min} = z_{min}$:

$$O_i = \begin{bmatrix} x_{min} & y_{min} & z_{max}; x_{min} & y_{min} & z_{min} \\ x_{min} & y_{max} & z_{max}; x_{min} & y_{max} & z_{min} \\ x_{max} & y_{min} & z_{max}; x_{max} & y_{min} & z_{min} \\ x_{max} & y_{max} & z_{max}; x_{max} & y_{max} & z_{min} \end{bmatrix} = \begin{bmatrix} 1131 & 1632 & 241; 1131 & 1632 & 23 \\ 1131 & 1703 & 241; 1131 & 1703 & 23 \\ 1209 & 1632 & 241; 1209 & 1632 & 23 \\ 1209 & 1703 & 241; 1209 & 1703 & 23 \end{bmatrix} \quad (22)$$

For example, an O whose original z_{max} is greater than the CPS height, and the bottom and top vertices are fully known, the pair of mid-points' two side faces denoted with f_1 and f_2 can be determined using Equations (23) and (24), respectively:

$$f_1 = \{\frac{x_{max} + x_{min}}{2}, y_{min}, z_{min}\}, \{\frac{x_{max} + x_{min}}{2}, y_{max}, z_{min}\} \quad (23)$$

$$f_2 = \{\frac{x_{max} + x_{min}}{2}, y_{min}, z_{max}\}, \{\frac{x_{max} + x_{min}}{2}, y_{max}, z_{max}\} \quad (24)$$

A similar procedure can be utilized to find the pair of vertices around all obstacles. After computing set X from the pertinent obstacles, we add both s and t in a set X and generate set Y of edges through two novel strategies and visibility-checks. Any two nodes u and v in X are inter-visible if a $\overline{uv}$ segment of line connecting them is collision free. A function named-line-of-sight (LOS) determines the visible segments among pairs of nodes through visibility analysis. The time complexity of this mechanism heavily relies on the function of LOS checking, and number of nodes. Meanwhile, in our work, we incorporate two additional strategies of far-reachability (FR) and direction-guidance (DG), thereby time complexity is significantly reduced. In addition, it only adds the edge between a vertices' pair that are as far as possible from each other and that guide to the t's direction. We set the visibility to off/false using coordinate values for those pairs of vertices that are on the same obstacle but do not favor the direction of the t. Hence, the visibility checking function has less time complexity in making a G. The time complexity of an ESWG formation is the $O((nfl)^2)$ time, where l denotes the number of levels, n denotes obstacles' count, and f represents the counts of obstacles' facets. However, the upper-bound of the l is constant (e.g., $l = 3$); therefore, ESWG's time complexity is $O((fn)^2)$. With the help of X and Y, a G is obtained that has reliable connectivity between s and t, and it encompasses all characteristic of a roadmap.

3.6. Path Finding from an ESWG and Enhancing Obtained Path Quality

Once an ESWG is modeled, a path searching algorithm is employed to search a P from it. In this paper, we used A^* algorithm for computing a P between s and t from an ESWG. The A* is reliable algorithms for extracting a P of low-cost. The evaluation function utilized by this algorithm is expressed in Equation (25):

$$f(n) = g(n) + h(n) \tag{25}$$

In Equation (25), the $f(n)$ denotes the estimated path cost in total between s and t via a node n, the $g(n)$ denotes the actual distance to reach node n, and the $h(n)$ is a heuristic function that computes the distance from node n to t. This algorithm was selected to make P's computing process fast. By exploring an ESWG using this algorithm, an abstract P is found. We consider both length and time of the obtained P for evaluating its quality. Meanwhile, in some cases, the P cannot be of minimal length, which needs post-processing to shorten it. We present the working of the A^* algorithm while finding a path between s and t in Figure 7.

In many UAV practical applications, the length of the P is paramount, and to preserve UAV resources, it should be minimum. To address this issue, we shorten P's length by including more nodes in close proximity of it, and refine the sharp turns. The path quality improvements is mainly carried out by determining the adjacent P's neighbor nodes, find the proximity between the adjacent neighbor nodes and P nodes, and in the close proximity of the P, we introduce new nodes with relatively denser resolution and add smooth edges. The reason to add more nodes closely is to retain visibility to improve P quality. After injecting additional nodes, a P of good quality is obtained by jointly using the newly added and the P nodes. This path-refining method has the potential to improve path quality significantly with reduced computing cost.

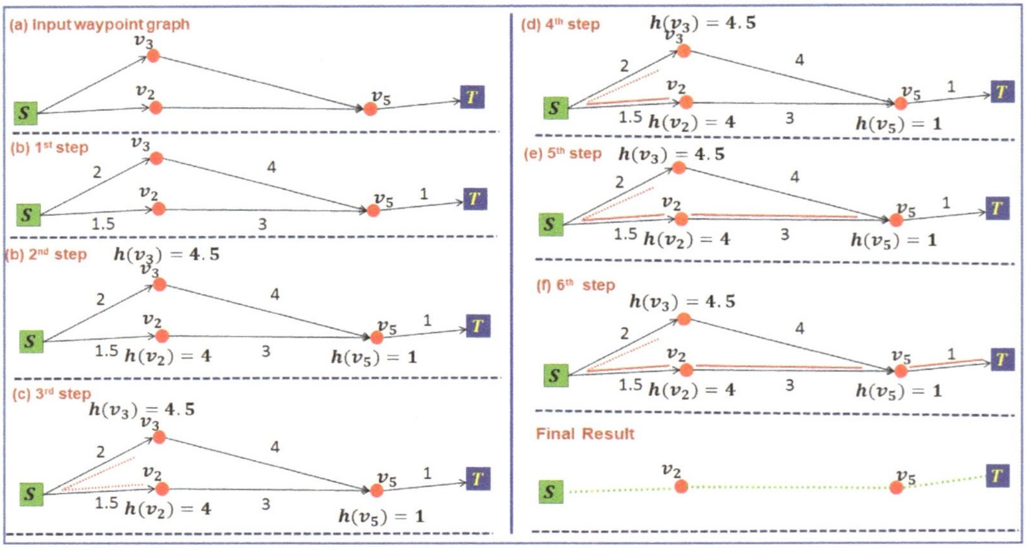

Figure 7. Example about the working of an A* algorithm for low-cost pathfinding.

4. Results and Discussion

This section explains the simulation experiments and corresponding results. The performance of the proposed PP method was analyzed using two criteria: time complexity and path length compared to prior studies. To make the proposed PP method a benchmark, we compared the simulation results with randomized motion planning and visibility graph–based algorithms. The simulation tests were performed and compared using Matlab v. 9.8.0.1451342 (R2020a) on a computer running Windows 10 with 8 GB of RAM and a 2.6 GHz CPU. In the tests, we assumed a 25 kg fixed-wing UAV similar to ones used in existing studies. We took into account both global and local constraints in the simulations. The parameters of the local constraints (i.e., on the UAV) were a 1 m wingspan and a maximum turning angle at a radius of $\pi/6$. The minimum and maximum flying altitudes were $h_{min} = 23$ m, $h_{max} = 155$ m. The global constraints belonged to the geometry of obstacles in W. We consider six complexity parameters that can significantly hinder the quality of a P and the UAV's safety while selecting a space size. We assumed that the UAV had enough battery power to complete the task in one flight. The safe distance to avoid collisions with obstacles was $10m$ ($d_{safe} = 10$ m). We assumed that wind was negligible during the flight. The proposed method finds P using an ESWG that respects both global and local constraints. The weights of space complexity parameters were $w_1 = 0.2$, $w_2 = 0.1$, $w_3 = 0.2$, $w_4 = 0.2$, $w_5 = 0.1$, and $w_6 = 0.2$. We assigned values to these weights by considering the significance and influence of each parameter on the accurate space selection for an optimal/quasi-optimal pathfinding. We tested our method with diverse combinations of the weight values, and analyzed the accuracy of space selection for optimal/quasi-optimal paths. Subsequently, we determined the best combination of these weight indexes' values that make accurate space selection consistently. Furthermore, these weight values were validated via numerical analysis by computing optimal paths from numerous maps using whole W, and analyzing the number of times optimal paths tend to lie in the selected space. The locations for s and t were chosen randomly during experiments. We compared our PP method's performance with two existing algorithms: the ApVL algorithm proposed by Guillermo et al. [60] and the RRT*-AB algorithm proposed by Noreen et al. [70]. Both comparison algorithms are state-of-the-art for PP. We tested them on our maps to compare the performance of our method with them. We show a sketch of the 3D maps employed in the experiments and three exemplary paths from each method in Figure 8.

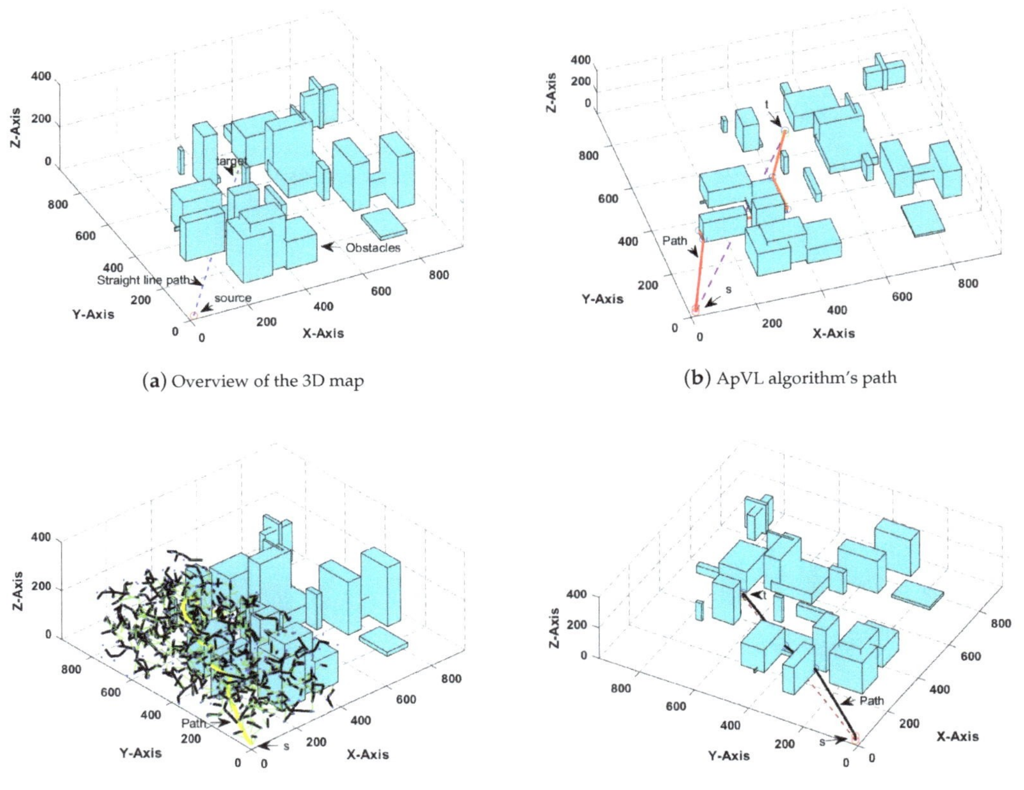

(**a**) Overview of the 3D map (**b**) ApVL algorithm's path

(**c**) RRT*-AB algorithm's path (**d**) Proposed method's path

Figure 8. Example of the paths produced by the proposed method, ApVL, and RRT*-AB algorithms.

The path produced by the proposed PP method was more smooth and shorter than the paths from the other methods. To analyze and compare the proposed PP method's results, we designed three distinct scenarios with sufficient obstacles in the 3D environment maps. Each scenario was tested with all three methods, and the results were analyzed. All obstacles had a random width, depth, and height with a rectangular-shaped base. Comprehensive details on maps counts, map sizes, s and t locations, numbers of obstacles and their geometric information, etc. are given in each scenario description below.

4.1. Comparison with the Existing Approaches by Varying Map Sizes and Obstacle Counts

This scenario is defined with W at sizes ranging between 100 m $\times$ 100 m $\times$ 300 m–1000 m $\times$ 1000 m $\times$ 400 m. It encompassed 50 maps with distinct obstacle counts (e.g., 5–50). For the sake of simplicity and rational comparisons, we categorized all maps into 10 distinct groups considering both map size and obstacle strengths, as given in Table 1. The locations for s and t were marked in alternate places for every test/map. Furthermore, the obstacle density in W varied on each map. The ApVL algorithm [60] processed only obstacles that were on the $\overline{l_o}$ and generated a dense graph to find P. However, the ApVL algorithm has higher complexity, and it produced a non-taut P in most scenarios due to the connected obstacles in urban environments, as shown in Figure 8b. The RRT*-AB algorithm restricts the space, but it explores many locations while finding P. The P generated by this algorithm had a longer length, and computing time was immense. In contrast, the proposed

PP method processed fewer obstacles and employed an ESWG to find a P with good quality. The complete information about maps utilized in this particular scenario and the mean computing time for task modeling by our PP method and its comparisons with the two previous algorithms are given in Table 1.

The computing time for modeling the UAV environment shown in Table 1 is the total time needed to constrain the space, an ESWG construction, and an ESWG's expansion for path-quality improvements. From Table 1, we can see that time surged with an increase in the map size and obstacle counts. Through extensive comparisons with prior PP algorithms, our method lowered computing time for task modeling by 15.05% on average. The pathfinding results (i.e., time needed and path length) and their comparisons with the two existing algorithms are depicted in Figure 9.

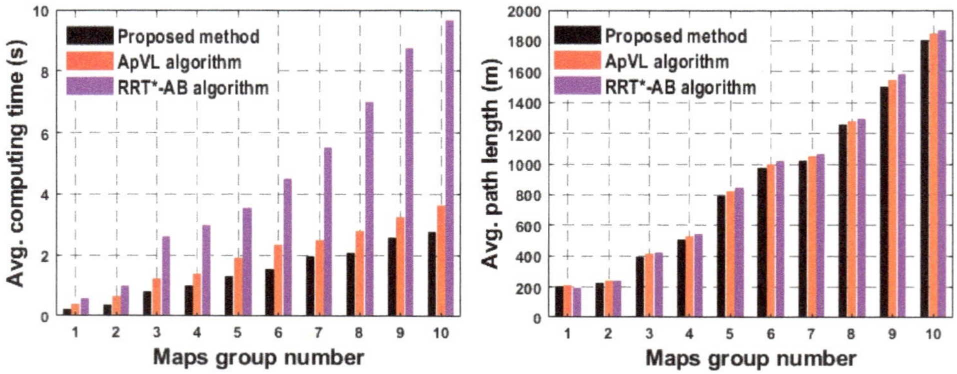

Figure 9. (**Left**) computing time: proposed method versus existing algorithms; (**Right**) path lengths: proposed method versus existing algorithms.

Both computing time and path length are the mean of five maps in each map's group (given in Table 1) with arbitrary obstacles' placement. The simulation results emphasize that, for each method, there is a surge in the computation time with the increase in W's complexity. Moreover, the proposed PP method shows 15.4% and 33.6% curtailment in mean computing time compared to the ApVL algorithm and the RRT*-AB algorithm, respectively. In path lengths, the proposed method shows 5.34% improvements compared to the ApVL algorithm. Moreover, average improvements in the path lengths compared to the RRT*-AB algorithm are 6.34%. These results highlight that the proposed PP method is superior in terms of both computing time and path length over prior algorithms.

Table 1. Details of the 3D maps used in simulations and comparisons of task modeling results.

Maps Group No.	3D Maps' Sizes (in m)	Obstacles Strength	Proposed PP Method	ApVL Algorithm [60]	RRT*-AB Algorithm [70]
	($x \times y \times z$)	(upto 50)	Avg. Computing Time (in s)	Avg. Computing Time (in s)	Avg. Computing Time (in s)
1.	100 × 100 × 300	5	0.98	1.46	2.97
2.	200 × 200 × 300	10	5.61	8.31	16.81
3.	300 × 300 × 300	15	17.01	23.54	34.20
4.	400 × 400 × 300	20	39.28	49.52	65.21
5.	500 × 500 × 300	25	76.26	98.81	115.04
6.	600 × 600 × 400	30	100.65	111.19	141.81
7.	700 × 700 × 400	35	134.03	137.599	185.93
8.	800 × 800 × 400	40	155.01	186.29	211.11
9.	900 × 900 × 400	45	180.21	201.56	225.13
10.	1000 × 1000 × 400	50	191.25	209.25	249.55

4.2. Comparison with the Existing Approaches by Varying the Number of Obstacles

This scenario is comprised of 10 maps with varying numbers of obstacles in a W of 1 km^2 to analyze the impact on our PP method's performance from increasing the number of obstacles. The obstacles were clustered between s and t in such a way that all methods would avoid them during pathfinding. Figure 10 presents the results of our proposed PP method and a comparison with the other algorithms from varying the number of obstacles. When W enclosed more obstacles, the suggested method could quickly determine a good-quality and safe P from W. It was better than the ApVL and RRT*-AB algorithms, based on the metrics, even when varying the number of obstacles in the CPS. The P determined by the proposed method was the shortest and smoother than the other two algorithms. Through simulation results and their comparison with previous methods on 10 obstacles' counts-based maps, the proposed method decreased the computing time for pathfinding by 27.06%, on average. For path lengths, the paths generated by our PP method, on average, were 4.6% shorter (i.e., produced at a lower cost) than the previous methods.

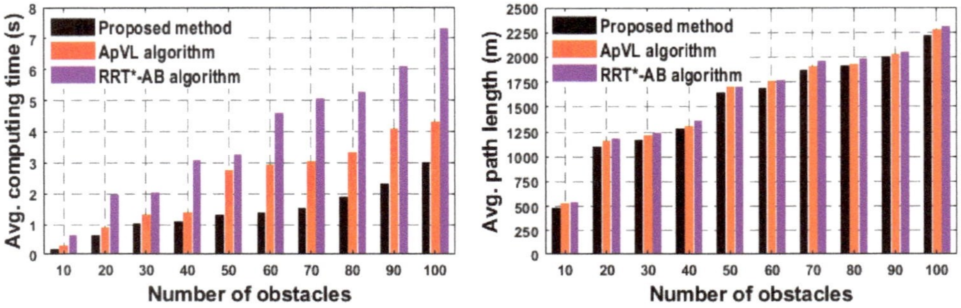

Figure 10. (**Left**) computing time: proposed method versus existing algorithms; (**Right**) path lengths: proposed method versus existing algorithms.

4.3. Comparison with the Existing Approaches by Varying Source and Target Locations

This scenario was tested using a W at 150 m × 150 m × 400–1000 m × 1000 m × 400 m. It encompassed five maps with obstacle counts of up to 25. We analyzed our method's performance through seven runs on every map with different coordinates for s and t in each run. By varying the positions of s and t, the number of obstacles to be modeled in each run/test can be distinct, and, accordingly, comparison metrics can vary with W complexity. The proposed method's averages, obtained from the seven runs on each map, are given in Table 2. The results indicate that the proposed method yielded comparable performance in all tests.

The proposed method yielded an average computing time of 0.71 s, compared to the ApVL and RRT*-AB algorithms, which had mean computing times of 1.01 s and 1.37 s, respectively. In addition, the proposed method lowered the path length, compared to both prior algorithms, by 5.05%. Although the proposed method gave better results, a relatively higher number of initial P nodes can degrade its performance in complex environments. The worst-case complexity with our method was $O(n^3)$. However, the test results revealed that time complexity did not accelerate like $O(n^3)$ in all test scenarios for finding good quality paths. The results obtained from all these scenarios showed that our method performed consistently better than the ApVL and the RRT*-AB algorithms. Aside from the path lengths and computing times, we analyzed and compared its performance against prior algorithms with respect to graph/tree nodes and path node counts. In Figure 11, the results of the proposed method in terms of average path nodes and graph/tree nodes for the above experiments are presented. As shown in Figure 11, both the graph nodes and path points of our ESWG method were lessened, compared to the prior methods.

Table 2. Proposed method pathfinding results' comparison with the ApVL and the RRT*-AB algorithms.

Maps Sizes (m)/Obstacles' Count	ApVL Algorithm [60]		RRT*-AB Algorithm [70]		Proposed PP Method	
	Avg. Time (s)	Avg. Path Length (m)	Avg. Time (s)	Avg. Path Length (m)	Avg. Time (s)	Avg. Path Length (m)
150 × 150 × 400/5	0.35	305.4	0.61	295.91	0.26	300.14
400 × 400 × 400/10	0.55	620.12	0.75	650.75	0.47	599.27
600 × 600 × 400/15	0.74	860.51	0.96	900.90	0.59	800.05
800 × 800 × 400/20	0.98	1020.01	1.81	1050.05	0.75	990.5
1000 × 1000 × 400/25	2.45	1100.05	2.75	1120.75	1.52	1070.21

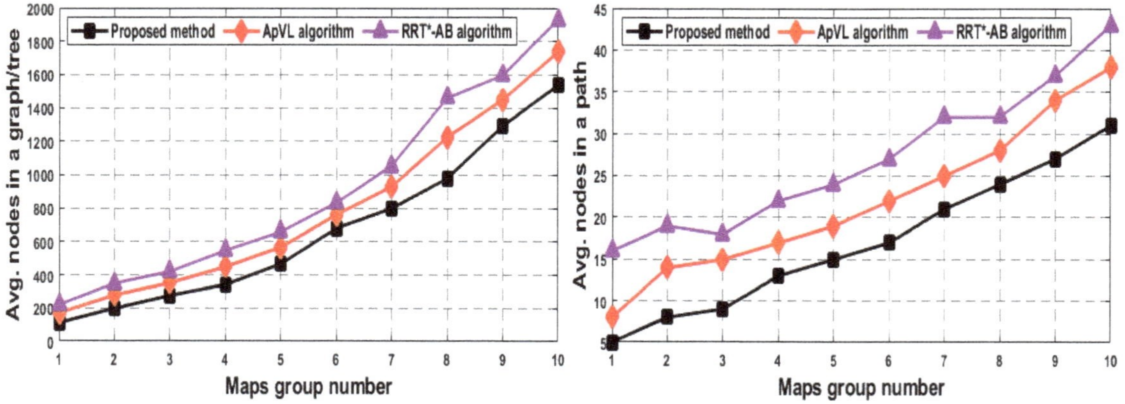

Figure 11. (**Left**) Graph/tree nodes: proposed method versus the ApVL algorithm and the RRT*-AB algorithm; (**Right**) path nodes: proposed method versus the ApVL algorithm and the RRT*-AB algorithm.

Analysis of the memory requirements: As shown in Figure 11a, the proposed PP method generates a WG with fewer vertices. In addition, it does not register visibility of the edges that contribute minimally in an optimal/quasi-optimal path due to less coverage in terms of distance or they are not in the same direction as the target location while generating an ESWG. Therefore, the memory requirements of proposed method are not high compared to the existing methods and a complete graph. However, most global PP methods keep all visible edges in the memory that significantly increase the memory requirements. Furthermore, the visibility check function is called a substantial number of times in visibility graph-based PP methods, thereby space complexity drastically increases. The proposed method resolves these space complexity related issues through reduction in search space, modeling tasks with an ESWG, producing far lower but relevant edges and vertices, and reducing the visibility checks between vertices by incorporating far-reachability and direction-guidance concepts while making an ESWG.

The proposed PP method is complete, and it can be used for many UAV practical applications in urban environments. The proposed method gives good performance due to two main concepts: (i) a new space reduction concept is proposed, which not only assists in lowering the time complexity by restricting the path exploration in the space of highest priority, but also assists in finding low-cost paths in most cases; and (ii) an ESWG, which models the tasks with far lower edges and vertices that curtails the computing time of path searching significantly by making a direction-guided search of the target location. It effectively resolves the trade-off between optimality and efficiency in pathfinding from an urban environment populated by various obstacles.

5. Conclusions and Future Directions

This article proposed a new PP method based on CPS and an ESWG to enable a UAV's safe navigation in 3D urban environments. The main objectives of the proposed PP method are to lower the time complexity in both task modeling and pathfinding without degrading the path quality for UAVs operating at lower elevations in urban environments with fixed, convex obstacles. The main contributions of this article are listed as follows:

- We propose a new PP method based on CPS and an ESWG that has the potential to find an optimal/quasi-optimal path with considerably reduced time complexity.
- We propose a new space reduction method that abstracts the full map into a 3D constrained polygonal space that guarantees a path for the UAV's mission.
- We analyze the effectiveness of CPS for low-cost paths considering six complexity parameters, including the ratio of free space, obstacle density in distinct regions of the CPS, the complexity in the options for avoiding obstacles, deviation in the length from the optimal path, the ratio of connected obstacles to pertinent obstacles, and obstacle tendencies in the CPS.
- The proposed method enlarges the CPS to the next level/space by including obstacles that are in close proximity to the first CPS if the first CPS fails to provide an opportunity for low-cost solutions owing to a higher complexity from obstacles in it.
- The proposed method generates an ESWG from the CPS, leveraging the principles of maximum distance reachability, having only a few nodes and edges, and direction guidance, and it computes an abstract path that is further improved with the assistance of more nodes and edges around it.
- This initial work makes use of obstacle information from an underlying environment in order to lower the computing overhead for pathfinding without compromising path quality in 3D urban environments.

The proposed method performance is substantiated through extensive tests, and, in most cases, it performs consistently better than prior PP methods. It lowers pathfinding time complexity considerably by restricting path exploration solely in the highest priority CPS that has the greatest chance of providing an optimal/sub-optimal path. While conducting the tests, we considered numerous parameters related to the underlying operating environment's complexity and UAV's safety. Meanwhile, during the PP at lower elevations in urban environments, we may need to consider hanged/thin obstacles (e.g., electrical wires and poles in the streets). Another group of evaluation parameters can be the wind/gust and wind/crosswind (e.g., wind's direction and speed), especially when passing through buildings. Hence, further testing with these parameters is yet to be investigated in future work. Finally, we intend to analyze the fidelity of our proposed PP method with other task modeling methods (e.g., Voronoi diagrams, grids, and navigation meshes, etc.).

Author Contributions: All authors contributed equally to this work. All authors have read and agreed to the published version of the manuscript.

Funding: This work was supported by the National Research Foundation of Korea (NRF) grant funded by the Korea government (MSIT) (2020R1A2B5B01002145).

Institutional Review Board Statement: Not applicable.

Informed Consent Statement: Not applicable.

Data Availability Statement: The data used in the experimental evaluation of this study are available within this article.

Conflicts of Interest: The authors declare no conflict of interest regarding the publication of this manuscript.

References

1. Song, B.D.; Park, K.; Kim, J. Persistent UAV delivery logistics: MILP formulation and efficient heuristic. *Comput. Ind. Eng.* **2018**, *120*, 418–428. [CrossRef]
2. Haidari, L.A.; Brown, S.T.; Ferguson, M.; Bancroft, E.; Spiker, M.; Wilcox, A.; Ambikapathi, R.; Sampath, V.; Connor, D.L.; Lee, B.Y. The economic and operational value of using drones to transport vaccines. *Vaccine* **2016**, *34*, 4062–4067. [CrossRef]
3. Ko, Y.D.; Song, B.D. Application of UAVs for tourism security and safety. *Asia Pac. J. Mark. Logist.* **2021**. [CrossRef]
4. Torresan, C.; Berton, A.; Carotenuto, F.; di Gennaro, S.F.; Gioli, B.; Matese, A.; Miglietta, F.; Vagnoli, C.; Zaldei, A.; Wallace, L. Forestry applications of UAVs in Europe: A review. *Int. J. Remote Sens.* **2017**, *38*, 2427–2447. [CrossRef]

5. Saloi, A. Drone in Libraries for Document Delivery: "Flying Documents". *Libr. Philos. Pract.* **2021**, 1–14. Available online: https://digitalcommons.unl.edu/libphilprac/4599/ (Accessed on: 20-02-2021).
6. Sarris, Z. Survey of uav applications in civil markets. In Proceedings of the 9th Mediterranean Conference on Control and Automation, Puglia, Italy, 22–25 June 2001; Volume 11.
7. Madridano, Á.; Al-Kaff, A.; Flores, P.; Martín, D.; de la Escalera, A. Software Architecture for Autonomous and Coordinated Navigation of UAV Swarms in Forest and Urban Firefighting. *Appl. Sci.* **2021**, *11*, 1258. [CrossRef]
8. Näsi, R.; Honkavaara, E.; Blomqvist, M.; Lyytikäinen-Saarenmaa, P.; Hakala, T.; Viljanen, N.; Kantola, T.; Holopainen, M. Remote sensing of bark beetle damage in urban forests at individual tree level using a novel hyperspectral camera from UAV and aircraft. *Urban For. Urban Green.* **2018**, *30*, 72–83. [CrossRef]
9. Hu, T.; Sun, X.; Su, Y.; Guan, H.; Sun, Q.; Kelly, M.; Guo, Q. Development and Performance Evaluation of a Very Low-Cost UAV-Lidar System for Forestry Applications. *Remote Sens.* **2021**, *13*, 77. [CrossRef]
10. Yuan, C.; Zhang, Y.; Liu, Z. A survey on technologies for automatic forest fire monitoring, detection, and fighting using unmanned aerial vehicles and remote sensing techniques. *Can. J. For. Res.* **2015**, *45*, 783–792. [CrossRef]
11. Neuville, R.; Bates, J.S.; Jonard, F. Estimating forest structure from UAV-mounted LiDAR point cloud using machine learning. *Remote Sens.* **2021**, *13*, 352. [CrossRef]
12. Kanistras, K.; Martins, G.; Rutherford, M.J.; Valavanis, K.P. A survey of unmanned aerial vehicles (UAVs) for traffic monitoring. In Proceedings of the IEEE 2013 International Conference on Unmanned Aircraft Systems (ICUAS), Atlanta, GA, USA, 28–31 May 2013; pp. 221–234.
13. Yoshii, T.; Matsumura, N.; Lin, C. Integrating UAV and Lidar Data for Retrieving Tree Volume of Hinoki Forests. In Proceedings of the IGARSS 2020—2020 IEEE International Geoscience and Remote Sensing Symposium, Waikoloa, HI, USA, 26 September–2 October 2020; pp. 4124–4127.
14. Stöcker, C.; Eltner, A.; Karrasch, P. Measuring gullies by synergetic application of UAV and close range photogrammetry—A case study from Andalusia, Spain. *Catena* **2015**, *132*, 1–11. [CrossRef]
15. Emilien, A.; Thomas, C.; Thomas, H. UAV & satellite synergies for optical remote sensing applications: A literature review. *Sci. Remote Sens.* **2021**, *3*, 100019.
16. Erdelj, M.; Natalizio, E.; Chowdhury, K.R.; Akyildiz, I.F. Help from the sky: Leveraging UAVs for disaster management. *IEEE Pervasive Comput.* **2017**, *16*, 24–32. [CrossRef]
17. Wang, W.; Jiang, B.; Yang, J.; Li, C. Research on UAV Application in Mountain Anti-terrorism Combat. *J. Phys. Conf. Ser.* **2021**, *1792*, 012079. [CrossRef]
18. Liao, K.-W.; Lee, Y.-T. Detection of rust defects on steel bridge coatings via digital image recognition. *Autom. Constr.* **2016**, *71*, 294–306. [CrossRef]
19. Balamuralidhar, N.; Tilon, S.; Nex, F. MultEYE: Monitoring System for Real-Time Vehicle Detection, Tracking and Speed Estimation from UAV Imagery on Edge-Computing Platforms. *Remote Sens.* **2021**, *13*, 573. [CrossRef]
20. Sujit, P.B.; Sousa, J.; Pereira, F.L. UAV and AUVs coordination for ocean exploration. In Proceedings of the IEEE Oceans 2009-Europe, Bremen, Germany, 11–14 May 2009; pp. 1–7.
21. Wang, Y.; Zhang, N.; Li, H.; Cao, J. Research on Digital Twin Framework of Military Large-scale UAV Based on Cloud Computing. *J. Phys. Conf. Ser.* **2021**, *1738*, 012052. [CrossRef]
22. Zikidis, K.C. Early Warning Against Stealth Aircraft, Missiles and Unmanned Aerial Vehicles. In *Surveillance in Action*; Springer: Cham, Switzerland, 2018; pp. 195–216.
23. Raja, P.; Pugazhenthi, S. Optimal path planning of mobile robots: A review. *Int. J. Phys. Sci.* **2012**, *7*, 1314–1320. [CrossRef]
24. Xue, Y.; Sun, J.-Q. Solving the path planning problem in mobile robotics with the multi-objective evolutionary algorithm. *Appl. Sci.* **2018**, *8*, 1425. [CrossRef]
25. Krishnan, J.; Rajeev, U.P.; Jayabalan, J.; Sheela, D.S. Optimal motion planning based on path length minimisation. *Robot. Auton. Syst.* **2017**, *94*, 245–263. [CrossRef]
26. Lv, T.; Zhao, C.; Bao, J. A global path planning algorithm based on bidirectional SVGA. *J. Robot.* 2017, *2017*, 8796531. [CrossRef]
27. Chen, Y.; Yu, J.; Mei, Y.; Wang, Y.; Su, X. Modified central force optimization (MCFO) algorithm for 3D UAV path planning. *Neurocomputing* **2016**, *171*, 878–888. [CrossRef]
28. Kala, R.; Shukla, A.; Tiwari, R. Robotic path planning in static environment using hierarchical multi-neuron heuristic search and probability based fitness. *Neurocomputing* **2011**, *74*, 2314–2335. [CrossRef]
29. Meng, B.; Gao, X. UAV path planning based on bidirectional sparse A* search algorithm. In Proceedings of the IEEE 2010 International Conference on Intelligent Computation Technology and Automation, Changsha, China, 11–12 May 2010; Volume 3, pp. 1106–1109.
30. Hwang, J.Y.; Kim, J.S.; Lim, S.S.; Park, K.H. A fast path planning by path graph optimization. *IEEE Trans. Syst. Manand Cybern.Part A Syst. Hum.* **2003**, *33*, 121–129. [CrossRef]
31. Chen, G.; Shen, D.; Cruz, J.; Kwan, C.; Riddle, S.; Cox, S.; Matthews, C. A novel cooperative path planning for multiple aerial platforms. In *Infotech@ Aerospace*; AIAA: Arlington, VA, USA, 2005; p. 6948.
32. Dijkstra, E.W. A note on two problems in connexion with graphs. *Numer. Math.* **1959**, *1*, 269–271. [CrossRef]
33. Imai, T.; Kishimoto, A. A Novel Technique for Avoiding Plateaus of Greedy Best-First Search in Satisficing Planning. In Proceedings of the AAAI Conference on Artificial Intelligence, San Francisco, CA, USA, 7–11 August 2011; Volume 25.

34. Hart, P.E.; Nilsson, N.J.; Raphael, B. A formal basis for the heuristic determination of minimum cost paths. *IEEE Trans. Syst. Sci. Cybern.* **1968**, *4*, 100–107. [CrossRef]
35. Korf, R.E. Depth-first iterative-deepening: An optimal admissible tree search. *Artif. Intell.* **1985**, *27*, 97–109. [CrossRef]
36. Nash, A.; Daniel, K.; Koenig, S.; Felner, A. Theta*: Any-Angle Path Planning on Grids. In Proceedings of the AAAI Conference on Artificial Intelligence, Vancouver, BC, Canada, 22–26 July 2007; pp. 1177–1183.
37. Nash, A.; Koenig, S.; Tovey, C. Lazy Theta*: Any-angle path planning and path length analysis in 3D. In Proceedings of the AAAI Conference on Artificial Intelligence, Atlanta, GA, USA, 11–15 July 2010; Volume 24.
38. Koenig, S.; Likhachev, M. Fast replanning for navigation in unknown terrain. *IEEE Trans. Robot.* **2005**, *21*, 354–363. [CrossRef]
39. Reyes, N.H.; Barczak, A.L.C.; Susnjak, T.; Jordan, A. Fast and Smooth Replanning for Navigation in Partially Unknown Terrain: The Hybrid Fuzzy-D* lite Algorithm. In *Robot Intelligence Technology and Applications 4*; Springer: Cham, Switzerland, 2017; pp. 31–41.
40. Bulitko, V.; Sturtevant, N.; Lu, J.; Yau, T. Graph abstraction in real-time heuristic search. *J. Artif. Intell. Res.* **2007**, *30*, 51–100. [CrossRef]
41. Harabor, D.; Grastien, A. Online graph pruning for pathfinding on grid maps. In Proceedings of the AAAI Conference on Artificial Intelligence, San Francisco, CA, USA, 7–11 August 2011; Volume 25.
42. Nussbaum, D.; Yörükçü, A. Moving target search with subgoal graphs. In Proceedings of the International Conference on Automated Planning and Scheduling, Jerusalem, Israel, 7–11 June 2015; Volume 25.
43. Aversa, D.; Sardina, S.; Vassos, S. Path planning with inventory-driven jump-point-search. In Proceedings of the AAAI Conference on Artificial Intelligence and Interactive Digital Entertainment, Santa Cruz, CA USA, 14–18 November 2015; Volume 11.
44. Sturtevant, N.R.; Felner, A.; Barrer, M.; Schaeffer, J.; Burch, N. Memory-based heuristics for explicit state spaces. In Proceedings of the Twenty-First International Joint Conference on Artificial Intelligence, Pasadena, CA, USA, 14–17 July 2009.
45. Strasser, B.; Botea, A.; Harabor, D. Compressing optimal paths with run length encoding. *J. Artif. Intell. Res.* **2015**, *54*, 593–629. [CrossRef]
46. Gonzalez, J.P.; Dornbush, A.; Likhachev, M. Using state dominance for path planning in dynamic environments with moving obstacles. In Proceedings of the 2012 IEEE International Conference on Robotics and Automation, Saint Paul, MN, USA, 14–18 May 2012; pp. 4009–4015.
47. Pochter, N.; Zohar, A.; Rosenschein, J.; Felner, A. Search space reduction using swamp hierarchies. In Proceedings of the AAAI Conference on Artificial Intelligence, Atlanta, GA, USA, 11–15 July 2010; Volume 24.
48. Amador, G.P.; Gomes, A.J.P. xTrek: An Influence-Aware Technique for Dijkstra's and A Pathfinders. *Int. J. Comput. Games Technol.* **2018**, *2018*, 5184605. [CrossRef]
49. Ninomiya, K.; Kapadia, M.; Shoulson, A.; Garcia, F.; Badler, N. Planning approaches to constraint-aware navigation in dynamic environments. *Comput. Animat. Virtual Worlds* **2015**, *26*, 119–139. [CrossRef]
50. Omar, R.; Gu, D.-W. Visibility line based methods for UAV path planning. In Proceedings of the IEEE 2009 ICCAS-SICE, Fukuoka, Japan, 18–21 August 2009; pp. 3176–3181.
51. Liang, X.; Meng, G.; Xu, Y.; Luo, H. A geometrical path planning method for unmanned aerial vehicle in 2D/3D complex environment. *Intell. Serv. Robot.* **2018**, *11*, 301–312. [CrossRef]
52. Sariff, N.; Buniyamin, N. An overview of autonomous mobile robot path planning algorithms. In Proceedings of the IEEE 2006 4th Student Conference on Research and Development, Shah Alam, Malaysia, 27–28 June 2006; pp. 183–188.
53. Kim, H.; Yu, K.-A.; Kim, J.-T. Reducing the search space for pathfinding in navigation meshes by using visibility tests. *J. Electr. Eng. Technol.* **2011**, *6*, 867–873. [CrossRef]
54. Lv, Z.; Yang, L.; He, Y.; Liu, Z.; Han, Z. 3D environment modeling with height dimension reduction and path planning for UAV. In Proceedings of the 2017 IEEE 9th International Conference on Modelling, Identification and Control (ICMIC), Kunming, China, 10–12 July 2017; pp. 734–739.
55. Liang, H.; Zhong, W.; Chunhui, Z. Point-to-point near-optimal obstacle avoidance path for the unmanned aerial vehicle. In Proceedings of the IEEE 2015 34th Chinese Control Conference (CCC), Hangzhou, China, 28–30 July 2015; pp. 5413–5418.
56. Hota, S.; Ghose, D. Optimal path planning for an aerial vehicle in 3D space. In Proceedings of the 49th IEEE Conference on Decision and Control (CDC) 2010; pp. 4902–4907.
57. Zhang, K.; Liu, P.; Kong, W.; Lei, Y.; Zou, J.; Liu, M. An improved heuristic algorithm for UCAV path planning. In *International Conference on Bio-Inspired Computing: Theories and Applications*; Springer: Singapore, 2016; pp. 54–59.
58. Plaku, E.; Plaku, E.; Simari, P. Direct path superfacets: An Intermediate representation for motion planning. *IEEE Robot. Autom. Lett.* **2016**, *2*, 350–357. [CrossRef]
59. Stenning, B.E.; Barfoot, T.D. Path planning with variable-fidelity terrain assessment. *Robot. Auton. Syst.* **2012**, *60*, 1135–1148. [CrossRef]
60. Frontera, G.; Martín, D.J.; Besada, J.A.; Gu, D.-W. Approximate 3D Euclidean shortest paths for unmanned aircraft in urban environments. *J. Intell. Robot. Syst.* **2017**, *85*, 353–368. [CrossRef]
61. Kavralu, L.E.; Svestka, P.; Latombe, J.-C.; Overmars, M.H. Probabilistic roadmaps for path planning in high-dimensional configuration spaces. *IEEE Trans. Robot. Autom.* **1996**, *12*, 566–580.
62. LaValle, S.M. *Rapidly-Exploring Random Trees: A New Tool for Path Planning*; TR 98-11; Department of Computer Science, Iowa State University: Ames, IA, USA, 1998.

63. Gammell, J.D.; Srinivasa, S.S.; Barfoot, T.D. Informed RRT*: Optimal sampling-based path planning focused via direct sampling of an admissible ellipsoidal heuristic. In Proceedings of the 2014 IEEE/RSJ International Conference on Intelligent Robots and Systems, Chicago, IL, USA, 14–18 September 2014; pp. 2997–3004.
64. Jaillet, L.; Cortés, J.; Siméon, T. Sampling-based path planning on configuration-space costmaps. *IEEE Trans. Robot.* **2010**, *26*, 635–646. [CrossRef]
65. Kuffner, J.J.; LaValle, S.M. RRT-connect: An efficient approach to single-query path planning. In Proceedings 2000 ICRA. Millennium Conference. IEEE International Conference on Robotics and Automation. Symposia Proceedings (Cat. No. 00CH37065), San Francisco, CA, USA, 24–28 April 2000; Volume 2, pp. 995–1001.
66. Nieuwenhuisen, D.; Overmars, M.H. Useful cycles in probabilistic roadmap graphs. In Proceedings of the IEEE International Conference on Robotics and Automation (ICRA'04), New Orleans, LA, USA, 26 April–1 May 2004; Volume 1, pp. 446–452.
67. Karaman, S.; Frazzoli, E. Sampling-based algorithms for optimal motion planning. *Int. J. Robot. Res.* **2011**, *30*, 846–894. [CrossRef]
68. Nasir, J.; Islam, F.; Malik, U.; Ayaz, Y.; Hasan, O.; Khan, M.; Muhammad, M.S. RRT*-SMART: A rapid convergence implementation of RRT. *Int. J. Adv. Robot. Syst.* **2013**, *10*, 299. [CrossRef]
69. Li, Y.; Wei, W.; Gao, Y.; Wang, D.; Fan, Z. PQ-RRT*: An improved path planning algorithm for mobile robots. *Expert Syst. Appl.* **2020**, *152*, 113425. [CrossRef]
70. Noreen, I.; Khan, A.; Ryu, H.; Doh, N.L.; Habib, Z. Optimal path planning in cluttered environment using RRT*-AB. *Intell. Serv. Robot.* **2018**, *11*, 41–52. [CrossRef]

applied sciences

MDPI

Article

Designing a Reliable UAV Architecture Operating in a Real Environment

Krzysztof Andrzej Gromada [1,*,†] and Wojciech Marcin Stecz [2,†]

1 The Institute of Automatic Control and Robotics, Warsaw University of Technology, 02-525 Warsaw, Poland
2 Faculty of Cybernetics, Military University of Technology, 00-908 Warsaw, Poland; Wojciech.Stecz@wat.edu.pl
* Correspondence: Krzysztof.Gromada.dokt@pw.edu.pl
† These authors contributed equally to this work.

Abstract: The article presents a method of designing a selected unmanned aerial platform flight scenario based on the principles of designing a reliable (Unmanned Aerial Vehicle) UAV architecture operating in an environment in which other platforms operate. The models and results presented relate to the medium-range aerial platform, subject to certification under the principles set out in aviation regulations. These platforms are subject to the certification process requirements, but their restrictions are not as restrictive as in the case of manned platforms. Issues related to modeling scenarios implemented by the platform in flight are discussed. The article describes the importance of Functional Hazard Analysis (FHA) and Fault Trees Analysis (FTA) of elements included in the hardware and software architecture of the system. The models in Unified Modeling Language (UML) used by the authors in the project are described, supporting the design of a reliable architecture of flying platforms. Examples of the transformations from user requirements modeled in the form of Use Cases to platform operation models based on State Machines and then to the final UAV operation algorithms are shown. Principles of designing system test plans and designing individual test cases to verify the system's operation in emergencies in flight are discussed. Methods of integrating flight simulators with elements of the air platform in the form of Software-in-the-Loop (SIL) models based on selected algorithms for avoiding dangerous situations have been described. The presented results are based on a practical example of an algorithm for detecting an air collision situation of two platforms.

Keywords: Unmanned Aerial Vehicle (UAV); collision avoidance; safety procedures; reliable architecture; Unified Modeling Language (UML)

Citation: Gromada, K.A.; Stecz, W.M. Designing a Reliable UAV Architecture Operating in a Real Environment. *Appl. Sci.* **2022**, *12*, 294. https://doi.org/10.3390/app12010294

Academic Editor: Seong-Ik Han

Received: 12 november 2021
Accepted: 27 December 2021
Published: 29 December 2021

1. Introduction

Unmanned aerial platforms for special tasks often move in an environment with an increasing number of other threatening objects, including aerial platforms. That can be a source of potential danger for UAVs. It should also be assumed that the air platform, primarily used in rescue operations, will move in a hostile environment. Such an environment can be understood as flight in conditions of GPS signal interference or flight in unfavorable weather conditions. Such environments may include operations where the platform might be destroyed due to intentional human activity (i.e., mainly due to military actions, etc.).

Designing a reliable UAV architecture operating in such an environment requires compliance with modern standards for the safety of the flying systems. It is insufficient to meet the requirements of a user who describes only his/her operational needs. Flight safety is the responsibility of the system builders, who must consider the guidelines for the safety of air systems that are in force in given countries.

Preparing a reliable and safe system is a comprehensive activity on many levels:

1. The development of hardware and software architecture that has the required reliability determined based on the so-called Fault Tree Analysis (FTA), in particular

designing the physical architecture of the system that ensures redundancy of the most critical subsystems,
2. Meeting the functional requirements in the context of ensuring an appropriate level of security, usually specified in the Functional Hazard Analysis (FHA) documentation, which presents the decomposition of critical functions in the system,
3. Designing algorithms for the operation of the air platform per the principles of Model-Based Systems Engineering (MBSE), which are verifiable with the use of selected formal techniques,
4. Designing and describing algorithms that support the occurrence of emergency situations during flight, such as loss of radio link with the Ground Control Station (GCS), loss of GPS signal, avoidance of platform collisions in the air,
5. Describing and proving the correctness of specific numerical algorithms that are used during the implementation of the mission (e.g., algorithms determining the change of the platform's course after detecting the possibility of a collision in air-collision avoidance algorithms),
6. Designing system testing procedures based on mission simulators and flight tests,
7. Developing documentation rules for the most critical procedures affecting flight safety and documentation of the tests performed.

The hardware architecture is first determined in designing an unmanned platform. The probability of damage to the elements is checked following the Fault Tree Analysis methodology for the developed architecture. The probability of damage to the elements is checked following the Fault Tree Analysis methodology for the developed architecture. The purpose of the activity is to check which of the elements involved in implementing a specific flight scenario is prone to failure and may lead to a potential system crash. The FTA methodology is described in literature in [1]. The general principle of proceeding in the construction of FTA trees is to arrange a series of devices that implement a given function. Then, for each of such devices, Mean Time Between Failures (MTBF) is determined, based on which the probability of failure of the entire system is verified. If the probability is above the acceptable threshold, then the system must not be allowed to operate. It is a fundamental step taken in designing an unmanned system. In a situation where the FHA shows that some system elements are too unreliable, these elements must be replaced before further design work because the platform will not meet the safety requirements.

After verifying the hardware architecture to be used in the designed unmanned system, the decomposition of key processes directly impacts the platform's safety in flight should be made. Based on pre-defined scenarios of platform operation (scenarios can be provided by the system contracting authority), a function decomposition called Functional Hazard Analysis [2] is prepared in the literature. It concerns the fulfilment of functional requirements in the context of ensuring an appropriate safety level, denoted in the ARP4754 methodology as Design Assurance Level (DAL) [3]. Each primary process (scenarios) is decomposed into a set of component subprocesses (scenarios) up to the point where atomic functions are defined (functions that are not worth decomposing because they describe a specific single operation of the system assigned to one of the components).

The Software Level, also known as the Design Assurance Level (DAL) as defined in ARP4754, is determined from the safety assessment process and hazard analysis by examining the effects of a failure condition in the system. The failure conditions are categorized by their effects on the aircraft, crew, and passengers.

(A) Catastrophic—Failure may cause deaths, usually also includes the destruction of the airplane.
(B) Hazardous—Failure has a sizeable negative impact on safety. It may reduce the ability of the crew to operate the aircraft due to physical distress or causes serious or fatal injuries among the passengers.
(C) Major—Failure significantly reduces the safety margin. It may increase crew workload.

(D) Minor—Failure slightly reduces the safety margin or slightly increases crew workload. Examples might include causing passengers inconvenience or a routine flight plan change.
(E) No Effect—Failure has no impact on safety, aircraft operation, or crew workload.

In practice, instead of specifying the DAL values directly, it is enough to define the most critical failure states that may occur during the execution of this function for each of the decomposed functions. In such a case, proving that the system is resistant to the occurrences of these emergency states means that its reliability level can be considered sufficient to perform specific tasks. It should be remembered that the DAL also depends on the hardware architecture, which significantly affects this parameter.

The preparation of a reliable and safe system requires designing following Model-Based System Engineering (MBSE) standards [4,5]. In the case of air platforms, it is required to introduce additional mechanisms to the design process, allowing for the preparation of a design that is easy to expand, maintain and verify. In recent years, the use of UAV functionalities based on the System Modeling Language (SysML) [6] and Unified Modeling Language (UML) [7] models have become widely used. UML also uses the Object Constraint Language (OCL), which allows for additional detailing of the system's functionality and defining constraints that must always be met. The OCL is a declarative language describing rules applying to UML. The Object Constraint Language provides a constraint on the metamodel that cannot otherwise be expressed by diagrammatic notation. OCL provides expressions that have neither the ambiguities of natural language nor the inherent difficulties of using complex mathematics.

However, formal system description languages such as SysML or UML alone do not guarantee the development of a secure aircraft platform architecture integrated with the Ground Control Station (GCS). For this purpose, dedicated metamodels should be developed to transform user requirements and the requirements of safety standards into technical models. In practice, this means the manufacturer needs to develop such systems. The set of metamodels describes the mapping of user requirements to system use cases. Use cases can be mapped to system state machines or, in simpler cases, directly to activity sequences (system function calls).

Due to the nature of the system, which is a close-to-real-time system, UML models primarily describe the transition from Use Case models to State Machines, the most commonly used system dynamics modeling mechanisms with multiple concurrent processes. Of course, the development of complete and consistent system models (GCS and UAV) does not guarantee that the prepared models do not contain any errors. In recent years, intensive work has been carried out on developing formal verification mechanisms for models prepared in SysML or UML [8].

Based on the UML-based scenario description methodology and the previous FHA decomposition, the process of designing and describing algorithms that support the occurrences of emergency situations during the flight takes place. Complex numerical algorithms are often used, the correctness of which must be proven. An example of such an algorithm is provided in the article [9]. The article presents the design of an algorithm for avoiding collisions between aerial platforms. The algorithms supporting safety also include algorithms for checking the correctness of the operation of GPS systems, algorithms for checking the possibility of a potential collision with terrain or other platforms, etc. Depending on the purpose of the air platform and the areas in which it can operate, the list of algorithms handling emergency situations can be very long. Each of these algorithms can have high computational complexity. Hence, appropriate onboard computers are selected depending on the class of the air platform. The article presents a description of the algorithm for avoiding collisions with another platform in the air, which is implemented on a medium-range UAV. Due to the size of the platform, the algorithm is implemented not only on GCS but also in the software of the platform itself. Due to its complexity, a separate thread of the onboard computer processor is allocated. However, it is not possible to assign one processor core to check the occurrence of each specified emergency situation (platforms of this type are too small). The article presents a formal way of modeling such an emergency situation from the stage of definition of action in the form of a Use Case in UML, through detailed State Machine models, to formal mathematical models.

The final step in the design process is to design system testing procedures based on mission simulators and air tests. It is also necessary to develop the principles of documentation of the most critical procedures affecting flight safety and the documentation of the tests performed. An example of system tests is shown in the article. Testing guidelines can be found in the DO-178 [10] methodologies.

This article presents the architecture concept for an unmanned aerial platform, which must operate in unfavorable environmental conditions, such as flight in an area with a large number of other air platforms.

The methodology of designing a reliable UAV architecture, which can be used for autonomous flight, was presented. It is possible thanks to integrated algorithms such as detections and avoidance of collisions with other UAVs or detection of collisions with terrain obstacles. A method of modeling an unmanned platform operation scenario was also presented, in which algorithms for detection and avoidance of situations threatening the platform's security are integrated. Formal methods based on UML notation are used to describe the problem presented like the method of transforming the requirements described in the so-called Use Cases in UML on diagrams describing the dynamics of the system. In this article, we mainly rely on State Machines, which are the basic method for modeling the operation of real-time systems.

A special case presented in the article is the automatic correction of the flight route to eliminate the possibility of a potential collision in the air with another platform. The article assumes that each of the platforms is equipped with the ADS-B (Automatic Dependent Surveillance Broadcast) system, which allows identifying the platforms' and the directions and speed of their movement. Algorithms of this type are usually built into unmanned platforms that fly long distances from the Ground Control Station, because flights at such a distance generate the risk of losing communication with GCS.

The article shows how to integrate the described algorithms with the platform management algorithms described in the form of State Machines. An exemplary method of managing the detected emergencies and UAV operations in the event of two situations co-occurring is also presented.

In Section 2, reference was made to works on a similar subject. The concept of UAV architecture modeling and the formal description of requirements for selected algorithms used in systems of this class were presented. Reference was also made to verifying the correctness of the developed models, although the discussion of these methods is not the subject of this article. The types of mathematical methods that are used to implement collision avoidance algorithms are also described. Particular attention was paid to geometric methods. Other optimization-based methods, including heuristic methods, are also mentioned.

Section 3 covers models for modeling the system architecture, from user requirements to modeling class diagrams. Particular attention was paid to modeling emergencies that may occur during the platform's flight. The principles of selecting emergencies in order to minimize their number are discussed. Methods of verifying the consistency of a set of states and the transitions between individual states in which the unmanned platform may be found are presented. Reference was also made to the very important but often overlooked topic of integrating formal optimization methods with UML or SysML models in unmanned systems. The methods of checking the completeness and adequacy of mathematical models for an exemplary emergency situation in flight are discussed.

The following part of the article shows an example of the physical architecture of the UAV, based on which the described algorithms for handling emergencies were designed and tested. The deterministic algorithm for handling collision avoidance to ensure separation between air platforms is presented in detail in Section 4. The algorithms are presented in the form of formal descriptions, which are verified on the simulator. Section 5 presents sample tests of the presented algorithms included in the system test plan and the results of testing the algorithms on the simulator. In the last section, further directions of the development of the presented models are discussed. Our experiences related to the preparation of formal models and their verification were also presented.

The innovative elements presented in the article are:

(a) The methodology of building advanced UAV flight algorithms that fly in an environment with a large number of obstacles, which has been adapted to designing algorithms for small and medium air platforms;
(b) A complete and tested collision avoidance algorithm of an unmanned platform equipped with ADS-B, for which methods of detecting situations in which the algorithm shows erroneous results have been defined (detection of a potential collision in a situation where there is no such collision);
(c) An example of a complete scenario that can be used as documentation in the certification process;
(d) A simulator for the verifying the UAV collision avoidance algorithm, the results of which are presented.

2. Related Works

In order to understand the importance of designing reliable UAVs, it is necessary to understand the principles of operation of these systems and their architecture. An excellent introduction to this is given in the article by Sanchez-Lopez, et al. [11]. The article describes the relationship between tasks related to mission planning and their implementation. The logical dependencies between the UAV control components and the payload control components were presented. Atyabi, et al. [12] introduce the reader to aspects of UAV mission planning and management systems and discuss selected future directions for the development of such systems. An extensive study was also presented on the assessment of UAV autonomy, including the provision of situational awareness and the development of decision-making methods.

Generally speaking, technical literature contains many articles discussing particular aspects of designing the hardware and software architecture of a safe UAV [13]. In fact, papers containing a comprehensive description of the software design process itself and the architecture of the unmanned system are difficult to access due to the complexity of problems encountered in aviation. In particular, available literature lacks proposals for methods that combine mathematical models with UAV operation models described in formal languages such as UML. Such methodologies are currently being developed, also in the form of methodologies such as DO-331, DO-332, and DO-333 [14–16].

Preparing a reliable and safe system is a comprehensive activity on many levels. The first level is the development of hardware and software architecture that has the required reliability determined based on the so-called FTA. This applies particularly to the design of the system's physical architecture that ensures redundancy of the most important subsystems. The approach to modeling hardware and software architecture is described, for example, in [1]. It is worth noting that in the case of designing unmanned platforms of the MALE class and larger, the operation of FCC flight computers is particularly carefully designed, the functionality of which covers most of the user's requirements of this platform class. However, it should be borne in mind that this element of the project is relatively rarely presented in scientific articles due to the information it contains. It constitutes the potential of the company producing unmanned systems. Hence, it is difficult to find studies strictly related to the topic of the architecture of the entire system. Rather, UAV elements such as FCC [13] flight computers are described.

Subsequently, the system to be designed must be verified in terms of meeting functional requirements in the context of ensuring the appropriate level of safety, usually defined in the FHA. Within the FHA, depending on the class of the air platform, several primary groups of functions are defined that are subject to functional decomposition. In the case of an unmanned aerial platform, the most important functional groups are:

- Mission planning
- Ensuring the stability and control of the platform
- Provision of platform navigation
- Data link management
- Payload management
- Flight systems management.

Each group is a set of UAV operation scenarios. In each of the groups, a number of emergency situations can be distinguished that should be taken into account when modeling the platform's behavior during the mission. The article describes the platform collision avoidance algorithm, which is an emergency situation in the group of scenarios "Ensuring the stability and control of the platform". The models in UML presented in the following sections of the article are elements of the mentioned group of scenarios.

Based on the precisely defined functional decomposition of the unmanned system and the developed hardware and software architecture of such a system, the next step is to design algorithms (scenarios) for the operation of the air platform in accordance with MBSE principles. Models compliant with the MBSE methodology, widely used in systems engineering, verify individual elements of the air platform flight scenario. The MBSE approach used in the design of the air platform is described quite extensively in the literature on the subject. Examples include [4,17].

Designing basic platform operation scenarios is not sufficient when designing a secure platform. The FHA defines the so-called emergency situations, i.e., potential failures that may occur during the flight in a given mode. Usually, for each function described in the decomposition process, potential emergency situations that may occur are indicated. The system designer's task is to design and describe the algorithms handling the occurrence of emergency situations during the flight so that the platform retains the possibility of at least a safe return in the event of one or several such events occurring in a short period. Examples of exceptional situations are the loss of a radio link with NSK, loss of a GPS signal, a potential collision of two platforms in the air, or a collision of a platform with a terrain obstacle. For many events of this type, there is literature that allows the analyst to design software to protect the platform against emergency situations. An example is the in-flight separation algorithm [9], the development of which is presented in this article.

In the case of building advanced unmanned platforms, functions related to the implementation of missions during the flight are designed. These often require optimization tasks to be solved. The UAV uses deep learning neural networks for image recognition, algorithms for route planning, avoiding obstacles, etc. These algorithms require detailed descriptions and formal proof of correctness. Among these specific numerical algorithms are the algorithms used during the mission, for example, algorithms determining a change of the platform's course after detecting the possibility of a collision in the air, i.e., algorithms ensuring the separation between [9,18] platforms.

In this article, we focus on the particular situation in flight related to the need to maintain the separation between two platforms whose flight trajectories intersect (also taking into account time). In such a situation, a deterministic algorithm should be designed to ensure that the appropriate minimum distance between platforms is maintained. At the same time, aircraft must be equipped with systems that communicate data about their position to others. The presence of ADS-B modules on the platform is assumed, working both as emitters and receivers. These devices broadcast flight parameters (plane position, velocity, and heading) to nearby vehicles. Having minimal information on the position and velocities of nearby airplanes allows for effective and suboptimal dodge maneuver prediction.

There are three main approaches used in solving the task of separating platforms in the air, described in the literature on the subject. The first approach, the most natural and efficient, is based on determining the change in the direction of the UAV flight as well as the speed and height of the UAV based on the available data using geometric methods. These methods are characterised by a high speed of determining the solution. However, they require multiple repetitions in the case of minor modifications of the flight parameters of one of the air platforms. Geometric approaches rely on the analysis of the geometric attributes of the UAV to ensure that minimum distances between the air platforms are maintained. This requires calculating the time to potential collision using the distance between UAVs and their speed.

This article is mainly based on the geometric algorithm presented by Park, et al. [9]. The method proposed by Park is tested to ensure flight safety in various situations. Different algorithm, described in [19], allows the UAV to avoid obstacles of various types (including other UAVs) in 3D. Depending on the obstacle type identified and the information available about the obstacle, the algorithm determines the time when the UAV should start avoiding the obstacle. When the UAV reaches the point where obstacle avoidance begins, the algorithm starts the avoidance operation for a specified period of time. The length of the time window in which the UAV modifies the flight direction, is flexible and depends on the size and distance from the obstacle. After completing the obstacle avoidance maneuvers, the algorithm searches for new route points that will allow the UAV to return to the planned route as quickly as possible. The work [20] also contains extensive literature review on collision avoidance algorithms. Peng et al. [21] present a geometric model for UAVs where horizontal maneuvers are only performed by varying the speed of the UAV. The direction of flight remains constant. The model predicts the separation to be achieved by a horizontal collision avoidance maneuver. In addition, they calculate the effects of the speed change time and autopilot response on the horizontal miss distance and the reserved time to the nearest potential collision point.

The second group of methods is methods based on optimization, in particular, methods based on the principles of optimal control. Their description is out of the scope of this article due to their complexity. The reader interested in the theory of optimization is referred to the works of Pytlak [22] and Betts [23], who present direct and indirect methods in optimal control. In the case of optimal control, it is worth emphasizing that the task to be solved consists in determining the UAV flight trajectory between the starting and ending point, assuming the knowledge of the air platform flight dynamics model. The dynamics model is represented in the task in the form of state equations, which are subject to discretization during the determination of feasible solutions using one of the aforementioned methods. Ikeda et al. [24] describe the collision avoidance problem as an optimal control task. The goal is to find a combination of safe maneuvers between the two UAVs that guarantees the longest minimum separation distance among possible escape maneuvers. The optimal control problem was formulated with quadratic performance criteria.

Many authors include among the set of algorithms that can be used in the process of determining the separation of platforms, UAV route planning algorithms based on the general VRP displacement planning task. For an example of the third group algorithm, see [25]. It seems, however, that the use of integer methods for this purpose does not guarantee a short time of obtaining the result, because the general VRP problem belongs to the NP-hard problems class. The authors of the article also believe that the use of heuristic methods in the form of Genetic Algorithms or Particle Swarm Optimization (a good introduction can be found in the article [26]) for this purpose, is also not a perspective path. For heuristic algorithms, it is not possible to determine convergence, which makes their use in control in the event of emergency situations very risky. In the case of system certification, it should rather be assumed that the certification authority may prevent the operation of the system with implemented algorithms of this class.

At this point, it is also worth noting that for several years there have been attempts to formalize the methods of testing models based on languages such as UML or SysML, as described, among others, in [8]. However, testing according to these methods is a design task to which a separate team should be assigned. This topic is not covered in the article. The interested reader is referred to the papers described in the article [8].

The final stage of the design work is the design of system testing procedures based on mission simulators and air tests. This includes the development of documentation rules for the most critical procedures affecting flight safety and the documentation of the tests performed. There are no clear guidelines for testing unmanned systems, but NATO and EU standards and norms are emerging [8]. The most important standard in the design of unmanned aerial vehicles is certainly STANAG 4586 [27], which presents general requirements for a UAV that would cooperate with various Ground Control Stations (GCS) and operate in a swarm with other platforms. Accompanying these standards are system design guidelines such as DO-1878C (System Building Guidelines [10]), DO-331 [14], DO-332 [15] or DO-333 [16]. In particular, the DO-331 standard-"Model-Based Development and Verification" is extremely important. The standard is an attachment to DO-178C and DO-278A. The equivalent of these guidelines is EMAR documents, which are also described in [8].

3. Modeling of UAV Architecture

This section describes a practical example of modeling a selected unmanned aerial platform flight scenario, which considers the system's response to selected emergency in flight situations. An example of a scenario will be the problem of maintaining the separation between two platforms, understood as a requirement not to exceed the minimum safe distance between platforms. The algorithm described in the article is implemented in the air platform Mission Computer and works automatically mainly when there is no communication with the GCS. When the radio link is active, the pilot has priority in making decisions on avoiding collisions with other platforms.

3.1. Selected Design Assumptions

The functional requirements and the requirements specified under the FHA are the bases for developing the system architecture and the detailed description of the aircraft flight scenarios. Each functional requirement is transformed, depending on the level of detail, into a scenario or a single function of the designed system. In the further part of the article, the operating models of the system in flight will be presented, taking into account the detection of a potential air collision of two air platforms equipped with ADSB systems. Figure 1 shows a Use Case diagram covering the scenario presented in the article. We assume that the functional analysis of the system operation (FHA) has shown that a collision with another aircraft equipped with ADSB may occur during the flight of the platform. We also assume that the UAV may not be in contact with GCS at the time preceding the collision. The UAV itself is equipped with ADSB with the option of receiving the signal generated by other air platforms. Otherwise, the system should only prompt the pilot on possible action, but the final decision must always be with the pilot. Based on these assumptions, the model shown in the article was developed.

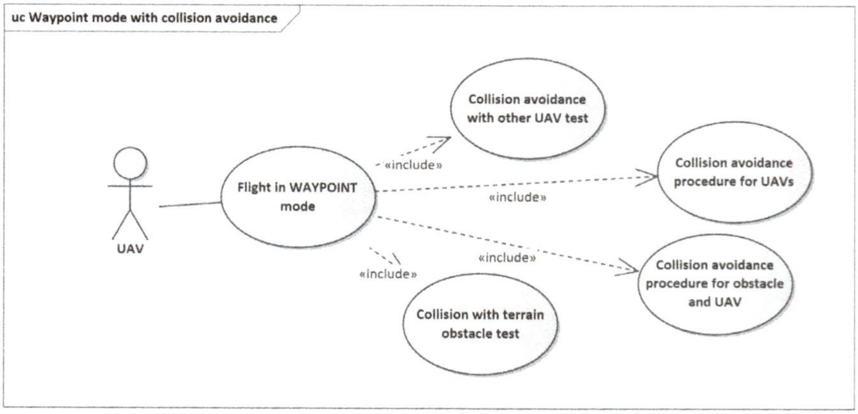

Figure 1. Use case scenario: UAV flight in WAYPOINT mode [27]. UAV can perform a flight in a WAYPOINT mode with the handling of exceptional situations.

Use Cases map to State Machine diagrams that best represent the interactions occurring in concurrent systems. This modeling approach requires the determination of the following data:

i. How many states are to be that model for the described scenarios?
ii. What are the transitions between states?
iii. What are the emergency situations in flight, when can they occur, and what do they depend on?
iv. How to minimize the number of states describing emergency situations in flight so that the pilot can easily manage the system (particularly UAV computers)?

3.2. The Modeling of an Emergency Situation

In this section, a practical example of modeling an emergency situation in the flight of an unmanned platform will be presented, which concerns the response to a potential air collision of two platforms.

Figure 1 shows an exemplary Use Case model that describes the flight of the UAV in the WAYPOINT mode (flight along an a priori pre-set route consisting of many points). The basic Use Case must include handling an emergency situation in flight, which concerns the occurrence of a potential collision of air platforms that inform themselves about their positions using the ADSB system. The same applies to the avoidance of collisions with a terrain obstacle.

Figure 2 shows a general diagram of a State Machine for a Use Case that includes testing for the possibility of collision between air platforms during flight. The state machine model is shown, which describes the system's operation during the flight in the WAYPOINT mode (automatic flight after the *a priori* setpoints). All state changes occur within a single thread, which allows the use of a Mission Computer with lower performance (and thus smaller dimensions of the device). However, critical functions do not run in parallel, so the method of implementation described by the model is unacceptable for larger platforms.

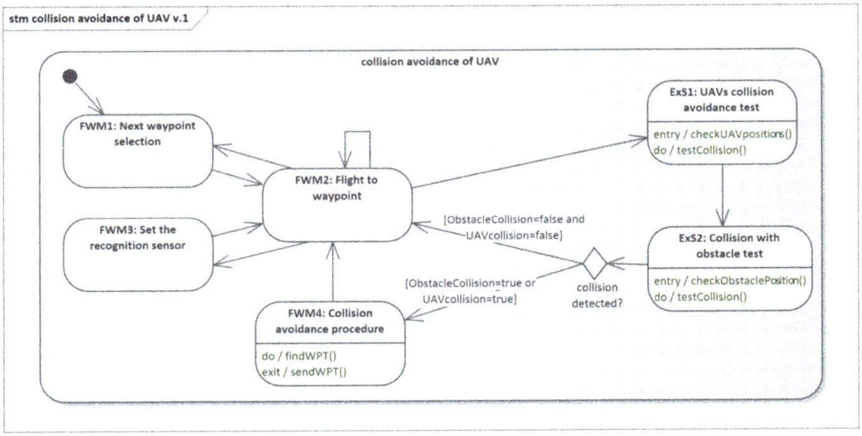

Figure 2. State machine diagram for a collision avoidance of UAV. Model for one core unit processor. Actions within the states are presented.

Figure 3 shows the State Machine model, which describes the operation of the system in flight in the same mode. The handling of the flight to the following points is separate from checking for the occurrence of emergency situations. In this model, there are two parallel threads, each affecting how the other works. The critical functions for testing emergencies are performed serially in a separate thread. This implementation method is acceptable for larger platforms and does not consume a significant amount of computer resources (in this case, the Mission Computer).

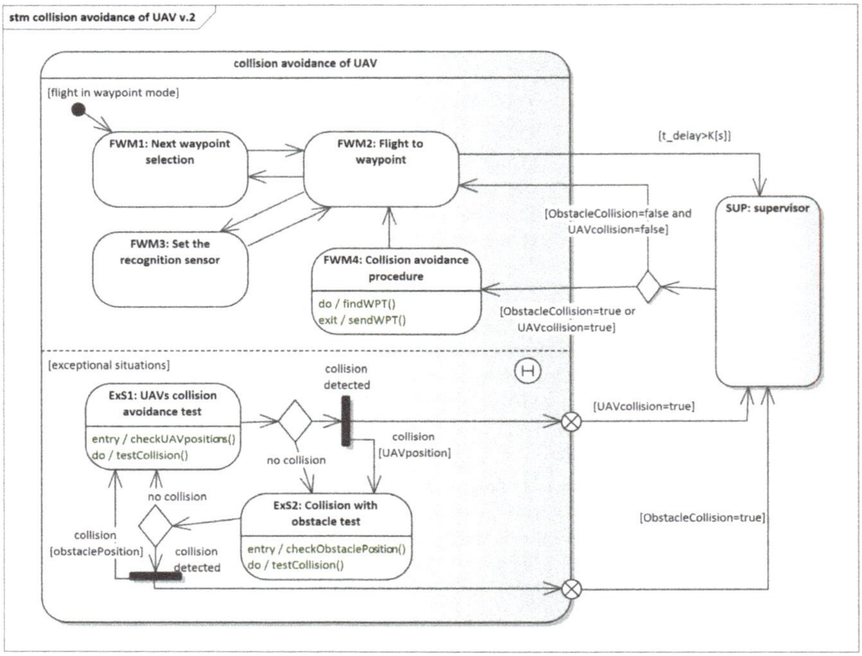

Figure 3. State machine diagram for collision avoidance of two UAVs. Model for two core units processor. Actions within the states are presented.

Figure 3 shows a fragment of the State Machine model with selected actions in selected states. From a formal point of view, each state must have an action with the stereotype *do:* that describes the processing that is performed in that state. This processing can be modeled as an activity or sequence diagram. In our case, the functions *do:testCollision()* and *do:findWPT()* are given numerical algorithms that are used to implement them. Collision testing is performed according to the Equations (1) and (2) (see Section 4). It is worth noting that the determination of the collision situation takes a short time so that individual tests can be performed sequentially without risk. Determining a new flight direction is based on the Equations (3) and (4).

The basic scenario carried out by the UAV, which performs a flight in the route mode, consists in going through the following states in sequence:

$FWM1 \rightarrow FWM2 \rightarrow FWM3 \rightarrow FWM2 \rightarrow SUP \rightarrow FWM1$ (selection of the next waypoint, flight to a point, the configuration of the recognition sensor, checking for an emergency, and going to the selection of the next waypoint).

An alternative processing scenario will occur when a parallel process described in the [exceptional situation] state will execute the $testCollision()$ function, which will determine the UAV in a potential collision with the UAV. At this point, an alternative scenario is realized:

$FWM1 \rightarrow FWM2 \rightarrow SUP \rightarrow FWM4 \rightarrow FWM2 \rightarrow SUP \rightarrow FWM1$ (following waypoint selection, flight to point, emergency test, emergency collision avoidance (*do:findWPT()* function, which determines the point to which the UAV must fly to avoid a collision), a continuation of the flight to a point, a test of the occurrence of an emergency, and selection of the next waypoint).

As part of handling emergency situations, potential collisions between aerial platforms and a collision with a terrain obstacle are tested. The reader may notice two entries from the state ExS2 to the state ExS1. It is not accidental. Depending on whether the possibility of a collision with a terrain obstacle was observed, the test of the potential collision of the aircraft may give different results. For example, in the event of a potential collision with a terrain obstacle, UAV algorithms must determine a route point above the obstacle, which ensures no collision with another UAV. By default, the platform can lower the flight altitude in certain situations, which is not possible in the event of a potential collision with a terrain obstacle. The processing itself is strictly dependent on the adopted rules and considerations that go beyond the article's scope. They are issues bordering on aviation law.

473 The system's physical architecture is designed after designing scenarios that take into account specific cases (emergency situations) in flight. A modern approach to the construction of unmanned platforms in line with the so-called Open architecture makes it possible to use commercial control systems even in military systems. Therefore, many modern UAVs use commercial Flight Control Computers (FCC), the functionality of which is extended by the use of a special Mission Computer (MC), which performs functions that cannot be performed with the use of FCC. A diagram of the architecture of the discussed unmanned system is shown in Figure 4. Systems of this class have redundant FCC and MC computers.

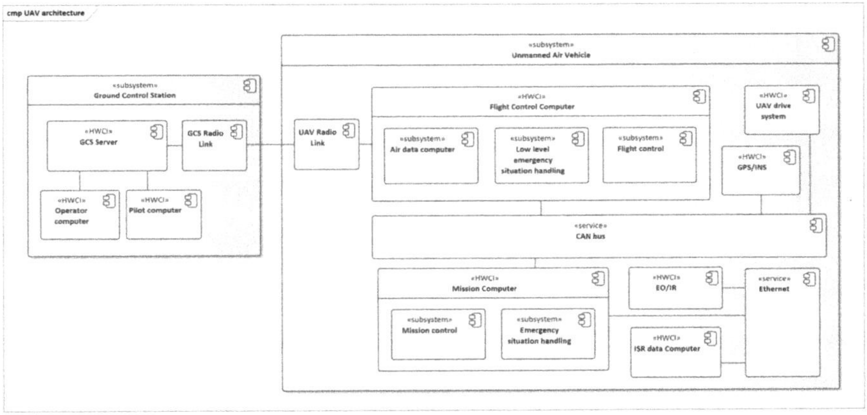

Figure 4. Physical architecture of UAV with logical units depicted (collision avoidance modules). The figure shows GCS components and UAV components with the radio data link marked.

Architecture refers to the physical components processing data in the system with indicating the types/roles of these elements («HWCI» stereotype means Hardware Configuration Item). The figure shows selected subsystems embedded on platform computers. In particular, the subsystems responsible for the specific situation described in the article, related to the avoidance of air platform collisions, were indicated. In classic systems equipped with ADSB, the FCC detects the collision situation, but the collision avoidance algorithm itself can be performed as part of MC processes. It depends on the computational complexity of the algorithm used.

3.3. System Architecture with SIL Elements

This section shows the extensions of the physical architecture that concern the preparation of additional components simulating the operation of those systems that cannot be run during system tests in the laboratory. Due to the scope of the simulation performed and the used UAV physical components (GPS / INS and ADSB), it was decided to use the Software-in-the-Loop (SIL) simulation scheme. Software-in-the-Loop represents the integration of a production code into a mathematical model simulation, providing engineers with a virtual simulation environment for the development and testing of complex systems. SIL makes it possible to test the software before the hardware prototyping phase and accelerates the development cycle. SIL enables the earliest detections of system-level defects or bugs.

Testing of complex systems, in particular those that can cause harm to humans, must be performed in conditions similar to reality. In this sense, simulators of real systems are built. If most or part of the critical UAV systems is simulated by software, it is referred to as SIL (Software-in-the-Loop) simulator testing. This approach has many disadvantages, the main of which is the need to prove that the software simulator corresponds precisely to the software embedded on the actual platform. In this respect, the most difficult thing is to simulate the computing power, including the processors' load. In addition, systems often use a different architecture associated with differences in the interpretation of variables and how some operations are performed. However, it is not easy to test a system other than based on SIL in some situations. Examples include simulating collisions of unmanned aerial vehicles, simulating a collision with a terrain obstacle, or simulating GPS/INS signal interference. In the case of GPS/INS systems, modern systems are so intelligent that it is impossible to substitute GPS position data to the device to simulate a flight. Therefore, in most cases, GPS/INS signal interference is tested based on SIL. The presented approach simulates the operation of GPS/INS spatial orientation systems and the ADSB system's operation, which generate information about UAVs whose flight trajectories may cause an air collision with our UAV. The SIL scheme is sufficient in this case to test the correctness of the designed algorithm. The SIL architecture integrated with GCS and UAV elements is shown in Figure 5.

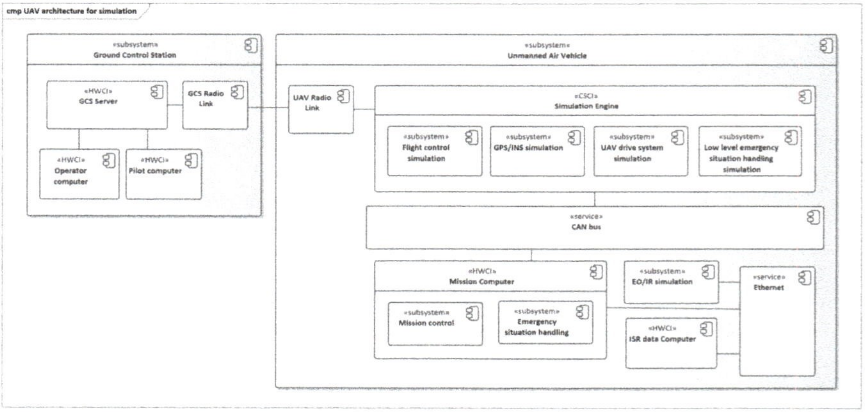

Figure 5. The physical architecture of UAV with additional simulation units. Simulation packages replace FCC. The remaining UAV components remained unchanged.

When testing the collision avoidance procedures, one can use the software embedded on the real Mission Computer (MC). The change in the platform's position is still simulated by software, but algorithms are used to determine new flight courses embedded in the MC. At this point, we can talk about the Hardware-in-the-Loop (HIL) class simulation, in which original elements of the air platform are used. However, in order to consider the simulation environment to meet the requirements of the HIL test environment fully, an FCC would have to be added to this environment, which would be fed with the necessary data about the platform position and the status of onboard equipment. However, it is not necessary to properly test the collision avoidance algorithm.

4. Collision Avoidance Algorithm

4.1. The General Model of Collision Avoidance

To present the algorithms supporting the UAV reaction to the occurrence of a critical situation, an algorithm for avoiding the collision of two air platforms was selected. The basic version of the algorithm is described in the article [9]. It defines a sequence of equations that allow the calculation of the minimum safe manoeuvres of two airplanes, which restrains the platforms from violating the protection zone.

In brief, the article assumes that two UAVs fly with a constant speed and direction (see Figure 6). Since the velocity vectors, $\vec{V_A}$ and $\vec{V_B}$ are known, the nearest approach vector is defined as:

$$\vec{r}_m = \hat{c} \times (\vec{r} \times \hat{c}), \tag{1}$$

where: $\vec{c} = \vec{V_B} - \vec{V_A}$, is defined in the Figure 6, $\hat{c}$ is its normalized representation ($\hat{c} = \frac{\vec{c}}{\|\vec{c}\|}$). If $\|\vec{r_m}\| < r_{safe}$, then separation distance will be violated. r_{safe} is the safe distance.

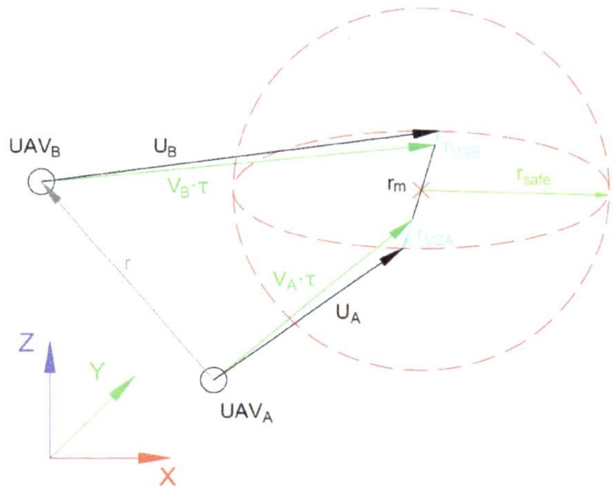

Figure 6. Graphical representation of variables of the algorithm in a 3D cartesian coordinate system. $\vec{V_A}$ and $\vec{V_B}$ are UAVs speed vectors. $\vec{U_A}$ and $\vec{U_B}$ are new UAVs speed vectors.

Next, the time of closest approach τ is calculated:

$$\tau = -\frac{\vec{r} \cdot \vec{c}}{\vec{c} \cdot \vec{c}}. \tag{2}$$

This enables to calculation of the positions of the UAVs for the closest approach. For UAV_A, these vectors can be calculated as presented in the Equation (3):

$$\vec{r}_{VSA} = k \cdot \frac{r_{safe} - \vec{r}_m}{\vec{r}_m}(-\vec{r}_m), \tag{3}$$

where $k = \frac{\vec{V_B}}{\|\vec{V_A}\| + \|\vec{V_B}\|}$ represents coefficient forcing a slower airplane to take a bigger turn, as higher speed reduces maneuverability. The resulting dodge vector $\vec{U_A}$ is calculated as follows:

$$\vec{U_A} = \vec{V_A} \cdot \tau + \vec{r}_{VSA}. \tag{4}$$

Park, et al. [9] propose to calculate $\vec{U_A}$ as unit vectors to define the direction of requested movement. However, for some reason, it is better to use the non normalized $\vec{U_A}$ on some systems to infer the collision avoidance waypoint position.

4.2. Visualization of the Results of the Collision Avoidance Algorithm

The results for this algorithm are shown in Figure 7. Each picture contains the current positions of the UAVs and their flight history and two markers defining the current target waypoint. The two-color scheme represents two UAVs. Both are surrounded by red ellipses (as an equirectangular projection of WGS84 is used for coordinates around N54deg). They represent the safety radius in the horizontal plane. Each UAV model of presents a vector of (horizontal) speed, while the number represents altitude at that point. The simulations assume very close speed values of both UAVs as it is the worst-case scenario.

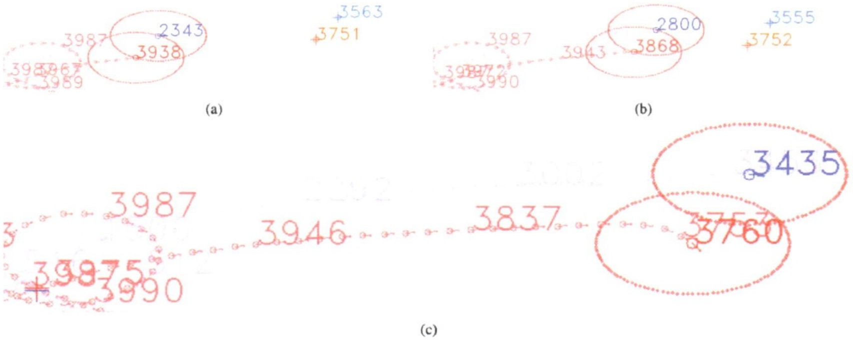

Figure 7. Example of the working principle of the algorithm with visualization of a dynamic environment. The red circles show the UAV's safety zone. The numerical values in the figures show the height of each UAV. (**a**) First detection of a long-distance collision of two UAVs flying in almost parallel trajectories. (**b**) Slight target waypoints adjustment after 30 iterations. The trajectories are corrected for the first time. (**c**) The end of the collision avoidance maneuver with visible target points. A change in the trajectory of each UAV is presented.

4.3. *Modifications to the Collision Avoidance Algorithm*

The article proposes an additional set of conditions to improve the response of systems in a realistic environment.

4.3.1. Time and Distance Limitations

The original algorithm does not consider the problem of detecting collisions of two UAVs at very long distances (Figure 8). If we consider the situation of one UAV flying in a straight line, while the other UAV turns with a constant turning radius around a single point very far from the first UAV but of the same altitude, the algorithm has a high probability of detecting a collision. This indicates a need to define an additional set of conditions to limit launch cases for emergency situations.

The following additional condition (5) is proposed as a solution to this issue.

$$\|r\| < (\max{(k_{mar} \cdot (UAV_A.r_{safe} + UAV_B.r_{safe}),} \\ \Delta T \cdot (\|V_A\| + \|V_B\|))\,), \tag{5}$$

where k_{mar} is the safety margin coefficient, proposed value is $k_{mar} = 2$, ΔT is assumed time of safety margin, proposed value is either 60 [s] or twice the time required for 180 [deg.] turn of the UAV. These proposed values are suggested minimal values. Higher distances will increase the number of algorithm launches and the smoothness of avoidance maneuvers.

566 To further limit the number of calls on the systems, additional conditions regarding time (τ) and distance ($V_A \cdot \tau$) to the collision can be added after the first calculation steps of the algorithm. A very long time to collision may suggest that no intervention is needed. However, the distance between UAVs must be monitored as there is a possibility that they will be very close and running in almost parallel straight lines. This can break safety zones, especially for airplanes with minimal vertical speed or low-altitude flight situations where lowering the flight level is impossible.

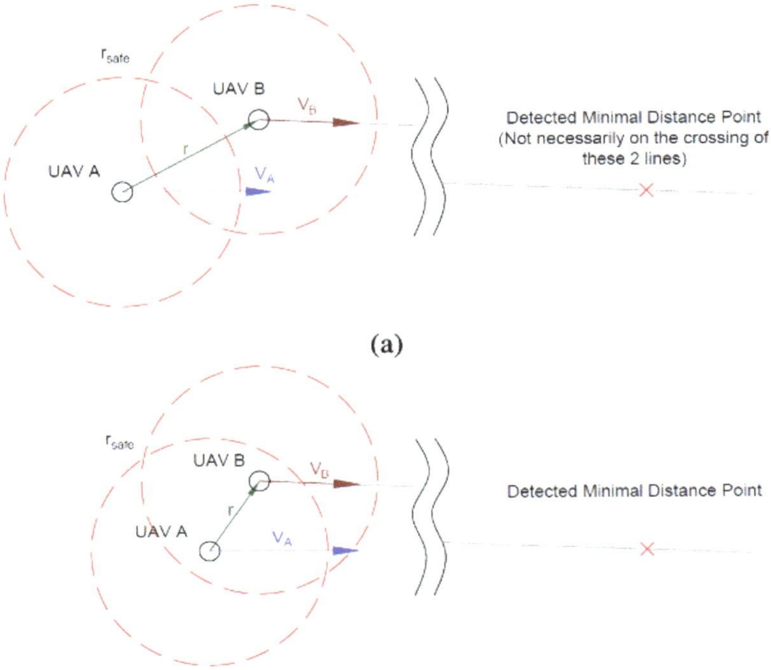

Figure 8. Example of the problematic situation after introduction of a collision time limit. (**a**) Detection of collision over a very long time, (**b**) safe area violation after a short period.

4.3.2. Spheroid Separation Areas

576 Most airplanes measure altitude positions with higher accuracy than other coordinates. Therefore most of the norms assume safe ellipsoid areas. For example, Reduced Vertical Separation Minimum (RVSM) [28] introduces lowered 1000 ft. vertical separation for flights under 41,000 ft. and 2000 ft. above that threshold. Meanwhile, horizontal separation is defined as more complex due to different definitions. Usually, horizontal separation requires a distance of 15 nautical miles (27.8 km—lateral separation) or 15 min of flight (longitudinal separation) [29].

There are many definitions for the safety radii, varying for countries, heights, speeds or weather conditions (e.g., [30], or [31] presented in Figure 9b). In this paper, separation values, due to low altitude flights, is 9260 m for horizontal distance and 600 m for vertical distance, defining spheroid presented in Figure 9a.

Additionally, the distances depend on object type (UAVs usually assume smaller collision spheroids) or flight conditions (IFR-Instrument flight rules versus VFR-visual flight rules).

To solve this problem, simple coordinate transformation can be used. The operations on the ellipsoid or spheroid are more complex than for sphere (for example, requiring iterative calculations for projections), and the original algorithm does not consider variable r_{safe} distance. It is the easiest to assume a spheroid separation zone (r_{safe-H}, r_{safe-V}) and stretch the Z-axis with coefficient $\frac{r_{safe-H}}{r_{safe-V}}$ before the appliance of the calculations for the sphere. Afterward, the Z-axis has to be brought back.

That yields few profits: dodge maneuvers in the Z-axis are smaller than initially, so it is simpler to test the algorithm and prove its efficiency. The r_m still represents the violation grade of the separation distance as it can be compared to r_{safe-H}.

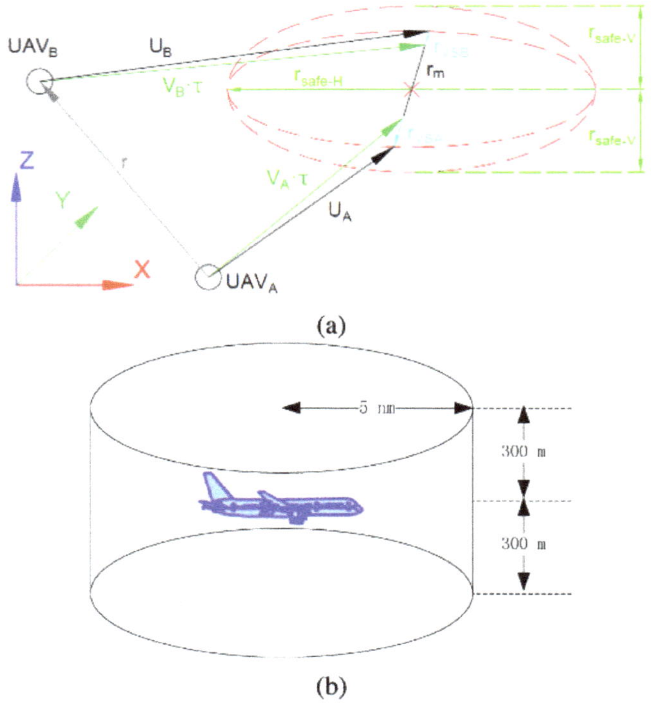

Figure 9. Examples of separation space definitions: (**a**) Spheroid safety zone and definition of used vectors in this paper, (**b**) Cylindrical safety zone from [31].

4.3.3. Multiple Obstacle Scenario

Precise analysis of multiple obstacle scenario is outside of the scope of this article. An exemplary approach to such situation is presented in an article by Lin et al. [19]. When the collision is detected with more than one object there might be 2 cases:

1. Only one object has overlap with initial path and possible dodge manoeuvre—a single dodge manoeuvre is needed in order to avoid safe area violation.
2. Two or more objects overlap the initial path—authors propose the expansion of the collision zone for new, single virtual target.

5. Results

5.1. Assumptions for the Testing Process

Several critical assumptions were made for the presented scenario of the system's operations. Firstly, it was assumed that it is not possible for more than two platforms to be in the area of a potential collision in a short period. Secondly, it was assumed that the algorithm determining a dodge maneuver for collision avoidance could be allotted any direction, i.e., there are no *no-fly* areas in the flight area. Otherwise (as is the case in the real system), a solution considering the above limitations should be determined at one stage of the algorithm. The state machine model for such a case would be beyond the scope of this article.

5.2. Testing of Emergency Situation in Flight

The article [9] shows the basic version of the algorithm for calculating a new UAV course to avoid a collision with another UAV that is on a collision course. Two prominent cases were considered: when the platforms are flying towards a head-on collision from two directions and when the courses of the platforms intersect in the near distance. For these cases, a method of determining a new course of both platforms was shown.

Additionally, the real-life application requires consideration over a non-zero sampling interval, which introduces delays and step changes of setpoints for the flight controllers.

The article presents the results of simulations of cases described in [9] and new cases that the basic algorithm developed by Jung-Woo did not correctly detect.

Testing scenarios:

(a) The platforms are on a collision course, and the collision will occur at a short distance (Figure 10a);
(b) The platforms are on a collision course, but the collision will occur at a considerable distance from the actual position of the platforms (Figure 10b);
(c) The platforms fly to points close to each other, almost parallel, which creates a risk of collision at any point in the route (Figure 10c,d).

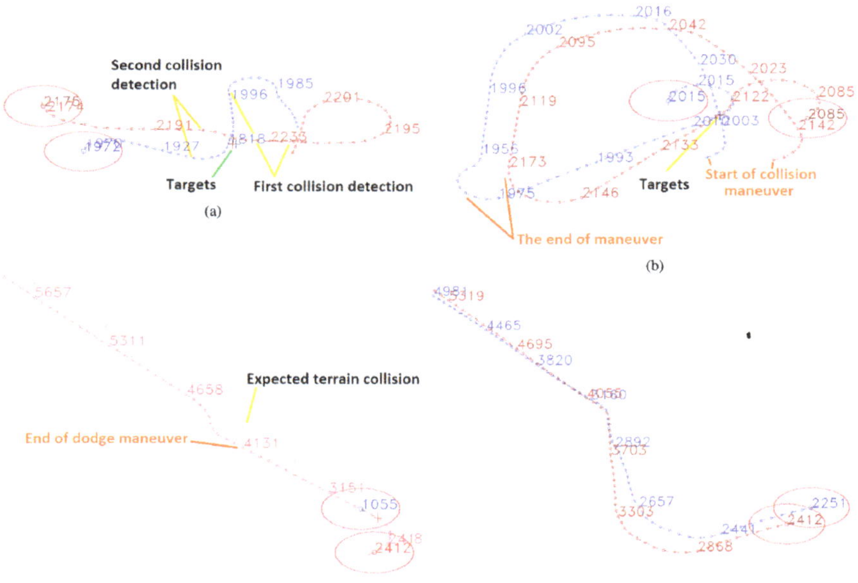

Figure 10. Output images from the simulation. (**a**) Dodge maneuvers of two crossing airplanes. (**b**) Dodge maneuvers of two semi-parallel airplanes. (**c**) Straight route with spherical safety zone—original algorithm. (**d**) Straight route with limited vertical movement and spheroid safety zone.

5.3. Results from the Simulator and Calculations

The result of the algorithm is shown in Figure 10. Figure 10a presents two dodge maneuvers need for two UAVs to dodge a safe area collision when given the same target waypoints. Their vectors of approach differ considerably. Figure 10b presents a similar situation, but the speed vectors of the UAVs are similar, which leads to more prolonged dodge maneuvers. Figure 10c presents a dodge for the original algorithm. This case leads to a terrain collision, as the lower UAV drops altitude to avoid conflict. Figure 10c presents a modified algorithm, which leads to much lower altitude changes, but a bigger and slower horizontal dodge.

6. Conclusions

The article discusses the methods of designing a reliable hardware and software architecture of unmanned aerial vehicles, that consider the modeling of emergency situations in the flight. The models and results are for a medium-range UAV system subject to certification regulations under current EU legislation. Therefore, it should be assumed that these platforms will be subject to a certification process similar to that to which airplanes are subject.

The article focuses primarily on the methods of modeling scenarios implemented by the platform in flight, using the technique of functional threat analysis and failure trees of elements included in the hardware and software architecture of the system. It will not be an overstatement to say that this approach will soon dominate companies producing flying systems.

Particular attention was paid to describing methods of modeling the collision avoidance behavior in a flight of two platforms equipped with ADSB devices. One of the better algorithms has been referred to, which gives a deterministic solution to the above problem. Ways to test this algorithm are provided. It was shown that the descriptions of the algorithm in the source article contained inaccuracies that could cause errors in the platform or its unexpected reactions in real-world conditions. The simulation tests proved it, and an analytical solution was given, an extension of the base algorithm. The article focuses on the principles of designing system test plans and designing individual test cases to verify the system's operation in the event of emergency situations in flight. We also present the air platform flight simulator that we use in practice in the form of a SIL model based on selected algorithms for avoiding dangerous situations.

Further work will go in two directions. First of all, the model should be extended with additional algorithms for reacting to emergencies, such as, for example, ensuring separation from a terrain obstacle or maintaining the UAV flight in the permitted area. Secondly, the described simulator should be expanded to a full HIL model using additional equipment. However, this requires the construction of simulators of mechanical components such as the engine, which is a separate project due to the complexity of the task.

Author Contributions: Methodology and system modeling in UML, W.M.S.; Collision avoidance algorithm and software, K.A.G.; Writing an original draft, W.M.S. and K.A.G. All authors have read and agreed to the published version of the manuscript.

Funding: This research received no external funding.

Institutional Review Board Statement: Not applicable.

Informed Consent Statement: Not applicable.

Data Availability Statement: Not applicable.

Conflicts of Interest: The authors declare no conflict of interest.

References

1. Ruijters, E.; Stoelinga, M. Fault tree analysis: A survey of the state-of-the-art in modeling, analysis and tools. *Comput. Sci. Rev.* **2015**, *15*, 29–62. [CrossRef]
2. Kritzinger, D. Assessments for Initial Airworthiness Certification. In *Aircraft System Safety*, 1st ed.; Woodhead Publishing: Sawston, UK, 2016.
3. SAE. *ARP4754A Guidelines for Development of Civil Aircraft and Systems*; SAE: Nashville, TN, USA, 2010.
4. Russell, M. Using MBSE to Enhance System Design Decision Making. *Proc. Comput. Sci.* **2012**, *8*, 188–193. [CrossRef]
5. Stecz, W.; Kowaleczko, P. Designing operational safety procedures for UAV according to NATO architecture framework. In Proceedings of the 16th International Conference Software Technology (ICSOFT), London, UK, 7–9 December 2021; pp. 135–142.
6. Steurer, M.; Morozov, A.; Janschek, K.; Neitzke, K.-P. SysML-based Profile for Dependable UAV Design. *IFAC-PapersOnLine* **2018**, *24*, 1067–1074. [CrossRef]
7. OMG Homepage—UML Specification File 2.5.1. Available online: https://www.omg.org\protect\discretionary{\char\hyphenchar\font}{}{}/spec/UML/2.5.1/\protect\discretionary{\char\hyphenchar\font}{}{}About-UML/ (accessed on 17 May 2020).
8. Stecz, W.; Bejtan, W.; Rulka, J. R&D Activities in the UAV Production and Certification Process. In Proceedings of the 37th Inter-national Business Information Management Association Conference, Cordoba, Spain, 30–31 May 2021.
9. Park, J.; Oh, H.; Tahk, M. UAV collision avoidance based on geometric approach. In Proceedings of the 2008 SICE Annual Conference, Tokyo, Japan, 20–22 August 2008; pp. 2122–2126. [CrossRef]

10. RTCA Homepage—DO178. Available online: https://standards.globalspec.com/std/\protect\discretionary{\char\hyphenchar\font}{}{}14369281/RTCA%20DO-178 (accessed on 28 September 2021).
11. Sanchez-Lopez, J.L.; Pestana, J.; de la Puente, P.; Campoy, P. A Reliable Open-Source System Architecture for the Fast Designing and Prototyping of Autonomous Multi-UAV Systems: Simulation and Experimentation. *J. Intel. Robot. Syst.* **2016**, *84*, 779–797. [CrossRef]
12. Atyabi, A.; MahmoudZadeh, S.; Nefti-Meziani, S. Current advancements on autonomous mission planning and management systems: An AUV and UAV perspective. *Ann. Rev. Control* **2018**, *46*, 196–215. [CrossRef]
13. Zhang, X.; Zhao, X. Architecture Design of Distributed Redundant Flight Control Computer Based on Time-Triggered Buses for UAVs. *IEEE Sens. J.* **2021**, *21*. [CrossRef]
14. RTCA. *DO-331 Model-Based Development and Verification Supplement to DO-178C and DO-278A*; RTCA: Washington, DC, USA, 2011.
15. RTCA. *DO-332 Object-Oriented Technology and Related Techniques Supplement to DO-178C and DO-278A*; RTCA: Washington, DC, USA, 2011.
16. RTCA. *DO-333 Formal Methods Supplement to DO-178C and DO-278A*; RTCA: Washington, DC, USA, 2011
17. Mitchell, S. Model-Based System Development for Managing the Evolution of a Common Submarine Combat System. In Proceedings of the AFCEA/GMU 2010 Symposium on Critical Issues in C4I, Fairfax, VR, USA, 18–19 May 2010.
18. Guan, X.; Lyu, R.; Shi, H.; Chen, J. A survey of safety separation management and collision avoidance approaches of civil UAS operating in integration national airspace system. *Chin. J. Aeronaut.* **2020**, *33*, 2851–2863. [CrossRef]
19. Lin, Z.; Castano, L.; Mortimer, E.; Xu, H. Fast 3D Collision Avoidance Algorithm for Fixed Wing UAS. *J. Intel. Robot. Syst.* **2020**, *97*, 577–604. [CrossRef]
20. Lin, Z.; Castano, L.; Xu, H. A Fast Obstacle Collision Avoidance Algorithm for Fixed Wing UAS. In Proceedings of the 2018 International Conference on Unmanned Aircraft Systems (ICUAS), Dallas, TX, USA, 12–15 June 2018; pp. 559–568. [CrossRef]
21. Peng, L.; Lin, Y. A closed-form solution of horizontal maneuver to collision avoidance system for UAVs. In Proceedings of the 2010 Chinese Control and Decision Conference, Xuzhou, China, 25–26 May 2010; pp. 4416–4421. [CrossRef]
22. Pytlak, R. Numerical Methods for Optimal Control Problems with State Constraints. In *Lecture Notes in Computer Science*; Springer: Berlin/Heidelberg, Germany,1999.
23. Betts, J.T. *Practical Methods for Optimal Control and Estimation Using Nonlinear Programming*; Cambridge University Press: Cambridge, UK, 2009.
24. Ikeda, Y.; Kay, J. An optimal control problem for automatic air collision avoidance. In Proceedngs of the 42nd IEEE International Conference on Decision and Control (IEEE Cat. No.03CH37475), Maui, HI, USA, 9–12 December 2003; Volume 3, pp. 2222–2227. [CrossRef]
25. Stecz, W.; Gromada, K. Determining UAV Flight Trajectory for Target Recognition Using EO/IR and SAR. *Sensors* **2020**, 20, 5712. [CrossRef] [PubMed]
26. Skrzypecki, S.; Tarapata, Z.; Pierzchala, D. Combined PSO Methods for UAVs Swarm modeling and Simulation. In *Modeling and Simulation for Autonomous Systems*; Springer: Berlin/Heidelberg, Germany, 2020; pp. 11–25.
27. NATO Standardization Agency (NSA). *STANAG 4586 Ed.4. Standard Interfaces of UAV Control System (UCS) for NATO UAV Interoperability*; NATO: Brussels, Belgium, 2017.
28. U.S. Department of Transportation Federal Aviation Administration. Authorization of Aircraft and Operators for Flight in Reduced Vertical Separation Minimum (RVSM) Airspace. Available online: https://www.faa.gov/air_traffic/separation_standards/rvsm/documents/\protect\discretionary{\char\hyphenchar\font}{}{}AC9_85B.pdf (accessed on 29 January 2019).
29. Skybrary Based on ICAO, Controled by EUROCONTROL Separation Standards. Available online: https://www.skybrary.aero/index.php/\protect\discretionary{\char\hyphenchar\font}{}{}Separation_Standards (accessed on 12 October 2021).
30. NAVAID Use Limitations. Available online: https://pointsixtyfive.com/xenforo/wiki/\protect\discretionary{\char\hyphenchar\font}{}{}711065_ch4/ (accessed on 12 October 2021).
31. Qian, X.; Mao, J.; Chen, C.; Chen, S.; Yang, C. Coordinated multi-aircraft 4D trajectories planning considering buffer safety distance and fuel consumption optimization via pure-strategy game. *Transport. Res. Part C Emerg. Technol.* **2017**, *81*, 18–35. [CrossRef]

applied sciences

MDPI

Article

Unmanned Aerial Traffic Management System Architecture for U-Space In-Flight Services

Carlos Capitán *, Héctor Pérez-León, Jesús Capitán, Ángel Castaño and Aníbal Ollero

Engineering School, University of Seville, Av. de los Descubrimientos, s/n, 41092 Seville, Spain; hectorperez@us.es (H.P.-L.); jcapitan@us.es (J.C.); castano@us.es (Á.C.); aollero@us.es (A.O.)
* Correspondence: ccapitan@us.es

Featured Application: This work can be applied to integrate autonomous unmanned aerial vehicles in civil airspace.

Abstract: This paper presents a software architecture for *Unmanned aerial system Traffic Management* (UTM). The work is framed within the U-space ecosystem, which is the European initiative for UTM in the civil airspace. We propose a system that focuses on providing the required services for automated decision-making during real-time threat management and conflict resolution, which is the main gap in current UTM solutions. Nonetheless, our software architecture follows an open-source design that is modular and flexible enough to accommodate additional U-space services in future developments. In its current implementation, our UTM solution is capable of tracking the aerial operations and monitoring the airspace in real time, in order to perform in-flight emergency management and tactical deconfliction. We show experimental results in order to demonstrate the UTM system working in a realistic simulation setup. For that, we performed our tests with the UTM system and the operators of the aerial aircraft located at remote locations with the consequent communication issues, and we showcased that the system was capable of managing in real time the conflicting events in two different use cases.

Keywords: UTM; system architecture; U-space; UAS

Citation: Capitán, C.; Pérez-León, H.; Capitán, J.; Castaño, Á.; Ollero, A. Unmanned Aerial Traffic Management System Architecture for U-Space In-Flight Services. *Appl. Sci.* **2021**, *11*, 3995. https://doi.org/10.3390/app11093995

Academic Editor: Sylvain Bertrand

Received: 30 March 2021
Accepted: 22 April 2021
Published: 28 April 2021

1. Introduction

In the last few years, there has been a clear trend to use *Unmanned Aircraft Systems* (UAS), or drones, for many commercial and civil applications. There are studies [1] estimating that up to 400,000 drones will be providing services in the airspace by 2050, with a total market value of 10 billion euros per year by 2035. Last-mile delivery [2], surveillance [3], infrastructure inspection [4], traffic monitoring [5], media production [6], or managing health emergency situations [7] are just a few examples of the wide spectrum of drone applications. Indeed, the integration of UAS in the civil airspace is probably one of the most revolutionary events for *Air Traffic Management* (ATM) since the beginning of its implementation. Although ATM has been traditionally based on voice communication through an *Air Traffic Control* (ATC) entity, its bounded workload and communication capacities turn this centralized resource into a bottleneck for system scalability. Therefore, the rise of UAS operations brings the need for a new paradigm for airspace management, where digital communication will play a key role, and the responsibilities will be shared among different stakeholders instead of a single central actor.

There are already some initiatives to integrate UAS into civil airspace and fulfill their operational requirements [8]. The *National Aeronautics and Space Administration* (NASA) has created the concept for *UAS Traffic Management* (UTM) [9] to enable safe, large-scale operations with UAS in low-altitude airspace [10]; whereas Europe has extended this UTM concept by proposing the *U-space* ecosystem [11]. More specifically, an overview of the

U-space ecosystem recently proposed by the *European Aviation Safety Agency* (EASA) [12] is depicted in Figure 1. The idea is to have a *U-space service Provider Platform*, which is a server running on the cloud, as the core component. There, the UTM system consists of a software architecture that provides U-space services to the different actors in the U-space ecosystem using as a bridge the *U-space Service Manager* (USM), which is a specific module of this UTM system.

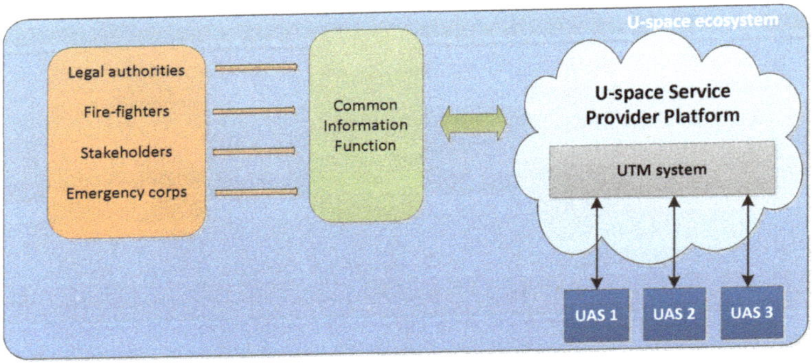

Figure 1. Overview of the U-space ecosystem proposed by EASA [12]. The UTM system offers U-space services to the different actors and runs on a remote server called U-space Service Provider Platform.

Currently, the community is in the process of further developing these U-space services. In this paper, we take a first step and propose a novel software architecture that aims to serve as a common framework for implementing and integrating U-space services. Our solution is being developed within the context of the European project GAUSS (https://projectgauss.eu, accessed on 26 April 2021), whose main objective is leveraging high-performance positioning functionalities provided by the Galileo ecosystem for U-space operations, including a validation phase with actual fixed-wing and rotary-wing UAS (see Figure 2). We present an architecture that is service-oriented and safety-centered, and that allows the airspace actors to abstract from specific UAS technologies. Besides, we implement a set of U-space services to manage complete UAS operations, but focusing on *in-flight* services (i.e., those required to handle the operations during the flight phase). Nonetheless, the architecture is modular and flexible enough to be extended with additional functionalities as new services become functional.

Figure 2. The Atlantic I (**left**) and DJI M600 (**right**) UAS will be used to validate the UTM functionalities developed in the GAUSS project.

Our main contributions are as follows. First, we introduce the main concepts and the roadmap for the U-space initiative, and we review other relevant works about UTM (Section 2). Second, we analyze the design properties for our UTM architecture (Section 3). Given a series of desired architectural guidelines (Section 3.1), we propose the open-source *Robot Operating System* (http://www.ros.org, accessed on 26 April 2021) as underlying middleware for our UTM system (Section 3.2). Third, we contribute with a new UTM system architecture implementing the U-space concept (Section 4). Our proposal represents a general framework for U-space services, which is modular, flexible, and technology-agnostic; but we describe our specific implementation for a set of core in-flight services dealing with unexpected UAS conflicts during their flight phase. Our software framework integrates automated decision-making procedures, which is one of the main gaps for current UTM solutions. Additionally, we show an actual realization of our UTM architecture that is available as open-source software for the community, and we demonstrate its capabilities (Section 5). In order to showcase the correct integration of all our components and services, we have defined use cases for UAS operations involving all the developed functionalities (Section 5.1); and we have assessed our results in terms of performance by running the whole system in a realistic simulation setup for multi-UAS operations (Sections 5.2 and 5.3). Finally, we draw the main conclusions of this work and point at future lines for further development (Section 6).

2. Background

In this section, we introduce the U-space initiative and its offered services, as well as its development roadmap. Then, we review the related work about UTM systems.

2.1. U-Space

U-space is a collaborative effort among researchers, industry, and regulators to enable the integration of UAS operations within the civil airspace, providing UAS situational awareness and digital communication with manned aviation, the ATM service providers, and the legal authorities. There exists a roadmap [11] to deploy U-space in Europe, consisting of the four phases depicted in Figure 3. Each phase will propose a new set of services with increasing complexity and integration level between UAS and manned aircraft, as well as an upgraded version of existing services in the previous phases.

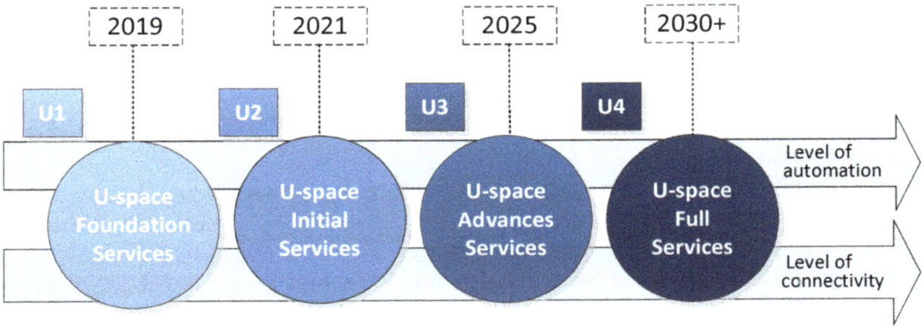

Figure 3. The implementation roadmap for the U-space initiative [13], consisting of 4 deployment phases.

The detailed functional system architecture is still under development, but there is already a list of services defined for each deployment phase [14], and a report with the current progress of their implementation and deployment [15]. Table 1 depicts these services and their current level of implementation in Europe.

Table 1. The U-space services for each development phase, together with their current implementation level. Our system focuses on in-flight services to handle unexpected events during the flight phase of the UAS.

Phase	Service	Overall Implementation Level	Covered in Our UTM System
U1 Foundation Services	E-registration	19%	
	E-identification	17%	
	Pre-tactical geofencing	23%	
U2 Initial Services	Tactical geofencing	13%	
	Flight planning management	6%	
	Weather information	3%	
	Tracking	4%	✓
	Monitoring	5%	✓
	Drone aeronautical information management	18%	
	Procedural interface with ATC	20%	
	Emergency management	9%	✓
	Strategic deconfliction	6%	
U3 Advanced Services	Dynamic geofencing	5%	
	Collaborative interface with ATC	8%	
	Tactical deconfliction	0%	✓
	Dynamic capacity management	4%	
U4 Full Services	To be defined	0%	

The U-space framework proposes a UTM system as the software architecture that provides services to the different U-space actors. A possible classification for the services is depending on whether they are activated in the UAS pre-flight phase or during the flight:

- Pre-flight services are those related with the functionalities needed to prepare and schedule a UAS operation. The vehicle and the operator need to register (*E-registration*), and the initial flight plan has to be handled before being accepted (*Flight planning management*). Then, the pilot may get assistance through information about predefined restricted areas (*Pre-tactical geofencing*) and the resolution of possible conflicts before flying (*Strategic deconfliction*).
- In-flight services deal with the functionalities required to handle the operation after the UAS flight has started. This means the possibility to update the operator (*Tactical geofencing*) or the UAS itself (*Dynamic geofencing*) with geofencing information during the flight, and to track the current position and predicted trajectory for each UAS (*Tracking*). This updated information is then used to create a situation of the airspace (*Monitoring*) and to generate warnings and contingency actions under possible threats (*Emergency management*). Alternative plans could also be suggested in-flight to maintain the required separation between aircraft and with geofences (*Tactical deconfliction*).
- There are other services that could be used either before flying or during the flight. These are functionalities that aim to provide identification (*E-identification*), weather forecasts (*Weather Information*), or more generic information (*Drone Aeronautical Information Management*), to create an interface with the ATC (*Procedural Interface with ATC* and *Collaborative interface with ATC*), and to control and manage the UAS density in the airspace (*Dynamic Capacity Management*).

According to Table 1 and to our study of the state of the art, in-flight services have been less addressed by UTM systems in general, with a notorious integration gap still existing. In this paper, we focus on in-flight functionalities to develop a UTM system, although our architecture is general enough to cover all kinds of services. In particular, we integrate those services related to the management of unexpected events while the UAS are flying, namely tracking, monitoring, emergency management, and tactical deconfliction. These services belong to the U2 and U3 implementation phases, which are scheduled to be developed between 2021 and 2029.

2.2. Related Work

The development of completely operational UTM systems is still at an early stage, even though it has recently become a growing field. The authors in [16] define what a UTM system is, and they give an overview of both a physical UTM architecture and a UTM software manager based on automated services. Big companies are one of the major parties interested in boosting the deployment of UTM. For instance, Google has proposed an ecosystem [17] where all UAS should be equipped with communication and *sense & avoid* technologies in order to perform cooperative flights when encountering other UAS or manned aircraft. In their proposal, the separation and planning services would be provided by an *Airspace Service Provider*. Furthermore, Amazon has put forward a one-operator-to-many-vehicle model [18], where the decision-making authority gets significantly distributed among the operators.

Additionally, there exist several commercial UTM system applications in the market. They implement most pre-flight services, but just partially a few in-flight services. For instance, Airmap [19] has its focus on UAS registration, geographic information systems, flight communication, traffic monitoring, and user interfaces. The Unifly platform [20] connects authorities with pilots to safely integrate UAS into the airspace. On the one hand, the authorities can visualize and approve the UAS flights, as well as manage *No Flight Zones* in real time. On the other hand, the pilots can manage their UAS (e.g., with the E-registration, E-identification, and Flight plan management services) and they can plan and receive flight approvals aligned with international and local regulations. Another framework is the Thales ECOsystem UTM [21], which integrates UAS and pilot registration. ECOsystem provides a flight planning functionality, using airspace rules and situational awareness as guidelines. It also includes tools to manage map overlays and 3D terrain views.

The aforementioned UTM applications offer pre-flight UTM services and some in-flight capabilities such as UAS tracking. Even though they are capable of publishing real-time information about the UAS, they do not manage operations autonomously during the flight phase. Moreover, it is important to highlight that all those applications are commercial products that are not available for the community as open software.

The scientific community has also been putting effort into functional UTM frameworks; a recent review of related works can be seen in [22]. A prototype UTM for flight surveillance has recently been proposed in Taiwan [23]. One of its core properties is the capability to monitor vehicles, being the ADS-B (*Automatic Dependent Surveillance Broadcast*) technology used for surveillance. There is a pre-flight procedure to schedule and approve flights, and then the UTM system can send surveillance alerts during the operation, though all the decisions for conflict resolution are up to the pilot. Another UTM system has been presented in Sweden [24]. It incorporates a complete toolkit to manage traffic, geofences, flight altitude segregation as in the general aviation, and complex visualization. This research has also identified problems that dense traffic in the low-level airspace will bring to the city users, by simulating the future urban airspace. In general, the functionalities of the aforementioned systems have only been demonstrated through simplistic simulations, and quite a few works have been devoted to field flight campaigns for preliminary tests [25,26]. We have also proposed in a previous work [27] a more realistic simulator for UAS operations based on the ROS middleware and the 3D simulation suite Gazebo (http://gazebosim.org, accessed on 26 April 2021). In that work, we introduced a preliminary definition of our in-flight services and a tool for mission validation. In the current paper, we go beyond by implementing a complete UTM architecture. We integrate in-flight services to handle unexpected conflicts that may occur while the UAS are flying, and we showcase the performance of our system through heterogeneous use cases.

Finally, regarding the implementation of particular in-flight services, there are different approaches for conflict resolution and emergency management. Many works [28–30] have focused on flight planning and scheduling at a strategic level, i.e., in the pre-flight phase; though in-flight automated decision-making has not been properly covered in UTM systems. In general, given the massive search space to find optimal resolutions for conflicts in *Very Low-Level* (VLL) airspace scenarios, approximate solutions based on heuristic solvers [28] or lane maneuvers [30] predominate over optimal deconfliction approaches. In [31], a probabilistic framework is proposed to formulate the risk involved in UAS operations. That methodology could be integrated for automated, real-time data analysis in an emergency management solution. We take methodological ideas from these previous works, in order to implement conflict resolution and emergency management in our system considering the specifics of UAS operations in a civil airspace. However, the focus of this paper is more on the architecture design and integration, rather than on the particular algorithms for conflict resolution.

3. Design Framework

This section settles the framework for our UTM architecture. First, we analyze the desired properties and requirements for a UTM architecture from a design perspective. Then, we introduce ROS, which is the open-source middleware that we use to implement our architecture. We justify this selection by discussing the main features in ROS and how they fit our UTM system requirements.

3.1. Guidelines for System Design

The *Global UTM Association* (GUTMA) is a non-profit consortium of worldwide UTM stakeholders, and it has promoted a discussion about which key properties should be present in future UTM systems [13]. After reviewing their technical report, we came up with a summary of these key features for UTM systems. We believe that the following aspects should be taken into account during the design phase of any UTM architecture:

- **Digital**. The process of system digitization consists of making the communication between the different actors and components digital, and introducing automated decision-making procedures. This is a key aspect in UTM to reduce the operators' workload in an efficient and secure manner. Moreover, it enables the real-time exchange of data between the relevant parties for situation awareness and an easier integration of the UTM services.
- **Flexible and modular**. A UTM architecture should be flexible and adaptable to incorporate new actors (e.g., stakeholders) and functionalities (e.g., services), as they appear. Besides, the system should be modular, i.e., made of composable and reusable modules, in order to ease the process of creating more complex functionalities.
- **Scalable**. A scalable architecture is needed to grow with new actors and services. In order to achieve that, not only is the aforementioned modularity desirable, but also a paradigm with distributed responsibilities, rather than the obsolete scheme with a centralized ATC.
- **Safe and secure**. These two features are top priorities in any UTM ecosystem. In this sense, the system should know who is flying each unmanned aircraft, where they are flying (or intend to fly) to, and whether they are conforming (or not) to mandatory operating requirements.
- **Automated**. A UTM system providing automated services to assist the UAS operators will be more efficient and secure. Therefore, the system should provide support through automated functionalities for flight planning, monitoring, and real-time deconfliction, in order to ensure safe operations for both manned and unmanned aircraft.

- **Open-source**. The use of open-source technologies is preferable, as they offer a global approach towards creating and evolving the necessary services and protocols for scalable operations. Moreover, open-source components can speed up the development and the deployment of UTM services.

3.2. Robot Operating System

Robot Operating System (ROS) is an open-source framework for robot software development. It consists of a collection of libraries, tools, and conventions to ease the creation of complex applications in robot systems; including hardware abstraction, low-level device control, implementation of commonly-used functionalities, message-passing between processes, and package management. ROS is also well known among the UAS community, as it allows drivers to communicate with a wide spectrum of both open-source and commercial autopilots and onboard sensors. The use of ROS for multi-UAS systems is extending fast, as it paves the way for integration of heterogeneous hardware and software systems. ROS is a framework based on multiple processes (so-called *nodes*) that run in a distributed fashion. These processes can be grouped into *packages*, and communicate with each other by passing *messages*, which are typed data structures. On the one hand, ROS implements asynchronous communication through a publish/subscribe paradigm where nodes can stream messages over different *topics*. On the other hand, synchronous communication is implemented through *services* for request/response interactions.

We decided to use ROS as middleware for our UTM architecture because it offers multiple features that fit our design guidelines. First, ROS is designed to create modular and reusable components, and its preferred development model is to write ROS-agnostic libraries with clean functional interfaces. Therefore, ROS yields flexible and scalable systems that can be adapted easily to incorporate new functionalities. Second, ROS is open-source and strongly supported by a large community. Its federated system of code repositories enables collaboration and fast development for UAS complex systems. Communication solutions and drivers for most popular autopilots (e.g., PX4, ArduPilot, DJI, etc.) are already available in ROS. Moreover, ROS provides remarkable tools for system integration and testing, and there exist multiple options for multi-UAS simulation, including *Software-In-The-Loop* (SITL) solutions for common autopilots [32].

ROS also presents some issues for multi-UAS systems. Mainly, its centralized nature due to the existence of a single *master* node that handles all the procedures for node registration, and its lack of proper *Quality of Service* (QoS) policies. However, there exist efficient solutions for these issues. Multi-master architectures have already been used for applications with multiple UAS [6]; and the adoption of ROS 2 is growing fast among the community, with a smooth transition from primary ROS. ROS 2 proposes a fully distributed scheme, where each node has the capacity to discover other nodes, without the need for a central master. Since it is built on top of the industrial standards DDS (*Data Distribution Service*) and RTPS (*Real-Time Publish-Suscribe*), ROS 2 is capable of offering multiple QoS options for improved communication.

Even though we have chosen ROS to implement our UTM architecture, mainly due to its advantages for system integration and realistic SITL simulation, it is important to remark that the proposed UTM architecture is a more general concept, and it could be adapted to alternative middleware solutions.

4. UTM System Architecture

This section describes our proposed UTM system architecture. Figure 4 depicts an overview of all the software modules involved, as well as their interactions. The modules in green implement specific U-space services. As it was explained in Section 2.1, we focus on those services that are required to address unexpected events during the flight operation of a UAS. In particular, we cover four services with their corresponding modules: *Tracking*, *Monitoring*, *Emergency Management* (EM), and *Tactical Deconfliction* (TD). Besides, our system includes additional software modules that provide support to the UTM architecture. First, there is a *Data Base* (DB) component that is in charge of handling all the relevant information about the state of the airspace, for instance, the current flight plans and tracks for all UAS operations (which are updated by the Tracking module) and the active geofences (which can be activated externally by auxiliary stakeholders like fire brigades or internally by the Emergency Management module). Second, the *U-space Service Manager* (USM) is a key module that acts as an interface between the UTM system and the rest of the U-space ecosystem. Basically, it receives state information and alerts from both the UAS and the external auxiliary stakeholders, and it communicates back recommended actions to deal with threatening events. These recommendations are generated by means of the interaction between the Tracking, Monitoring, Emergency Management, and Tactical Deconfliction modules.

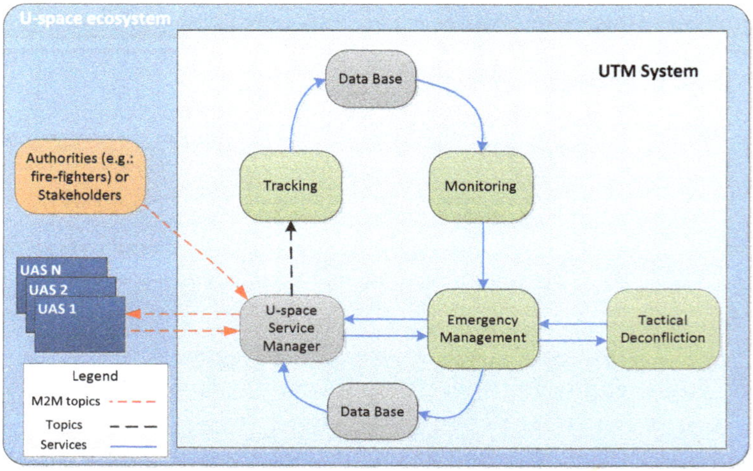

Figure 4. Overview of the proposed UTM system architecture. This system would be running in a remote server named U-space Service Provider Platform. The red arrows indicate remote communication links with other machines in the ecosystem.

Our system is built upon ROS (Section 3.2) and hence, each module consists of a software process implemented as a ROS node. The communication between modules takes place through ROS topics and services. In particular, the system is designed to use services in a preferable manner, as they provide the possibility of acknowledging message reception, which is crucial to reliably manage many of the UTM interactions. In those cases, one of the modules acts as a server while others act as clients, which results in an asynchronous communication between the modules. Upon a client request, the server module will carry out the requested activity and then it will reply, indicating whether the result was successful or not. Nevertheless, there are also a few cases where ROS topics are needed. Topics provide a synchronous communication, and they are used by modules that need to publish information at a constant rate.

In the following sections, we will provide a more detailed description of the different modules in our UTM system. For each module, we describe its functionality and interactions with other modules, as well as the methodology that we have used to implement them.

4.1. U-Space Service Manager

The U-space Service Manager is a key module in the UTM system, as it provides an interface with the rest of the actors in the U-space ecosystem, i.e., the UAS operators and auxiliary stakeholders like the airspace authorities, the fire-fighters or the police.

First, the USM receives positioning measurements from the control station of each UAS, which is transmitted by their onboard telemetry and ADS-B transceivers (if available). This information is forwarded to the Tracking module in order to keep updated a list of tracks for all the operational UAS. Second, the USM receives warning information that may be relevant for the UTM system, coming from external stakeholders (e.g., a declaration of a wildfire by the fire-fighters) or from the UAS (e.g., the detection of a jamming attack or a technical failure due to a lack of power). A jamming attack consists of an attempt to jeopardize the GNSS (*Global Navigation Satellite System*) signal of a UAS. These previous events are treated as possible threats by the system and are forwarded to the EM, which is in charge of processing them. Last, the USM communicates back to the UAS operators any action determined by the EM (e.g., an immediate landing or an alternative flight plan). Due to regulatory restrictions, the actions involving the variation of a UAS flight plan are just recommendations that must be confirmed or rejected by the corresponding pilot. In case of acceptance, the USM would notify the DB to update the state of that operation and its flight plan.

4.2. Data Base

The function of the Data Base module is to handle a digital data base with the required information to represent the situation of the current UAS operations, in the airspace managed by the UTM system. Basically, this information is made up of active geofences and UAS operations. The DB works as a server for the rest of the UTM system and hence, other modules can read the database in order to carry out their tasks (e.g., the Monitoring module uses the UAS predicted trajectories to detect events of lack of separation); or they can write the database to update the airspace situation (e.g., the USM can notify new accepted flight plans and the new EM geofences).

The DB manages two types of objects internally: geofences and UAS operations. Tables 2 and 3 depict the data structures for each of these objects. A geofence is a 4D portion of the airspace (a 3D geometrical space with an activation period of time) which has special restrictions for UAS, like flight prohibition. In the UTM context, the term *dynamic* geofence is used for those created during the UAS operation, while the *static* geofences are set in a pre-flight phase. The DB stores for each geofence in the airspace the following information: a unique identifier, its type (cylindrical or polygonal), its geometry definition, its minimum and maximum altitude, and its starting and finishing time instants. Besides, the DB stores each UAS operation, which consists of the following data: a unique identifier for the UAS, given by its ICAO (*International Civil Aviation Organization*) address; the priority level of the operation; its associated flight plan; the next waypoint assigned to the UAS; the predicted trajectory of the UAS; a brief description of the UAS operation; and the sizes of the Flight Geometry and the Operational Volume.

Table 2. Attributes of a geofence object.

Attribute	Data Type	Description
Identifier	Integer	Unique number for geofence identification
Type	Enum	Cylindrical or polygonal
Geometry	List of 2D waypoints	Definition of the horizontal shape, defined by a circle or a polygon
Min/max altitude	Float	Altitude range where the geofence is active
Start/end time	Float	Time period in which the geofence is active

Table 3. Attributes of a UAS operation object.

Attribute	Data Type	Description
Identifier	Integer	Unique identification of the aircraft
Priority	Enum	Priority of the operation in the airspace
Flight plan	List of waypoints (x, y, z, t)	Reserved 4D trajectory for the operation
Next waypoint	Integer	Waypoint index that the UAS is currently targeting
Predicted trajectory	Float	Prediction of the future UAS trajectory
ConOps	String	Description of the concept of the operation
Flight Geometry	Float	Radius of the cylindrical volume where the UAS is intended to remain during its operation
Operational Volume	Float	Radius of the outer cylindrical volume to account for environmental or performance uncertainties

4.3. Tracking

The Tracking module implements the U-space service with the same name. According to the U-space definition (Section 2.1), the main functionality of this service is to track the operational UAS in the airspace. These tracks contain information updated in real time about the UAS current position and its predicted trajectory within a certain time horizon. The module computes the tracks by fusing information from different sources that it receives through the USM. In particular, measurements from the UAS telemetry and ADS-B transceivers (when available) are integrated to achieve a more accurate estimation of the UAS positions. Moreover, the future trajectory of each UAS is predicted given its current position and velocity, as well as its flight plan. The tracking component keeps updated the UAS tracks in the DB module, so that this information is available for the rest of the system.

Mathematically, the Tracking module implements a stochastic filter that maintains a list of objects to estimate the state of each UAS, as depicted in Figure 5. This filter allows the system to cope with noisy and delayed measurements, as well as irregular sensor rates. The state of each UAS consists of its 3D position and velocity (expressed in *Universal Transverse Mercator* coordinates), and its current waypoint, i.e., the next waypoint of the flight plan that the UAS is targeting. The continuous variables are estimated through a *Kalman Filter* that integrates the measurements coming from the UAS telemetry and the onboard ADS-B transceivers. These data are previously transformed from geographic to Universal Transverse Mercator coordinates.

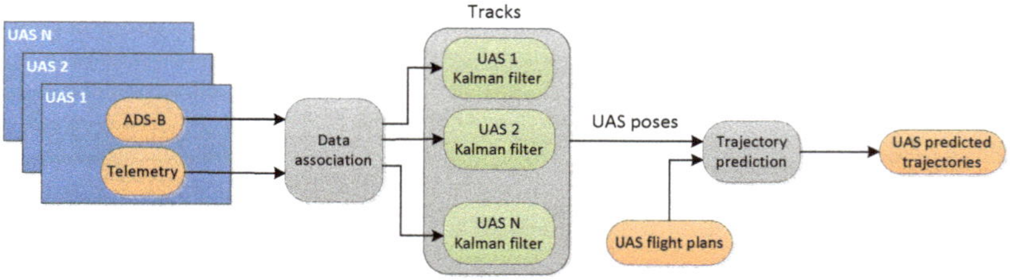

Figure 5. Scheme with the internal components of the Tracking module. The *data association* component matches the measurements from the UAS with their tracks, to update the corresponding Kalman filters. The future UAS trajectories are predicted using the tracks and the flight plans.

The procedures is as follows. At a constant rate, the list of operations is read from the DB in order to identify the active UAS. The state of all those UAS is predicted and then updated with the received observations. Each observation can be easily associated with its corresponding track, since they all come with a unique UAS identifier. The observations with *unknown* identifiers are ignored by the filter, as they are considered non-cooperative aircraft. Moreover, the current waypoint for each UAS is computed by searching for the waypoint in its flight plan that is closest to its current position. The future trajectory within a given time horizon is also predicted for each track. If the current position of the UAS is close enough to its current waypoint (according to a given distance threshold), the prediction of the future trajectory sticks to the flight plan. Otherwise, the Kalman Filter is used to predict a trajectory given the current UAS position and velocity. Finally, after each step, the Tracking module updates all the information about the tracks in the DB module.

4.4. Monitoring

The functionality of the Monitoring module is to monitor the state of the airspace and to detect potential conflicts or threats that need to be managed by the UTM system. In particular, the module deals with conflicts related with UAS trajectories. Thus, it detects: (i) whether a UAS gets out of its reserved flight volume; (ii) whether it is in conflict with a geofence; or (iii) whether two UAS lose a minimum required separation. For that, the Monitoring module periodically reads information from the DB about the UAS tracks and the geofences, and it analyzes that information to determine when a threatening situation should be reported to the EM. When the Monitoring notifies the EM, it indicates the type of the detected threat, a prediction of the time instant when the event will occur and a snapshot with the current predicted trajectories of the involved UAS. This last piece of information is sent so that the modules resolving the conflicts use exactly the same data to evaluate the situation and hence, time glitches and incoherent solutions are avoided.

The first type of issue that is evaluated by the Monitoring module is related to the *Operational Volume* that is reserved by each UAS operation (see Figure 6). The Operational Volume is a 4D space that consists of a 3D volume around the flight plan with a temporal component representing the time that the volume, as part of an operation, will be reserved in the U-space ecosystem. The Operational Volume is composed by: the *Flight Geometry*, which defines the volume of airspace where the UAS is intended to remain during its operation; and the *Contingency Volume*, which is an outer surrounding volume to account for environmental or performance uncertainties. The closest distance between the current UAS position and its flight plan is computed to determine whether the UAS is out of its Operational Volume.

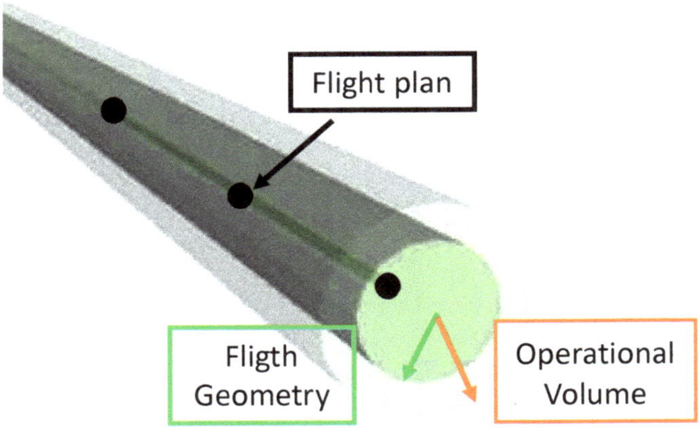

Figure 6. Graphical representation of the Operational Volume of a UAS operation (the orange arrow represents its radius). Given a flight plan, the green cylindrical volume around would represent its Flight Geometry (the green arrow indicates its radius), whereas the outer volume is the Contingency Volume.

In addition, this module monitors possible intrusions in geofences. For that, every waypoint belonging to the predicted trajectory of each UAS is compared against the active geofences, to determine whether the UAS is already intruding a geofence or it is estimated to enter one in a short future time. This check is carried out in 4D, i.e., the 3D volume of the geofence and its activation time are taken into account. More specifically, apart from checking the waypoint altitude with the minimum and maximum altitudes of the geofence, an evaluation on the horizontal plane is done depending on the shape of the geofence. If it is cylindrical, the distance of the given waypoint to the cylinder center is computed and compared with the geofence radius. If the geofence is defined by a polygonal shape, the *signed angle* method is applied. This method computes the sum of the angles between the segments that connect the test waypoint and each pair of points in the polygon. If this sum is 360°, the waypoint is within the polygon, whereas it is outside if the sum is 0°. Figure 7 depicts an example.

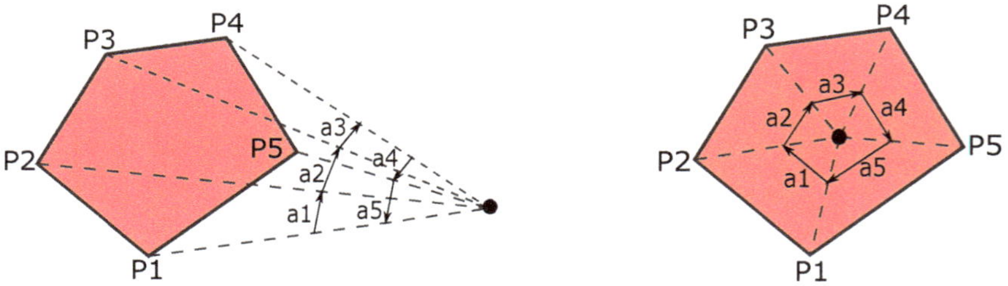

Figure 7. The signed angle method is used to evaluate whether a tested waypoint (black dot) is inside or outside a polygonal geofence. (**Left**), an example where the angles of an external waypoint sum up to 0°. (**Right**), an interior waypoint whose angles sum up to 360°.

Finally, the Monitoring module checks whether there is any loss of separation between UAS that needs to be notified. This check is done with a geometrical approach whose details can be seen in [33]. Basically, the idea is to discretize the airspace to model it as a 4D grid (see Figure 8), where each cell represents a 4D volume in space and time (dX, dY, dZ, dT) and stores a list of all the UAS whose trajectory is estimated to be inside. Thus, each waypoint of a UAS trajectory only needs to be compared with other waypoints within the neighboring cells (space and time neighborhood). For each waypoint in the 4D grid, the distances to the waypoints in the lists of its neighboring cells are calculated. If any of these distances is shorter than a safe distance, a threatening event of loss of separation will be reported.

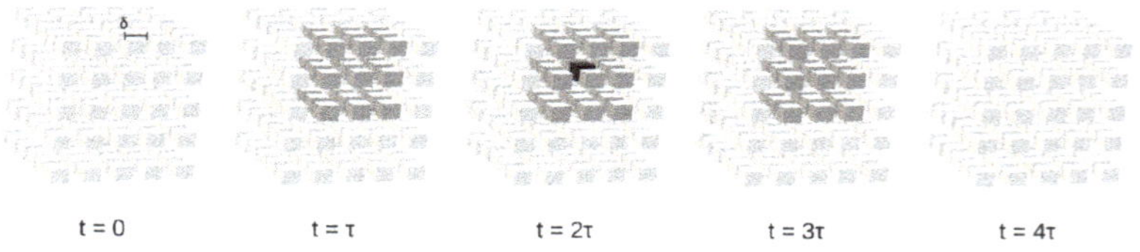

Figure 8. A 4D grid representation of the airspace. The dark grey cells would be the neighboring cells of the black cell.

4.5. Emergency Management

The Emergency Management module is the component of the UTM system that handles the threatening or unexpected situations in the U-space ecosystem. The module centralizes all the information related to the events that may become a threat, either due to conflicting UAS operations or to external warnings (e.g., a jamming attack or a bad weather situation). After analyzing the threatening events, the EM determines which are the recommended actions to resolve the conflicts, and it sends them to the corresponding UAS operators.

The EM is a central module in the UTM architecture and, as such, it interacts with the Monitoring, the USM, the DB, and the TD. The possible threats or conflicts are notified to the EM by the Monitoring or the USM modules. The former reports about conflicts related with the UAS flight plans, as it was explained in Section 4.4. The latter reports about external warnings coming from UAS technical issues, UAS operators or auxiliaries stakeholders in the U-space. For instance, this is the case of a jamming attack, a bad weather forecast, the declaration of a wildfire by the fire brigades or any other threatening event notified by emergency corps.

Depending on the severity of each threat, the EM executes a decision-making procedure to determine the best possible actions to solve the conflict [34]. In this procedure, the EM takes into account the current flight plans for the involved UAS, the priority of their operations, and other restrictions in the airspace like the geofences. As output, the EM can decide to take three different types of actions: (i) to send a specific command to a particular UAS to terminate the flight, to go back to the flight plan, etc.; (ii) to create a geofence to isolate the detected threat; and (iii) to propose an alternative flight plan to one or several UAS to resolve the conflict.

In the first type of action, the EM acts, sending a notification to the corresponding UAS operator through the USM. In the second type of action, the EM creates a geofence and it interacts with the DB in order to update the database with geofences. In the third type of action, the EM sends the alternative recommended flight plans to the USM, which is in charge of forwarding them to the corresponding UAS. For the computation of these alternative plans, the EM receives the support of the TD module, which is requested to compute a series of alternative routes for the involved UAS, depending on the situation. The TD generates these routes by applying a set of predefined maneuvers for each UAS (see Section 4.6). Then, the EM selects which are the best alternative routes for all the UAS in conflict by minimizing the following value function:

$$\sum_{i=1}^{N}\sum_{j=1}^{M}\alpha \cdot c_{ij} + \beta \cdot r_{ij}\ ; \tag{1}$$

where N and M represent the number of conflicting UAS and the number of available maneuvers for each UAS, respectively; c_{ij} is the cost incurred if the UAS i executes the route j; r_{ij} is the riskiness associated with the route j executed by the UAS i; and $\alpha, \beta \in [0,1]$ are the optimization weights. Each type of UAS maneuver considered by the TD will generate an alternative route for the UAS, with an associated cost and riskiness. The former is related to the additional time that the UAS has to travel to execute the route, while the latter measures the risk level of the route, e.g., how close it comes to other existing flight plans or geofences. The values of the weights assigned to the two terms need to be tuned by a human designer. In general, the system should favor safety over efficiency, so higher values for β than for α are expected.

Finally, it is important to remark that all the actions sent by the EM to the UAS are just recommendations. According to the current regulation of the U-space ecosystem, the UTM can only suggest automatically possible correction actions, but those must be accepted or rejected by each UAS operator eventually. Nonetheless, our approach would be able to accommodate a UTM system where the whole process is executed autonomously without the need for human intervention, which is the final objective in the U-space framework.

4.6. Tactical Deconfliction

The Tactical Deconfliction module provides support to compute alternative flight plans for UAS that need to resolve a potentially threatening or conflicting situation. The TD receives requests from the EM indicating the necessary information related to the event to solve, i.e., the type of threatening situation and the data of the affected operations and the active geofences. Depending on the situation, the TD will attempt different types of maneuvers to generate a list of alternative flight plans for the involved UAS. For each possible solution, the TD will compute an associated cost and riskiness level, which will be reported back to the EM, together with the generated alternative flight plans. Then, as it was explained in Section 4.5, the EM is the module that makes a final decision about which the best solution to resolve the conflict is.

The TD uses two different approaches to compute the alternative routes, depending on whether the threat is a conflict between different UAS or a situation with a single UAS involved. The first case occurs when the flight plans of several UAS are in conflict, e.g., due to a loss of separation. In that case, a geometric approach based on repulsive forces is used to modify the original flight plans. The details of the implemented algorithm can be seen in [35], but it basically models the UAS trajectories as cords with electrical charges that repel each other, in order to increase their separation. By applying vertical or horizontal separation maneuvers between the involved UAS trajectories in an iterative procedure (see Figure 9), the TD can generate several alternative solutions. The priorities of the conflicting flight plans are also considered. The algorithm tends not to modify the flight plans of those UAS whose operations present a higher priority in the U-space. For each computed solution, its cost is calculated as the total distance traveled by the UAS, whereas its riskiness is the length of the UAS routes that still get in conflict with other geofences. Even though these types of conflicts are solved in an iterative manner, by applying the tactical deconfliction procedure for each pair of UAS sequentially, the final solution could still produce additional conflicts with geofences. In this case, the Monitoring module would report those new pending conflicts in subsequent iterations.

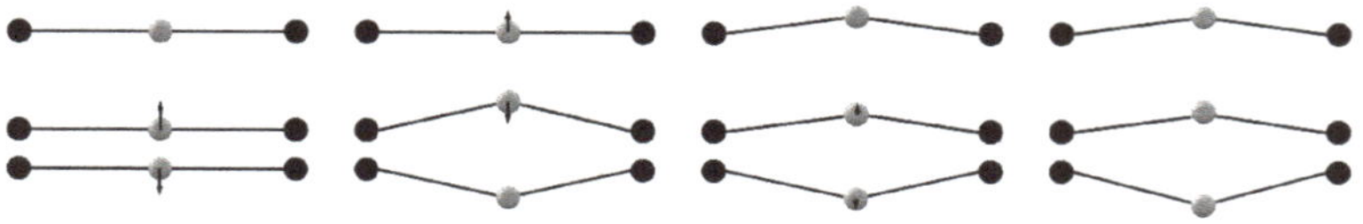

Figure 9. Iterative procedure to solve a conflict in the case of a loss of separation (from **left** to **right**). The flight plans of the two lower UAS are in conflict and need to be separated. Then, the middle UAS enters in conflict with the upper UAS, so these two get separated again. As the plan of the middle UAS gets modified, the lowest UAS is also adapted to achieve a final solution without loss of separation.

A second approach is used to solve situations with a single UAS involved. This is the case of a UAS that presents a technical problem, that is out of its Operational Volume, or that has a conflict with a geofence. In all those cases, a heuristic path planner based on the well-known A* algorithm is used. First, if the UAS flight plan goes through a geofence, the path planner generates an alternative route avoiding that geofence (see Figure 10, left). Second, if the UAS is already within a geofence, it gets out of the geofence through an *escape* point, and then it avoids the geofence to resume with its flight plan afterwards (see Figure 10, right). The TD also computes an alternative route from the current UAS position to the last waypoint in its flight plan, in order to skip the conflicting part of the plan and fly directly to the final goal. Third, if a UAS is out of its Operational Volume, two alternative routes are computed: one from the current UAS position to the closest point of its flight plan; and another from the current UAS position to its next waypoint in the flight plan, regardless of how long the UAS remains out of its Operational Volume.

In the three cases, an alternative route to return back to the home station is also computed. The EM could select this option if all the other solutions to continue with the operation are too risky. In all the generated solutions, the cost is determined by the total distance traveled by the UAS. The riskiness is determined by the minimum distance between the alternative route and any geofence, or by the length of the route portions that remain within a geofence, in case that the solution goes through any geofence partially. In case of a UAS out of its Operational Volume, the riskiness of the solution is determined by the length of the route portion where the UAS stays out of its Operational Volume.

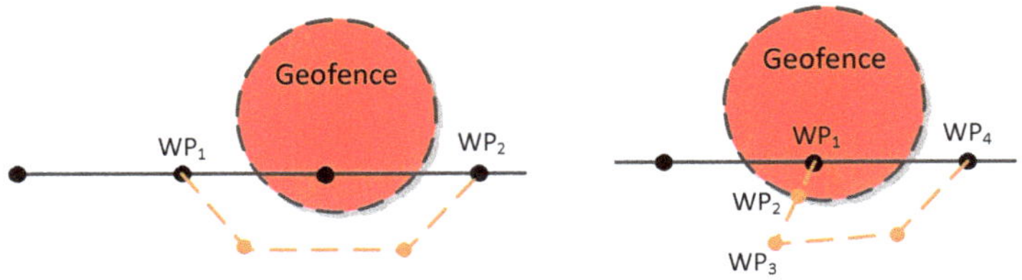

Figure 10. (**Left**), a UAS with a flight plan crossing a geofence. The last waypoint of its flight plan before entering the geofence (WP_1) and the first waypoint after leaving it (WP_2) are obtained, and this portion of the flight plan is replaced by an alternative route (dashed line). (**Right**), a UAS that is inside a geofence. The escape point (WP_2) is that on the geofence's border closest to the UAS (WP_1). From WP_3, which is already at a safety distance from the geofence, to the first point of the flight plan after leaving the geofence (WP_4), an alternative route avoiding the geofence is inserted to modify the original flight plan.

Finally, an alternative route where the UAS travels to its closest landing site can also be computed in some cases, for instance, if the UAS presents a technical problem like a lack of battery. In those cases, the riskiness is determined by the distance of the route that goes through any geofence in the airspace.

4.7. Discussion

In this section, we discuss the functionalities implemented by the U-space services of our UTM architecture, when compared to those expected in the current definition of the U-space ecosystem. For that, we have summarized in Table 4 the expected functionalities to be covered by each of the U-space services included in our system, according to the bibliography studied in Section 2.1. In the following, we discuss which capabilities are already covered by our system and the missing points for future implementations.

Table 4. Summary of the functionalities to be covered by the U-space services included in our UTM system.

U-Space Service	Functionalities	Covered in Our UTM System
Tracking	Cooperative UAS tracking	✓
	Non-cooperative UAS tracking	✗
	Capability to exchange data with other services	✓
	Real-time tracking with data fusion from multiple sources	✓
	Tracking data recording	✓
Monitoring	Air situation monitoring	✓
	Non-cooperative UAS identification	✗
	Flight non-conformance detection	✓
	Restricted area infringement detection	✓
	Provision of traffic information for UAS operators	✗
	Conflict alerts	✓
Emergency Management	Emergency alerts	✓
	Provision of assistance information for UAS operators	✓
Tactical Deconfliction	Transmission of deconfliction information from the USM to the UAS	✗
	Transmission of deconfliction information in real time	✓

- *Tracking*. This service is supposed to consider cooperative and non-cooperative UAS, but our current implementation only manages cooperative UAS. This is because we have focused on enabling automated decision-making for the operating UAS, which makes no sense for non-cooperative vehicles. Those should be treated as uncontrollable intruders (i.e., threats) in the airspace. However, our Tracking module does have the capability to update and record data in real time from different sources. Other services can also access these data through the DB module if needed.
- *Monitoring*. As in the previous case, our current implementation does not consider non-cooperative UAS. We did not establish a specific communication link to provide traffic information to the UAS operators either, though this could be easily done through the USM. However, our Monitoring module does accomplish all the other expected functionalities, i.e., it detects and alerts in real time about conflicts related to flight non-conformances, geofences, and inter-UAS separation.
- *Emergency Management*. This service is expected to provide the UAS operators with notifications about alerts and any other emergency assistance. Besides, our EM module includes automated decision-making capabilities, in order to manage threats in real time by proposing safe and optimal actions to the UAS.
- *Tactical Deconfliction*. Although this service is supposed to provide deconfliction information to the UAS operators through the USM, in our scheme this role is played by the EM module. This is because the automated decision-making capability is implemented in the EM module, which uses the TD module to get support generating possible alternative plans. Then, the EM is the one in charge of deciding the best option for real-time deconfliction.

5. Experiments

This section contains experimental results to showcase the capabilities of the proposed UTM system. The objectives of these experiments are twofold: (i) we show the integration of the complete architecture, with all its functional modules interacting together to accomplish the specified UAS operations; and (ii) we demonstrate our system operating in real time in a realistic setup, to test its capabilities to solve different types of conflicts in an automated manner. For that, we have defined two use cases (Section 5.1) involving heterogeneous UAS and several types of conflicts, in order to validate all the modules in our UTM system. The tested use cases are realistic both in terms of the UAS operational parameters and the experimental setup (Section 5.2). Our experiments were carried out by means of *Hardware-In-The-Loop* (HITL) simulations where the UAS operators and the UTM framework ran at different physical locations, with a real long-distance communication link in between. All of the results of the tests are described in Section 5.3.

5.1. Use Cases Definition

We defined two use cases using the heterogeneous UAS that were depicted in Figure 2: the multirotor DJI M600 and the fixed-wing Atlantic I. These UAS are used in the GAUSS project to run tests integrating aircraft with different maneuverability and different proprietary autopilots. Both use cases involve a pair of UAS performing operations with different or equal priorities, and both require the interaction of all the modules of the proposed UTM system.

Figure 11 depicts a top view of each use case, with the corresponding initial flight plans. Table 5 summarizes the operational parameters for the use case 1. UAS_1 is a multirotor performing an operation for precision agriculture, while UAS_2 is a fixed-wing aircraft that has to inspect an electrical power line. Given its easier maneuverability, the priority of the UAS_1 operation is set lower. The initial flight plans (see Figure 11, left) are such that the UAS do not coincide in space and time throughout their operations. However, we simulated an unexpected delay in the start of the UAS_1 operation, which resulted in a later violation of the minimum safety distance between both UAS. Thus, this use case is used to test how the UTM is able to detect a loss of separation between the UAS and to perform real-time tactical deconfliction for an inter-vehicle conflict, deciding new flight plans for both UAS.

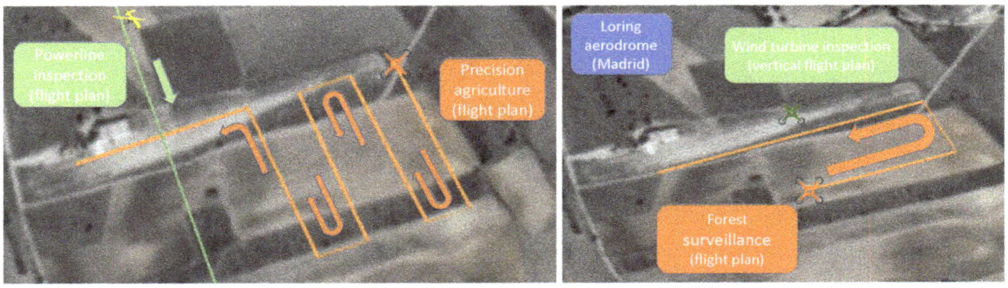

Figure 11. Top views including the initial flight plans of the use case 1 (**left**) and the use case 2 (**right**). All the operations were planned in an area of the Loring aerodrome in Madrid (Spain).

Table 5. Operational parameters for the use case 1.

	Operation 1.1	Operation 1.2
ConOps	Precision agriculture	Powerline inspection
UAS type	M600 (UAS_1)	Atlantic I (UAS_2)
Cruising speed	3.3 m/s	30 m/s
Altitude (Above Ground Level)	40 m	100 m
Operation priority	Low	High
Events involved	Loss of separation	Loss of separation

Table 6 summarizes the operational parameters for the use case 2. In this case, both UAS_1 and UAS_2 are multi-rotors, performing two operations with equal priority. In their initial flight plans (see Figure 11, right), UAS_1 moves on a vertical line to accomplish the inspection of a wind turbine, while UAS_2 has to fly on a horizontal plane to survey a nearby forest. During the operation, a jamming attack is simulated over UAS_1. The objective of this use case is to test how the UTM is able to react in an automated manner to an emergency generated by an external source, creating a new geofence and adapting to the conflicting flight plans.

Table 6. Operational parameters for the use case 2.

	Operation 2.1	Operation 2.2
ConOps	Wind turbine inspection	Forest surveillance
UAS type	M600 (UAS_1)	M600 (UAS_2)
Cruising speed	1 m/s	1 m/s
Altitude (Above Ground Level)	30–90 m	70 m
Operation priority	High	High
Events involved	Jamming attack	Geofence conflict

5.2. Experimental Setup

We have developed our UTM system architecture in ROS (Kinetic version), and the software is available online (https://github.com/grvcTeam/gauss, accessed on 26 April 2021). First, we used an airspace SITL simulation based on Gazebo [27] for system integration and preliminary tests. Then, we setup a realistic environment to run experiments in real time with HITL simulations. These experiments were carried out within the framework of the GAUSS project, with the configuration depicted in Figure 12.

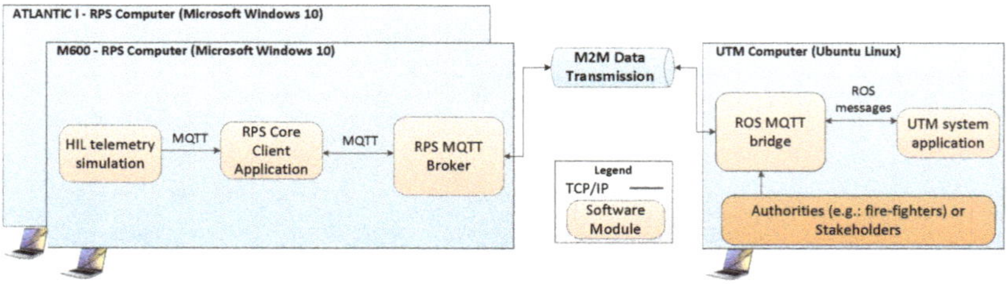

Figure 12. Setup for the experiments. The computers running the RPS for the two UAS and the UTM system were placed at remote locations and communicated through the Internet via the MQTT protocol.

The company EVERIS (https://www.everis.com/global/en, accessed on 26 April 2021) ran on its headquarters in Madrid (Spain) a *Remote Pilot Station* (RPS) for each type of UAS. Each RPS has an integrated HITL simulation producing real-time telemetry data for the operating UAS, a graphical user interface to show this telemetry, and the operational information to the safety pilot (*RPS Client Application*), and an *RPS MQTT Broker* to communicate data over the Internet. The RPS Client Application was developed by the company SATWAYS (https://www.satways.net, accessed on 26 April 2021) and it can be seen in Figure 13). Simultaneously, we ran our UTM system on a server located in Seville (Spain), connected to the Internet via a *ROS MQTT bridge*. The UAS RPS communicated with the remote UTM system exchanging JSON (*JavaScript Object Notation*) messages sent over the MQTT (*Message Queuing Telemetry Transport*) transport protocol (We used the open-source Apache Active MQ broker). Moreover, the time synchronization for the exchanged data between the remote computers was achieved thanks to an NTP (*Network Time Protocol*) server. It is important to highlight that this experimental setup is close to the real U-space ecosystem, where the UTM system would be running on a server located at a remote distance of the UAS operators.

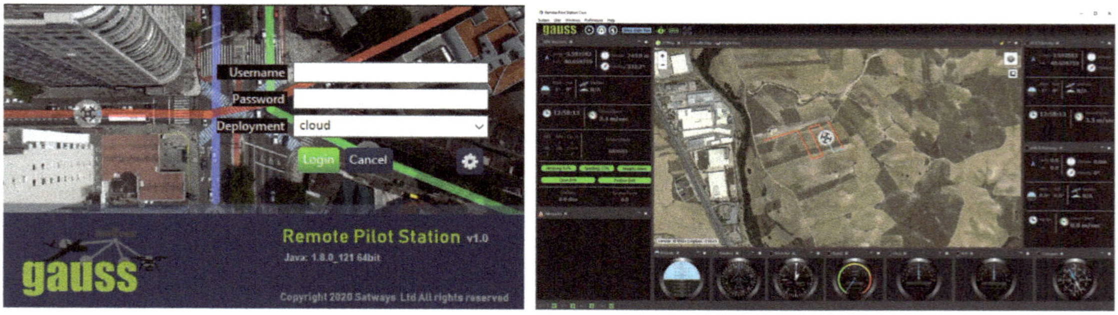

Figure 13. Screenshot of the graphical user interface developed by SATWAYS running on the RPS Client Application.

5.3. Results

In this section, we present results of the experimental tests for the two proposed use cases (an illustrative video with the use cases can be seen at https://grvc.us.es/downloads/videos/UTM_System.mp4, accessed on 26 April 2021), with all the modules in our UTM system working together. It is important to highlight that the experiments were carried out in real time, with the UTM system monitoring the operations and managing the unexpected events properly. Moreover, the proposed solutions to solve the conflicts were executed in an automated manner by the simulated UAS, and supervised by human safety pilots.

Figure 14 shows a timeline for the experiment of the use case 1. Both UAS were supposed to start their operations simultaneously ($t = 0$ s) according to their pre-flight generated plans, without conflicts. However, we simulated a delay of 3 s in the start of the UAS_1 operation. The Tracking module received periodically positioning information from both UAS and it updated the DB accordingly. The Monitoring module checked for conflicts periodically using the updated tracks from the DB and, at $t = 24$ s, it detected a future loss of separation conflict between the UAS. This was communicated to the EM, which ran an automated decision-making process (supported by the TD) to propose the optimal conflict resolution. In this case, an alternative flight plan was sent to UAS_1 through the USM module. Figure 15 shows the initial flight plans for the UAS and their reserved Operational Volume. Despite not having conflicts initially, the delay in the UAS_1 operation provoked an eventual loss of separation in the last part of its operation, which was resolved with an alternative flight plan. Figure 16 depicts the three options generated by the TD module and the optimal solution (in terms of risk and traveled distance) selected by the EM. In the experiment, the conflict was detected by the UTM system well in advance, and the total time between the detection and the communication of a solution to the USM took 0.13 s.

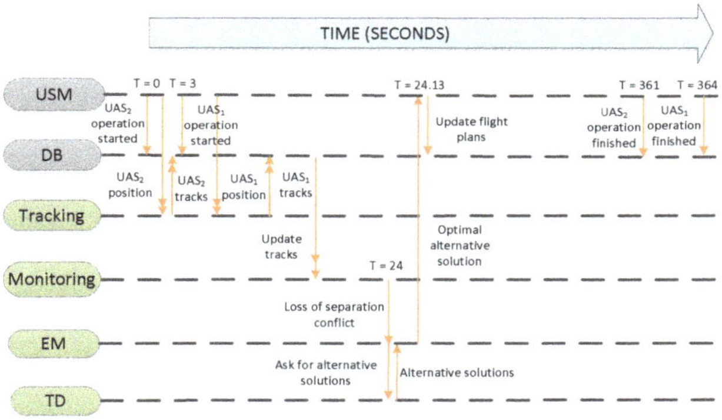

Figure 14. Timeline of the experiment of the use case 1, where a loss of separation event is resolved. Single arrows indicate isolated interactions between modules, whereas double arrows indicate periodic communication.

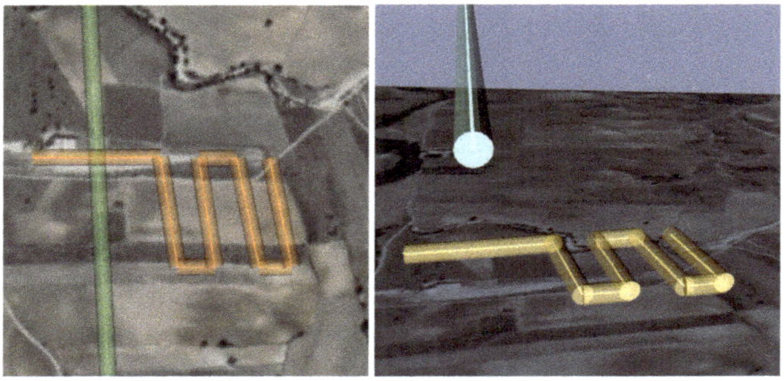

Figure 15. A top (**left**) and a perspective view (**right**) of the initial flight plans in the use case 1. The Operational Volumes are shown for both UAS. There are no conflicts given the UAS 4D trajectories.

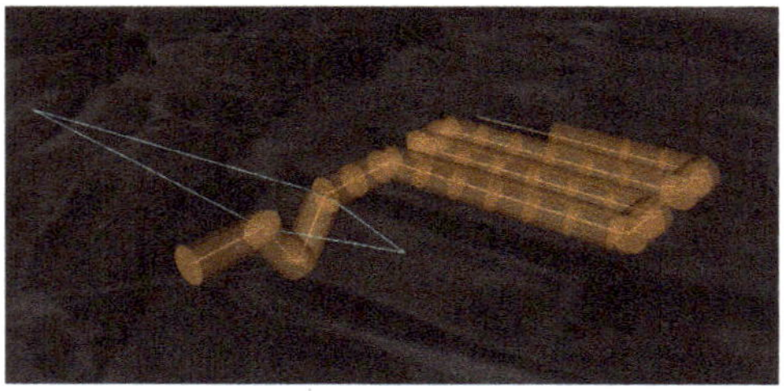

Figure 16. A perspective view of the conflict resolution in the use case 1. A new flight plan for UAS_1 (with a final *go down* maneuver) was selected to keep the safety distance with UAS_2. The other alternative maneuvers generated by the TD module (*go left* and *go right*) are also shown.

Figure 17 shows a timeline for the experiment of the use case 2. Both UAS started their operations simultaneously ($t = 0$ s) following pre-flight plans without conflicts. The Tracking module received periodically positioning information from both UAS, and it updated the DB accordingly. The Monitoring module checked for conflicts periodically using the updated tracks from the DB. We simulated a jamming attack over UAS_1 ($t = 12$ s) that was notified by the USM to the EM, which ran an automated decision-making process. In this type of threat, due to the involved risks, the EM decided to suspend the UAS_1 operation (notifying the USM) and to create a geofence around (updating the DB). Then, the Monitoring module detected ($t = 15$ s) a future geofence conflict with the UAS_2 flight plan, which was resolved by the EM (with the support of the TD) with an alternative plan avoiding the geofence. Again, the time between the detection of the conflict and the communication of the optimal solution to the USM was less than 1 s. Figure 18 shows the initial flight plans for the UAS and their reserved Operational Volumes, and the situation right after the jamming attack. Despite not having conflicts initially, the creation of a new geofence provoked an eventual conflict, which was resolved with an alternative flight plan for UAS_2 (see Figure 19).

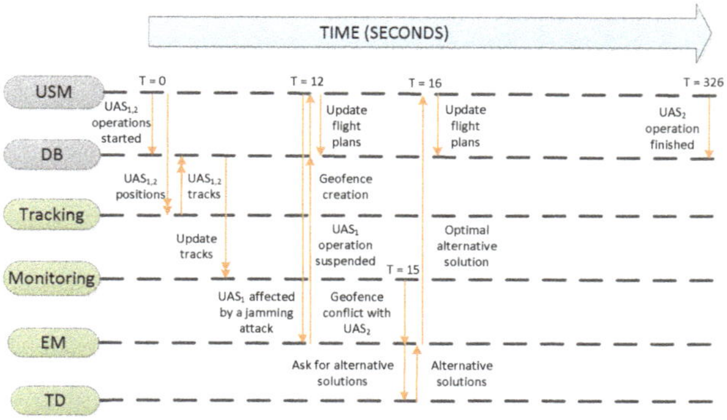

Figure 17. Timeline of the use case 2, where a jamming attack and a geofence conflict are resolved. Single arrows indicate isolated interactions between modules, whereas double arrows indicate periodic communication.

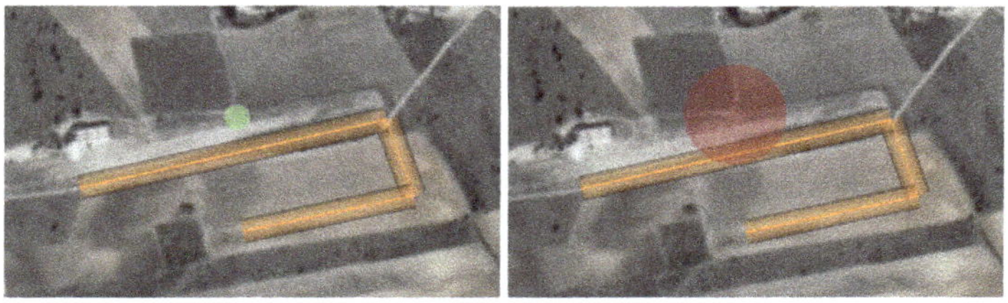

Figure 18. (**Left**), top view with the initial flight plans in the use case 2. The Operational Volumes without conflicts are also shown. (**Right**), situation after the detection of the jamming attack. A geofence (in red) is created around the attacked UAS, which generates a conflict with the flight plan of the other UAS.

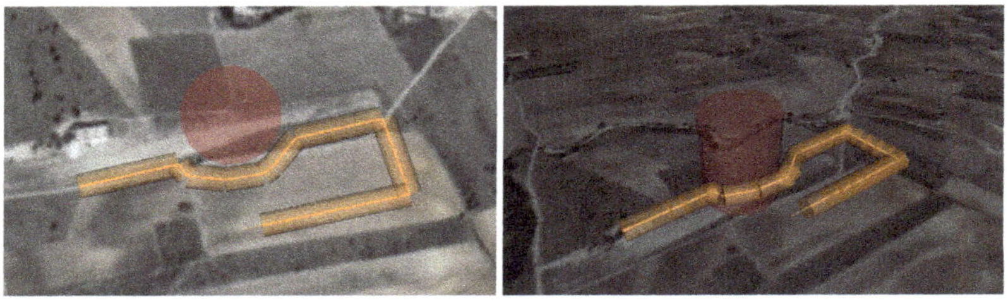

Figure 19. A top (**left**) and a perspective (**right**) view of the optimal solution in the use case 2. An alternative flight plan for UAS_2 is generated to avoid the geofence.

Finally, it is important to recall that the experiments were carried out with a setup where the UTM system ran at a remote distance of the UAS stations. Despite that, the communication delays and response times by the UTM system were adequate for a real-time resolution of the unexpected conflicts. In particular, we measured a reception of the UAS telemetry data at the USM of an average rate of 1 Hz with a maximum delay of 40 ms.

6. Conclusions

In this paper, we have presented a UTM system architecture framed within the U-space ecosystem. Our software architecture is flexible and general, and it is built as an open-source solution that could be easily extended with additional U-space functionalities. Nonetheless, we have focused on in-flight services for automated threat management and conflict resolution, which is a major gap in the current state of the art. In our realistic experimental setup, with the involved systems running HITL simulations communicated through a remote link with the UTM system, we have demonstrated that the proposed UTM solution is capable of managing unexpected events in real time, proposing solutions in an automated manner. In our experiments, the system was able to detect and resolve different types of conflicts, reasoning about 4D UAS trajectories and Operational Volumes. Besides, we have tested the feasibility of the system for the future U-space, integrating heterogeneous types of UAS (fixed and rotary wing), heterogeneous positioning technologies (ADS-B and telemetry from different autopilots), and a database to keep track in real time of the different UAS operations and geofences.

Our system has still some practical limitations. First, it relies on a centralized UTM server that requires continuous communication with the other actors. This bottleneck could be addressed by splitting the UTM system into a set of distributed and interconnected servers. Second, our approach does not consider non-cooperative vehicles in the VLL airspace, such as ultralight planes, nor pre-flight services. However, the architecture is flexible enough to integrate additional services, e.g., for flight operation pre-planning. Besides, non-cooperative vehicles could be tackled by working with see&avoid systems on board the UAS. As future work, we plan to develop further on the emergency management functionality, analyzing the possible threats that could appear in the VLL airspace, and quantifying the involved risks of the alternative solutions proposed by our system. Thus, we will be able to improve the capabilities of the system to solve more conflicts safely and efficiently, and to test it in more varied use cases. Furthermore, we plan to adapt our UTM system to ROS 2 and to validate it in field trials within the framework of the GAUSS project, which will be a significant step toward a totally automated U-space environment.

Author Contributions: Conceptualization, C.C., Á.C. and J.C.; software and validation, C.C. and H.P.-L.; writing and editing, C.C., H.P.-L. and J.C.; supervision, Á.C., J.C. and A.O.; funding acquisition, A.O. All authors have read and agreed to the published version of the manuscript.

Funding: This work has received funding from the European Union's Horizon 2020 Research and Innovation Programme under grant agreement No 776293 (GAUSS).

Institutional Review Board Statement: Not applicable.

Informed Consent Statement: Not applicable.

Data Availability Statement: The original contributions presented in the study are included in the article; further inquiries can be directed to the corresponding author.

Conflicts of Interest: The authors declare no conflict of interest.

abbreviations

The following abbreviations are used in this manuscript:

ADS-B	Automatic Dependent Surveillance Broadcast
ATC	Air Traffic Control
ATM	Air Traffic Management
DB	Data Base
DDS	Data Distribution Service
EASA	European Aviation Safety Agency
EM	Emergency Management
GNSS	Global Navigation Satellite System
GUTMA	Global UTM Association
HITL	Hardware-In-The-Loop
ICAO	International Civil Aviation Organization
JSON	JavaScript Object Notation
MQTT	Message Queuing Telemetry Transport
NASA	National Aeronautics and Space Administration
NTP	Network Time Protocol
QoS	Quality of Service
ROS	Robot Operating System
RPS	Remote Pilot Station
RTPS	Real-Time Publish-Subscribe
SITL	Software-In-The-Loop
TD	Tactical Deconfliction
UAS	Unmanned Aircraft System
USM	U-space Service Manager
UTM	Unmanned aerial system Traffic Management
VLL	Very Low Level

References

1. SESAR. *European Drones Outlook Study*; Technical report; SESAR: Brussels, Belgium, 2016. [CrossRef]
2. Aurambout, J.P.; Gkoumas, K.; Ciuffo, B. Last mile delivery by drones: An estimation of viable market potential and access to citizens across European cities. *Eur. Transp. Res. Rev.* **2019**, *11*, 1–21. [CrossRef]
3. Capitan, J.; Merino, L.; Ollero, A. Cooperative Decision-Making Under Uncertainties for Multi-Target Surveillance with Multiples UAVs. *J. Intell. Robot. Syst.* **2016**, *84*, 371–386. [CrossRef]
4. Sanchez-Cuevas, P.J.; Gonzalez-Morgado, A.; Cortes, N.; Gayango, D.B.; Jimenez-Cano, A.E.; Ollero, A.; Heredia, G. Fully-Actuated Aerial Manipulator for Infrastructure Contact Inspection: Design, Modeling, Localization, and Control. *Sensors* **2020**, *20*, 4708. [CrossRef] [PubMed]

5. Garcia-Aunon, P.; Roldán, J.J.; Barrientos, A. Monitoring traffic in future cities with aerial swarms: Developing and optimizing a behavior-based surveillance algorithm. *Cogn. Syst. Res.* **2019**, *54*, 273–286. [CrossRef]
6. Alcantara, A.; Capitan, J.; Torres-Gonzalez, A.; Cunha, R.; Ollero, A. Autonomous Execution of Cinematographic Shots with Multiple Drones. *IEEE Access* **2020**, *8*, 201300–201316. [CrossRef]
7. Kramar, V. UAS (drone) in Response to Coronavirus. In Proceedings of the Conference of Open Innovations Association (FRUCT), Trento, Italy, 9–11 September 2020; pp. 90–100. [CrossRef]
8. Peinecke, N.; Kuenz, A. Deconflicting the urban drone airspace. In Proceedings of the IEEE/AIAA Digital Avionics Systems Conference (DASC), St. Petersburg, FL, USA, 17–21 September 2017; pp. 1–6. [CrossRef]
9. Kopardekar, P. *Unmanned Aerial System (UAS) Traffic Management (UTM): Enabling Civilian Low-Altitude Airspace and Unmanned Aerial System Operations*; Technical Report; NASA: Washington, DC, USA, 2015.
10. Kopardekar, P.; Rios, J.; Prevot, T.; Johnson, M.; Jung, J.; Robinson, J.E. Unmanned Aircraft System Traffic Management (UTM) Concept of Operations. In Proceedings of the AIAA Aviation Technology, Integration, and Operations Conference, Washington, DC, USA, 13–17 June 2016. [CrossRef]
11. SESAR. *U-Space Blueprint*; Technical report; SESAR: Brussels, Belgium, 2017. [CrossRef]
12. European Aviation Safety Agency (EASA). *High-Level Regulatory Framework for the U-Space*; Technical Report 18; EASA: Cologne, Germany, 2020.
13. Global UTM Association. *Designing UTM for Global Success*; Technical Report; Global UTM Association: Lausanne, Switzerland, 2020.
14. Barrado, C.; Boyero, M.; Brucculeri, L.; Ferrara, G.; Hately, A.; Hullah, P.; Martin-Marrero, D.; Pastor, E.; Rushton, A.P.; Volkert, A. U-Space Concept of Operations: A Key Enabler for Opening Airspace to Emerging Low-Altitude Operations. *Aerospace* **2020**, *7*, 24. [CrossRef]
15. Eurocontrol. *U-Space Services Implementation Monitoring Report*; Technical Report; Eurocontrol: Brussels, Belgium, 2020.
16. Jiang, T.; Geller, J.; Ni, D.; Collura, J. Unmanned Aircraft System traffic management: Concept of operation and system architecture. *Int. J. Transp. Sci. Technol.* **2016**, *5*, 123–135. [CrossRef]
17. NASA. *Google UAS Airspace System Overview*; NASA: Washington, DC, USA, 2015.
18. Amazon Prime Air. *Revising the Airspace Model for the Safe Integration of Small Unmanned Aircraft Systems*; NASA: Washington, DC, USA, 2015; pp. 2–5.
19. AirMap. *Five Critical Enablers or Safe, Efficient, and Viable UAS Traffic Management (UTM)*; Technical report; AirMap: Santa Monica, CA, USA, 2018.
20. Unifly. Unifly UTM Platform. Available online: https://www.unifly.aero/solutions/unmanned-traffic-management (accessed on 26 April 2021).
21. Thales. *Thales Launches ECOsystem UTM and Joins Forces with Unifly to Facilitate Drone Use*; Technical Report; Thales: La Defense, France, 2017.
22. Rumba, R.; Nikitenko, A. The wild west of drones: A review on autonomous- UAV traffic-management. In Proceedings of the International Conference on Unmanned Aircraft Systems (ICUAS), Athens, Greece, 1–4 September 2020; pp. 1317–1322. [CrossRef]
23. Lin, C.E.; Chen, T.; Shao, P.; Lai, Y.; Chen, T.; Yeh, Y. Prototype Hierarchical UAS Traffic Management System in Taiwan. In Proceedings of the Integrated Communications, Navigation and Surveillance Conference (ICNS), Washington, DC, USA, 9–11 April 2019; pp. 1–13. [CrossRef]
24. Lundberg, J.; Palmerius, K.L.; Josefsson, B. Urban Air Traffic Management (UTM) Implementation In Cities—Sampled Side-Effects. In Proceedings of the IEEE/AIAA 37th Digital Avionics Systems Conference (DASC), London, UK, 23–27 September 2018; pp. 1–7. [CrossRef]
25. Aweiss, A.; Homola, J.; Rios, J.; Jung, J.; Johnson, M.; Mercer, J.; Modi, H.; Torres, E.; Ishihara, A. Flight Demonstration of Unmanned Aircraft System (UAS) Traffic Management (UTM) at Technical Capability Level 3. In Proceedings of the IEEE/AIAA Digital Avionics Systems Conference (DASC), San Diego, CA, USA, 8–12 September 2019; pp. 1–7. [CrossRef]
26. Alarcon, V.; Garcia, M.; Alarcon, F.; Viguria, A.; Martinez, A.; Janisch, D.; Acevedo, J.J.; Maza, I.; Ollero, A. Procedures for the Integration of Drones into the Airspace Based on U-Space Services. *Aerospace* **2020**, *7*, 128. [CrossRef]
27. Millan-Romera, J.A.; Acevedo, J.J.; Perez-Leon, H.; Capitan, C.; Castaño, A.R.; Ollero, A. A UTM simulator based on ROS and GAZEBO. In Proceedings of the International Workshop on Research, Education and Development of Unmanned Aerial Systems (RED-UAS), Cranfield, UK, 25–27 November 2019.
28. Tan, Q.; Wang, Z.; Ong, Y.; Low, K.H. Evolutionary Optimization-based Mission Planning for UAS Traffic Management (UTM). In Proceedings of the International Conference on Unmanned Aircraft Systems (ICUAS), Atlanta, GA, USA, 11–14 June 2019; pp. 952–958. [CrossRef]
29. Ho, F.; Geraldes, R.; Gonçalves, A.; Rigault, B.; Oosedo, A.; Cavazza, M.; Prendinger, H. Pre-Flight Conflict Detection and Resolution for UAV Integration in Shared Airspace: Sendai 2030 Model Case. *IEEE Access* **2019**, *7*, 170226–170237. [CrossRef]
30. Sacharny, D.; Henderson, T.C.; Cline, M. Large-Scale UAS Traffic Management (UTM) Structure. In Proceedings of the IEEE International Conference on Multisensor Fusion and Integration for Intelligent Systems (MFI), Karlsruhe, Germany, 14–16 September 2020; pp. 7–12. [CrossRef]

31. Rubio-Hervas, J.; Gupta, A.; Ong, Y.S. Data-driven risk assessment and multi-criteria optimization of UAV operations. *Aerosp. Sci. Technol.* **2018**, *77*, 510–523. [CrossRef]
32. Real, F.; Torres-Gonzalez, A.; Ramon-Soria, P.; Capitan, J.; Ollero, A. Unmanned aerial vehicle abstraction layer: An abstraction layer to operate unmanned aerial vehicles. *Int. J. Adv. Robot. Syst.* **2020**, *17*, 1–13. [CrossRef]
33. Acevedo, J.J.; Castaño, A.R.; Andrade-Pineda, J.L.; Ollero, A. A 4D grid based approach for efficient conflict detection in large-scale multi-UAV scenarios. In Proceedings of the International Workshop on Research, Education and Development of Unmanned Aerial Systems (RED-UAS), Cranfield, UK, 25–27 November 2019; pp. 18–23. [CrossRef]
34. Capitán, C.; Castaño, A.R.; Capitán, J.; Ollero, A. A framework to handle threats for UAS operating in the U-space. In Proceedings of the International Workshop on Research, Education and Development on Unmanned Aerial Systems (RED-UAS), Cranfield, UK, 25–27 November 2019.
35. Acevedo, J.J.; Capitán, C.; Capitán, J.; Castaño, A.R.; Ollero, A. A Geometrical Approach based on 4D Grids for Conflict Management of Multiple UAVs operating in U-space. In Proceedings of the International Conference on Unmanned Aircraft Systems (ICUAS), Athens, Greece, 1–4 September 2020; pp. 263–270.

applied sciences

MDPI

Article

Airspace Geofencing and Flight Planning for Low-Altitude, Urban, Small Unmanned Aircraft Systems

Joseph Kim [1,*] and Ella Atkins [2]

1 Robotics Institute, University of Michigan, Ann Arbor, MI 48109-2106, USA
2 Department of Aerospace Engineering, Robotics Institute, University of Michigan, Ann Arbor, MI 48109-2106, USA; ematkins@umich.edu
* Correspondence: jthkim@umich.edu

Abstract: Airspace geofencing is a key capability for low-altitude Unmanned Aircraft System (UAS) Traffic Management (UTM). Geofenced airspace volumes can be allocated to safely contain compatible UAS flight operations within a fly-zone (keep-in geofence) and ensure the avoidance of no-fly zones (keep-out geofences). This paper presents the application of three-dimensional flight volumization algorithms to support airspace geofence management for UTM. Layered polygon geofence volumes enclose user-input waypoint-based 3-D flight trajectories, and a family of flight trajectory solutions designed to avoid keep-out geofence volumes is proposed using computational geometry. Geofencing and path planning solutions are analyzed in an accurately mapped urban environment. Urban map data processing algorithms are presented. Monte Carlo simulations statistically validate our algorithms, and runtime statistics are tabulated. Benchmark evaluation results in a Manhattan, New York City low-altitude environment compare our geofenced dynamic path planning solutions against a fixed airway corridor design. A case study with UAS route deconfliction is presented, illustrating how the proposed geofencing pipeline supports multi-vehicle deconfliction. This paper contributes to the nascent theory and the practice of dynamic airspace geofencing in support of UTM.

Keywords: geofencing; unmanned aircraft systems; UAS traffic management; air traffic control; UAS; low-altitude airspace; computational geometry; path planning; route deconfliction; separation assurance; map processing

Citation: Kim, J.; Atkins, E. Airspace Geofencing and Flight Planning for Low-Altitude, Urban, Small Unmanned Aircraft Systems. *Appl. Sci.* **2022**, *12*, 576. https://doi.org/10.3390/app12020576

Academic Editor: Sylvain Bertrand

Received: 17 November 2021
Accepted: 3 January 2022
Published: 7 January 2022

1. Introduction

Small Unmanned Aircraft System (UAS) operations are expected to proliferate [1,2] for applications such as small package delivery, surveillance, and the visual inspection of assets including wind turbines, construction sites, bridges, and agricultural products. Several challenges must be overcome to enable routine small UAS operations. The aviation community has proposed UAS Traffic Management (UTM) [3–5] to safely and efficiently manage low-altitude airspace where small UAS are expected to operate. UTM services are expected to be based on web apps and datalinks which facilitate the efficient definition and coordination of UAS flight plans.

Airspace geofencing is one of the key capabilities required for UTM [3]. The envisioned geofencing system will enable safe flight operations by dividing airspace into available fly-zone (keep-in geofence) and no-fly zone (keep-out geofence) volumes with statically and dynamically adjusted virtual barriers or "fences" designed to assure UAS separation from obstacles, sensitive areas, and each other. Geofencing will facilitate safety management (i.e., Situational Awareness (SA) for trajectory monitoring, trajectory deviation alerts/geofence breaches, and contingency plans) and flight management (i.e., route-planning, the selection of take-off/landing sites, and mission priority adjustment) for UTM. Figure 1 illustrates airspace geofence examples for UAS flight operations near the One World Trade Center in Manhattan (left) and for wind turbine inspection (right).

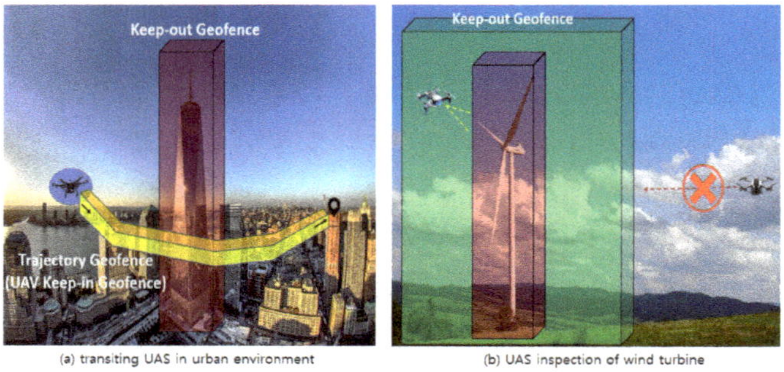

Figure 1. UAS airspace geofencing examples. The left figure shows a keep-out geofence (red) around One World Trade Center in New York City. A transiting UAS keeps clear of this geofence with a path wrapped by a trajectory or keep-in geofence (yellow). The right figure shows a wind turbine being inspected by a small UAS. During inspection, the usual wind turbine keep-out geofence (red) is expanded as depicted in green to also enclose the inspection UAS. Any other nearby UAS will keep clear of this expanded keep-out geofence (green) during inspection activities. This geofence design assures separation between the two illustrated UAS.

Researchers have previously investigated airspace geofencing systems for UTM. A two-dimensional keep-in/keep-out geofence construction algorithm was developed in [6]. Real-time geofence violation detection capabilities have been developed using Ray Casting [7] and Triangle Weight Characterization with Adjacency (TWCA) [8] methods. Potential intersections of 2-D geofences can be rapidly detected using a convex hull approach [9]. A constrained control scheme was developed using an Explicit Reference Governor (ERG) in [10]; this approach ensures a UAS does not violate geofence boundaries. This previous research primarily focused on geofence definition, boundary violation detection, and UAS avionics augmentation to support geofencing. Our work's focus on 3D path planning with geofence volumes in a realistically mapped urban environment is complementary.

Cooperative UAS flight tests were also evaluated using "separation by segregation" geofencing features in [11]. To define a local geofence volume for applications such as crop inspection, the maximum flight distance a UAS can travel after flight termination was calculated using vehicle dynamics and position sensors in [12] to define geofence geometry. This research demonstrated that a UAS stays within its prescribed keep-in geofence in both nominal and off-nominal (e.g., flight termination) conditions.

A three-dimensional dynamic geofencing volumization solution was proposed using "operational" and "inverse" volumization functions in [13]. Per [13], *airspace operational volumization* is the process by which a requested flight plan is "wrapped" with a geofence reserved over an approved flight time window. *Inverse volumization* is the opposite process in which a flight is planned to always remain within a designated airspace geofence volume. This paper extends our work in [13] in several ways. First, we integrate the individual airspace volumization algorithms into a geofencing pipeline described in Section 3. This geofencing pipeline is shown in Figure 2. We also construct simplified keep-in/keep-out 3-D geofencing boundaries based on buildings and UAS flight plans, as illustrated in Figure 1. This algorithm uses parameters such as vehicle speed, geofence boundary safety buffer size, and polygon simplification parameters to generate a flight plan that does not violate keep-in/keep-out geofences in the surrounding region. We define a trajectory keep-in geofence as the airspace volume surrounding the planned flight path with constant ceiling and floor safety buffers. Pathfinding logic is developed for different start and desired end locations of a vehicle in the flight plan. The algorithms are built on computational geometry, where obstacles, buildings, and flight path keep-in geofences are represented as sets of 3-D polygons. Path planning modules are computed efficiently based on a visibility graph approach and set operations in a 3-D environment.

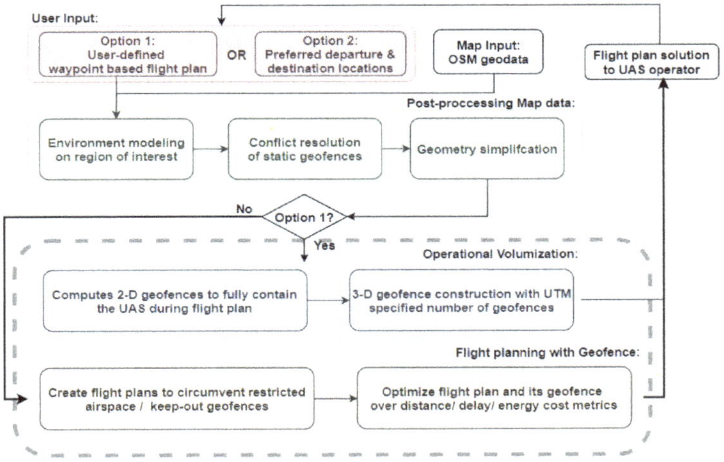

Figure 2. Airspace and environment geofencing functionality and data flow.

This work is unique in its joint consideration of low-altitude mapped obstacles and geofence volume requirements in 3D multicopter sUAS flight planning. Urban terrain and building maps necessarily create more complex flight paths and safety constraints [14]. As an example, consider package delivery UAS in an urban canyon environment. Safe operation requires obstacle-free path planning for all sUAS operating in this shared low-altitude airspace. Planned sUAS paths must therefore treat both physical obstacles (e.g., buildings, power lines, and terrain) and keep-out geofences as impenetrable obstacles that must be circumvented in a safe flight plan. Our work bridges the gap in the existing geofencing literature by focusing on path planning solutions that assure the satisfaction of keep-in/keep-out geofenced airspace volume constraints.

The contributions of this work are:

- The specification of formal algorithms to define keep-in/keep-out geofences for obstacles to plan UAS paths with separation assurance;
- The integration of airspace and environmental geofencing processing pipelines with user inputs to construct geofences and geofence-wrapped path plans in a real-world urban environment;
- Map data processing to generate keep-out geofences around buildings and terrain and a process to simplify a detailed map dataset to support a more compact representation and improved path planning efficiency;
- A benchmark comparison of our geofenced path planning solutions with a fixed sUAS airway flight corridor design, and a case study of sUAS route deconfliction in shared airspace.

The remaining structure of this paper is organized as follows. Section 2 summarizes previous work in UAS Traffic Management (UTM), sUAS and robotic path planning, and computational geometry methods used in geofencing algorithms. Section 3 defines an airspace geofence, states assumptions made in this work, and introduces sUAS geofencing pipeline algorithms used in the generation of flight trajectory solutions. Section 4 describes OpenStreetMap (OSM) data processing steps to minimize computational time in generating solutions. Section 5 describes Monte Carlo simulation setups that integrate pipeline algorithms with map data processing. Section 6 presents statistics comparing results from our airspace volumization algorithm with a fixed airway flight corridor solution for a region of Manhattan in New York City. Section 7 describes a case study for sUAS route deconfliction, while Section 8 concludes the paper.

2. Literature Review

This section presents related work in UTM, computational geometry, and path planning, all of which are relevant to our geofencing algorithms.

2.1. Unmanned Traffic Management and Geofencing

UTM has been identified as a critical capability for future small UAS operations due to their unique operating profiles at low altitude, near complex infrastructure, and likely in mixed-use airspace [3]. UTM-like concepts have been investigated by industry, government, and academia across the globe. As an example, Single European Sky ATM Research (SESAR) recommended UTM to the European Union (EU) to safely coordinate UAS [15]. Centralized and distributed UTM with airspace volumes distinguished by altitude layer was investigated to deconflict UAS traffic in Sweden [16]. UTM was modeled using a multiplayer network of nodes and airways at low-altitude airspace in Luxembourg [5]. The National Aeronautics and Space Administration (NASA) perhaps first coined the term UTM as a system architecture necessary to accommodate UAS in a low-altitude National Airspace System (NAS) layer not frequently occupied by legacy manned aircraft [3]. Representatives from industry have worked to establish adequate security protocols for managing UTM datalinks [17]. NASA, in cooperation with industry, has pursued a series of flight test events to evaluate cooperative UAS operations in beyond visual line of sight (BVLOS) conditions with a "separation by segregation" geofence design [11]. Airspace capacity estimation was analyzed using keep-in and keep-out geofences in [18]. A roadmap for geofence implementation in urban areas with 5G networks and blockchain was introduced in [19].

Dynamic airspace geofencing algorithms are novel to UTM. Two different but equally important perspectives (i.e., local/global) exist in geofencing designs. One perspective is a classical guidance/navigation/control (GNC) approach, where geofence layering is only generated for the individual UAS that has full knowledge of its control system. This vehicle-centered geofence perspective focuses on controlling UAS to ensure that the vehicle does not violate the geofence boundaries (given expected trajectory tracing errors) [10,20]. In this work, each UAS monitors its real-time state vector relative to geofence boundaries to detect and react to potential breaches given uncertainties due to sensor errors and wind disturbances.

Vehicle-centered geofencing research is important but does not consider properties of the operating area airspace or the ground-based environment. Geofencing has also been researched from an *airspace system* perspective. With this viewpoint, geofences are managed by UTM to organize airspace structure and improve Situational Awareness (SA). UTM will not model individual UAS capabilities and uncertainties in detail, but it can conservatively monitor UAS travel through an approved geofence to offer impending breach warnings to the UAS and actual boundary violations to all traffic per [21].

SA is a fundamental requirement for all flight operations, autonomous or human-piloted [22,23]; while legacy air Traffic Management (ATM) will remain distinct from UTM in the near term, advanced air mobility (AAM) supporting increasingly to fully autonomous flight will motivate the integration of ATM and UTM over the long term. UTM calls for the automation of airspace management tasks. Airspace organization and protection through geofencing can improve SA and in turn safety. Our algorithms can be integrated into both GNC (onboard) and airspace system (UTM) geofencing realizations.

AAM operations, including but not limited to Urban Air Mobility (UAM), are envisioned to have higher altitudes than 400 AGL, where current UTM is designed to serve. Researchers at NASA and Uber investigated the applicability of UTM to coordinate UAM routes safely and efficiently [24]. In their case studies, "Transit-Based Operational Volumes (TBOVs)" were used to wrap the UAM flight path, a notion analogous to the trajectory keep-in geofence discussed in this paper. Inspired by the static "UAM-authorized airspace" active over a fixed duration [24] as an airspace management alternative to geofencing, in our case studies, we designed fixed flight corridors and simulated sUAS flight missions operating in these flight corridors. This alternative solution offers a benchmark with which our dynamic geofence volumization and path planning solutions are compared (Section 5).

2.2. Computational Geometry

Computational geometry has been used to construct and deconflict airspace geofence volumes and to detect/prevent airspace boundary violations (onboard). A scaling algorithm was developed for two-dimensional keep-in/keep-out concave polygon geofences in [6]. This paper uses vehicle performance constraints and steady wind conditions to generate scaled "warning" and "override" geofence boundaries. Once a UAS crosses one of these boundaries, onboard GNC can trigger a corrective response [25] or flight termination. In [26–28], algorithms for polygon set operations (i.e., polygon intersections and unions), polygon clipping, convex hull, and point-in-polygon were developed. We use these algorithms to detect and resolve geofence boundary conflicts and generate new geofence volumes by merging conflicting boundaries. A UAS geofence violation detection method was defined in [7] using Ray Casting [29]. A Triangle Weight Characterization with Adjacency (TWCA) algorithm was developed as a faster real-time geofence violation detection method in [8]. TWCA divides geofence into a finite number of triangles and then finds UAS location in a pre-generated adjacency graph. In [9], a 3-D dynamic geofence ("moving geofence") was constructed using maximum cruise time, speed, and range of the UAS as a pre-departure flight planning algorithm. This paper also proposes a convex hull approach to find conflicts between current and newly submitted flight plans.

2.3. Path Planning

Determining a collision-free geofence-based flight trajectory is central to the design of our geofencing volumization work. A variety of path planning algorithms were considered. Grid-based path planning methods overlay a fixed-resolution grid on top of the configuration space and find discretized line segment paths connecting start state to destination. This search is fast in low-dimensional space but quickly becomes computationally intractable with high-resolution maps and appreciable travel distance. The most notable grid-based path planning algorithms are $\mathcal{A}^*$ [30] and $\mathcal{D}^*$ [31]. A family of roadmap-based path planning algorithms have been developed to offer a more compact search space optimizing a specific cost metric. For example, a visibility graph [32] minimizes travel distance, while a Voronoi diagram maximizes obstacle clearance distance [33]. The application of cell decomposition [34] offers a compact map for discrete search path planning in an obstacle field. Other path planning methods include potential-field algorithms [35] that efficiently build plans with gradient descents but are subject to local minima issues. Sampling-based path planning algorithms [36] have also been developed and are particularly well suited to planning in uncertain environments. Our work utilizes a visibility graph approach to path planning. This approach allows us to directly translate geofence volumes generated with computational geometry into visibility graph roadmaps. As is discussed below, we scale keep-out geofences to assure safe separation is maintained. Note that a visibility graph does not require a rasterized map, enabling geofences to be represented without distortion or approximation.

3. Definitions and Algorithms

The term airspace geofence was formally defined in [37] to support a common framework for airspace volume reservation in UTM. Our work follows this definition:

Definition 1. *A Geofence* $g = \{n, v[], z_f, z_c, m, h[]\}$ *is a volume defined by a list of n vertices in the horizontal plane* $v = [(x_1, y_1), (x_2, y_2), \cdots, (x_n, y_n)]$, *where* $n \geq 3$, *and an altitude floor* z_f *and ceiling* z_c. *The volume is defined relative to a set of home locations,* $h_i = (x_i, y_i, z_i, t_i)$, *where* $h[]$ *is a list of length* $m \geq 2$. *Lateral home positions can be represented as latitude/longitude pairs* (ϕ_i, λ_i) *or locally referenced Cartesian coordinates* (x_i, y_i). z_i *is the altitude of the home location above Mean Sea Level (MSL).* t_i *is the activation time for home location i where* $1 \leq i \leq m$. t_m *is* *the deactivation time for geofence g. For consistency, Cartesian coordinates and altitudes are defined in meters and activation/deactivation times are in seconds.*

This data structure supports geofence types: static, durational, and dynamic. A *static geofence* has a permanent fixed home location $h[\,]$ and typically surrounds physical obstacles such as buildings or sensitive areas (i.e., no-fly zones). A *durational geofence* is active over a finite time interval with a fixed home location $h[\,]$. A *dynamic geofence* is active over a specific time interval; its *home* location can move over time.

The following simplifying assumption is made in this paper to facilitate path planning and eliminate the need for traffic deconfliction.

Assumption 1. *One aircraft (e.g., UAS) is allocated to each local geofence volume. No other UAS is permitted to cross into this volume. UTM efficiency therefore relies on minimizing each reserved geofence volume and its duration.*

Dynamic airspace volumization for geofencing will enhance safety by wrapping a UAS in an airspace volume that assures separation from other traffic. The below subsections describe our geofencing algorithm pipeline for UTM, where flight plans are designed with keep-in/keep-out geofencing volumes on a low-altitude airspace map. Three-dimensional trajectory keep-in geofence volumes safely wrapping UAS flight paths are described in Section 3.1, keep-out geofence construction for a low-altitude urban map is described in Section 3.2, and geofence-based path planning solutions are illustrated in Section 3.3.

3.1. Airspace Operational Volumization

Operational volumization constructs a trajectory keep-in geofence overlaid on a user-defined 3-D flight trajectory. Climb and descent segments are first generated with vehicle dynamics inputs such as velocity and desired time to climb/descend. Then, three-dimensional cruise operational volumes are created between the climb and descent geofence pair. This assures a geofence volume always encloses the flight trajectory with the prescribed safety buffer $\delta_{vehicle}$. This algorithm integrates 2-D flight trajectory operational volumization with the Multiple Staircase Geofence (MSG) algorithm per [13]. Three-dimensional trajectory volumization is shown in Algorithm 1. Figure 3 shows an example of a 3-D trajectory with its corresponding three-dimensional geofence volume. A sequence of geofence volumes is constructed by connecting climb, cruise, and descent geofences with user-specified safety buffers.

Algorithm 1 3D Flight Trajectory Operational Volumization (***3dOperVol***).

Inputs: 2-D Trajectory waypoints $\mathcal{W}$, Velocity $\mathcal{V}$, Time to Climb t_{climb}, Time to Descent t_{desc}, Number of Geofence $\mathcal{N}_{geo}$, UAS Safety Buffer $\delta_{vehicle}$, Cruise Altitude h_{cruise}
Outputs: 3-D Flight Trajectory $\mathcal{P}_{traj}$, 3-D Geofence for 3-D Flight Trajectory $\mathcal{G}$
Algorithm:

1: $[\mathcal{P}_{climb}, \mathcal{G}_{climb}] \leftarrow MSG(\mathcal{W}[1:2], \mathcal{V}, t_{climb}, \mathcal{N}_{geo}, \delta_{vehicle})$ $\lhd$ generate climb geofence
2: $[\mathcal{P}_{desc}, \mathcal{G}_{desc}] \leftarrow MSG(\mathcal{W}[end-1:end], \mathcal{V}, t_{desc}, \mathcal{N}_{geo}, \delta_{vehicle})$ $\lhd$ descent geofence
3:
4: $\mathcal{P}_{cruise} \leftarrow [\mathcal{P}_{climb}[end-1:end], \mathcal{W}[3:end-2], \mathcal{P}_{desc}[1:2]]$ $\lhd$ 3-D Cruise flight
5: $[\mathcal{G}_{cruise}] \leftarrow MSG(2dOperVol(\mathcal{P}_{cruise}, \delta_{vehicle}), h_{cruise})$ $\lhd$ Generate cruise geofence
6: $\mathcal{P}_{traj} \leftarrow [\mathcal{P}_{climb}; \mathcal{P}_{cruise}; \mathcal{P}_{desc}]$
7: $\mathcal{G} \leftarrow [\mathcal{G}_{climb}; \mathcal{G}_{cruise}; \mathcal{G}_{desc}]$
8: **return** $[\mathcal{P}_{traj}, \mathcal{G}]$

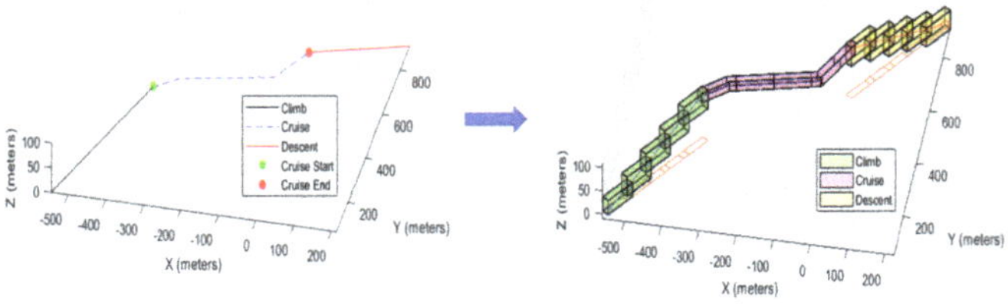

Figure 3. Example application of Algorithm 1. A sample 3-D flight path is shown on the left. A corresponding flight trajectory keep-in geofence is shown on the right.

To minimize airspace volume reservation duration, we utilize the shrinking durational geofence (SDG) and multi-stage durational geofence (MDG) algorithms in [13] for the cruise segment. A shrinking durational geofence (SDG) removes a previously occupied geofence volume at each time update in UTM. A multi-stage durational geofence (MDG) has multiple volumes generated over the flight trajectory with temporal or spatial overlap. For transitions between MDG regions, either temporal or spatial overlap is used to guarantee the UAS is always enclosed by at least one MDG. Overlap offers a buffer in case the UAS flies faster or slower than expected. Note that climb and descent segments utilize multiple staircase geofences so that previously occupied staircase geofences can be removed sequentially.

3.2. Constructing a Geofence Volume from an Urban Map

Keep-out geofences are constructed around obstacles (i.e., buildings) to assure separation between UAS and obstacles or no-fly airspace zones. The construction of keep-out geofence volumes from a building and terrain map must be efficiently carried out to constrain the computation time needed to generate geofence-based path planning solutions. For this work, we utilize a visibility graph approach to path planning, as illustrated in Section 3.3. The time complexity of visibility graph generation is $O(n^2 log(n))$, where n is the total number of vertices in all polygons. In a real-world environment, the number of keep-out geofences in the urban environment can be significant (i.e., 14,000 building cluster geofence polygons in the southern Manhattan map). We utilize two algorithms to achieve map simplification. First, we downsample geofence vertices in the map as shown in Algorithm 2 per [38] with user-defined parameters $n_{maxVert}$ and $p_{dwnSmple}$. This updated set of keep-out geofences is then used to construct a region of interest (ROI) visibility graph. The ROI in the map is first constructed as a rectangular box surrounding departure and destination points. Then, polygon intersection, point-in-polygon, and convex hull operations are used to define the actual region of interest for which geofence-based path planning solutions are generated. Generation of the flight planning visibility graph ROI is shown in Algorithm 3. Figure 4 shows an example of polygon vertex set downsampling. Figure 5 illustrates an initial rectangular ROI P_{recROI} example.

Algorithm 2 Reduce Map Geofence Vertex Set.

Inputs: Set of Keep-out Geofences $\mathcal{S}_{geo}$, Downsample Threshold $n_{maxVert}$, Downsample Tolerance In Percentage $p_{dwnSmple}$
Outputs: Set of Downsampled Keep-out Geofences $\mathcal{S}_{ds}$
Algorithm:

$\mathcal{S}_{ds} \leftarrow []$ ◁ initialize the output set

for $\mathcal{S} \in \mathcal{S}_{geo}$ **do**
  if $len(\mathcal{S})/2 > n_{maxVert}$ **then**
    $\mathcal{S}_{out} \leftarrow DecimatePoly(\mathcal{S}, p_{dwnSmple})$ ◁ downsample polygon vertices
    $k \leftarrow 1$
    for $j = 1 : len(\mathcal{S}_{out})$ **do**
      $\mathcal{G}[k : k+1] \leftarrow \mathcal{S}_{out}[j, 1:2]$ ◁ obtain geofence data structure
      $k = k + 2$
    end for
  end if
  $\mathcal{S}_{ds} \leftarrow \mathcal{S}_{ds}.add(\mathcal{G})$
end for
return $\mathcal{S}_{ds}$

Algorithm 3 Compute Visibility Graph ROI.

Inputs: Departure Point $\mathcal{P}_{start}$, Destination Point $\mathcal{P}_{end}$, ROI Inital Buffer δ_{ROI}, Keep-out Geofence Set $\mathcal{S}_{geo}$
Outputs: Keep-out Geofences in ROI $\mathcal{S}_{ROI}$
Algorithm:

$P_{recROI} \leftarrow getRecROI(\mathcal{P}_{start}, \mathcal{P}_{end}, \delta_{ROI})$ ◁ get Rectangular ROI vertices

//get convexhull ROI where geofencing solutions are generated
$S_{intersect} \leftarrow []$ ◁ initialize the intersecting geofence set
for $\mathcal{S} \in \mathcal{S}_{geo}$ **do**
  if $searchIntersect(S, P_{recROI}) \neq \emptyset$ **then**
    $\mathcal{S}_{intersect} \leftarrow \mathcal{S}_{intsct}.add(S)$ ◁ Append intersecting geofence
  end if
end for

//Search keep-out geofences inside the convex hull P_{ROI}
$S_{ROI} \leftarrow []$ ◁ initialize S_{ROI}
$P_{ROI} \leftarrow convexHull(S_{intersect})$ ◁ ROI where geofencing solutions are generated
for $\mathcal{S} \in \mathcal{S}_{geo}$ **do**
  if $searchIntersect(S, P_{ROI}) \neq \emptyset$ **then**
    $\mathcal{S}_{ROI} \leftarrow \mathcal{S}_{intsct}.add(S)$ ◁ Append intersecting geofence
  end if
end for
return S_{ROI}

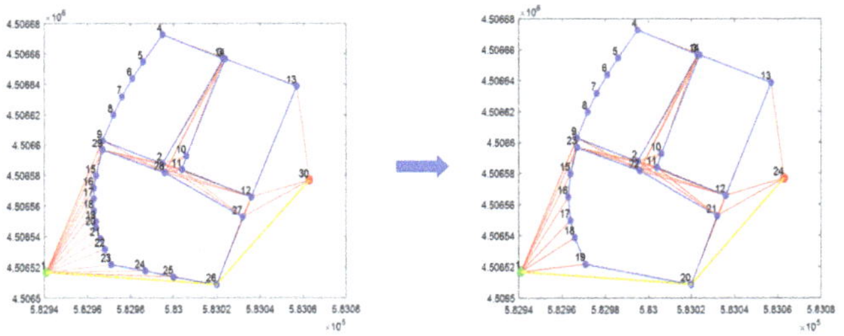

Figure 4. Example of reducing number of vertices to simplify the associated visibility graph. The left illustration shows three original polygons. The right illustration shows the polygons after applying the vertex downsampling algorithm. $n_{maxVert}$ and $p_{dwnSmple}$ are 15 and 60%, respectively. The time complexity of visibility graph generation is $O(n^2 log(n))$, where n is the total number of vertices in all polygons. The number of vertices in the lower polygon illustrated here is reduced from 15 to 9.

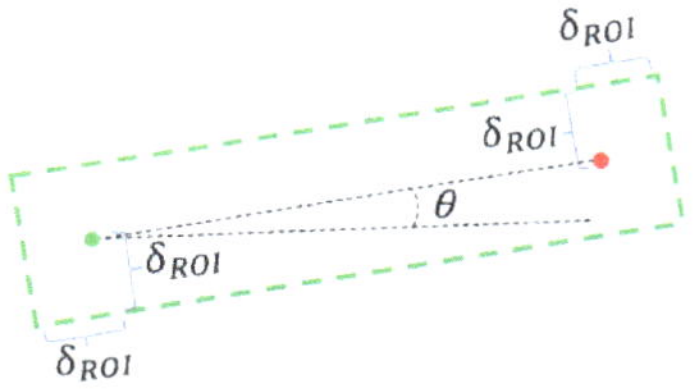

Figure 5. Illustration of rectangular ROI generation. Start point, destination point, and ROI initial buffer size δ_{ROI} are used to initialize the rectangular ROI per Algorithm 3.

3.3. UAS Flight Planning in a Geofenced UTM Airspace

Flight plans are typically optimized over distance, energy usage, and flight time (delay) cost metrics. A UAS configuration space is first obtained from user-defined safety buffers $\delta_{vehicle}$, $\delta_{building}$ around the vehicle and obstacles, respectively. The UAS can then be treated as a point mass in configuration space with obstacle boundaries expanded for safety by:

$$\delta_{sb} = \delta_{vehicle} + \delta_{building}. \tag{1}$$

This safety buffer ensures the vehicle maintains sufficient clearance from any obstacles. $\delta_{vehicle}$ and $\delta_{building}$ are user-specified parameters in this work.

Our proposed geofencing pipeline applies three inverse volumization options per [13] based on user-specified departure and destination locations. The first option is a "turn" solution that calculates climb, cruise, and descent flight trajectories that turn away from nearby obstacles, maintaining a minimum-distance path from start to end. For this module, a low-dimensional visibility graph search with Dijkstra's algorithm [39] plans paths around obstacles (i.e., polygons) defined in a local Cartesian frame. We modeled keep-out geofences on obstacles as open set 3-D polygons extruded from 2-D obstacles with fixed heights. Per Section 2.3, an obstacle-free visibility graph or roadmap space can be constructed from geofence and obstacle polygons without rasterization [32,40].

The second path planning option is a "constant cruise altitude climb" module for which the UAS climbs over no-fly and obstacle volumes until a direct-heading route to the destination is obstacle-free. For this option, a vehicle first climbs to a pre-determined cruise altitude greater than the highest building en route to the destination. Then, the vehicle flies directly to the destination at cruise altitude. As the vehicle approaches the end of its cruise segment, it descends to the destination free of obstacles along the path. The third path planning option is a "vertical terrain follower" module, where a UAS follows the terrain altitude profile en route to the destination, flying as low as possible. This solution minimizes the time a UAS will spend at a high altitude potentially in conflict with other transiting traffic, but it adds complexity to the altitude profile. Figure 6 shows examples of turn, constant cruise altitude, and terrain follower climb solutions per [13].

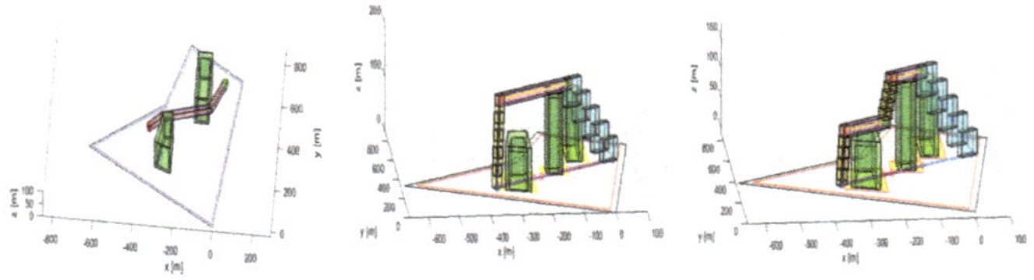

Figure 6. Three candidate flight planning solutions respecting keep-out airspace geofence and obstacle "no-fly" volumes. A turn solution uses a visibility graph to define a constant-altitude path around no-fly zones (**left**). A cruise altitude solution climbs to an altitude greater than the highest building enroute to the destination (**center**). The terrain follower defines an altitude profile maintaining minimum safe clearance or greater from no-fly zones (**right**).

To determine which of three solutions is best, a weighted cost function over time, distance, and energy is defined:

$$C = \alpha * d_{travel} + \beta * P_{travel} + \gamma * t_{wait}. \quad (2)$$

where d_{travel}, P_{travel}, and t_{wait} are distance traveled, power consumption, and time delay until durational geofences disappear, respectively. Weighting factors $\alpha, \beta, and \gamma$ are user-defined. The path planning solution with minimum cost is then suggested to an operator and/or automation. The flight planning process with geofencing is shown in Algorithm 4. In this algorithm, the departure point, destination point, cruise altitude, and keep-out geofence boundary coordinates are input along with cruise velocity and climb/descent times. For the turn solution, a Rotational Plane Sweep (RPS) algorithm is used to find all straight-line segments connecting line-of-sight vertices to form a visibility graph map. Then, Dijkstra's algorithm finds the minimum distance path from departure to destination point. For constant altitude climb and terrain follower solutions, points of intersection between a straight line solution path and obstacles are found using a polygon-line intersection operator. Then, obstacle height at the intersection points are extracted from keep-out geofence data. Three-dimensional flight trajectory "turn", "constant cruise altitude", and "terrain follower" solutions are wrapped with geofences using Algorithm 1. The best solution is the minimum cost module based on Equation (2). Note that geofence segment duration is not explicitly considered in this paper. Instead, it is assumed the flight trajectory keep-in geofence generated using Algorithm 4 remains active from UAS launch to landing.

Algorithm 4 Flight Planning With Geofencing.

Inputs: Departure Point $\mathcal{R}_{start}$, Destination Point $\mathcal{R}_{end}$, Cruise Altitude h_{cruise}, Keep-out Geofence Boundaries $\mathcal{S}_{geo}$, Aircraft Velocity $\mathcal{V}$, Time to Climb t_{climb}, Time to Descend t_{desc}, Number of Geofences $\mathcal{N}_{geo}$, UAS Safety Buffer $\delta_{vehicle}$
Outputs: Planned Flight Trajectory $\mathcal{P}_{traj}$, Trajectory-wrapping 3-D Geofence Volumes $\mathcal{G}$
Algorithm:

//turn solution module
$\mathcal{R}_{VG} \leftarrow [\mathcal{R}_{start}; \mathcal{S}_{geo}; \mathcal{R}_{end}]$ ◁ Vertices of Visibility Graph
$[edges, vert\_ID] \leftarrow RPS(\mathcal{R}_{VG})$ ◁ get visibility graph edges on the map using RPS
$[R_{turn}] \leftarrow dijkstraPath(\mathcal{R}_{start}, \mathcal{R}_{end}, edges, vert\_ID)$ ◁ get min. distance path
$[\mathcal{P}_{turn}, \mathcal{G}_{turn}] \leftarrow 3dOperVol(\mathcal{R}_{turn}, \mathcal{V}, [t_{climb}, t_{desc}], \mathcal{N}_{geo}, \delta_{vehicle}, h_{cruise})$
$\mathcal{D}_{turn} \leftarrow getDist(\mathcal{P}_{turn})$ ◁ get turn module flight distance

//climb solution modules
$\mathcal{R}_{intersect} \leftarrow searchIntersect(\mathcal{R}_{VG})$ ◁ get intersections from $[\mathcal{R}_{start}; \mathcal{R}_{end}]$ to $\mathcal{S}_{geo}$
if $\mathcal{R}_{intersect} \neq \emptyset$ **then**
  $h_{intersect} \leftarrow extractHeight(\mathcal{R}_{intersect}, \mathcal{S}_{geo})$ ◁ get heights at intersections
  $h_{max} \leftarrow max(h_{intersect})$

  //constant cruise altitude
  $[\mathcal{P}_{const}, \mathcal{G}_{const}] \leftarrow 3dOperVol(\mathcal{R}_{intersect}, \mathcal{V}, [t_{climb}, t_{desc}], \mathcal{N}_{geo}, \delta_{vehicle}, h_{max})$
  $\mathcal{D}_{const} \leftarrow getDist(\mathcal{P}_{const})$ ◁ get constant altitude cruise flight distance

  //terrain follower
  $[\mathcal{P}_{terr}, \mathcal{G}_{terr}] \leftarrow 3dOperVol(\mathcal{R}_{intersect}, \mathcal{V}, [t_{climb}, t_{desc}], \mathcal{N}_{geo}, \delta_{vehicle}, h_{intersect})$
  $\mathcal{D}_{terrain} \leftarrow getDist(\mathcal{P}_{terr})$ ◁ get terrain follower flight distance
end if

//cost comparison
$[\mathcal{C}_{min}, opt] \leftarrow costCompare(\mathcal{D}_{turn}, \mathcal{D}_{const}, \mathcal{D}_{terrain})$
if opt == 1 **then**
  $[\mathcal{P}_{traj}, \mathcal{G}_{traj}] \leftarrow [p_{turn}, \mathcal{G}_{turn}]$ ◁ best sol: turn module
else if opt == 2 **then**
  $[\mathcal{P}_{traj}, \mathcal{G}_{traj}] \leftarrow [p_{const}, \mathcal{G}_{const}]$ ◁ best sol: constant cruise altitude module
else
  $[\mathcal{P}_{traj}, \mathcal{G}_{traj}] \leftarrow [p_{terr}, \mathcal{G}_{terr}]$ ◁ best sol: terrain follower module
end if
return $[\mathcal{P}_{traj}, \mathcal{G}_{traj}]$

4. Environment Modeling

Map Data Processing

To evaluate the proposed geofencing capability in a complex low-altitude environment, we processed OpenStreetMap (OSM) data for the Manhattan Borough of New York City (USA). OSM is a collaborative global mapping project that creates geographical data and information [41]. OSM is frequently updated and provides map entities including airways, roads, buildings, and more. To minimize map processing overhead for this work, we used pre-processed georeferenced OSM Manhattan building data as described in Ref. [42]. This raw data contain building coordinates represented as polygon vertices, building heights, and street level in WGS 84/UTM zone 18N [43], where units are in meters with East, North, Up (ENU) axes. We applied a combination of set and convex hull [32] operations to simplify geofence geometry for flight planning. Figure 7 shows the flowchart for map data post-processing. After post-processing, the dataset was partitioned into four categories: buildings with heights greater than 20 m, 60 m, 122 m, and 400 m. Depending on sUAS start and end altitude (i.e., roof of building, ground), flight planning utilizes one of these four datasets to generate plans and associated geofence volumes.

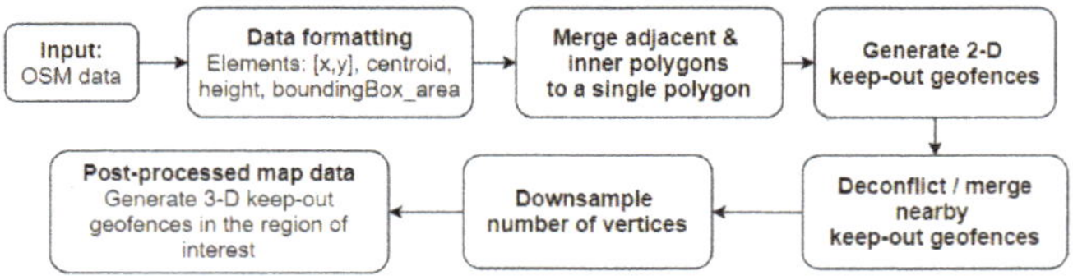

Figure 7. Flowchart of post-processing map data. OSM data were converted to a MATLAB format, then processed using polygon set convex hull operators to reduce the number of keep-out geofences in the region of interest (ROI), the area between departure and destination points. If the number of vertices in a geofence is greater than threshold $n_{maxVert}$, it is downsampled to $p_{dwnSmple}$. $n_{maxVert}$ and $p_{dwnSmple}$ are user-defined parameters set to 15 and 60%, respectively, in this work. Algorithms 2 and 3 are used in finding ROI and reducing number of map vertices. Three-dimensional keep-out geofences around buildings are generated with safety buffer $\delta_{building}$.

Figure 8 shows a map of southern Manhattan with closely spaced building clusters each enclosed by a single keep-out geofence to simplify the Manhattan urban canyon map. Figure 9 shows an example of post-processed georeferenced data and its 3-D keep-out geofence.

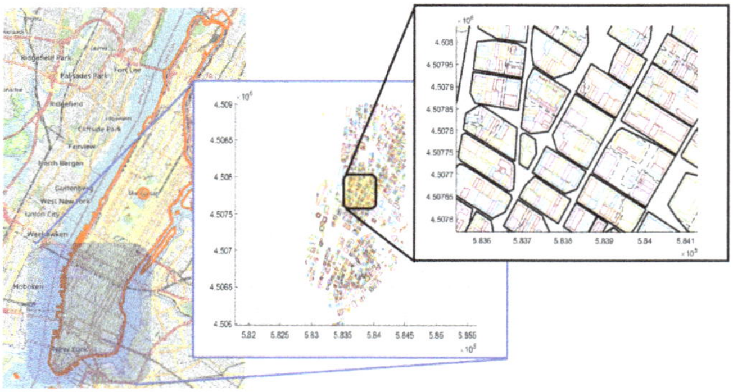

Figure 8. Post-processing map data for southern Manhattan. Buildings with heights greater than 20 m are shown. The rightmost plot shows keep-out geofences enclosing building clusters (black solid lines), individual building keep-out geofences (black dashed lines), and building outlines (colored lines). Geofence maps for 60 m, 122 m, and 400 m altitude cross-sections are constructed in the same manner.

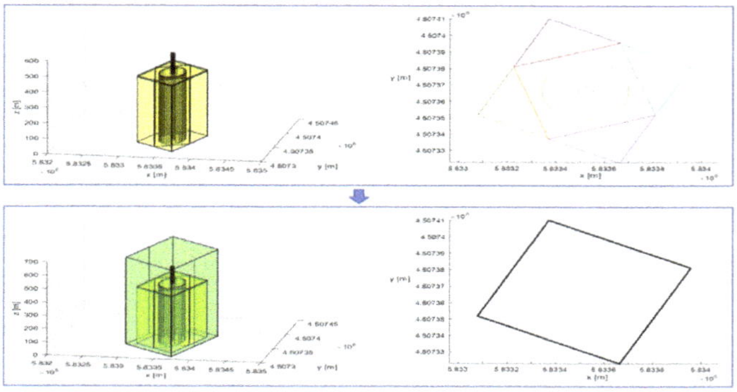

Figure 9. Post-processed georeferenced data for the One World Trade Center building in Manhattan. The top left and right show raw OSM data side and top views, respectively. The bottom left and right show post-processed keep-out geofence data (shaded in green) side and top views, respectively.

A southern Manhattan, New York City map was defined by 14,000 building cluster geofence polygons using the above procedure. To further simplify the map, we downsampled geofence vertices and construct an updated set of keep-out geofences from the ROI visibility graph per Algorithms 2 and 3 in Section 3.2. Figure 10 shows an example of the rectangular ROI, ROI obstacle polygon, and visibility graph generation pipeline. The "turn" flight planning visibility graph was constructed from keep-out geofences inside the ROI along with departure and destination locations.

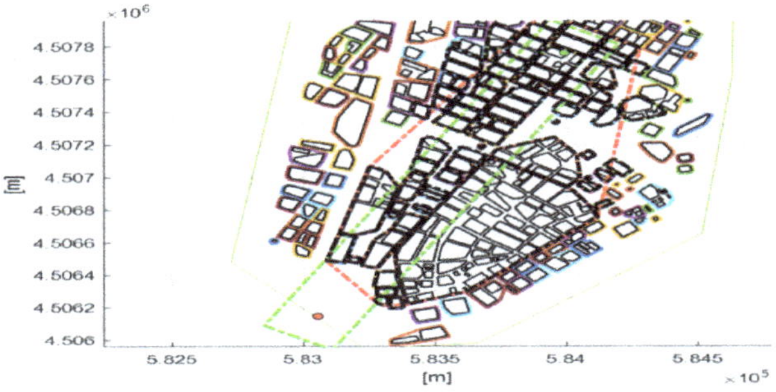

Figure 10. Keep-out geofence polygon extraction for UAS flight planning. The initial ROI (green dashed line) is a rectangular box per Figure 5. Keep-out geofences (solid black lines) inside or intersecting the rectangular ROI box are found using polygon intersection and point-in-polygon operations. The final ROI (red dashed line) is the convex hull around these keep-out geofences. For our simulation, $\delta_{ROI} = 150$ m.

5. Simulation Setup

Monte Carlo simulations were used to evaluate proposed airspace volumization strategies on the Manhattan map. Figure 11 shows the flowchart of pathfinding logic in our simulation setup. Pathfinding logic comprises four solution modules for the airspace geofencing algorithm. Once the start and end locations were defined, the keep-out geofence ROI polygons (Figure 10) were extracted from post-processed map data. Constant cruise altitude and terrain follower modules were generated by searching the intersection points between the buildings' keep-out geofences and the line that connects UAS start and end waypoints. A pure turn solution was generated if both start and end locations were on the ground. If either start or end location was on the roof of the building (i.e., inside of the keep-out geofence), a constant cruise altitude algorithm was first used to find the flight path from the start/end point to the outside of the keep-out geofence, and the turn module solution was used to calculate the remaining flight path, creating a combined solution.

Control parameters are shown in Table 1. To offer an experimentally grounded dataset, a prototype quadplane's power consumption model [44] was used per Table 2 to compute P_{travel} in climb, cruise, and descent segments. A quadplane is a hybrid quadrotor/fixed-wing UAS designed to vertically takeoff and land in an urban environment. For our simulations, the quadrotor motors were active in all phases of flight; cruise power would otherwise be lower. Cost function weighting factors $\alpha = 0.6$, $\beta = 0.2$, $\gamma = 0.0$ were chosen to prioritize minimum-distance solutions. Note that γ was set to zero because building obstacles have static or permanent geofences.

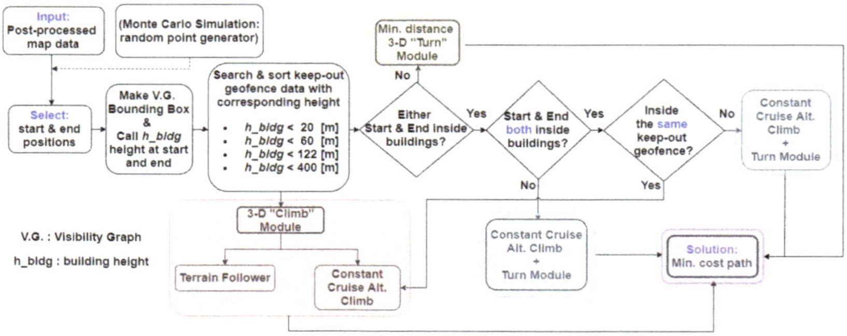

Figure 11. Flow chart of pathfinding logic for different start and end locations. In the chart, V.G. abbreviates visibility graph, and h_{bldg} is the height of a geofence around a cluster of buildings. If the departure/destination is not inside the keep-out geofence ROI box, h_{bldg} at start/end point is set to street/terrain altitude.

Table 1. Control parameters for geofenced flight planning case studies.

$v_{vehicle}$	$\delta_{vehicle}$	$\delta_{building}$	$N_{geofence}$	z_{cruise}
5 (m/s)	2 (m)	5 (m)	5	50 (m)

Table 2. Flight power consumption data from [44].

Climb	Descent	Forward Flight
312 (J/s)	300 (J/s)	328 (J/s)

6. Simulation Results

A total of 1010 Monte Carlo simulations were run with our Manhattan maps. For each case, start and destination points were randomly defined. Selected start/end altitudes ranged from 20 m above ground level to the highest building roof. The 20 m value represents an above-ground vertical climb to hover waypoint to ensure the multicopter is well clear of people on the ground when it begins executing its lateral flight plan. If both start and end points had altitudes less than 50 m, the cruise altitude for the turn solution was set at 50 m. Otherwise, cruise altitude was adjusted based on the following condition:

$$z_{cruise} = max\{h_{start}, h_{end}\} if\ h_{start} > 50 \text{ m} \mid\mid h_{end} > 50 \text{ m}. \tag{3}$$

Our airspace volumization algorithm used this condition to choose one of the fixed-altitude datasets described in Section 4. As z_{cruise} becomes larger, fewer obstacles were present, so fewer calculations were needed to generate and plan a flight through the visibility graph. For each case, cost values of the four planning options ("turn", "constant cruise alt.", "terrain follower", "combined (constant cruise altitude + turn)") were calculated using Equation (2), and the minimum cost solution was selected as the best solution. Note in the Manhattan data the "wait" solution was never used because buildings are permanent, resulting in static geofence obstacles only.

Monte Carlo results offer an opportunity to compare our airspace volumization solutions against a manual fixed airway or "flight corridor" airspace design. A conventional fixed-altitude airway is permanently designated on a map to enable traffic "queues" to organize in a way that can be managed by human air traffic controllers. It is unclear whether UTM will benefit from this legacy design practice, motivating our comparison of path costs for our airspace volumization and fixed airway solutions. Unlike our airspace volumization, fixed airway/flight corridor maps only require a local search for the closest airway to join. The UAS then follows fixed airway routes until exiting over a short final segment to the end state. We generated a pair of low-altitude horizontal and vertical airways through our Lower Manhattan map to illustrate the airways concept and support our evaluation.

The designed vertical airway in Lower Manhattan follows Broadway, the north–south main thoroughfare, from its origin at Bowling Green to Houston Street. The horizontal airway follows Chambers Street from River Terrance in the west to Municipal Plaza in the east, and then follows the Brooklyn Bridge until it reaches the East River. We provided two sets of the same cross airways at 150 m and 500 m to offer each UAS an altitude choice since more obstacles are present at 150 m but the climb will be more substantial to 500 m. Figure 12 shows our manually defined airway corridors. To offer a practical comparison, only randomly generated start and end points that do not lie in the same quadrants (i.e., 712 out of 1010 simulation examples) were considered. If randomly-generated start and end points were located in the same quadrant, the airways were unused, thus offering no benefit to efficiency or airspace organization.

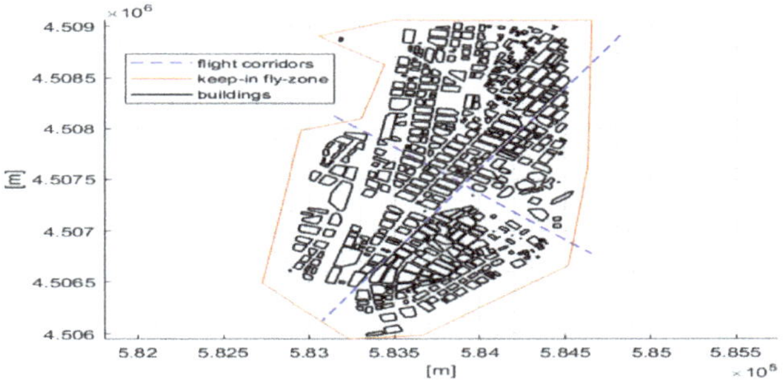

Figure 12. Example horizontal and vertical airway corridors in Manhattan.

Figure 13 shows a top-down route view comparing our airspace volumization and flight corridor solutions. Cost weights $\alpha = 0.6$, $\beta = 0.2$, $\gamma = 0.0$ were again used, so d_{travel} was prioritized in minimizing overall cost. For the illustrated case, the "turn" solution is best. Flight corridor solution cost was in fact typically higher than any of our airspace volumization solutions. For the same example, altitude vs. time plots for each solution are shown in Figure 14. Examples of geofencing solutions are shown in Figures 15–17, where three alternative trajectory solutions are generated, ensuring the avoidance of no-fly zones. Building keep-out geofences are shown in green.

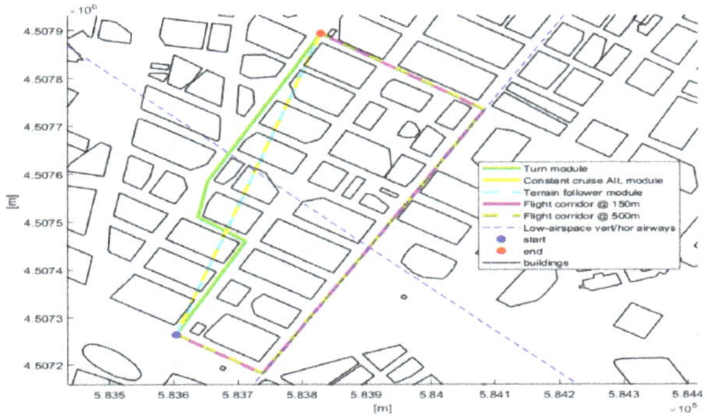

Figure 13. Top-down view of example flight paths for airspace volumization and fixed flight corridor solutions. Distances traveled are 770 m (turn), 1051 m (constant cruise), 1139 m (terrain follower), 1528 (150 m flight corridor), and 1977m (500 m flight corridor).

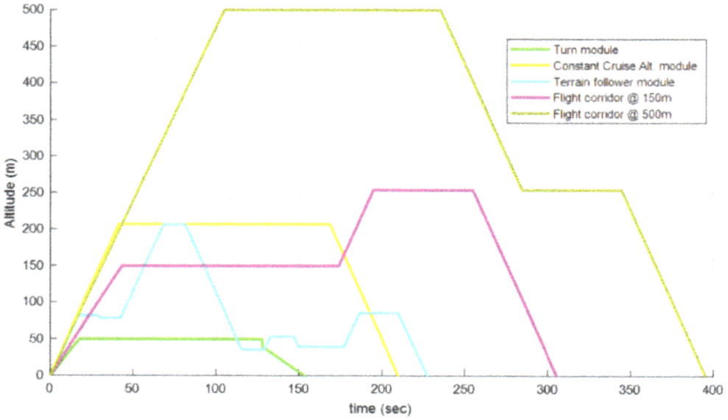

Figure 14. Flight altitude time histories for airspace volumization and flight corridor solutions for Figure 13 example.

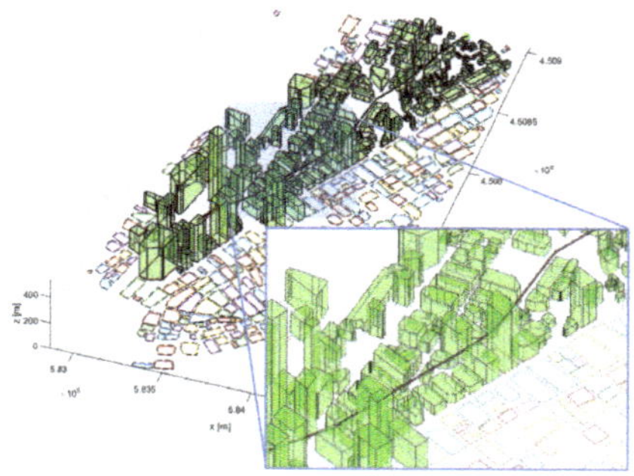

Figure 15. Example of a 3-D geofence wrapping a "turn" flight plan solution. Polyhedra in green denote keep-out geofences around buildings near the trajectory's keep-in geofence. The remaining 2-D polygons denote keep-out geofences around buildings that are more distant from the sUAS flight path.

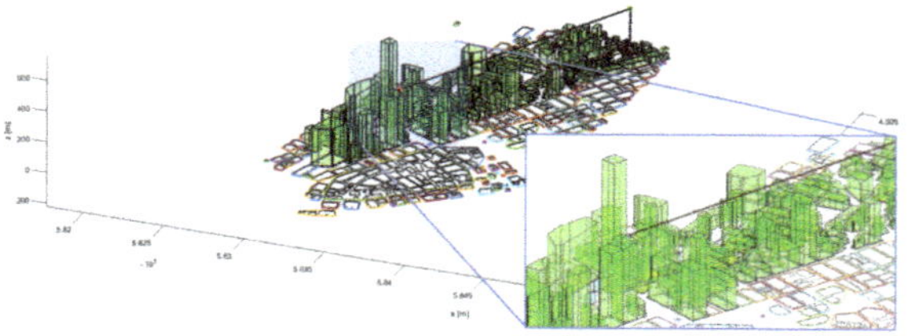

Figure 16. Example 3-D geofencing solution for a "constant cruise altitude" flight plan solution. Polyhedra in green denote keep-out geofences around buildings near the trajectory's keep-in geofence. The remaining 2-D polygons denote keep-out geofences around buildings that are more distant from the sUAS flight path.

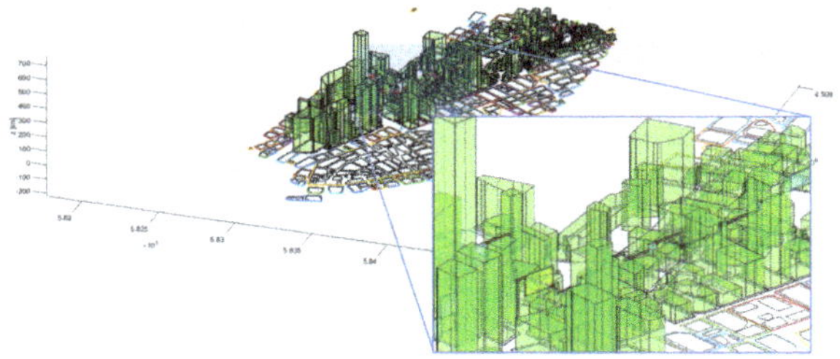

Figure 17. Example 3-D geofencing solution for a "terrain follower" flight plan solution. Polyhedra in green denote keep-out geofences around buildings near the trajectory's keep-in geofence. The remaining 2-D polygons denote keep-out geofences around buildings that are more distant from the sUAS flight path.

For each Monte Carlo simulation, the minimum-cost C solution was compared to flight corridor solution costs at 150 m and 500 m per Table 3. Since the flight corridor at 150 m was almost always better than the flight corridor at 500 m, benchmark data compare the best solution obtained using dynamic airspace volumization with the flight corridor at 150 m. The results indicate our airspace geofencing volumization solutions generally have lower cost than flight corridors at 150 m or 500 m do.

The average distance and power consumption of the two-dimensional straight-line path between each start and destination location are shown in Table 4.

Table 3. Number of cases where airspace volumization *vol* has minimum cost (left) and number of cases where the flight corridor at 150 m has lower cost than the corridor at 500 m.

$\#\{C_{vol.method} < C_{150m}\}$	$\#\{C_{150m} < C_{500m}\}$
698 out of 712 cases	702 out of 712 cases

Table 4. Average distance (d), power consumption (P), and minimum and maximum distances of 2D straight-line paths between start and destination states for the Monte Carlo simulations.

$\mu_{d_{2D\ path}}$	$\mu_{P_{2D\ path}}$	$min\{d_{2D\ path}\}$	$max\{d_{2D\ path}\}$
1391 (m)	91259 (J)	189 (m)	3003 (m)

The mean and standard deviation for d_{travel}, p_{travel} for the minimum cost airspace volumization solution are summarized in Table 5. The percent frequency distributions of the four solution options are shown in Figure 18.

Table 5. Mean μ and standard deviation σ of the minimum-cost airspace volumization solution.

$\mu_{d_{travel}}$	$\sigma_{d_{travel}}$	$\mu_{P_{travel}}$	$\sigma_{P_{travel}}$	$min\{d_{travel}\}$	$max\{d_{travel}\}$
1595 (m)	606 (m)	94,338 (J)	39,609 (J)	254 (m)	3349 (m)

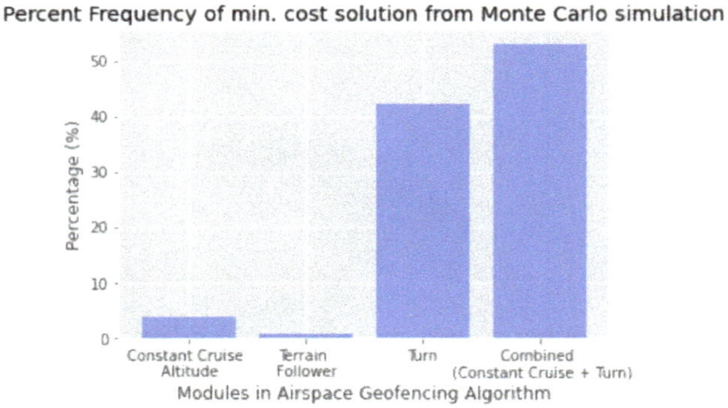

Figure 18. Percent frequency distribution of minimum-cost solutions over Monte Carlo simulations.

A similar analysis was performed to compute travel distance and power consumption statistics for the flight corridor solutions at 150 m and 500 m, as shown in Tables 6 and 7.

Table 6. Mean μ and standard deviation σ of 150 m flight corridor solutions.

$\mu_{d_{150m}}$	$\sigma_{d_{150m}}$	$\mu_{P_{150m}}$	$\sigma_{P_{150m}}$	$min\{d_{150m}\}$	$max\{d_{150m}\}$
2303 (m)	820 (m)	149,084 (J)	53,449 (J)	479 (m)	4464 (m)

Table 7. Mean μ and standard deviation σ of 500 m flight corridor solutions.

$\mu_{d_{500m}}$	$\sigma_{d_{500m}}$	$\mu_{P_{500m}}$	$\sigma_{P_{500m}}$	$min\{d_{500m}\}$	$max\{d_{500m}\}$
2796 (m)	788 (m)	179,363 (J)	51,502 (J)	1142 (m)	4836 (m)

Dynamic airspace volumization and flight corridor solutions at 150 m are normalized by the two-dimensional straight-line path parameters, indicating the percent increase in average travel distance and power consumption. A normalized benchmark comparison is shown in Table 8. On average, our 3-D airspace geofencing solution increased travel distance by 15% and power consumption by 3% compared to 2-D straight-line paths from start states to destination states. On the other hand, the travel distance increased by 66 % and power increases by 63% when comparing minimum-cost 3-D geofencing solutions with 150 m flight corridor solutions. This analysis indicates our airspace geofencing algorithm generates routes that offer nontrivial distance (time) and power (energy) reductions relative to flight corridor paths, at least for Manhattan.

Table 8. Normalized travel distance comparison between airspace geofencing and 150 m flight corridor solutions.

$\mu_{d_{travel}}/\mu_{d_{2D\ path}}$	$\mu_{P_{travel}}/\mu_{P_{2D\ path}}$	$\mu_{d_{150m}}/\mu_{d_{2D\ path}}$	$\mu_{P_{150m}}/\mu_{P_{2D\ path}}$
115 (%)	103 (%)	166 (%)	163 (%)

All simulations were executed on a standard laptop PC using uncompiled MATLAB code. The mean runtime and standard deviation over all 1010 Monte Carlo simulations were computed. The average runtime was 10.98 s with $\sigma = 12.68$. The minimum runtime was 0.13 s, and the maximum runtime was 90.66 s. As the number of obstacles inside the fly-zone increase, runtime also increased, as might be expected. A more computationally efficient visibility graph algorithm could be implemented in future work [45], particularly with a large obstacle set. Migration from uncompiled MATLAB to a compiled code (e.g., in C++) will also improve performance.

A Monte Carlo simulation generated a suite of random launch (start) and landing (end) points for a single sUAS flying in Lower Manhattan. Start and end points were either located on the ground or a flat building roof to simulate the diverse sUAS flight cases that might be encountered in a densely populated urban environment. Keep-out geofences were generated at each building or around blocks of clustered buildings, representing no-fly zones for the sUAS. Our airspace geofencing pipeline successfully generated flight plans and enclosing geofence volumes for four flight trajectory solution options for all 1010 Monte Carlo simulations. The minimum distance and energy cost was chosen as the best solution. Our geofence-based path planning solutions outperformed a more traditional fixed flight corridor routing option.

Our Monte Carlo simulations did not limit the maximum altitude for UAS flight, so the trajectories for some solutions had cruising altitudes greater than 400 ft AGL, beyond the UTM and sUAS ceiling. Our Monte Carlo results showed the "combined" solution option (i.e., constant cruise and turn) was preferred most often. A maximum altitude constraint would eliminate all solutions that climbed above UTM-managed airspace, likely resulting in the more frequent use of visibility graph "turning" solutions. The results in Table 8 showed that our algorithm generates solutions that are 51% and 60% more efficient than flight corridor solutions at 150 m altitude in terms of normalized average flight distance and power consumption, respectively. It is likely that for AAM airspace corridors accessible to sUAS, above 400 ft AGL will be designated. For longer-distance flights, a flight plan might use an efficient dynamically geofenced route to/from a high-altitude transit tube, potentially requiring a hybrid combination of dynamic flight planning and geofencing at UTM-managed altitudes and fixed corridor transit at altitudes managed by legacy ATM.

7. Case Study with sUAS Route Deconfliction

The above results describe single geofenced sUAS routes through a complex urban landscape. In general, UTM will manage multiple sUAS in shared airspace. This section presents a case study illustrating how the proposed geofencing pipeline supports multiple-sUAS deconfliction. For this study, we assume airspace is allocated first-come-first-served. Suppose $sUAS_1$ and $sUAS_2$ request flight plans each defined by departure and destination coordinates (WGS 84/UTM zone 18N), cruise speed, and targeted cruise altitude as defined in Table 9. Further, suppose $sUAS_1$ receives approval for its flight plan and associated geofence volume before $sUAS_2$ contacts UTM. $sUAS_2$ will then need to plan a flight that avoids the Manhattan terrain and building geofences as well as the flight trajectory geofence wrapping the $sUAS_1$ route. Figure 19 shows the resulting flight plans for $sUAS_2$ as a top-down route view comparing our airspace volumization and flight corridor solutions. For this example, altitude vs. time plots for the $sUAS_2$ solutions are shown in Figure 20.

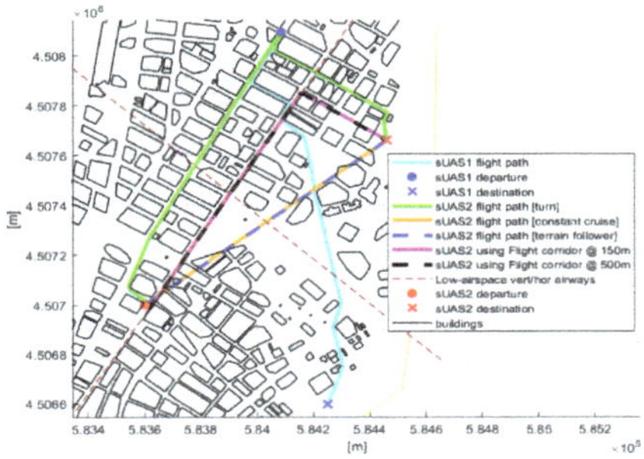

Figure 19. Top-down view of $sUAS_2$ sample solutions. Five flight trajectory solutions are generated for $sUAS_2$. Each solution provides route deconfliction from Manhattan terrain and building geofences and from the $sUAS_1$ flight trajectory geofence. Distances traveled are 2008 m (turn), 1585 m (constant cruise), 1634 (terrain follower), 1983 (150 m flight corridor), and 2395 (500 m flight corridor). The minimum-cost solution for $sUAS_2$ is the constant cruise altitude option.

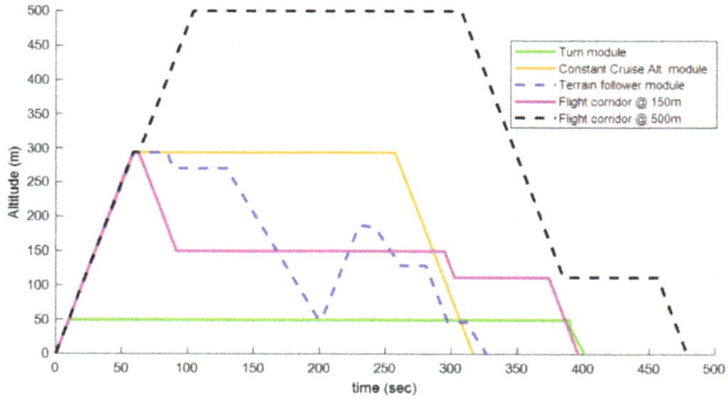

Figure 20. Flight altitude time histories for airspace volumization and flight corridor solutions for $sUAS_2$ in Figure 19 example.

Table 9. Flight plan parameters for $sUAS_1$ and $sUAS_2$.

	$P_{Departure}$ (m)	$P_{Destination}$ (m)	V_{UAS} (m/s)	$h_{targetCruise}$ (m)
$sUAS_1$	[584,085; 4,508,093; 0]	[584,248; 4,506,598; 0]	30	50
$sUAS_2$	[583,600; 4,507,000; 0]	[584,460; 4,507,660; 0]	20	50

Since $sUAS_1$ and $sUAS_2$ have the same target cruise altitude, a maneuver was required for $sUAS_2$ to deconflict its "turn" route from the $sUAS_1$ flight trajectory, making this the longest distance solution option. On the other hand, the "constant cruise altitude" and "terrain follower" solutions were not influenced by the $sUAS_1$ trajectory because the minimum building height along the straight line path from departure to destination for $sUAS_2$ was greater than $sUAS_1$'s target cruise altitude. If building height placed $sUAS_2$ at $sUAS_1$'s cruise altitude, $sUAS_2$ would also need to climb over the $sUAS_1$ geofence. Note that if $sUAS_1$'s airspace volume reservation duration was minimized using SDG or MDG, $sUAS_2$'s path had a lower probability of being impacted. Example 3-D $sUAS_2$ in "turn", "constant cruise altitude", "terrain follower" solutions are shown in Figures 21–23.

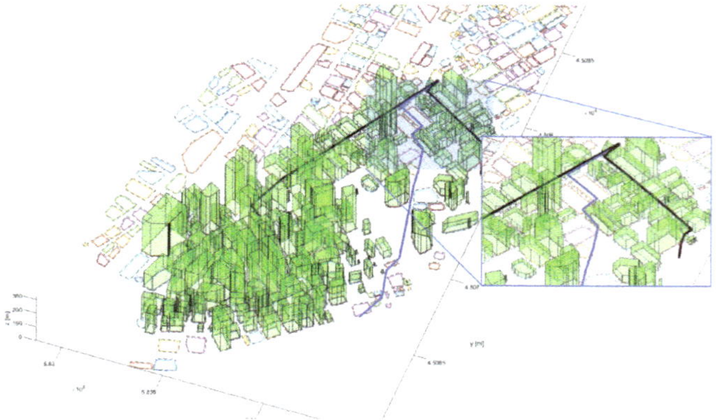

Figure 21. Example of a 3-D geofence wrapping a "turn" flight plan for $sUAS_2$. The $sUAS_2$ trajectory is shown in black, and the $sUAS_1$ trajectory is shown in blue. Polyhedra (green) denote keep-out geofences around buildings. The remaining 2-D polygons denote keep-out geofences around buildings that are outside the combined ROI.

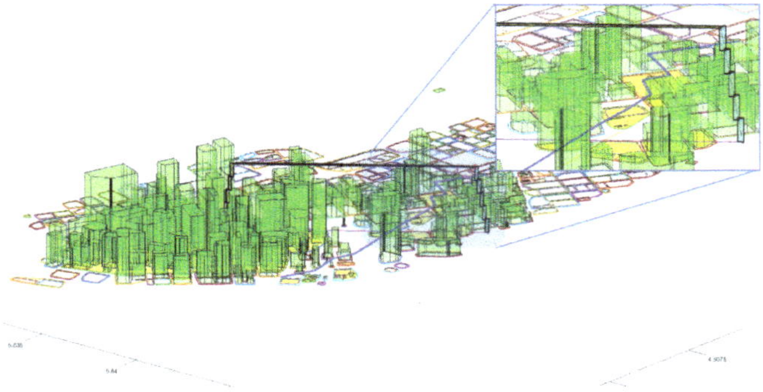

Figure 22. Example of a 3-D geofence wrapping a "constant cruise altitude" flight plan for $sUAS_2$. The $sUAS_2$ trajectory is shown in black, and the $sUAS_1$ trajectory is shown in blue. Polyhedra (green) denote keep-out geofences around buildings. The remaining 2-D polygons denote keep-out geofences around buildings that are outside the combined ROI.

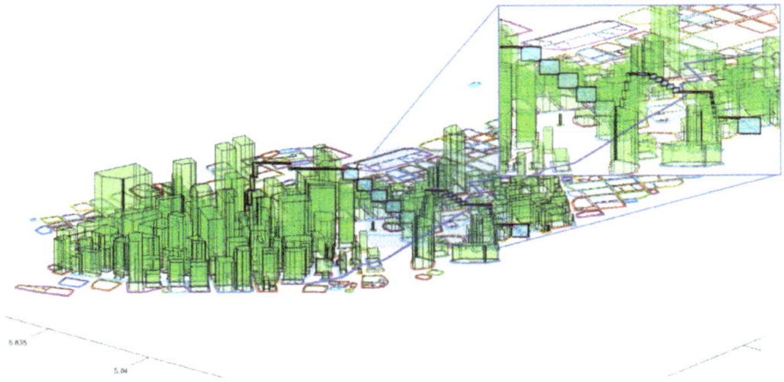

Figure 23. Example of a 3-D geofence wrapping a "terrain follower" flight plan for $sUAS_2$. The $sUAS_2$ trajectory is shown in black, and the $sUAS_1$ trajectory is shown in blue. Polyhedra (green) denote keep-out geofences around buildings. The remaining 2-D polygons denote keep-out geofences around buildings that are outside the combined ROI.

8. Conclusions and Future Work

This paper applied airspace geofencing volumization and path planning to support UTM management of low-altitude airspace. Layered durational geofences wrapping flight trajectories ensure the UAS will fly without conflict in designated or reserved airspace volumes. Our airspace volumization algorithms generated four conflict-free paths for any keep-in/keep-out geofence volume set based on turn, constant cruise, terrain follower and combination turn/cruise options. The algorithm ranked these paths using a weighted distance, energy, and time cost function, then selected the minimum-cost solution. A city map data of Lower Manhattan was used to construct keep-out geofences around buildings. Monte Carlo simulation studies validate our geofence algorithms and support the statistical characterization of performance including run time. A benchmark comparison of our dynamically geofenced flight plans and conventional flight corridor solutions is provided, showing that our solutions reduce flight distance and power compared to fixed corridor solutions. A case study of two sUAS flight planning demonstrated how the proposed geofencing pipeline supports multiple sUAS deconfliction. Algorithms and definitions from this paper can contribute to future UTM dynamic airspace geofencing operational standards.

This work simplifies flight planning to geometric paths. Future work will incorporate aircraft dynamics into flight plans and geofence layer sizing, extend airspace volumization to enclose cooperative groups of sUAS. Additionally, the altitude constraint and other factors such as day/night local population density, GPS dependency, air traffic volume, and vehicle-specific parameters should be incorporated in the geofenced path planning algorithm to generate solutions that are more realistic for UTM-specific applications. We hope to apply machine learning to large-scale flight track data and urban maps to generalize and optimize geofencing volume designs based on area topology, day/night occupancy, infrastructure, and existing air traffic patterns. We will also explore auto-code generation and Python/C++ implementations to improve path planning computational performance.

Author Contributions: Conceptualization, J.K. and E.A.; methodology, J.K. and E.A.; software, J.K.; validation, J.K. and E.A.; writing—original draft preparation, J.K.; writing—review and editing, J.K. and E.A.; visualization, J.K.; supervision, E.A. All authors have read and agreed to the published version of the manuscript.

Funding: This research was funded in part by the Collins Aerospace (Funder grant number: AWD014974) and by NASA (Funder grant number: T18-601046- UMICH).

Institutional Review Board Statement: Not applicable.

Informed Consent Statement: Not applicable.

Data Availability Statement: Not applicable.

Acknowledgments: The authors gratefully acknowledge the following people who supported this research: Mia Stevens for geofencing simulation software and support. Nicholas Liberko for the conceptualization of operational and inverse volumization, Akshay Mathur for the conceptualization of the trajectory geofence and flight power consumption data, and Cosme Ochoa for pre-processed Manhattan map data and documentation.

Conflicts of Interest: The authors declare no conflict of interest.

Abbreviations

The following abbreviations are used in this manuscript:

AAM	Advanced Air Mobility
AGL	Above Ground Level
ATC	Air Traffic Control
ATM	Air Traffic Management
BVLOS	Beyond Visual Line of Sight
d_{travel}	Vehicle travel distance
ERG	Explicit Reference Governor
GNC	Guidance Navigation and Control
IoT	Internet of Things
MDG	Multi-staged Durational Geofence
MSG	Multiple Staircase Geofence
NAS	National Airspace System
$n_{maxVert}$	Allowable maximum number of vertices in a geofence
OSM	OpenStreetMap
$p_{dwnSmple}$	Downsampling percentage of the number of vertices in a geofence
P_{travel}	Power consumption over d_{travel}
ROI	Region of Interest
RPS	Rotational Plane Sweep
SA	Situational Awareness
SBG	Single Big Geofence
SDG	Shrinking Durational Geofence
sUAS	small Unmanned Aerial System
TBOV	Transit Based Operational Volumnes

TWCA	Triangle Weight Characterization with Adjacency
t_{wait}	Wait time until a geofence disappears
UAS	Unmanned Aircraft System
UTM	UAS Traffic Management
UAM	Urban Air Mobility
V_{UAS}	UAS flight speed
$\delta_{building}$	Safety buffer around a building
δ_{sb}	Total safety buffer
δ_{ROI}	Safety buffer of initial ROI
$\delta_{vehicle}$	Safety buffer of vehicle

References

1. Joshi, D. Drone Technology Uses and Applications for Commercial, Industrial and Military Drones in 2020 and the Future. December 2019. Available online: https://www.businessinsider.in/tech/news/drone-technology-uses-and-applications-for-commercial-industrial-and-military-drones-in-2020-and-the-future/articleshow/72874958.cms (accessed on 3 January 2021).
2. Doole, M.; Ellerbroek, J.; Hoekstra, J. Estimation of traffic density from drone-based delivery in very low level urban airspace. *J. Air Transp. Manag.* **2020**, *88*, 101862. [CrossRef]
3. Kopardekar, P.; Rios, J.; Prevot, T.; Johnson, M.; Jung, J.; Robinson, J.E. Unmanned aircraft system traffic management (UTM) concept of operations. In Proceedings of the 16th AIAA Aviation Technology, Integration, and Operations Conference, Washington, DC, USA, 13–17 June 2016; pp. 1–16.
4. Barrado, C.; Boyero, M.; Brucculeri, L.; Ferrara, G.; Hately, A.; Hullah, P.; Martin-Marrero, D.; Pastor, E.; Rushton, A.P.; Volkert, A. U-Space Concept of Operations: A Key Enabler for Opening Airspace to Emerging Low-Altitude Operations. *Aerospace* **2020**, *7*, 24. [CrossRef]
5. Samir Labib, N.; Danoy, G.; Musial, J.; Brust, M.R.; Bouvry, P. Internet of unmanned aerial vehicles—A multilayer low-altitude airspace model for distributed UAV traffic management. *Sensors* **2019**, *19*, 4779. [CrossRef] [PubMed]
6. Stevens, M.N.; Atkins, E.M. Generating Airspace Geofence Boundary Layers in Wind. *J. Aerosp. Inf. Syst.* **2020**, *17*, 113–124. [CrossRef]
7. Fu, Q.; Liang, X.; Zhang, J.; Qi, D.; Zhang, X. A Geofence Algorithm for Autonomous Flight Unmanned Aircraft System. In Proceedings of the IEEE 2019 International Conference on Communications, Information System and Computer Engineering (CISCE), Haikou, China, 5–7 July 2019; pp. 65–69.
8. Stevens, M.N.; Rastgoftar, H.; Atkins, E.M. Geofence boundary violation detection in 3D using triangle weight characterization with adjacency. *J. Intell. Robot. Syst.* **2019**, *95*, 239–250. [CrossRef]
9. Zhu, G.; Wei, P. Low-altitude uas traffic coordination with dynamic geofencing. In Proceedings of the 16th AIAA Aviation Technology, Integration, and Operations Conference, Washington, DC, USA, 13–17 June 2016; p. 3453.
10. Hermand, E.; Nguyen, T.W.; Hosseinzadeh, M.; Garone, E. Constrained control of UAVs in geofencing applications. In Proceedings of the IEEE 2018 26th Mediterranean Conference on Control and Automation (MED), Zadar, Croatia, 19–22 June 2018; pp. 217–222.
11. Johnson, M.; Jung, J.; Rios, J.; Mercer, J.; Homola, J.; Prevot, T.; Mulfinger, D.; Kopardekar, P. Flight test evaluation of an unmanned aircraft system traffic management (UTM) concept for multiple beyond-visual-line-of-sight operations. In Proceedings of the 12th USA/Europe Air Traffic Management Research and Development Seminar (ATM2017), Seattle, WA, USA, 26–30 June 2017.
12. Dill, E.T.; Young, S.D.; Hayhurst, K.J. SAFEGUARD: An assured safety net technology for UAS. In Proceedings of the 2016 IEEE/AIAA 35th Digital Avionics Systems Conference (DASC), Sacramento, CA, USA, 25–29 September 2016; pp. 1–10.
13. Kim, J.T.; Mathur, A.; Liberko, N.; Atkins, E. Volumization and Inverse Volumization for Low-Altitude Airspace Geofencing. In Proceedings of the AIAA AVIATION 2021 FORUM, Virtual, 2–6 August 2021; p. 2383.
14. Prevot, T.; Rios, J.; Kopardekar, P.; Robinson, J.E., III; Johnson, M.; Jung, J. UAS traffic management (UTM) concept of operations to safely enable low altitude flight operations. In Proceedings of the 16th AIAA Aviation Technology, Integration, and Operations Conference, Washington, DC, USA, 13–17 June 2016.
15. Sesar, J. *European Drones Outlook Study Unlocking the Value for Europe*; SESAR: Brussels, Belgium, 2016.
16. Sedov, L.; Polishchuk, V. Centralized and distributed UTM in layered airspace. In Proceedings of the 8th International Conference on Research in Air Transportation, Barcelona, Spain, 26–29 June 2018.
17. Jiang, T.; Geller, J.; Ni, D.; Collura, J. Unmanned Aircraft System traffic management: Concept of operation and system architecture. *Int. J. Transp. Sci. Technol.* **2016**, *5*, 123–135. [CrossRef]
18. Cho, J.; Yoon, Y. How to assess the capacity of urban airspace: A topological approach using keep-in and keep-out geofence. *Transp. Res. Part C Emerg. Technol.* **2018**, *92*, 137–149. [CrossRef]
19. Dasu, T.; Kanza, Y.; Srivastava, D. Geofences in the sky: Herding drones with blockchains and 5G. In Proceedings of the 26th ACM SIGSPATIAL International Conference on Advances in Geographic Information Systems, Seattle, WA, USA, 6–9 November 2018; pp. 73–76.

20. Yoon, H.; Chou, Y.; Chen, X.; Frew, E.; Sankaranarayanan, S. Predictive runtime monitoring for linear stochastic systems and applications to geofence enforcement for uavs. In *International Conference on Runtime Verification*; Springer: Cham, Switzerland, 2019; pp. 349–367.
21. Stevens, M.N.; Rastgoftar, H.; Atkins, E.M. Specification and evaluation of geofence boundary violation detection algorithms. In Proceedings of the IEEE 2017 International Conference on Unmanned Aircraft Systems (ICUAS), Miami, FL, USA, 13–16 June 2017; pp. 1588–1596.
22. Endsley, M.R. Design and evaluation for situation awareness enhancement. In Proceedings of the Human Factors Society Annual Meeting, Anaheim, CA, USA, 24–28 October 1988; Sage Publications: Los Angeles, CA, USA, 1988; Volume 32, pp. 97–101.
23. Endsley, M.R.; Rodgers, M.D. Situation awareness information requirements analysis for en route air traffic control. In *Proceedings of the Human Factors and Ergonomics Society Annual Meeting*; SAGE Publications: Los Angeles, CA, USA, 1994; Volume 38, pp. 71–75.
24. Verma, S.A.; Monheim, S.C.; Moolchandani, K.A.; Pradeep, P.; Cheng, A.W.; Thipphavong, D.P.; Dulchinos, V.L.; Arneson, H.; Lauderdale, T.A.; Bosson, C.S.; et al. Lessons learned: Using UTM paradigm for urban air mobility operations. In Proceedings of the 2020 AIAA/IEEE 39th Digital Avionics Systems Conference (DASC), San Antonio, TX, USA, 11–16 October 2020; pp. 1–10.
25. Stevens, M.N.; Atkins, E.M. Multi-mode guidance for an independent multicopter geofencing system. In Proceedings of the 16th AIAA Aviation Technology, Integration, and Operations Conference, Washington, DC, USA, 13–17 June 2016; p. 3150.
26. Weiler, K. Polygon comparison using a graph representation. In Proceedings of the 7th Annual Conference on Computer Graphics and Interactive Techniques, Seattle, WA, USA, 14–18 July 1980; pp. 10–18.
27. Sklansky, J. Finding the convex hull of a simple polygon. *Pattern Recognit. Lett.* **1982**, *1*, 79–83. [CrossRef]
28. Haines, E. Point in Polygon Strategies. *Graph. Gems* **1994**, *4*, 24–46.
29. Hormann, K.; Agathos, A. The point in polygon problem for arbitrary polygons. *Comput. Geom.* **2001**, *20*, 131–144. [CrossRef]
30. Duchoň, F.; Babinec, A.; Kajan, M.; Beňo, P.; Florek, M.; Fico, T.; Jurišica, L. Path planning with modified a star algorithm for a mobile robot. *Procedia Eng.* **2014**, *96*, 59–69. [CrossRef]
31. Stentz, A. Optimal and efficient path planning for partially known environments. In *Intelligent Unmanned Ground Vehicles*; Springer: Boston, MA, USA, 1997; pp. 203–220.
32. De Berg, M.; Van Kreveld, M.; Overmars, M.; Schwarzkopf, O. Computational geometry. In *Computational Geometry*; Springer: Berlin/Heidelberg, Germany, 1997; pp. 1–17.
33. Latombe, J.C. *Robot Motion Planning*; Springer Science & Business Media: New York, NY, USA, 2012; Volume 124.
34. Lingelbach, F. Path planning using probabilistic cell decomposition. In *ICRA'04, Proceedings of the IEEE International Conference on Robotics and Automation, New Orleans, LA, USA, 26 April–1 May 2004*; IEEE: New York, NY, USA, 2004; Volume 1, pp. 467–472.
35. Hwang, Y.K.; Ahuja, N. A potential field approach to path planning. *IEEE Trans. Robot. Autom.* **1992**, *8*, 23–32. [CrossRef]
36. Burns, B.; Brock, O. Sampling-based motion planning using predictive models. In Proceedings of the 2005 IEEE International Conference on Robotics and Automation, Barcelona, Spain, 18–22 April 2005; pp. 3120–3125.
37. Stevens, M.; Atkins, E. Geofence Definition and Deconfliction for UAS Traffic Management. *IEEE Trans. Intell. Transp. Syst.* **2021**, *22*, 5880–5889. [CrossRef]
38. Semechko, A. Decimate 2D Contours/Polygons. 2018. Available online: https://github.com/AntonSemechko/DecimatePoly (accessed on 3 January 2021).
39. Bondy, J.A.; Murty, U.S.R. *Graph Theory with Applications*; Macmillan: London, UK, 1976; Volume 290.
40. Huang, H.P.; Chung, S.Y. Dynamic visibility graph for path planning. In Proceedings of the 2004 IEEE/RSJ International Conference on Intelligent Robots and Systems (IROS) (IEEE Cat. No. 04CH37566), Sendai, Japan, 28 September–2 October 2004; Volume 3, pp. 2813–2818.
41. Haklay, M.; Weber, P. Openstreetmap: User-generated street maps. *IEEE Pervasive Comput.* **2008**, *7*, 12–18. [CrossRef]
42. Ochoa, C.A.; Atkins, E.M. Urban Metric Maps for Small Unmanned Aircraft Systems Motion Planning. *arXiv* **2021**, arXiv:2102.07218.
43. Kumar, M. World geodetic system 1984: A modern and accurate global reference frame. *Mar. Geod.* **1988**, *12*, 117–126. [CrossRef]
44. Mathur, A.; Atkins, E.M. Design, Modeling and Hybrid Control of a QuadPlane. In Proceedings of the AIAA Scitech 2021 Forum, Virtual, 11–22 January 2021; p. 374.
45. Hershberger, J.; Suri, S. An optimal algorithm for Euclidean shortest paths in the plane. *SIAM J. Comput.* **1999**, *28*, 2215–2256. [CrossRef]

applied sciences

MDPI

Article

Drone Ground Impact Footprints with Importance Sampling: Estimation and Sensitivity Analysis †

Jérôme Morio [1,‡], Baptiste Levasseur [2,‡] and Sylvain Bertrand [2,*,‡]

1 ONERA/DTIS, Université de Toulouse, F-31055 Toulouse, France; jerome.morio@onera.fr
2 ONERA/DTIS, Université Paris Saclay, CEDEX, F-91123 Palaiseau, France; baptiste.levasseur@onera.fr
* Correspondence: sylvain.bertrand@onera.fr

† This paper is an extended version of our paper published in IEEE Aerospace Conference held in Big Sky, MT, USA, 2–9 March 2019.

‡ These authors contributed equally to this work.

Abstract: This paper addresses the estimation of accurate extreme ground impact footprints and probabilistic maps due to a total loss of control of fixed-wing unmanned aerial vehicles after a main engine failure. In this paper, we focus on the ground impact footprints that contains 95%, 99% and 99.9% of the drone impacts. These regions are defined here with density minimum volume sets and may be estimated by Monte Carlo methods. As Monte Carlo approaches lead to an underestimation of extreme ground impact footprints, we consider in this article multiple importance sampling to evaluate them. Then, we perform a reliability oriented sensitivity analysis, to estimate the most influential uncertain parameters on the ground impact position. We show the results of these estimations on a realistic drone flight scenario.

Keywords: UAV; probabilistic maps of impact; ground footprints; Monte Carlo; importance sampling; sensitivity analysis

Citation: Morio, J.; Levasseur, B.; Bertrand, S. Drone Ground Impact Footprints with Importance Sampling: Estimation and Sensitivity Analysis. *Appl. Sci.* **2021**, *11*, 3871. https://doi.org/10.3390/app11093871

Academic Editor: Massimiliano Mattei

Received: 1 April 2021
Accepted: 20 April 2021
Published: 25 April 2021

1. Introduction

Assessing the risks and feasibility of unmanned aerial vehicle (UAV) operations for outdoor inspection or monitoring missions has become a major challenge for regulatory authorities and drone operators. This evaluation relies on risk analysis methods that can be helpful in the process of flight authorization, but also in the design and the preparation of the mission. Two main types of methods are classically used. The first one relies on the qualitative evaluation of risks by applying some predefined methodologies or guides [1]. This is, for example, the case of classical methods such as failure modes and effects analysis (FMEA) or, more recently and more specifically developed for UAV operations, SORA (specific operation risk assessment) [2]. The second type of methods relies on the quantitative evaluation of risks, based on the use of models developed to represent the UAV behavior, its environment, etc. This is the case of model-based probabilistic risk assessment (PRA) approaches that have recently gained a huge interest for UAVs, see e.g., [1–5]. With these approaches, the accuracy of models used for risk probabilities' computations is of paramount importance. Indeed, being too conservative may prevent or restrict some operational uses of UAVs, while not being conservative enough may lead to uses with uncontrolled risks. A fundamental keystone in these methods, when considering ground risk evaluation, is the computation of probabilities of impact of the UAV at ground level. Accurate models should be developed to be able to compute representative predictions of impact points' locations and probabilities. Works from the literature have focused on computing impact point locations, enabling to obtain estimates of impact footprints on the ground level. In [6], impact footprints are computed by reachability analysis, considering a gliding descent model for a fixed-wing UAV, composed of a turning and a straight line phase. Different modes of failure (engine, engine+rudder+ailerons) are considered, as well as the effect of the altitude of the vehicle at the failure instant. This type of impact footprints has also been obtained in [3], considering a 6 degrees-of-freedom dynamic model of a fixed-wing aircraft. The effect of wind on impact footprints has been investigated in [7]. Computation of impact locations and footprints has also been performed in [8], for both fixed-wing and multi-rotor UAVs considering different modes of failure. Some level sets are computed to provide some insights on the distribution of the impact points inside the footprint.

The generation of probability maps has been investigated in other works to provide more information on the impact distributions on the ground that could be useful and reduce conservatism in risk analysis or decision making. In [9], a ballistic model with drag force is considered to represent the descent of a fixed-wing UAV and generate probability density functions of impact points. Uncertainties on drag, initial speed at the instant of failure and external wind are accounted for. Full flight dynamics of a Cesna 182 aircraft are used in [10] to compute ground impact probability maps by Monte Carlo simulations. Total loss of power is assumed and uncertainties on the initial conditions of the UAV at failure instant as well as on the deflection of unactuated control surfaces are considered. A 6 degrees-of-freedom flight mechanics model is also used in [11] for a fixed wing UAV to estimate ground impact probability maps, taking into account the influence of wind direction and speed. Real flight data have been used to model uncertainties on the turn rate and flight path angle of the vehicles for cruise-like mode at constant altitude and straight line. These uncertainties along with the ones on the actuators deflections at the instant of failure are used in the Monte Carlo process. Influence of initial altitude, speed and wind (speed and direction) are analyzed, and a full data basis has been obtained containing impact probability maps for a sampled set of values for these quantities. This data basis can be useful for risk evaluation along a given UAV flight trajectory, e.g., for mission preparation.

Since Monte Carlo simulations can be time-consuming, more recent works have been dedicated to the development of surrogate models for the generation of ground impact probability maps. *K*-Nearest neighbors models have been considered in [10] to approximate impact probability distribution. Other techniques such as Krigging have been investigated [7] regarding impact footprints or neural networks for both generation of impact footprints [7] and probability maps [11].

In all these works, assumptions are made on the uncertainties on the variables used as inputs of the computations. Uncertainty representations are mainly based on statistic models (probability distributions) and/or bounds (intervals with no statistical assumptions). Accuracy of the resulting outputs (ground impact locations, footprints, probability maps) strongly relies on the representativeness of these assumptions. Another important aspect in these approaches is the computation method itself and the choice of its hyper-parameters. For example, choice of simulation budgets in Monte Carlo approaches is crucial, as it may strongly influence the probability density estimation and its confidence.

Moreover, Monte Carlo methods with a low number of samples lead to an underestimation of extreme ground impact footprints, which may be of interest to provide more confidence in the risk assessment process for flight preparation and authorization. Knowledge of UAV's extreme fallout zones can also help defining safety levels at very low thresholds, which can be critical for certain high-risk infrastructures.

This paper therefore addresses the estimation problem of accurate extreme ground impact probability maps and footprints containing 95%, 99% and 99.9% of the impacts. Multiple importance sampling (MIS) is considered to estimate density minimum volume sets associated with these extreme quantiles.

In addition, a study of the sensitivity of hazard parameters is proposed to estimate the most influential uncertain parameters on ground impact positions. This analysis may enable both operators and drone constructors to better understand, design and anticipate fallout zones in the event of a failure. All the results in this paper are obtained for the case of a fixed wing UAV after main engine failure.

This paper is organized as follows. The first section focuses on the simulation approach used to compute the ground impact points coordinates. The characterization of uncertainties that are taken into account is discussed in Section 3. The estimation of ground footprints by Multiple Importance Sampling is then presented in Section 4 and an associated sensitivity analysis is proposed in Section 5.

2. Ground Impact Simulation

In this paper, we focus on impacts on the ground due to a loss of control of the UAV (unmanned aerial vehicle) after a main engine failure. It is assumed that immediately after the failure, the engine thrust becomes zero and the control surfaces remain stuck in their equilibrium positions. The objective of this section is to present the models and approach that are used to compute the impact points at ground by simulation, based on previous studies by the authors in [7].

2.1. UAV Dynamics

The model used to simulate the trajectory of the UAV to the ground is a six degrees of freedom (6DOF) dynamic model, including full flight mechanics, and hence enabling one to incorporate the influence of wind from a dynamical point of view (and not kinematical compared to some approaches developed in the literature [12]). The model considered here is a fixed wing aircraft such as the one presented in [13]. The control input vector $u = \begin{bmatrix}\delta_a & \delta_e & \delta_r & \delta_T\end{bmatrix}^\top$ is composed of ailerons, elevators, rudder deflections, and thrust command. The state of the dynamical system to be simulated is defined as $\chi = \begin{bmatrix}X^\top & V^\top & \eta^\top & \Omega^\top\end{bmatrix}^\top$ where $X = \begin{bmatrix}x & y & z\end{bmatrix}^\top$ is the position vector defined in a local NED (north east down) frame, V and Ω are the translation and angular velocity vectors in the aircraft body-frame, and $\eta = \begin{bmatrix}\phi & \theta & \psi\end{bmatrix}^\top$ is the vector of Euler angles (roll-pitch-yaw) describing the attitude of the UAV. The origin of the (inertial) local NED frame is chosen to correspond to z=0 (ground) and is arbitrarily chosen for x and y-components, as we are only interested in the description of the motion of UAV during its descent to the ground with respect to the vehicle position at the instant of failure (considered to be $x = y = 0$).

The rigid-body dynamics of the UAV is described as

$$\begin{cases} \dot{X} = R_\eta V \\ \dot{V} = -\Omega \times V + \frac{1}{m}F \\ \dot{\eta} = T_\eta \Omega \\ \dot{\Omega} = J^{-1}(-\Omega \times J\Omega + M) \end{cases} \tag{1}$$

where R_η is the orientation matrix parametrized in terms of Z-Y-X Euler angles given by

$$R_\eta = \begin{bmatrix} c_\theta c_\psi & s_\phi s_\theta c_\psi - s_\psi c_\psi & s_\phi s_\psi + s_\theta c_\phi c_\psi \\ c_\theta s_\psi & c_\phi c_\psi + s_\theta s_\phi s_\psi & s_\theta c_\phi s_\psi - s_\phi c_\psi \\ -s_\theta & c_\theta s_\phi & c_\theta c_\phi \end{bmatrix} \tag{2}$$

and T_η is the transformation matrix defined by

$$T_\eta = \begin{bmatrix} 1 & s_\phi t_\theta & c_\phi t_\theta \\ 0 & c_\phi & -s_\phi \\ 0 & s_\phi / c_\theta & c_\phi / c_\theta \end{bmatrix} \tag{3}$$

whith the notations $c_\alpha = \cos(\alpha)$, $s_\alpha = \sin(\alpha)$ and $t_\alpha = \tan(\alpha)$ for any given angle α. The inertia matrix of the UAV is denoted by J and its mass by m. Values used for the UAV parameters are given in Appendix A.

The resulting force F expressed in the aircraft body-frame

$$F = F_{eng}(\delta_T) + F_g(\eta) + F_a(V, V_w, \eta, \Omega, \delta_a, \delta_e, \delta_r) \tag{4}$$

is composed of the thrust, due to the engine ($F_{eng}(\delta_T)$) which is zero after engine failure, the gravity force ($F_g(\eta)$) and the resulting aerodynamic force F_a, which depends on the true air speed of the UAV and then on the wind speed vector V_w. To compute this force, a full aerodynamic model is used, such as the one described in [13] and is presented in Appendix A. Similarly, the resulting torque expressed in the aircraft body-frame

$$M = M_{eng}(\delta_T) + M_a(V, V_w, \eta, \Omega, \delta_a, \delta_e, \delta_r) \quad (5)$$

is composed of the torque component $M_{eng}(\delta_T)$, due to the thrust of the main engine (equals to zero after engine failure) and the aerodynamic torque M_a, which also depends on the wind speed vector V_w. To compute this torque, a full aerodynamic model of the aircraft is also considered (see Appendix A). The dynamic model of the UAV can be summarized with the following state-space representation

$$\dot{\chi} = f(\chi, u, V_w) \quad (6)$$

The simulation of the UAV descent trajectory is performed from an initial condition χ_0, defined at the engine failure instant t_0, to ground impact, that corresponds to instant t_f such that the altitude $h(t_f) = -z(t_f) = 0$.

During a steady flight (coordinated turn, straight flight, pull-up/pull-over etc.), the accessible space through the initial condition (χ_0, u_0) is considerably reduced. Defining the initial condition then consists of zeroing numerically the dynamic part of Equation (6), while simultaneously considering kinematic constraints related to flight mode [14]. In this case, the control vectors and the dynamic part of the state vector are entirely defined by these constraints. This method is called trim algorithm. A simple way to represent a trajectory is to consider the two following parameters:

- the turn rate $R = d\psi/dt$, where ψ is the heading angle
- the flight path angle $\gamma = \dot{z}/V_a$, where V_a is the aerodynamic speed of the aircraft.

Therefore, the trim algorithm can be run to determine the initial condition (χ_0, u_0), by assigning values R_0, γ_0, V_{a_0} and h_0 to the turn rate, flight path angle, aerodynamic speed and altitude, which are representative of the UAV flight conditions. A straight cruise flight mode at constant altitude can, for example, be considered by choosing $R_0 = 0$ and $\gamma_0 = 0$. Note that the wind is not considered in the trim algorithm.

From this initial condition (χ_0, u_0), the trajectory of the UAV is simulated until the impact time t_f, by considering zero thrust (main engine failure) and taking into account the wind speed V_w. The complete simulation process is represented in Figure 1.

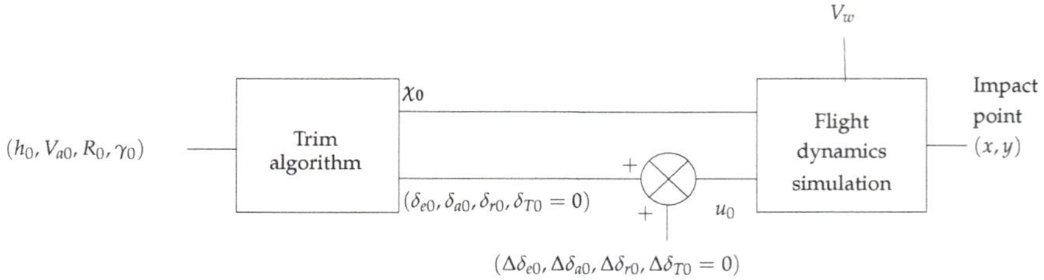

Figure 1. Simulation flowchart used to generate ground impact points.

Using this approach, one can simulate a single trajectory and compute the coordinates of the ground impact point. An example of the trajectory is provided on Figure 2, corresponding to $\gamma_0 = 0$ deg, $R_0 = 0.15$ rad/s, $V_{a_0} = 30$ m/s, $h_0 = 150$ m and no wind.

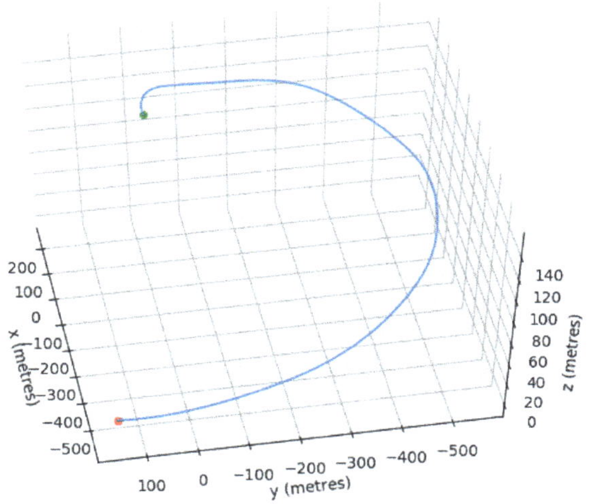

Figure 2. Example of simulated descent trajectory to ground (initial point in green corresponding to main engine failure instant, ground impact point in red) [7].

Nevertheless, impact point coordinates are not deterministic in the sense that their computation will suffer from different sources of uncertainties on the parameters of the problem. The next section covers the characterization of these uncertainties, as well as the Monte Carlo approach to handle them.

3. Uncertainties and Monte Carlo approach

The computation of an impact point involves a simulation relying on several parameters. Some of them will be considered as fixed values, such as the initial ground altitude h_0 and aerodynamic speed V_{a_0}. Note that these quantities are affected by some uncertainties, since the reference altitude and velocities commanded for the UAV are not exactly flown in practice. However, the influence of their respective incertitude levels on the impact point location is negligible. Uncertainties on the parameters of the UAV model (aerodynamic coefficients) are also not taken into account in this paper. A robustness analysis with regard to them should be carried out, especially since the descent phase to ground may lead to aerodynamic behaviors different from the ones that can be identified. This is beyond the scope of this paper and will be considered in future work. Uncertainties taken into account in this article are described in the following subsections.

3.1. Uncertainties on R_0 and γ_0

Experimental data have been recorded on flights realized by Altametris, the drone subsidiary of SNCF Réseau (French Railway Network) (see [7]). These data correspond to a cruise-like flight mode in a straight line and constant ground altitude. This is the flight mode of interest for the study in this paper. For this flight mode, R and γ should be zero, which is not the case in practice.

For simplicity reason, a bi-variate normal distribution has been fitted on these experimental data, after rejection of the outliers (see [7]). Its mean is given by $\mu = \begin{bmatrix}\mu_R & \mu_\gamma\end{bmatrix}^T$ with $\mu_R = 7.47e-5$ rad/s and $\mu_\gamma = 1.03e-1$ deg and its covariance matrix by:

$$\Sigma = \begin{bmatrix} 7.40e-4 & 1.75e-3 \\ 1.75e-3 & 8.53e-1 \end{bmatrix} \tag{7}$$

This distribution will be used to sample points for R_0 and γ_0

3.2. Uncertainties on Control Surface Deflections

As previously mentioned, a main engine failure is considered in this paper. A constant zero thrust command is therefore assumed for the descent trajectory simulation, that is $\delta_T(t) = \delta_{T_0} = 0, \forall t \in [t_0, t_f]$.

For the deflection of the control surfaces (elevators, ailerons and rudder), it is also assumed that they remain stuck in their trim position $(\delta_{e_0}, \delta_{a_0}, \delta_{r_0})$ during the descent trajectory. Some noise $\Delta\delta_{i_0}, i \in \{e, a, r\}$, is nevertheless added to these trim values, since in practice, a flapping behavior of these control surfaces has been observed on the UAV. It is defined as a zero-mean Gaussian noise of variance $\sigma_i^2 = \rho_i/30$, where ρ_i stands for the amplitude range of the control surface i. The coefficient 30 has been arbitrarily chosen, but to define a variance small enough to make the new initial condition $(\chi_{d_0}, u_0 + \Delta u_0)$ not to deviate too much from the computed trim point (χ_{d_0}, u_0), which is an equilibrium point for the UAV dynamics. The notation Δu_0 is used to define the noise vector $\begin{bmatrix}\Delta\delta_{e_0} & \Delta\delta_{a_0} & \Delta\delta_{r_0} & 0\end{bmatrix}^\top$.

3.3. Monte Carlo Simulations

Let us bring in the same vector U, the 5 uncertain variables R_0, γ_0 and $\Delta\delta_{i_0}, i \in \{e, a, r\}$, with joint density $f : \mathbb{R}^5 \to \mathbb{R}^+$ with respect to Lebesgue measure. The computation of the impact points is then done with the deterministic process described in Section 2.1 synthesized by a scalar continuous function $\mathcal{M} : \mathbb{R}^5 \to \mathbb{R}^2$. The impact position vector Z is such that $Z = [x, y] = \mathcal{M}(U)$. As U is a random vector, Z is also a random vector of unknown density $g : \mathbb{R}^2 \to \mathbb{R}^+$ with respect to the Lebesgue measure. If we consider N independent and identically distributed (iid) samples U_i, $i = 1..N$, with density f, we can generate N iid samples Z_i of density g thanks to $\mathcal{M}$. Figure 3 shows, for instance, 2000 iid samples of Z_i depending on the tuning of the wind in the function $\mathcal{M}$.

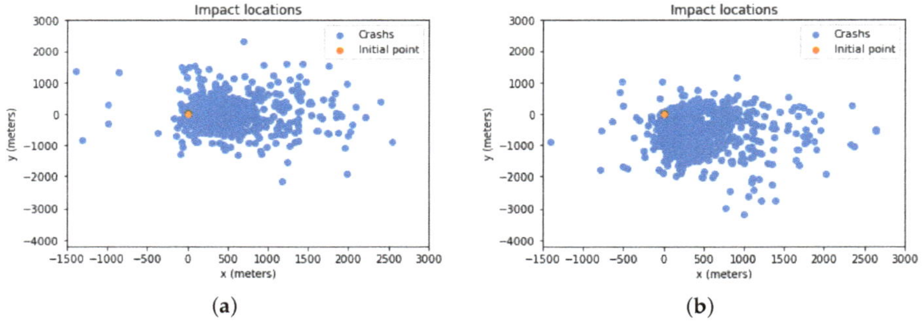

Figure 3. (**a**) 2000 Monte Carlo ground impact points with no wind (**b**) 2000 Monte Carlo ground impact points with a wind of 5 m.s $^{-1}$ with angle $-90°$.

4. Density Minimum Volume Set Estimation for the Analysis of Ground Impact Footprints

A volume set is a mathematical tool that enables one to analyse the density of drone ground impacts. In this section, we describe how to define a multidimensional density minimum volume set and how to estimate them in practice, with multiple importance sampling to focus on rare events.

4.1. Definition of a Density Minimum Volume Set

The t-level set $\mathbf{L}(t)$ of the multivariate probability density g of Z is defined as follows:

$$\mathbf{L}(t) = \left\{ z \in \mathbb{R}^2 : g(z) \geq t \right\} \tag{8}$$

for $t \geq 0$. The level sets of the probability density function (PDF) g are defined as the mapping ϵ:

$$\begin{aligned}\epsilon : [0, \sup g] &\rightarrow [0,1] \\ t &\rightarrow \int_{\mathbf{L}(t)} g(z)dz = P(Z \in \mathbf{L}(t)) = P(g(Z) \geq t) = \alpha\end{aligned}$$

The t-level set $\mathbf{L}(t)$ of density g is the minimum volume set of probability α under regularity conditions [15]. A density level set can be viewed in fact as a multidimensional α-quantile estimation.

$$V(t) = \inf_{A \in \mathbb{R}^r} \{\lambda(A) : P(A) \geq \alpha\},\ \alpha \in [0,1] \tag{9}$$

where A is a subset of $\mathbb{R}^r$ and λ is a real-valued function defined on A. If λ is the Lebesgue measure, V is a minimum volume set of probability α.

4.2. Statistical Estimation of a Density Minimum Volume Set with MIS

In this article, we want to estimate the t-level set $\mathbf{L}(t)$ of density g for a given probability α. The estimation principle is based on the following steps:

1. Propose an estimate $\hat{g}$ of g from a given set of samples $(Z_1 = \mathcal{M}(U_1),...,Z_N = \mathcal{M}(U_N)$.
2. Estimate the threshold $\hat{t} = \epsilon^{-1}(\alpha)$ with a simple binary search and determine the level set
$$\mathbf{L}(\hat{t}) = \left\{z \in \mathbb{R}^2 : \hat{g}(z) \geq \hat{t}\right\} \tag{10}$$

This estimator $\mathbf{L}(\hat{t})$ is a plug-in estimator of a minimum volume set [16]. To apply this 2-step procedure, it is necessary first to estimate the unknown density g. This can be done with classical Monte Carlo from samples Z_i, $i = 1, \ldots, N$, distributed with the unknown density g, but also with importance sampling. The principle of importance sampling is to modify the sampling distribution of Z_i, in order to improve the accuracy of the estimation of g on some part of its support. A comparison between Monte Carlo and classical importance sampling estimates of g is indeed performed in [17]. Depending on the value of α, a trade-off should be made. For this purpose, we consider in this article multiple importance sampling [18] that behaves well in the heart of the distribution g, because Monte Carlo and importance sampling samples can be taken into account in the estimation of the density g. Moreover, the estimation of $\hat{t}$ with binary search is often intractable and cannot be applied in practice, since it requires the estimation of integrals over a multidimensional domain. To avoid this difficulty, one also considers another plug-in estimator of t described in [19], based on density quantile.

To estimate the density level set $\mathbf{L}(\hat{t})$ with multiple importance sampling, the following computational steps are considered in this article:

1. Generate a set of N independent and identically (iid) distributed samples $(U_{11}, ..., U_{1N})$ of density f, and apply the function $\mathcal{M}$ on these samples to determine a set of samples $(Z_{11} = \mathcal{M}(U_{11}), ..., Z_{1N} = \mathcal{M}(U_{1N}))$.
2. Estimate the output density $\hat{g}^1$ from the samples $(Z_{11}, ..., Z_{1N})$ with multivariate kernel density estimate [20].
3. Estimate the density h of the samples $\{U_{1i} | \hat{g}^1(Z_{1i}) < \gamma\}$ for $i = 1, \ldots, N$ where γ is set by the user.
4. Generate a set of N iid samples $(U_{21}, ..., U_{2N})$ from density h, and applies the function $\mathcal{M}$ on these samples to determine a set of samples $(Z_{21}, ..., Z_{2N})$ [20].
5. Estimate the density $\hat{g}^{MIS}$ from the samples $(Z_{11}, ..., Z_{1N})$ and $(Z_{21}, ..., Z_{2N})$ with weighted multivariate kernel density estimate. The weight of each Z_{ij} is $\frac{f(U_{ij})}{\frac{1}{2}(f(U_{ij}+h(U_{ij}))}$.

6. Estimate the threshold $\hat{t}^{MIS}$ as the $(1-\alpha)$-quantile of the weighted samples $(Z_{11}, ..., Z_{1N}, Z_{21}, ..., Z_{2N})$.
7. The level set with MIS is then estimated with

$$\hat{\mathbf{L}}^{MIS}(t) = \left\{ \tilde{x} \in \mathbb{R}^2 : \hat{g}^{MIS}(z) \geq \hat{t}^{MIS} \right\} \tag{11}$$

The choice of γ can be made with a quantile of the samples $\hat{g}^1(Z_{1i}), ..., \hat{g}^1(Z_{1N})$. In this article, γ is quantile of level 0.1 as, from our experience, it corresponds to a good trade-off between Monte Carlo samples and extreme samples.

4.3. Application to Drone Ground Impact

An MIS algorithm for density minimum level set has been applied to the estimation of drone ground impacts with $N = 1000$. In Figure 4, we present the estimation of minimum density volume set for different probabilities α with MIS. Importance sampling with density h has consequently increased the frequency of impacts with a high distance from the aim without requiring a large number of simulations, and thus extreme level sets are more accurate. A similar analysis is also performed in Figure 5, when a wind of 5 m.s $^{-1}$ with angle $-90°$ is considered in the drone ground impact simulations.

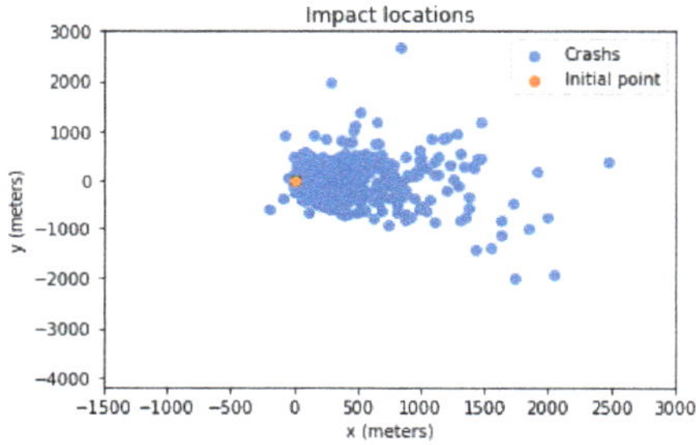

(a)

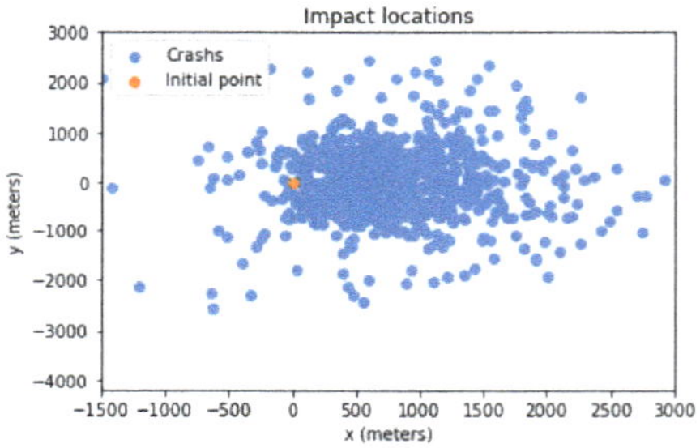

(b)

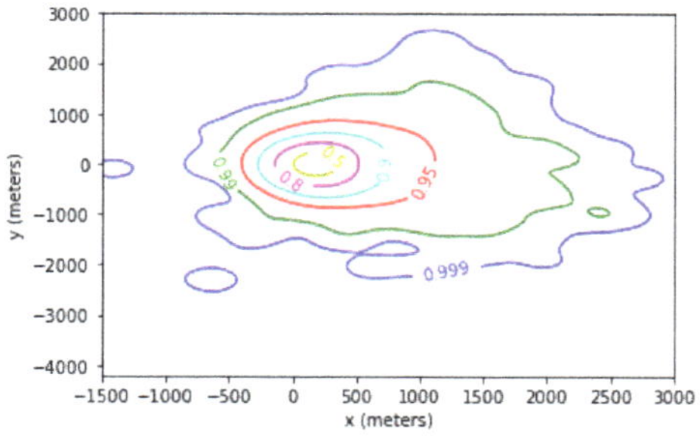

(c)

Figure 4. (**a**) 1000 Samples generated with the density f of MIS (**b**) 1000 Samples generated with the density h of MIS (**c**) Minimum density volume set estimation for different probability values α (0.5; 0.8; 0.9; 0.95; 0.99; 0.999) with MIS.

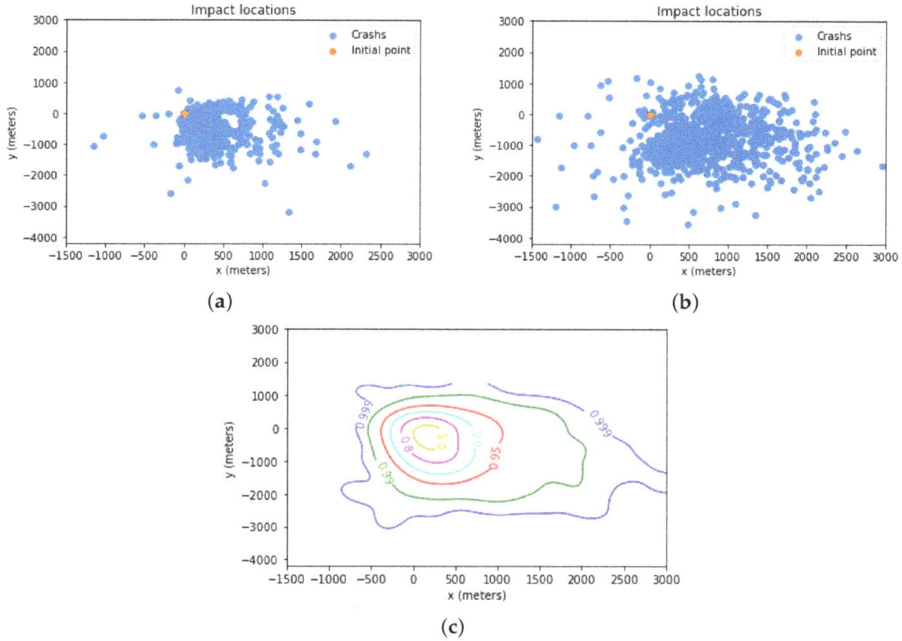

Figure 5. (**a**) 1000 Samples generated with the density f of MIS (**b**) 1000 Samples generated with the density h of MIS (**c**) Minimum density volume set estimation for different probability values α (0.5; 0.8; 0.9; 0.95; 0.99; 0.999) with MIS (wind of 5 m.s $^{-1}$ with angle $-90°$).

5. Reliability-Oriented Sensitivity Analysis

5.1. Definition of ROSA Sensitivity Indices

Reliability-oriented sensitivity analysis (ROSA) differs from classical sensitivity analysis in the nature of the output quantity of interest under study. Indeed, sensitivity analysis focuses on the model output, whereas ROSA analyses the impact of the input uncertainty on a reliability measure. Two kinds of ROSA indices can be computed in practice [21]:

- first, target sensitivity analysis evaluates the impact of inputs over a function of the output, typically the indicator function of a critical domain. In the drone impact application, it answers the question: which uncertain inputs lead to extreme drone impact?
- second, conditional sensitivity analysis, which aims at studying the impact of inputs exclusively within the critical domain, namely, conditionally to the failure event. In the drone application, this indice determines, conditionally to an extreme impact, which uncertain inputs are the most influential.

In this article, we consider two recent target and conditional ROSA moment-independent indices $\bar{\eta}_i$ and δ_i^f [22] to analyse the influence of the ith component of U, $U^{(i)}$, on the scalar output quantity $\tilde{Z} = ||Z||_2$ for a given failure event. We propose to define here the failure event as $Z \notin \mathbf{L}(t)$, that is, the ground impact is outside a given volume set and is thus an extreme impact. The two ROSA indices are defined by the following equations for the proposed drone fallout test case with $i = 1, \ldots, 5$ as there are 5 random inputs:

$$\bar{\eta}_i = \frac{1}{2}\left\| f_{U^{(i)}} - f_{U^{(i)}|Z\notin\mathbf{L}(t)} \right\|_{L^1(\mathbb{R})}. \tag{12}$$

and
$$\delta_i^f = \frac{1}{2}\left\| f_{(U^{(i)},\tilde{Z})|Z\notin\mathbf{L}(t)} - f_{U^{(i)}|Z\notin\mathbf{L}(t)} f_{\tilde{Z}|Z\notin\mathbf{L}(t)} \right\|_{L^1(\mathbb{R}^2)} = \frac{1}{2}\|c_i - 1\|_{L^1(\mathbb{R}^2)}. \quad (13)$$

where $f_{U^{(i)}}$ is the density of $U^{(i)}$, $f_{U^{(i)}|Z\notin\mathbf{L}(t)}$ is the density of $U^{(i)}$ conditionally to $Z \notin \mathbf{L}(t)$, $f_{\tilde{Z}|Z\notin\mathbf{L}(t)}$ is the density of $\tilde{Z}$ conditionally to $Z \notin \mathbf{L}(t)$, $f_{(U^{(i)},\tilde{Z})|Z\notin\mathbf{L}(t)}$ is the density of the couple $(U^{(i)}, \tilde{Z})$ conditionally to $Z \notin \mathbf{L}(t)$ and finally c_i the copula density $(U^{(i)}, \tilde{Z})$ conditionally to $Z \notin \mathbf{L}(t)$. The ROSA indices $\bar{\eta}_i$ and δ_i^f take values in $[0,1]$ where the low values of these indices mean this i^{th} component of U is not influential on the failure event analysis and conversely. The computation of these indices can be done with the samples generated for density level set estimation (see Section 4) and thus requires no additional calls to $\mathcal{M}$. Moreover, this methodology can be applied even if the random inputs U are dependent contrary to ROSA variance based indices [23].

5.2. Statistical Estimation of ROSA Sensitivity Indices

To practically estimate the ROSA indices $\bar{\eta}_i$ and δ_i^f, the following steps are required [22]:

1. Obtain $(V_1, \ldots, V_n)$ approximately i.i.d. from $f_{U|Z\notin\mathbf{L}(t)}$ and their corresponding value $Z_k = \mathcal{M}(V_k)$ by $\mathcal{M}$. From Z_k, the value of $\tilde{Z}_k$ is then easily computed with $\tilde{Z}_k = ||Z_k||_2$.
2. Use the sample $((V_k^{(i)}, \tilde{Z}_k), k = 1, \ldots, n, i = 1, \ldots 5)$ where $V_k^{(i)}$ is the ith component of V_k, to get estimates $\hat{f}_{U^{(i)}|Z\notin\mathbf{L}(t)}$ and $\hat{c}_i$ of the density $f_{U^{(i)}|Z\notin\mathbf{L}(t)}$ and of the copula density c_i respectively. In this article, they are both estimated with the non-parametric method [20,24], but any other efficient density and copula estimation techniques can be chosen.
3. Use the estimates $\hat{f}_{U^{(i)}|Z\notin\mathbf{L}(t)}$ and $\hat{c}_i$ to compute $\bar{\eta}_i$ and δ_i^f as follows:
   - for $\bar{\eta}_i$, estimate the one-dimensional integral $\|f_{U^{(i)}|Z\notin\mathbf{L}(t)} - f_{U^{(i)}}\|_{L^1(\mathbb{R})}$ either by direct numerical approximation, or if $f_{U^{(i)}}$ can be sampled from, by Monte Carlo method via
   $$\hat{\bar{\eta}}_i = \frac{1}{N'} \sum_{k=1}^{N'} \left| \frac{\hat{f}_{U^{(i)}|Z\notin\mathbf{L}(t)}(U_k^{(i)})}{f_{U^{(i)}}(U_k^{(i)})} - 1 \right| \quad (14)$$
   where the $U_k^{(i)}$ are i.i.d. with common distribution $f_{U^{(i)}}$;
   - for δ_i^f, generate $((H_{1k}, H_{2k}), k = 1, \ldots, N')$ i.i.d. uniformly distributed on $[0,1]^2$ and estimate δ_i^f by
   $$\hat{\delta}_i^f = \frac{1}{2N'} \sum_{k=1}^{N'} |\hat{c}_i(H_{1k}, H_{2k}) - 1| \,. \quad (15)$$

The estimates $\hat{\bar{\eta}}_i$ $\hat{\delta}_i^f$ can be computed for different failure events $Z \notin \mathbf{L}(t)$ for different values of $t = t_\alpha$ that correspond to several minimum volume sets of probability α. N' can be taken as large as possible, as it does not imply any calls to $\mathcal{M}$ and thus we set to $N' = 10^4$.

5.3. Application to Drone Ground Impact Sensitivity Analysis

The algorithm proposed in the previous section has been applied with MIS samples and thus without any supplementary calls to $\mathcal{M}$ to determine the influence on the reachability of extreme drone impacts of the different components of the random vector U. The ROSA indices are computed in Table 1 for three different level sets of probability $\alpha = 0.5, 0.8, 0.99$. The most influential variables are the third and fourth components of U, that is, the noise uncertainty on the UAV elevators and ailerons. The positions of the drone ground impact are less sensitive to the other uncertain simulation parameters in the heart of the impact position distribution. Nevertheless, when we consider more extreme impacts ($t_{0.99}$), these observations have to be mitigated. A parameter alone does not explain an extreme fallout, as the ROSA indices decrease for $U^{(3)}$ and $U^{(4)}$. A combination of parameters leads to an extreme fallout. The comparison between $\hat{\eta}_i$ $\hat{\delta}_i^f$ here is not really relevant and does not provide much information. The sensitivity analysis gives similar results when wind is taken into account in the simulation.

Table 1. ROSA indices for ground impact analysis. Bold numbers correspond to values greater than 0.1.

(a) No Wind			
ROSA indices	$t_{0.5}$	$t_{0.8}$	$t_{0.99}$
$\hat{\eta}_1$	0.07	0.06	0.06
$\hat{\delta}_1^f$	0.04	0.04	0.03
$\hat{\eta}_2$	0.05	0.03	0.03
$\hat{\delta}_2^f$	0.04	0.03	0.03
$\hat{\eta}_3$	**0.34**	**0.31**	**0.11**
$\hat{\delta}_3^f$	0.07	0.07	**0.15**
$\hat{\eta}_4$	**0.41**	**0.17**	0.06
$\hat{\delta}_4^f$	**0.16**	**0.18**	**0.16**
$\hat{\eta}_5$	0.06	0.04	0.04
$\hat{\delta}_5^f$	0.06	0.04	0.04
(b) Wind of 5 m.s $^{-1}$ **with angle** $-90°$.			
ROSA indices	$t_{0.5}$	$t_{0.8}$	$t_{0.99}$
$\hat{\eta}_1$	0.06	0.07	0.07
$\hat{\delta}_1^f$	0.05	0.03	0.03
$\hat{\eta}_2$	**0.13**	0.08	0.06
$\hat{\delta}_2^f$	0.02	0.04	0.03
$\hat{\eta}_3$	**0.40**	**0.24**	0.05
$\hat{\delta}_3^f$	0.16	**0.22**	**0.23**
$\hat{\eta}_4$	**0.34**	**0.13**	0.06
$\hat{\delta}_4^f$	**0.30**	**0.24**	**0.21**
$\hat{\eta}_5$	**0.10**	0.06	0.05
$\hat{\delta}_5^f$	0.04	0.03	0.03

6. Conclusions

The generation of extreme ground impact footprints map has been addressed in this paper for fixed-wing UAVs failure. In the proposed approach, the computation of impact points is based on simulation of a full dynamic model, including aerodynamics of the UAV and wind effect. Uncertainties accounted for in these simulations have been characterized, based on some real flight data. Monte Carlo simulations have been performed to generate footprints; however, it is not satisfying when we focus on extreme ground footprints. For this purpose, we have presented a rare-event simulation technique called multiple importance sampling to answer the issue of extreme drone ground impacts. We also show that at low computational cost, it is also possible to derive sensitivity indices to interpret the cause of extreme impacts.

Future work will include these characterizations of extreme drone impacts for the risk analysis of UAV missions. Sensitivity and robustness analysis with regard to uncertainties on some parameters of the UAV (aerodynamic coefficients) will also be considered.

Author Contributions: Conceptualization, J.M., B.L. and S.B.; methodology, J.M.; software, B.L.; validation, J.M., B.L. and S.B.; manuscript writing, J.M., B.L. and S.B. All authors have read and agreed to the published version of the manuscript.

Funding: This work has been supported by French DGAC in the context of the research partnership PHYDIAS with ONERA for safety improvement of UAVs.

Conflicts of Interest: The authors declare no conflict of interest.

Appendix A. UAV Model

The aerodynamic force F_a in (4) and torque M_a in (5) can be expressed in the aerodynamic reference frame (related to the aerodynamic speed of the drone) as:

$$F_a^{(w)} = q_d S \begin{bmatrix} C_X & C_Y & C_Z \end{bmatrix}^\top \tag{A1}$$

$$M_a^{(w)} = q_d S L \begin{bmatrix} C_l & C_m & C_n \end{bmatrix}^\top \tag{A2}$$

where $q_d = \frac{1}{2}\rho V_a^2$ is the dynamic pressure, ρ the air density, V_a the airspeed, S the reference surface and $L = diag(L_a, L_o, L_a)$ a matrix with lateral and longitudinal reference lengths L_a (wingspan) and L_o (mean aerodynamic chord).

In case of a non-zero wind speed vector V_w, the airspeed is $V_a = \|V - V_w\|$.

A linearized aerodynamic model is used in this paper, where the lift (C_L), lateral (C_Y) and drag (C_D) coefficients are computed by

$$\begin{aligned} C_L &= C_{L_0} + C_{L_\alpha}\alpha + C_{L_{\dot\alpha}}\dot\alpha + C_{L_q}\frac{q}{V_a} + C_{L_{\delta_e}}\delta_e \\ C_Y &= C_{Y_\beta} + C_{Y_P}\frac{p}{V_a} + C_{Y_r}\frac{r}{V_a} + C_{Y_{\delta_a}}\delta_a + C_{Y_{\delta_r}}\delta_r \\ C_D &= C_{D_0} + C_{D_{C_L}} C_L + C_{D_{C_{L_2}}} C_L^2 + C_{D_{\delta_e}}\delta_e \end{aligned} \tag{A3}$$

with $\begin{bmatrix} C_X & C_Y & C_Z \end{bmatrix}^\top = \begin{bmatrix} -C_D & C_Y & -C_L \end{bmatrix}^\top$.
Similarly, the aerodynamic coefficients regarding the torque are computed according to the following linearized model

$$\begin{aligned} C_l &= C_{l_\beta}\beta + \frac{L_a}{V_a}(C_{l_p}p + C_{l_r}r) + C_{l_{\delta_a}}\delta_a + C_{l_{\delta_r}}\delta_r \\ C_m &= C_{m_0} + C_{m_\alpha}\alpha + C_{m_{\dot\alpha}}\dot\alpha + \frac{L_0}{V_a}C_{m_q}q + C_{m_{\delta_e}}\delta_e \\ C_n &= C_{n_\beta}\beta + \frac{L_a}{V_a}(C_{n_p}p + C_{n_r}r) + C_{l_{\delta_a}}\delta_a + C_{l_{\delta_r}}\delta_r \end{aligned} \tag{A4}$$

with α the angle of attack, β the slideslip angle, $\Omega = [p, q, r]^T$ the angular velocity vector between the NED and body frames and $(\delta_a, \delta_e, \delta_r)$ the ailerons, elevators and rudder deflections.

Numerical values of the aerodynamic coefficients and other UAV model parameters used in this paper are given in Tables A1 and A2 below.

Table A1. Numerical values of aerodynamic coefficients.

C_{L_0}	0.3243	C_{l_β}	−0.0113
C_{L_α}	6.0204	C_{l_p}	−1.2217
$C_{L_{\dot\alpha}}$	1.93	C_{l_r}	0.015
C_{L_q}	6.0713	$C_{l_{\delta_a}}$	0.3436
$C_{L_{\delta_e}}$	0.9128	$C_{l_{\delta_r}}$	0.0076
C_{Y_β}	−0.3928	C_{m_0}	0.0272
C_{Y_p}	0	C_{m_α}	−1.9554
C_{Y_r}	0	C_{m_q}	−5.286
$C_{Y_{\delta_r}}$	0.1982	$C_{m_{\delta_e}}$	−2.4808
C_{D_0}	0.0251	C_{n_β}	0.0804
$C_{D_{C_L}}$	−0.0241	C_{n_p}	−0.0557
$C_{D_{C_{L_2}}}$	0.0692	C_{n_r}	−0.1422
$C_{D_{\delta_e}}$	0.1	$C_{n_{\delta_a}}$	−0.0165
$C_{n_{\delta_r}}$	−0.0598		

Table A2. UAV model parameters.

L_a	0.264 m
L_o	2.410 m
S	0.6360 m^2
m	10.0 kg
J	diag[1.00, 0.87, 1.40] kg·m^2

References

1. Washington, A.; Clothier, R.A.; Silva, J. A review of unmanned aircraft system ground risk models. *Prog. Aerosp. Sci.* **2017**, *95*, 24–44. [CrossRef]
2. la Cour-Harbo, A. The value of step-by-step risk assessment for unmanned aircraft. In Proceedings of the International Conference on Unmanned Aircraft Systems (ICAUS), Dallas, Texas, USA, 12–15 June 2018; pp. 149–157.
3. Wu, P.; Clothier, R. The development of ground impact models for the analysis of the risks associated with Unmanned Aircraft Operations over inhabited areas. In Proceedings of the International Probabilistic Safety Assessment and Management Conference and the 2012 Annual European Safety and Reliability Conference, Helsinki, Finland, 25–29 June 2012.
4. Melnyk, R.; Schrage, D.; Volovoi, V.; Jimenez, H. A Third-Party Casualty Risk Model for Unmanned Aircraft System Operations. *Reliab. Eng. Syst. Saf.* **2014**, *124*, 105–116. [CrossRef]
5. Bertrand, S.; Raballand, N.; Viguier, F.; Muller, F. Ground risk assessment for long-range inspection missions of railways by UAVs. In Proceedings of the International Conference on Unmanned Aircraft Systems (ICUAS), Miami, FL, USA, 13–16 June 2017; pp. 1343–1351.
6. Poissant, A.; Castana, L.; Xu, H. Ground Impact and Hazard Mitigation for Safer UAV Flight Response. In Proceedings of the International Conference on Unmanned Aircraft Systems (ICUAS), Dallas, TX, USA, 12–15 June 2018.
7. Levasseur, B.; Bertrand, S.; Raballand, N.; Viguier, F.; Goussu, G. Accurate Ground Impact Footprints and Probabilistic Maps for Risk Analysis of UAV Missions. In Proceedings of the IEEE Aerospace Conference, Big Sky, MT, USA, 2–9 March 2019 ; pp. 1–10.
8. Haartsen, Y.; Aalmoes, R.; Cheung, Y. Simulation of Unmanned Aerial Vehicles in the Determination of Accident Locations. In Proceedings of the International Conference on Unmanned Aircraft Systems, Arlington, VA, USA, 7–10 June 2016.
9. la Cour-Harbo, A. Ground impact probability distribution for small unmanned aircraft in ballistic descent. In Proceedings of the International Conference on Unmanned Aircraft Systems, Athens, Greece, 1–4 September 2020.
10. Rudnick-Cohen, E.; Herrmann, J.W.; Azarm, S. Modeling Unmanned Aerial System (UAS) Risks via Monte Carlo Simulation. In Proceedings of the International Conference on Unmanned Aircraft Systems, Atlanta, GA, USA, 11–14 June 2019.
11. Levasseur, B.; Bertrand, S.; Raballand, N. Efficient Generation of Ground Impact Probability Maps by Neural Networks for Risk Analysis of UAV Missions. In Proceedings of the International Conference on Unmanned Aircraft Systems, Athens, Greece, 1–4 September 2020.
12. La Cour-Harbo, A. Quantifying risk of ground impact fatalities for small unmanned aircraft. *J. Intell. Robot. Syst.* **2019**, *93*, 367–384. [CrossRef]
13. Lesprier, J.; Biannic, J.M.; Roos, C. Modeling and robust nonlinear control of a fixed-wing UAV. In Proceedings of the 2015 IEEE Conference on Control Applications (CCA), Sydney, Australia, 21–23 September 2015; pp. 1334–1339.

14. De Marco, A.; Duke, E.; Berndt, J. A general solution to the aircraft trim problem. In Proceedings of the AIAA Modeling and Simulation Technologies Conference and Exhibit, Reston, VA, USA, 20–23 August 2007; p. 6703.
15. Garcia, J.N.; Kutalik, Z.; Cho, K.H.; Wolkenhauer, O. Level sets and minimum volume sets of probability density functions. *Int. J. Approx. Reason.* **2003**, *34*, 25–47. [CrossRef]
16. Molchanov, I.S. Empirical estimation of distribution quantiles of random closed sets. *Theory Proba. Appli.* **1990**, *35*, 594–600. [CrossRef]
17. Morio, J.; Pastel, R. Plug-in estimation of d-dimensional density minimum volume set of a rare event in a complex system. *Proc. Inst. Mech. Eng. Part O J. Risk Reliab.* **2012**, *226*, 337–345. [CrossRef]
18. Owen, A.B. Importance Sampling. Monte Carlo Theory, Methods and Examples. 2013. Available online: http://statweb.stanford.edu/~owen/mc/Ch-var-is.pdf (accessed on 21 April 2021).
19. Park, C.; Huang, J.; Ding, Y. A computable Plug-In Estimator of Minimum Volume Sets for Novelty Detection. *Oper. Res.* **2010**, *59*, 1469–1480. [CrossRef]
20. Silverman, B., Density Estimation for statistics and data analysis. In *Monographs on Statistics and Applied Applied Probability*; Chapman and Hall: London, UK, 1986.
21. Marrel, A.; Chabridon, V. Statistical Developments for Target and Conditional Sensitivity analysis: Application on Safety Studies for Nuclear Reactor. 2020. Available online: https://hal.archives-ouvertes.fr/hal-02541142/document (accessed on 21 April 2021).
22. Derennes, P.; Morio, J.; Simatos, F. Simultaneous estimation of complementary moment independent and reliability-oriented sensitivity measures. *Math. Comput. Simul.* **2021**, *182*, 721–737. [CrossRef]
23. Perrin, G.; Defaux, G. Efficient evaluation of reliability-oriented sensitivity indices. *J. Sci. Comput.* **2019**, *79*, 1433–1455. [CrossRef]
24. Nagler, T.; Schellhase, C.; Czado, C. Nonparametric estimation of simplified vine copula models: Comparison of methods. *Depend. Model.* **2017**, *5*, 99–120. [CrossRef]

applied sciences

MDPI

Article

A Comparative Performance Evaluation of Routing Protocols for Flying Ad-Hoc Networks in Real Conditions

Antonio Guillen-Perez *, Ana-Maria Montoya, Juan-Carlos Sanchez-Aarnoutse and Maria-Dolores Cano *

Department of Information Technologies and Communications, Universidad Politécnica de Cartagena, 30203 Murcia, Spain; anamaria.montoya.osete@gmail.com (A.-M.M.); juanc.sanchez@upct.es (J.-C.S.-A.)
* Correspondence: antonio.guillen@edu.upct.es (A.G.-P.); mdolores.cano@upct.es (M.-D.C.)

Abstract: Unmanned aerial vehicles (UAVs) are widely used in our modern society and their development is rapidly accelerating. Flying Ad Hoc Networks (FANETs) have opened a new window of opportunity to create new value-added services. However, the characteristics that make FANETs unique, such as node mobility, node distance, energy constraints, etc., imply that several guidelines need to be considered for their successful deployment. Although numerous routing protocols have been proposed for FANETs, due to the wide range of applications in which FANETs can be applied, not all routing protocols can be used. Due to this challenge, after breaking down and classifying the different types of existing routing protocols for FANET, this paper analyzes and compares the performance of several routing protocols (Babel, BATMAN-ADV, and OLSR) in terms of throughput and packet loss in a real deployment composed of several UAV nodes using 2.4 and 5 GHz WiFi networks. The results show that Babel achieves better performance in the studied metrics than OLSR and BATMAN-ADV, while BATMAN-ADV delivers significantly lower performance. This experimental study confirms the importance of choosing the proper routing protocol for FANETs and their performance evaluation, something that will be extremely important in a few years when this type of network will be common in our day-to-day life.

Keywords: ad hoc networks; experimental study; Flying Ad Hoc Networks; FANET; practical case; routing protocols; testbed; unmanned aerial vehicles; UAV; WiFi

Citation: Guillen-Perez, A.; Montoya, A.-M.; Sanchez-Aarnoutse, J.-C.; Cano, M.-D. A Comparative Performance Evaluation of Routing Protocols for Flying Ad-Hoc Networks in Real Conditions. *Appl. Sci.* **2021**, *11*, 4363. https://doi.org/10.3390/app11104363

Academic Editor: Sylvain Bertrand

Received: 30 March 2021
Accepted: 10 May 2021
Published: 11 May 2021

1. Introduction

Ad hoc networks are becoming an essential part of our modern technological infrastructure, expanding the range of available applications and their characteristics. On the other hand, thanks to technological advances, unmanned aerial vehicles (UAVs) offer a wide range of possibilities (extending wireless coverage, use in agriculture, search and rescue, fire surveillance, etc.). The creation of an ad hoc network with UAVs also offers a significant advantage over other networks due to the high mobility of its nodes and their great versatility. This type of network is known as a Flying Ad Hoc Network (FANET) and is considered a subtype of Mobile Ad Hoc Networks (MANETs) or Vehicular Ad Hoc Networks (VANET). Due to the particular characteristics of UAVs, they bring new challenges for obtaining node mobility models, routing protocols, energy management, etc. Indeed, FANETs present high mobility of their nodes, both in speed and 3D mobility, frequent changes of network topology, or intense energy and weight restrictions. Consequently, it is essential to carry out a study of the different protocols for FANETs to identify the best ones to minimize the impact of the UAVs' characteristics and create a dynamic, agile, and efficient FANET network. A survey of mobility models, positioning algorithms, and propagation models can be found in [1–3].

In this work, after analyzing the types of routing protocols designed for FANETs, we focus on comparing, through a real deployment of a FANET using the IEEE 802.11 (WiFi) standard in the 2.4 GHz and 5 GHz bands, the performance offered by several routing protocols. We believe that a real deployment can provide much more reliable results than those obtained by simulators. Therefore, we consider that the highlight of this article is the actual deployment of the FANET, something that, as we will see, very few articles do, which limits their investigation to simulated experiments. The chosen protocols are Optimized Link-State Routing (OLSR), Better Approach to Mobile Ad Hoc Networking Advanced (BATMAN-ADV), and Babel. They are compared in terms of throughput and packet losses in a network composed of several UAVs and an intermediate relay node. We chose these proactive routing protocols because of their high mobility range, low latency, widespread use in ad hoc networks, and good power consumption; similarly, they have relatively low complexity and computational demands, allowing us to simplify and automate some of the challenges mentioned in [1–3]. The results highlight the importance of the correct choice of routing protocol for FANETs, showing that Babel achieves higher performance in the studied metrics whereas BATMAN-ADV and OLSR show lower performance.

The rest of the article is organized as follows: in Section 2, we include a review of state-of-the-art case studies with FANETs, and we break down the different types of routing protocols proposed for ad hoc networks and FANETs. The FANET deployed for this study and the tools used are detailed in Section 3. The results are discussed in Section 4. Finally, conclusions and future works are presented in Section 5.

2. State of the Art

In this section, we will look at proposed works for FANETs. First, we will look at the proposed studies that perform a real FANET deployment, and then we will see the different routing protocols proposed for ad hoc networks and FANETs.

2.1. Real Experimental Studies

In this subsection, we will see a compilation of papers that analyze, by means of real experimental studies, the performance of FANETs deployed by UAVs. Due to the high cost and complexity of building large-scale networks with variable topologies, and the difficulties related to the repetition of scenarios, it is challenging to find works that include real deployments of FANETs. The vast majority of the work completed expects the simulators used to be capable of simulating real conditions to ensure that the results they obtain or the algorithms they propose can resemble what could be obtained. However, we believe that a real deployment, controlling as many variables as possible and repeating the experiments several times, can provide greater assurance. The results obtained are, by comparison, more trustworthy. That is why we consider this work very interesting, helping the reader to understand the impact of choosing a routing protocol on the FANET performance.

Analyzing works that focus on real FANET deployments, Rosati et al. [4] studied the performance offered by Predictive-OLSR (P-OLSR) and OLSR in a FANET composed of two small fixed-wing UAVs (called "eBee"). Their results showed that, because P-OLSR uses GPS information, the performance obtained both in simulation and in experimental tests improves that obtained with OLSR, reducing the number of communication interruptions.

Furthermore, the P-OLSR performance was compared against OLSR and Babel in [5]. The results showed that P-OLSR, for the scenario they proposed (up to three UAVs with a highly dynamic ad hoc network), could provide more reliable multi-hop communication than Babel and OLSR.

In [6], the authors compared two modes of operation of the 802.11 WiFi standard (access point and ad hoc using BATMAN-ADV) in a real experimental scenario in terms of coverage, throughput, and energy efficiency with up to two quadcopter UAVs. The results revealed a better performance of the access point mode in terms of received signal strength and throughput but worse performance in terms of power consumption than the ad hoc mode.

The work presented by Lee et al. [7] proposed an approach similar to P-OLSR, in which the UAV nodes employed GPS information to improve the routing protocol, called Ground Control System-Routing (GCS-R). A real experiment was conducted with up to six UAVs in a network-coverage application scenario. The results showed that their proposed routing algorithm outperformed OLSR and DSDV in terms of throughput, stability, and network outage time. However, given that it suggests a centralized algorithm, it has a single point of failure that could be critical because of the instability of the UAV nodes and presents serious scalability problems.

On the other hand, in [8], the authors showed the deployment of a FANET using BATMAN as the routing protocol, showing the capabilities that this type of network can offer in terms of coverage and throughput. They used a UAV in three different scenarios, and the results showed that the maximum distance they could transmit without packet loss was 117 m.

Finally, the authors of [9] proposed a security protocol for FANETs called SUAP, which incorporates geographical leashes, hash chains, and public-key cryptography into the AODV routing protocol. Although SUAP proved to be effective in encrypting messages exchanged between nodes against various attacks, it did not provide a robust mechanism to recover from disconnections between nodes. According to the experimental result performed with three nodes, the delay to re-establish a new route when a node failure occurred was considerable, especially for real-time applications such as video capture, monitoring, and aerial photography. Moreover, its performance should be evaluated in a network with a high density of nodes. A summary of the reviewed papers can be seen in Table 1.

Table 1. Related works that included a real FANET deployment. Also detailed are the routing protocols employed and the number of UAVs.

Work	Routing Protocols	Number of UAVs
Rosati et al. [4]	OLSR, P-OLSR	2
Rosati et al. [5]	Babel, OLSR, P-OLSR	3
Guillen et al. [6]	BATMAN-ADV	2
Lee et al. [7]	GCS-R, DSDV, OLSR	6
Kaysina et al. [8]	BATMAN	1
Maxa et al. [9]	SUAP	2

2.2. Routing Protocols

Routing protocols are responsible for guaranteeing the delivery of a message from a source node to a destination node. Routing protocols must adapt to the essential characteristics of FANETs: high mobility, energy efficiency, constant changes in the topology, etc. In addition, due to the wide variety of application fields, the requirements of routing protocols in FANETs can be very diverse. For example, a constant jitter would be necessary for real-time video applications, whereas a high level of reliability would be necessary in applications that extend telecommunication coverage in case of disasters (regardless of delay and jitter). Most FANETs' routing protocols are extensions of well-known MANET protocols, such as AODV, OLSR, or DSR [10]. However, there is still a lack of routing protocols that fit all FANETs' needs.

In this subsection, we will examine the different routing protocols proposed for FANETs and group them into five main categories: (i) topology-based, (ii) position-based, (iii) clustering/hierarchical, (iv) swarm-based, and (v) delay-tolerant network (DTN).

2.2.1. Topology-Based Routing Protocols

Topology-based routing protocols base their operation on using the links information for data forwarding. Within this category, four types of protocols can be distinguished: (i) static, (ii) proactive, (iii) reactive, and (iv) hybrid.

Static Routing Protocols

The nodes in this network have static routing tables which are configured at the beginning of a task and do not change. Networks that implement these protocols must have a constant topology, and therefore, being unable to adapt dynamically to changes, are susceptible to failures. The static routing protocols include Load Carry and Deliver Routing (LCAD) [11], Multi-Level Hierarchical Routing (MLHR) [12], and Data-Centric Routing (DCR) [13]. LCAD is based on the Store-Carry-Forward (SCF) paradigm [14]. The nodes capture the data and transport it, physically moving the message to a relay or destination node. It is usually used in DTN networks. MLHR solves the problem of scalability by clustering the nodes of the network and allowing a head node to communicate with the other head nodes of other clusters. Thus, the size and area of operation increase, but this head node can be a bottleneck and is a single point of failure [15]. DCR bases the routing on the information contained in the message. Thus, this protocol sends the information to several nodes that want specific data.

Proactive Routing Protocols

Each node has its routing table which is periodically updated and shared with the other nodes. The routing tables contain the routes to send a message from a source node to any destination node in the network. The main advantage of these types of protocols is their low delay in sending messages because the path is known beforehand. However, if there are many nodes in the network, the periodic updating of the routing tables severely consumes bandwidth and energy. In addition, proactive protocols react slowly to changes in topology, having to update the routing tables of all nodes in the network. Due to these disadvantages, these protocols are not usually used in applications with FANETs with a large number of UAVs with high mobility or when energy consumption must be low. The most important protocols proposed or adapted for FANETs are Optimized Link-State Routing (OLSR) [16], Destination Sequence Distance Vector (DSDV) [17], Better Approach to Mobile Ad Hoc Network (BATMAN) [18], and Directional OLSR (DOLSR) [19]. The OLSR protocol is currently the most widely used routing protocol in ad hoc networks. The approach of this protocol is to put a *cost* to each link (link-state) in the network. Each node evaluates the *cost* of sending a message through a link that is directly connected and shares it with its neighboring nodes using a flooding strategy. This *cost* can be the distance, delay, losses, bandwidth, etc. Once the view of the network has been updated, the route is searched applying the shortest path algorithm (*Dijkstra*). DSDV is a protocol based on the Bellman-Ford algorithm. It adds two parameters to vector-distance routing, which are a sequence number, to avoid loops and determine the freshness of the routes and the *Dampling* parameter. In addition, it uses two types of packages for route updates: full-dump and incremental. The full-dump packages contain all the information in the routing table, and due to its size, it is not usually transmitted. The incremental packet is used to update the routes in the last full-dump packet. DSDV has been used extensively in the field of FANETs, as can be seen in [20–22]. The BATMAN routing protocol is a relatively new proactive protocol that is used in MANETs. The BATMAN protocol does not discover the entire network; the protocol only learns from the nodes directly connected to them. In addition, BATMAN-Advanced (BATMAN-ADV) [23] is an improvement of the original BATMAN protocol in terms of performance due to its integration in the protocol stack. Numerous comparative studies of BATMAN-ADV performance in ad hoc networks and FANETs have been carried out [23,24]. DOLSR was specifically designed for FANETs and is a variant of OLSR. One of the most important performance factors in OLSR is the selection of multipoint relay (MPR) nodes. Thus, in DOLSR the number of MPRs is reduced to decrease the overhead. Each node selects a set of MPRs so that it can cover two-hop neighbors. DOLSR has the advantage of minimizing end-to-end delay, which is crucial for real-time applications and offers security improvements as it is resistant to jamming. In addition to the protocols explained, there are a variety of protocols with variants that are used in FANETs, such as Predictive-OLSR (P-OLSR) [5], Mobility and Load-Aware OLSR (ML-OLSR) [25], Contention-Based OLSR (COLSR) [26], Modified-OLSR (M-OLSR) [27], Cartography-Enhanced OLSR (CE-OLSR) [28], Topology Broadcast Based on Reverse-Path Forwarding (TBRPF) [29], Fisheye State Routing (FSR) [30], and Babel [31]. Babel builds on the ideas of DSDV, AODV, and other routing protocols to derive a loop-avoiding distance vector routing protocol that is designed to be robust and efficient in both relatively stable and highly dynamic networks.

Reactive Routing Protocols

Due to the large bandwidth consumption that periodically occurs in the network discovery process of proactive protocols, reactive protocols use on-demand network discovery processes, which makes reactive protocols bandwidth-efficient. These protocols have a great advantage over proactive protocols as it gives them a great dynamism, something necessary in FANETs, but it has a great disadvantage in the latency produced by the route-search process. The most important reactive routing protocols are Dynamic Source Routing (DSR) [32], Ad Hoc On-Demand Distance Vector (AODV) [33], and Time-Slotted AODV (TS-AODV) [34]. DSR is a source-routing protocol. The complete route that a packet must follow is indicated in the header and only the source node can indicate the route. AODV is the most popular reactive routing protocol and is typically used in MANETs, VANETs, and FANETs. During the route lookup process, the source node searches its routing table to see if a route to the destination node has been established in the past. If no route exists, a route request process is initiated. TS-AODV is a time slot routing protocol centered on FANETs and based on the AODV protocol. Due to the large number of UAVs that may exist in FANETs, TS-AODV uses the time-division mechanism to avoid collisions in the transmission of information, significantly reducing packet losses and increasing the available bandwidth. In addition to the reactive protocols explained above, there are other routing protocols used in the literature on ad hoc networks and FANETs, such as Multicast AODV (MAODV) [35] and AODV Security (AODVSEC) [36].

Hybrid Routing Protocols

To solve the problems of bandwidth consumption of control messages and the low dynamism of proactive protocols, as well as the long delay in the route search of reactive protocols, hybrid routing protocols were introduced. These hybrid protocols are particularly suitable for large networks since they base their operation on the division of the network into sub-networks or zones. Thus, a proactive routing protocol operates within each zone and a reactive routing protocol is used for communication between zones. For hybrid solutions, the Zone Routing Protocol (ZRP) [37] is the most popular. ZRP divides the network into clusters of nodes in which their maximum separation distance is predefined. Within the clusters, a proactive routing protocol is applied, and nodes in different zones are routed to a subset that is common to both zones. There are other hybrid protocols designed or modified for FANETs, such as the Temporarily Ordered Routing Algorithm (TORA) [38], Rapid-reestablish TORA (RTORA) [39], Hybrid Wireless Mesh Protocol (HWMP) [40], Sharp Hybrid Adaptive Routing Protocol (SHARP) [41], and Hybrid Routing Protocol (HRP) [42].

2.2.2. Position-Based Routing Protocols

These routing protocols base their operation on the knowledge of the geographical position of the nodes. Therefore, they are the most suitable routing protocols for FANETs with high mobility. Protocols in this category can be divided into (i) reactive-based, (ii) greedy-based, and (iii) heterogeneous.

Reactive-Based Routing Protocols

These types of protocols are based on a reactive technique and use the position of the nodes to obtain higher performance. In this category, we highlight the Reactive-Greedy-Reactive (RGR) algorithm [43]. RGR is based on the reactive routing protocol AODV [33] for the on-demand route lookup process, and for message delivery, it is based on the Greedy Geographic Forwarding (GGF) protocol [44]. Other reactive-based protocols are Ad Hoc Routing Protocol for Aeronautical Mobile Ad Hoc Networks (ARPAM) [45] and Multipath Doppler Routing (MUDOR) [46].

Greedy-Based Routing Protocols

This set of protocols searches the path with the least number of hops between nodes through the Greedy Forwarding approach [47], minimizing message delay. One of the most prominent protocols of this category is Geographic Position Mobility-Oriented Routing (GPMOR) [48]. GPMOR was designed considering some characteristics of FANETs and bases its operation on the predictive approach. Based on the Gauss–Markov mobility model [1] and the geographical positions of the nodes, GPMOR predicts the node positions and selects the next closest forwarding node to the receiver optimally, following the Greedy approach. Other greedy-based routing protocols are Mobility Prediction-Based Geographic Routing (MPGR) [49], Geographic Load Share Routing (GLSR) [50], Geographic Greedy Perimeter Stateless Routing (GPSR) [51], Greedy-Hull-Greedy (GHG) [52], Greedy-Random-Greedy (GRG) [53], Greedy Distributed Spanning Tree Routing 3D (GDSTR-3D) [54], and UAV Search Mission Protocol (USMP) [51].

Heterogeneous Routing Protocols

This set of protocols is applied to networks where the nodes that form them are of different natures, i.e., the network can be formed by fixed or mobile nodes on the ground and the aerial nodes of a FANET. This architecture has several advantages including increased coverage and performance. Within this set of protocols, Connectivity-Based Traffic Density Aware Routing Using UAVs for VANETs (CRUV) [55] stands out. CRUV relies on the existence of a DTN VANET supported by a FANET, which allows the interconnection of the DTN. In addition, there are also the following: UAV-Assisted VANET Routing Protocol (UVAR) [56], Position-Aware Secure and Efficient Routing (PASER) [57], Cross-Layer Link Quality and Geographical-Aware Beaconless (XLinGo) [58], and Secure UAV Ad Hoc Routing Protocol (SUAP) [9].

2.2.3. Clustering/Hierarchical Routing Protocols

In this category, UAV nodes are grouped into clusters, in a hierarchical fashion, and within each cluster, there is a node (cluster head) that enables inter-cluster communication via other cluster heads. In these routing protocols, the selection of the node acting as cluster-head is a critical task since the overall performance of the network will depend in part on this node. Within this category of protocols, the Clustering Algorithm of UAV Networking (CAUN) [59], and Mobility Prediction Clustering Algorithm for UAV Networking (MPCA) [60] may be mentioned. The operation of CAUN is quite simple: the initial cluster is built on the ground depending on the mission application, and once the network is deployed, the cluster adapts according to real-time conditions. On the other hand, MPCA predicts network topology updates, and, in turn, determines cluster formation based on UAV mobility, allowing for more stable clusters. In addition to the protocols described above, there are other protocols belonging to the category of clustering/hierarchical routing protocols, such as Landmark Ad Hoc Routing (LANMAR) [61], Multi-Meshed Tree Protocol (MMT) [62], Cluster-Based, Location-Aided DSR (CBLADSR) [63], and Disruption Tolerant Mechanism (DTM) [64].

2.2.4. Swarm-Based Routing Protocols

Swarm-based routing protocols are inspired by the behavior of animals and nature. Thus, for example, as swarm-based routing protocols that are created for FANETs, we highlight the BeeAdHoc [65] algorithm that bases its operation on the behavior of bees, as well as AntHocNet [66], and APAR [67] algorithms, both inspired by the behavior of ants.

2.2.5. Delay-Tolerant Network (DTN) Routing Protocols

Delay-tolerant routing protocols are proposed for networks with constant outages, partitions, and topology changes, such as FANETs. Thus, these protocols make use of the Store-Carry-and-Forward (SCF) technique when nodes lose connectivity. This technique eliminates the overhead since no control messages are transmitted but greatly increases the communication delay. The most widespread protocol of this group is Location-Aware Routing for Opportunistic Delay Tolerant (LAROD) [68]. LAROD is based on the combination of two approaches: SCF and Greedy Forwarding techniques, depending on the situation, and uses the beaconless strategy to reduce the overload of the network. In addition, within this category, it is worth highlighting the following algorithms: AeroRP [69], Geographic Routing Protocol for Aircraft Ad Hoc Network (GRAA) [70], Epidemic [71], Maxprop [72], Spray and Wait [73], and Prophet [74].

Table 2 shows a summary of the different classes of routing algorithms, their subclasses in this section, and their algorithms.

Table 2. A summary of routing algorithms and their classes and subclasses.

Type	Subtype	Routing Algorithm
Topology-Based	Static	LCAD [11], MLHR [12], DCR [13]
	Proactive	OLSR [16], DSDV [17], BATMAN [18], BATMAN-ADV [23], DOLSR [19], P-OLSR [5], ML-OLSR [25], COLSR [26], M-OLSR [27], CE-OLSR [28], TBRPF [29], FSR [30], Babel [31]
	Reactive	DSR [32], AODV [33], TS-AODV [34], MAODV [35], AODVSEC [36]
	Hybrid	ZRP [37], TORA [38], RTORA [39], HWMP [40], SHARP [41], HRP [42]
Position-Based	Reactive-Based	RGR [43], GGF [44], ARPAM [45], MUDOR [46]
	Greedy-Based	GPMOR [48], MPGR [49], GLSR [50], GPSR [51], GHG [52], GRG [53], GDSTR-3D [54], USMP [51]
	Heterogeneous	CRUV [55], UVAR [56], PASER [57], XLinGo [58], SUAP [9]
Clustering/Hierarchical		CAUN [59], MPCA [60], LANMAR [61], MMT [62], CBLADSR [63], DTM [64]
Swarm-Based		BeeAdHoc [65], AntHocNet [66], APAR [67]
Delay-Tolerant Network		LAROD [68], AeroRP [69], GRAA [70], Epidemic [71], Maxprop [72], Spray and Wait [73], Prophet [74]

3. Materials, Methods, and Scenario

In this section, we will explain the testbed and tools employed in this study. By using this deployment, we demonstrate the communication capabilities that FANETs can offer, comparing the performance achieved with various proactive routing protocols, specifically OLSR, BATMAN-ADV, and Babel. The 3 protocols were selected based on the criteria of "state of the art" technology, experience working with them, suitability for the proposed scenario, and compatibility with the communications module.

For this purpose, a WiFi network composed of several UAVs was deployed. Each UAV had a specific communication module (WiTi [75]); see Figure 1a. These communication modules had a dual 2.4/5 GHz communication band, using the IEEE 802.11g standard for the 2.4GHz band and the IEEE 802.11a standard for the 5GHz band, and 2 antennas for each frequency band. The tool *iperf3* [76] was employed to evaluate the performance of the different routing algorithms, analyzing throughput and packet losses. To analyze packet losses in an efficient way, the UDP communication protocol was used. In the deployed scenario (see Figure 1b), 5 communication nodes were placed: specifically, 2 UAVs (quadcopters), 2 PCs, and 1 ground node. The 2 PCs were located under the UAVs to simulate devices in a coverage extension scenario. They also allowed monitoring of the network status. On the other hand, the ground node operated as a relay base station, acting as an intermediate node between the drones when the distance between the nodes required its use. The relay node was located at 35 m from PC1 (UAV1). Whereas UAV1 was kept at a fixed location, UAV2 flew horizontally, stopping every 10 m to take performance measurements. Each measurement had a duration of 60 s and consisted of sending data between the 2 PCs through the UAVs. Both UAVs flew at an altitude of 10 m, calibrating all distances, positions, and heights with the UAVs' GPS modules. A representation of the deployed scenario can be seen in Figure 2a,b. For instance, this scenario could simulate a situation where a coverage extension is required due to an emergency, to solve a specific moment of overload of the communications network, or for remote surveillance.

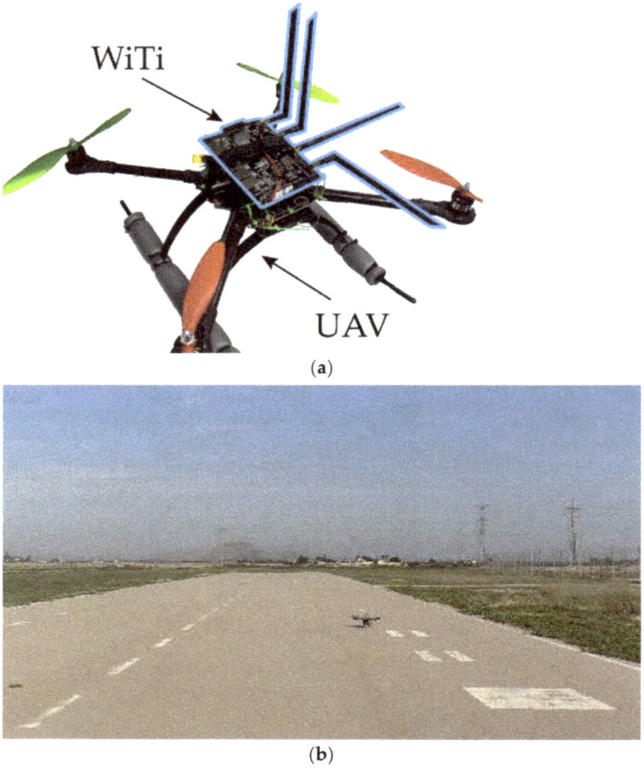

Figure 1. (**a**) UAV with WiFi communication module (WiTi). (**b**) Flying field, "Los Halcones de la Rambla", Murcia, Spain.

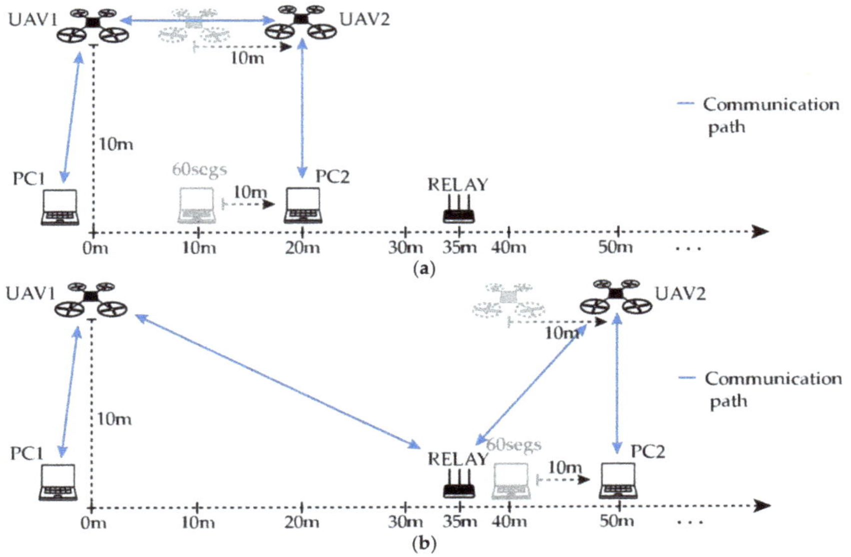

Figure 2. Real testbed FANET with relay node. (**a**) Communication before 35-m mark. (**b**) Communication after 35-m mark. Note that after the 35-m mark, communication between the 2 UAVs (UAV1 and UAV2) was performed through the relay node. Before this mark, the communication was completed through a direct path. Performance measurements were taken between the 2 UAVs every 10 m step and each measurement had a duration of 60 s. The PCs were used to monitor the network and followed their UAV. The height of the UAVs was 10 m. The first UAV (UAV1) was located at the 0-meter mark. The relay node was located at the 35-m mark. The second UAV (UAV2) was positioned for 60 s at the measurement points located 10 m from the 0-meter mark. That is, the first measuring point was located at 10 m, the second at 20 m, and so on.

4. Results

This section will analyze the results obtained in the tests detailed in the previous section, comparing in terms of throughput and packet losses the different routing algorithms presented.

The results showed that the Babel and OLSR protocols obtained higher throughput in both the 2.4 GHz and 5 GHz bands. Besides, the expected inverse relationship between packet losses and throughput was demonstrated since one increases while the other decreases. It was also observed that the 5 GHz band presented a faster throughput decay. Because the received power is inversely proportional to frequency and distance, the 2.4 GHz band will present a higher received power at equal transmitted power and communication distance, which will allow the maintenance of a more stable and robust communication. Only at 10 m, 40 m, and 50 m, the performance of both bands is comparable. For the 10 m point, the communication between the UAVs was direct on both bands, so propagation losses were low and communication was very stable. For the 40 m and 50 m points, the communication for the 2.4 GHz band was still direct. However, for the 5 GHz band, the communication made an intermediate hop through the relay node, decreasing the distance between the communication nodes. Throughput results can be seen in Table 3 and Figure 3. Likewise, packet loss results are shown in Table 4 and Figure 4.

Table 3. Average throughput (Mbps) obtained between PCs-UAVs at different gap distances for OLSR, BATMAN-ADV, and Babel, for 2 frequency bands: 2.4 and 5 GHz. Each measurement had a duration of 60 s.

Routing Protocol	Frequency Band	Distance						
		10 m	20 m	30 m	40 m	50 m	60 m	70 m
OLSR	2.4 GHz	1.02	1.05	1.05	0.75	0.91	0.84	0.00
	5 GHz	0.98	0.73	0.51	0.99	0.82	0.00	0.00
BATMAN-ADV	2.4 GHz	1.05	1.05	0.90	0.82	0.51	0.27	0.75
	5 GHz	0.89	0.16	0.03	0.82	0.58	0.00	0.00
Babel	2.4 GHz	1.05	1.05	1.05	1.05	0.93	0.67	0.86
	5 GHz	0.96	0.66	0.49	0.88	0.71	0.00	0.00

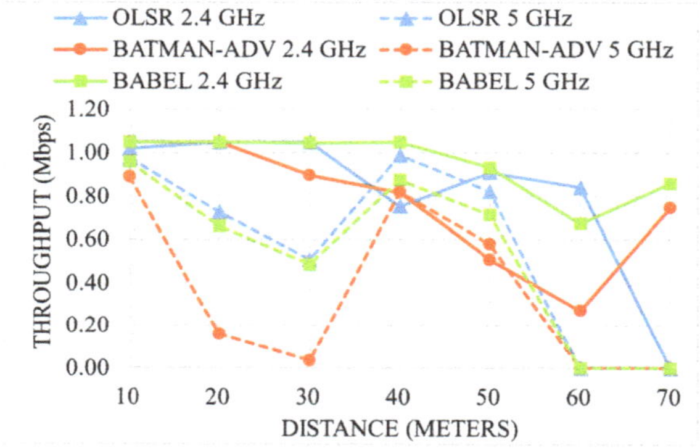

Figure 3. Average throughput (Mbps) vs. distance (meters) for each protocol and frequency band.

Table 4. Average packet loss (%) obtained between PCs-UAVs at different gap distances for OLSR, BATMAN-ADV, and Babel, for 2 frequency bands: 2.4 and 5 GHz. Each measurement had a duration of 60 s.

Routing Protocol	Frequency Band	Distance						
		10 m	20 m	30 m	40 m	50 m	60 m	70 m
OLSR	2.4 GHz	3.3%	0%	0%	27%	11%	19%	100%
	5 GHz	6.6%	30%	44%	6%	20.4%	100%	100%
BATMAN-ADV	2.4 GHz	0%	0%	13%	22.1%	44%	73%	27%
	5 GHz	14%	84%	97%	21.3%	43%	100%	100%
Babel	2.4 GHz	0%	0%	0%	0%	10%	35%	18%
	5 GHz	8%	36%	45.9%	15%	29.7%	100%	100%

For the 2.4 GHz band, as shown in the results, especially in Figures 3 and 4, the routing protocols present a different behavior. OLSR switches from direct communication to relay communication between the measurement points located at 40 m and 50 m. However, both BATMAN-ADV and Babel switch between 60 m and 70 m. This can be verified by the fact that there is a change of trend in the metrics studied at these two measurement points and the corresponding change in the routing table.

If we now focus on the 5 GHz band, the trend change occurs between the measurement points located between 30 and 40 m. Between these two points, communication switched from direct UAVs communication to relay mode. After this change in communication, the metrics improved, to rapidly decline again with distance. Additionally, for the 5 GHz frequency band, the maximum distance at which we were able to establish communication between two nodes was 40 m. This is again reflected above 40 m, which is the distance at which the intermediate hop to the relay node existed. If the relay node had been located more than 40 m away, it would have been impossible to use it as an intermediate communication node in the 5 GHz band.

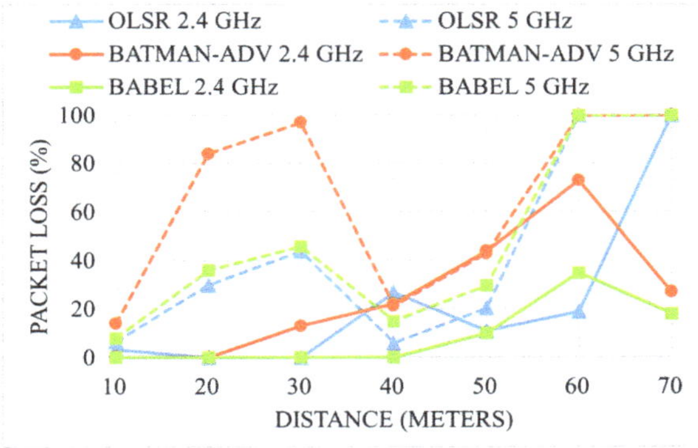

Figure 4. Average packet loss (%) vs. distance (meters) for each protocol and frequency band.

We can conclude that Babel and OLSR obtained better results than BATMAN-ADV, given the results obtained. This may have been expected because BATMAN-ADV has a too-high routing update period that was originally not intended for FANETs. Finally, between Babel and OLSR, Babel offered the best overall performance, guaranteeing more stable results in both frequency bands.

5. Conclusions

The potential of FANETs can increase in the coming years with the advent of new communication technologies and standards such as 6G, the Internet of Things (IoT), and connected autonomous vehicles (CAVs). The benefits that FANETs can offer are immeasurable (coverage extension, application in emergency communications, search and rescue, agriculture, etc.). When deploying FANETs, it is essential to be able to find the path that a packet must follow to reach its destination. Due to the characteristic constant changes in the network topology of FANETs, the selection of the routing protocol is a crucial task for effective deployment and successful operation. After detailing the different routing protocols that have been proposed for FANETs, this paper performed a case study comparing the performance offered, in terms of throughput and packet loss, of three proactive routing protocols: OLSR, BATMAN-ADV, and Babel. The scenario studied could simulate a coverage extension scenario in emergencies, search and rescue, or remote surveillance. The results obtained showed that Babel achieved higher performance in the studied metrics, outperforming OLSR and BATMAN-ADV. Besides, it was shown that BATMAN-ADV performed significantly worse than OLSR and Babel due to its low frequency of routing-table updating. In particular, the sending period of messages in charge of finding alternative routes in BATMAN-ADV is 1 s (OGM interval). In FANETs, the network topology can vary a lot in 1 s, hence discouraging its use for FANETs. As future work, we consider that an extension of the routing protocols to be evaluated would be a logical next step, employing a more significant number of UAVs with different characteristics and comparing experimental results and propagation models such as those presented in [77].

Author Contributions: Conceptualization, M.-D.C.; methodology, M.-D.C. and J.-C.S.-A.; software and tests measurements, A.G.-P., A.-M.M., and J.-C.S.-A.; formal analysis, A.-M.M. and M.-D.C.; writing, A.G.-P. and M.-D.C. All authors have read and agreed to the published version of the manuscript.

Funding: This research was funded by the AEI/FEDER, UE project grant TEC2016-76465-C2-1-R (AIM), project SPID202000X116746SV0 (AriSe2: FINe), and 20740/FPI/18. Fundación Séneca. Región de Murcia (Spain).

Conflicts of Interest: The authors declare no conflict of interest. The funders had no role in the design of the study; in the collection, analyses, or interpretation of data; in the writing of the manuscript, or in the decision to publish the results.

Abbreviations

AeroRP	Aeronautical Routing Protocol.
AODV	Ad Hoc On-Demand Distance Vector.
AODVSEC	AODV Security.
APAR	Ant Colony Optimization-Based Polymorphism Aware Routing Algorithm.
ARPAM	Ad Hoc Routing Protocol for Aeronautical Mobile Ad Hoc Networks.
BATMAN	Better Approach to Mobile Ad Hoc Network.
BATMAN-ADV	BATMAN-Advanced.
CAUN	Clustering Algorithm of UAV Networking.
CBLADSR	Cluster-Based Location-Aided DSR.
CE-OLSR	Cartography-Enhanced OLSR.
COLSR	Contention-Based OLSR.
CRUV	Connectivity-Based Traffic-Density Aware Routing Using UAVs for VANETs.
DCR	Data-Centric Routing.
DOLSR	Directional OLSR.
DSDV	Destination-Sequenced Distance Vector.
DSR	Dynamic Source Routing.
DTM	Disruption-Tolerant Mechanism.
DTN	Delay-Tolerant Network.
FANET	Flying Ad Hoc Network.
FSR	Fisheye-State Routing.
GCS-R	Ground Control System-Routing
GDSTR-3D	Greedy Distributed Spanning Tree Routing 3D.
GGF	Greedy Geographic Forwarding.
GHG	Greedy-Hull-Greedy.
GLSR	Geographic Load-Share Routing.
GPMOR	Geographic Position Mobility-Oriented Routing.
GPSR	Geographic Greedy Perimeter Stateless Routing.
GRAA	Geographic Routing Protocol for Aircraft Ad Hoc Network.
GRG	Greedy-Random-Greedy.
HRP	Hybrid-Routing Protocol.
HWMP	Hybrid Wireless Mesh Protocol.
LANMAR	Landmark-Routing Protocol.
LAROD	Location-Aware Routing for Opportunistic Delay Tolerant.
LCAD	Load Carry and Deliver Routing.
MANET	Mobile Ad Hoc Network.
MAODV	Multicast AODV.
MLHR	Multi-Level Hierarchical Routing.
ML-OLSR	Mobility and Load-Aware OLSR.
MMT	Multi-Meshed Tree Protocol.
M-OLSR	Modified-OLSR.
MPCA	Mobility Prediction Clustering Algorithm.
MPGR	Mobility Prediction-Based Geographic Routing.
MUDOR	Multipath Doppler Routing.
OLSR	Optimized Link-State Routing.
PASER	Position-Aware Secure and Efficient Routing Approach.
P-OLSR	Predictive-OLSR.
RGR	Reactive-Greedy-Reactive.

RTORA	Rapid-Reestablish TORA.
SHARP	Sharp Hybrid Adaptive Routing Protocol.
SUAP	Secure UAV Ad Hoc Routing Protocol.
TBRPF	Topology Broadcast Based on Reverse-Path Forwarding.
TORA	Temporarily Ordered Routing Algorithm.
TS-AODV	Time-Slotted AODV.
UAV	Unmanned Aerial Vehicle.
USMP	UAV Search Mission Protocol.
UVAR	UAV-Assisted VANET Routing Protocol.
VANET	Vehicular Ad Hoc Network.
XLinGo	Cross-Layer Link Quality and Geographical-Aware Beaconless.
ZRP	Zone-Routing Protocol.

References

1. Guillen-Perez, A.; Cano, M.-D. Flying Ad Hoc Networks: A New Domain for Network Communications. *Sensors* **2018**, *18*, 3571. [CrossRef] [PubMed]
2. Bekmezci, I.; Sahingoz, O.K.; Temel, Ş. Flying Ad-Hoc Networks (FANETs): A survey. *Ad Hoc Netw.* **2013**, *11*, 1254–1270. [CrossRef]
3. Oubbati, O.S.; Atiquzzaman, M.; Lorenz, P.; Tareque, H.; Hossain, S. Routing in Flying Ad Hoc Networks: Survey, Constraints, and Future Challenge Perspectives. *IEEE Access* **2019**, *7*, 81057–81105. [CrossRef]
4. Rosati, S.; Kruzelecki, K.; Heitz, G.; Floreano, D.; Rimoldi, B. Dynamic Routing for Flying Ad Hoc Networks. *IEEE Trans. Veh. Technol.* **2016**, *65*, 1690–1700. [CrossRef]
5. Rosati, S.; Kruzelecki, K.; Traynard, L.; Rimoldi, B. Speed-aware routing for UAV ad-hoc networks. In Proceedings of the 2013 IEEE Globecom Workshops (GC Wkshps), Atlanta, GA, USA, 9–13 December 2013; pp. 1367–1373.
6. Guillen-Perez, A.; Sanchez-Iborra, R.; Cano, M.-D.; Sanchez-Aarnoutse, J.C.; Garcia-Haro, J. WiFi networks on drones. In Proceedings of the 2016 ITU Kaleidoscope: ICTs for a Sustainable World (ITU WT), Bangkok, Thailand, 14–16 November 2016; pp. 1–8.
7. Lee, J.; Kim, K.; Yoo, S.; Chung, A.Y.; Lee, J.Y.; Park, S.J.; Kim, H. Constructing a reliable and fast recoverable network for drones. In Proceedings of the 2016 IEEE International Conference on Communications (ICC), Kuala Lumpur, Malaysia, 22–27 May 2016.
8. Kaysina, I.A.; Vasiliev, D.S.; Abilov, A.; Meitis, D.S.; Kaysin, A.E. Performance evaluation testbed for emerging relaying and coding algorithms in Flying Ad Hoc Networks. In Proceedings of the 2018 Moscow Workshop on Electronic and Networking Technologies (MWENT), Moscow, Russia, 14–16 March 2018.
9. Maxa, J.-A.; Ben Mahmoud, M.S.; Larrieu, N. Joint Model-Driven design and real experiment-based validation for a secure UAV Ad hoc Network routing protocol. In Proceedings of the 2016 Integrated Communications Navigation and Surveillance (ICNS), Herndon, VA, USA, 19–21 April 2016; pp. 1–16.
10. Maxa, J.A.; Mahmoud, M.S.B.; Larrieu, N. Survey on UAANET routing protocols and network security challenges. *Ad-Hoc Sens. Wirel. Netw.* **2017**, *37*, 231–320.
11. Cheng, C.-M.; Hsiao, P.-H.; Kung, H.T.; Vlah, D. Maximizing throughput of UAV-relaying networks with the load-carry-and-deliver paradigm. In Proceedings of the 2007 IEEE Wireless Communications and Networking Conference, Hong Kong, China, 11–15 March 2007; pp. 4417–4424. [CrossRef]
12. Sahingoz, O.K. Networking Models in Flying Ad-Hoc Networks (FANETs): Concepts and Challenges. *J. Intell. Robot. Syst.* **2014**, *74*, 513–527. [CrossRef]
13. Ko, J.; Mahajan, A.; Sengupta, R. A network-centric UAV organization for search and pursuit operations. In Proceedings of the IEEE Aerospace Conference, Big Sky, MT, USA, 9–16 March 2002; Volume 6, pp. 2697–2713.
14. Kolios, P.; Friderikos, V.; Papadaki, K. Store carry and forward relay aided cellular networks. In Proceedings of the Seventh International Workshop on Vehicular Ad Hoc Networks, VANET 2010, Chicago, IL, USA, 24 September 2010; pp. 71–72.
15. Abolhasan, M.; Wysocki, T.; Dutkiewicz, E. A review of routing protocols for mobile ad hoc networks. *Ad Hoc Netw.* **2004**, *2*, 1–22. [CrossRef]
16. Jacquet, P.; Muhlethaler, P.; Clausen, T.; Laouiti, A.; Qayyum, A.; Viennot, L. Optimized link state routing protocol for ad hoc networks. In Proceedings of the IEEE International Multi Topic Conference, IEEE INMIC 2001, Technology for the 21st Century, Lahore, Pakistan, 30–30 December 2001; pp. 62–68.
17. Perkins, C.E.; Bhagwat, P. Highly dynamic Destination-Sequenced Distance-Vector routing (DSDV) for mobile computers. *ACM SIGCOMM Comput. Commun. Rev.* **1994**, *24*, 234–244. [CrossRef]
18. Johnson, D.L.; Ntlatlapa, N.; Aichele, C. Simple pragmatic approach to mesh routing using BATMAN. In Proceedings of the 2nd IFIP International Symposium on Wireless Communications and Information Technology in Developing Countries, CSIR, Pretoria, South Africa, 6–7 October 2008; pp. 10–20.
19. Alshbatat, A.I.; Dong, L. Cross layer design for mobile Ad-Hoc Unmanned Aerial Vehicle communication networks. In Proceedings of the 2010 International Conference on Networking, Sensing and Control (ICNSC), Chicago, IL, USA, 10–12 April 2010; pp. 331–336. [CrossRef]

20. Singh, K.; Verma, A.K. Experimental analysis of AODV, DSDV and OLSR routing protocol for flying adhoc networks (FANETs). In Proceedings of the 2015 IEEE International Conference on Electrical, Computer and Communication Technologies (ICECCT), Coimbatore, India, 5–7 March 2015; pp. 1–4.
21. Yadav, M.; Gupta, S.K.; Saket, R.K. Multi-hop wireless ad-hoc network routing protocols- a comparative study of DSDV, TORA, DSR and AODV. In Proceedings of the 2015 International Conference on Electrical, Electronics, Signals, Communication and Optimization (EESCO), Visakhapatnam, India, 24–25 January 2015; pp. 1–5. [CrossRef]
22. Broch, J.; Maltz, D.A.; Johnson, D.B.; Hu, Y.-C.; Jetcheva, J. A performance comparison of multi-hop wireless ad hoc network routing protocols. In Proceedings of the Fourth Annual ACM/IEEE International Conference on Mobile Computing and Networking (MobiCom'98), Dallas, TX, USA, 25–30 October 1998; pp. 85–97.
23. Pojda, J.; Wolff, A.; Sbeiti, M.; Wietfeld, C. Performance analysis of mesh routing protocols for UAV swarming applications. In Proceedings of the 2011 8th International Symposium on Wireless Communication Systems, Aachen, Germany, 6–9 November 2011; pp. 317–321.
24. Sanchez-Iborra, R.; Cano, M.-D.; Garcia-Haro, J. Performance Evaluation of BATMAN Routing Protocol for VoIP Services: A QoE Perspective. *IEEE Trans. Wirel. Commun.* **2014**, *13*, 4947–4958. [CrossRef]
25. Zheng, Y.; Jiang, Y.; Dong, L.; Wang, Y.; Li, Z.; Zhang, H. A mobility and load aware OLSR routing protocol for UAV mobile ad-hoc networks. In Proceedings of the 2014 International Conference on Information and Communications Technologies (ICT 2014), Nanjing, China, 15–17 May 2014; pp. 1–7.
26. Li, Y.; Luo, X. Cross layer optimization for cooperative mobile ad-hoc UAV network. *Int. J. Digit. Content Technol. Appl.* **2012**, *6*, 367–375.
27. Paul, A.B.; Nandi, S. Modified Optimized Link State Routing (M-OLSR) for Wireless Mesh Networks. In Proceedings of the 2008 International Conference on Information Technology, Bhubaneswar, India, 17–20 December 2008; pp. 147–152.
28. Belhassen, M.; Belghith, A.; Abid, M.A. Performance evaluation of a cartography enhanced OLSR for mobile multi-hop ad hoc networks. In Proceedings of the 2011 Wireless Advanced, London, UK, 20–22 June 2011; pp. 149–155. [CrossRef]
29. Bellur, B.; Ogier, R.G.; Templin, F.L. Topology broadcast based on reverse-path forwarding routing protocol (tbrpf). In *ETF Internet Draft*; RFC: Fremont, CA, USA, 2003.
30. Pei, G.; Gerla, M.; Chen, T.-W. Fisheye state routing: A routing scheme for ad hoc wireless networks. In Proceedings of the 2000 IEEE International Conference on Communications, ICC 2000, Global Convergence through Communications, Conference Record, New Orleans, LA, USA, 18–22 June 2000; pp. 70–74. [CrossRef]
31. Chroboczek, J. The Babel Routing Protocol. 2011. Available online: https://www.hjp.at/(en)/doc/rfc/rfc6126.html (accessed on 11 May 2021).
32. Johnson, D.B.; Maltz, D.A. Dynamic source routing in ad hoc wireless networks. In *Mobile Computing*; Springer: Boston, MA, USA, 2007; pp. 153–181.
33. Perkins, C.E.; Royer, E.M. Ad-hoc on-demand distance vector routing. In Proceedings of the 2nd IEEE Workshop on Mobile Computing Systems and Applications (WMCSA 1999), New Orleans, LA, USA, 25–26 February 1999; pp. 90–100.
34. Forsmann, J.H.; Hiromoto, R.E.; Svoboda, J. A time-slotted on-demand routing protocol for mobile ad hoc unmanned vehicle systems. In Proceedings of the SPIE 6561, Unmanned Systems Technology IX, Orlando, FL, USA, 2 May 2007; Volume 6561, p. 65611P. [CrossRef]
35. Royer, E. Multicast ad hoc on-demand distance vector (MAODV) routing. In *IETF Internet Draft*; RFC: Fremont, CA, USA, 2000.
36. Aggarwal, A. AODVSEC: A Novel Approach to Secure Ad Hoc On-Demand Distance Vector (AODV) Routing Protocol from Insider Attacks in MANETs. *Int. J. Comput. Netw. Commun.* **2012**, *4*, 191–210. [CrossRef]
37. Haas, P.; Pearlman, Z.J.; Samar, M.R. The Zone Routing Protocol (ZRP) for Ad Hoc Networks. In *IETF Draft*; RFC: Fremont, CA, USA, 2002; p. 11.
38. Park, V.; Corson, S. Temporally-Ordered Routing Algorithm (TORA) Version 1 Functional Specification. In *IETF MANET Work. Gr. INTERNET-DRAFT*; RFC: Fremont, CA, USA, 2002; pp. 1–23.
39. Zhai, Z.; Du, J.; Ren, Y. The Application and Improvement of Temporally Ordered Routing Algorithm in Swarm Network with Unmanned Aerial Vehicle Nodes. In Proceedings of the ICWMC 2013: The Ninth International Conference on Wireless and Mobile Communications, Nice, France, 21–26 July 2013; pp. 7–12.
40. Wei, Y.; Blake, M.B.; Madey, G.R. An Operation-Time Simulation Framework for UAV Swarm Configuration and Mission Planning. *Procedia Comput. Sci.* **2013**, *18*, 1949–1958. [CrossRef]
41. Ramasubramanian, V.; Haas, Z.J.; Sirer, E.G. SHARP: A Hybrid Adaptive Routing Protocol for Mobile Ad Hoc Networks. In Proceedings of the 4th ACM International Symposium Mobile ad hoc Network Computer—MobiHoc '03, Annapolis, MD, USA, 1–3 June 2013; p. 303.
42. Pei, G.; Gerla, M.; Hong, X.; Chiang, C.-C. A wireless hierarchical routing protocol with group mobility. In Proceedings of the 1999 IEEE Wireless Communications and Networking Conference (Cat. No.99TH8466), New Orleans, LA, USA, 21–24 September 1999; Volume 3, pp. 1538–1542. [CrossRef]
43. Shirani, R. Reactive-Greedy-Reactive in Unmanned Aeronautical Ad-Hoc Networks: A Combinational Routing Mechanism. Master's Thesis, Carleton University, Ottawa, ON, Canada, August 2011.

44. Shirani, R.; St-Hilaire, M.; Kunz, T.; Zhou, Y.; Li, J.; Lamont, L. The Performance of Greedy Geographic Forwarding in Unmanned Aeronautical Ad-Hoc Networks. In Proceedings of the 2011 Ninth Annual Communication Networks and Services Research Conference, Ottawa, ON, Canada, 2–5 May 2011; pp. 161–166.
45. Iordanakis, M.; Yannis, D.; Karras, K.; Bogdos, G.; Dilintas, G.; Amirfeiz, M.; Colangelo, G.; Baiotti, S. Ad-hoc routing protocol for aeronautical mobile ad-hoc networks. In Proceedings of the Fifth International Symposium on Communication Systems, Networks and Digital Signal Processing (CSNDSP), Achaia, Greece, 19–21 July 2006; pp. 1–5.
46. Ni, M.; Zhong, Z.; Wu, H.; Zhao, D. A New Stable Clustering Scheme for Highly Mobile Ad Hoc Networks. In Proceedings of the 2010 IEEE Wireless Communication and Networking Conference, Sydney, NSW, Australia, 18–21 April 2010; pp. 1–6. [CrossRef]
47. Dora, D.P.; Kumar, S.; Kaiwartya, O. Efficient dynamic caching for geocast routing in VANETs. In Proceedings of the 2015 2nd International Conference on Signal Processing and Integrated Networks (SPIN), Noida, India, 19–20 February 2015; pp. 979–983.
48. Lin, L.; Sun, Q.; Li, J.; Yang, F. A novel geographic position mobility oriented routing strategy for UAVs. *J. Comput. Inf. Syst.* **2012**, *8*, 709–716.
49. Lin, L.; Sun, Q.; Wang, S.; Yang, F. A geographic mobility prediction routing protocol for Ad Hoc UAV Network. In Proceedings of the 2012 IEEE Globecom Workshops, Anaheim, CA, USA, 3–7 December 2012; pp. 1597–1602. [CrossRef]
50. Medina, D.; Hoffmann, F.; Rossetto, F.; Rokitansky, C.-H. A Geographic Routing Strategy for North Atlantic In-Flight Internet Access Via Airborne Mesh Networking. *IEEE/ACM Trans. Netw.* **2011**, *20*, 1231–1244. [CrossRef]
51. Lidowski, R.L.; Mullins, B.E.; Baldwin, R.O. A novel communications protocol using geographic routing for swarming UAVs performing a Search Mission. In Proceedings of the 2009 IEEE International Conference on Pervasive Computing and Communications, Galveston, TX, USA, 9–13 March 2009; pp. 1–7.
52. Liu, C.; Wu, J. Efficient Geometric Routing in Three Dimensional Ad Hoc Networks. In Proceedings of the IEEE INFOCOM 2009—The 28th Conference on Computer Communications, Rio de Janeiro, Brazil, 19–25 April 2009; pp. 2751–2755.
53. Flury, R.; Wattenhofer, R. Randomized 3D geographic routing. In Proceedings of the IEEE INFOCOM 2008—The 27th Conference on Computer Communications, Phoenix, AZ, USA, 13–18 April 2008; pp. 1508–1516.
54. Zhou, J.; Chen, Y.; Leong, B.; Sundaramoorthy, P.S. Practical 3D geographic routing for wireless sensor networks. In Proceedings of the 8th ACM Conference on Web Science, Zurich, Switzerland, 3–5 November 2010; pp. 337–350.
55. Oubbati, O.S.; Lakas, A.; Lagraa, N.; Yagoubi, M.B. CRUV: Connectivity-based traffic density aware routing using UAVs for VANets. In Proceedings of the 2015 International Conference on Connected Vehicles and Expo (ICCVE), Shenzhen, China, 19–23 October 2015; pp. 68–73. [CrossRef]
56. Oubbati, O.S.; Lakas, A.; Lagraa, N.; Yagoubi, M.B. UVAR: An intersection UAV-assisted VANET routing protocol. In Proceedings of the 2016 IEEE Wireless Communications and Networking Conference, Doha, Qatar, 3–6 April 2016; pp. 1–6. [CrossRef]
57. Sbeiti, M.; Goddemeier, N.; Behnke, D.; Wietfeld, C. PASER: Secure and Efficient Routing Approach for Airborne Mesh Networks. *IEEE Trans. Wirel. Commun.* **2016**, *15*, 1950–1964. [CrossRef]
58. Rosário, D.; Zhao, Z.; Braun, T.; Cerqueira, E.; Santos, A.; Alyafawi, I. Opportunistic routing for multi-flow video dissemination over Flying Ad-Hoc Networks. In Proceedings of the IEEE International Symposium on a World of Wireless, Mobile and Multimedia Networks 2014, Sydney, NSW, Australia, 19 June 2014; pp. 1–6. [CrossRef]
59. Liu, K.; Zhang, J.; Zhang, T. The clustering algorithm of UAV Networking in Near-space. In Proceedings of the 2008 8th International Symposium on Antennas, Propagation and EM Theory, Kunming, China, 2–5 November 2008; pp. 1550–1553.
60. Zang, C.; Zang, S. Mobility prediction clustering algorithm for UAV networking. In Proceedings of the 2011 IEEE GLOBECOM Workshops (GC Wkshps), Houston, TX, USA, 5–9 December 2011; pp. 1158–1161. [CrossRef]
61. Pei, G.; Gerla, M.; Hong, X. LANMAR: Landmark routing for large scale wireless Ad Hoc Networks with group mobility. In Proceedings of the 2000 First Annual Workshop on Mobile and Ad Hoc Networking and Computing, MobiHOC (Cat. No.00EX444), Boston, MA, USA, 11 August 2000; pp. 11–18.
62. Martin, N.; Al-Mousa, Y.; Shenoy, N. An Integrated Routing and Medium Access Control Framework for Surveillance Networks of Mobile Devices. *Comput. Vis.* **2011**, *6522*, 315–327. [CrossRef]
63. Shi, N.; Luo, X. A Novel Cluster-Based Location-Aided Routing Protocol for UAV Fleet Networks. *Int. J. Digit. Content Technol. Appl.* **2012**, *6*, 376–383. [CrossRef]
64. Fu, B.; Dasilva, L.A. A mesh in the sky: A routing protocols for airbone networks. In Proceedings of the MILCOM 2007—IEEE Military Communications Conference, Orlando, FL, USA, 29–31 October 2007; pp. 1–7.
65. Leonov, A.V. Application of bee colony algorithm for FANET routing. In Proceedings of the 2016 17th International Conference of Young Specialists on Micro/Nanotechnologies and Electron Devices (EDM), Erlagol, Russia, 30 June–4 July 2016; pp. 124–132. [CrossRef]
66. Leonov, A.V. Modeling of bio-inspired algorithms AntHocNet and BeeAdHoc for Flying Ad Hoc Networks (FANETs). In Proceedings of the 2016 13th International Scientific-Technical Conference on Actual Problems of Electronics Instrument Engineering (APEIE), Novosibirsk, Russia, 3–6 October 2016; Volume 2, pp. 90–99. [CrossRef]
67. Yu, Y.; Ru, L.; Chi, W.; Liu, Y.; Yu, Q.; Fang, K. Ant colony optimization based polymorphism-aware routing algorithm for ad hoc UAV network. *Multimedia Tools Appl.* **2016**, *75*, 14451–14476. [CrossRef]
68. Kuiper, E.; Nadjm-Tehrani, S. Geographical Routing with Location Service in Intermittently Connected MANETs. *IEEE Trans. Veh. Technol.* **2010**, *60*, 592–604. [CrossRef]

69. Jabbar, A.; Sterbenz, J.P.G. AeroRP: A Geolocation Assisted Aeronautical Routing Protocol for Highly Dynamic Telemetry Environments. In Proceedings of the International Telemetering Conference, Las Vegas, NV, USA, 26–29 October 2009; pp. 1–10.
70. Hyeon, S.; Kim, K.-I. A new geographic routing protocol for aircraft ad hoc networks. In Proceedings of the 29th Digital Avionics Systems Conference, Salt Lake City, UT, USA, 3–7 October 2010; pp. 1–8. [CrossRef]
71. Whitbeck, J.; Conan, V. HYMAD: Hybrid DTN-MANET routing for dense and highly dynamic wireless networks. *Comput. Commun.* **2010**, *33*, 1483–1492. [CrossRef]
72. Burgess, J.; Gallagher, B.; Jensen, D.; Levine, B.N. MaxProp: Routing for Vehicle-Based Disruption-Tolerant Networks. In Proceedings of the IEEE INFOCOM 2006, 25th IEEE International Conference on Computer Communications, Barcelona, Spain, 23–29 April 2006; Volume 6, pp. 1–11. [CrossRef]
73. Spyropoulos, T.; Psounis, K.; Raghavendra, C.S. Spray and wait: An efficient routing scheme for intermittently connected mobile networks. In Proceedings of the ACM SIGCOMM 2005 Work, Delay-Tolerant Networking, WDTN 2005, Philadelphia, PA, USA, 22–26 August 2005; pp. 252–259.
74. Lindgren, A.; Doria, A.; Schelén, O. Probabilistic routing in intermittently connected networks. *ACM SIGMOBILE Mob. Comput. Commun. Rev.* **2003**, *7*, 19–20. [CrossRef]
75. mqmaker. WiTi Board. Available online: https://goo.gl/bfSvM8 (accessed on 4 December 2017).
76. Iperf. Available online: https://iperf.fr/ (accessed on 20 May 2016).
77. Khawaja, W.; Guvenc, I.; Matolak, D.W.; Fiebig, U.-C.; Schneckenburger, N. A Survey of Air-to-Ground Propagation Channel Modeling for Unmanned Aerial Vehicles. *IEEE Commun. Surv. Tutor.* **2019**, *21*, 2361–2391. [CrossRef]

applied sciences

MDPI

Article

Hybrid Direction of Arrival Precoding for Multiple Unmanned Aerial Vehicles Aided Non-Orthogonal Multiple Access in 6G Networks

Laura Pierucci

Department of Information Engineering, University of Florence, 50139 Firenze, Italy; laura.pierucci@unifi.it

Abstract: Unmanned aerial vehicles (UAV) have attracted increasing attention in acting as a relay for effectively improving the coverage and data rate of wireless systems, and according to this vision, they will be integrated in the future sixth generation (6G) cellular network. Non-orthogonal multiple access (NOMA) and mmWave band are planned to support ubiquitous connectivity towards a massive number of users in the 6G and Internet of Things (IOT) contexts. Unfortunately, the wireless terrestrial link between the end-users and the base station (BS) can suffer severe blockage conditions. Instead, UAV relaying can establish a line-of-sight (LoS) connection with high probability due to its flying height. The present paper focuses on a multi-UAV network which supports an uplink (UL) NOMA cellular system. In particular, by operating in the mmWave band, hybrid beamforming architecture is adopted. The MUltiple SIgnal Classification (MUSIC) spectral estimation method is considered at the hybrid beamforming to detect the different direction of arrival (DoA) of each UAV. We newly design the sum-rate maximization problem of the UAV-aided NOMA 6G network specifically for the uplink mmWave transmission. Numerical results point out the better behavior obtained by the use of UAV relays and the MUSIC DoA estimation in the Hybrid mmWave beamforming in terms of achievable sum-rate in comparison to UL NOMA connections without the help of a UAV network.

Keywords: hybrid precoding; millimeter wave; non-orthogonal multiple access scheme; massive MIMO; unmanned aerial vehicles; direction of arrivals (DoA); MUSIC algorithm

Citation: Pierucci, L. Hybrid Direction of Arrival Precoding for Multiple Unmanned Aerial Vehicles Aided Non-Orthogonal Multiple Access in 6G Networks. *Appl. Sci.* **2022**, *12*, 895. https://doi.org/10.3390/app12020895

Academic Editors: Sylvain Bertrand and Juan-Carlos Cano

Received: 29 November 2021
Accepted: 15 January 2022
Published: 16 January 2022

1. Introduction

The explosive number of devices demands data traffic growth and new radio spectrum resources in future 6G systems and the IoT context. Key enabling technologies are (i) the underutilized millimeter wave (mmWave) band (between 30 GHz and 300 GHz), which could be valuable for its wide bandwidth and higher spectral efficiency, and (ii) NOMA, which could be valuable for simultaneously supporting multiple users on the same time-frequency resources. The joint use of the large spectrum available in mmWaves together with massive multiple-input-multiple-output (MIMO) strategies allows ultra-high data rates to be guaranteed through spatial directional transmissions compensating for the high propagation loss of mmWaves communications [1]. This directional nature of mmWaves transmissions needs the support of one radio frequency (RF) chain for each user on the same time-frequency resource. Therefore, the hardware complexity and costs of the mmWave MIMO system increase with an increase in the number of users. Hybrid architectures, which combine phase shifters based on analog precoding and digital precoding, reduce the costs practically by selecting a reduced number of RF chains. However, even if hybrid beamforming structures are implemented, the user's channels are highly correlated in mmWave communications, and thus, the users cannot be separated by linear operations. Such a correlation facilitates integration with NOMA technology. NOMA can simultaneously serve multiple users on the same time-frequency resource by converting their channel gains into multiplexed gains in the power domain and by using successive interference cancellation (SIC) at the receiver to remove intra-channel interference with a decoding order based on the channel conditions [2–4].

In the literature, different papers analyze the spectral efficiency (SE) and the energy efficiency (EE) maximization problem mainly for a downlink mmWave MIMO with hybrid beamforming [5,6]. In [3], a power allocation (PA) problem to optimize EE is considered for an uplink NOMA-assisted mmWave MIMO system under users' quality-of-service (QoS) and quality-of-experience requirements [7]. In [4], power allocation and beamforming are jointly considered to maximize the sum rate of a pair of users in a mmWave NOMA system by using an analog beamforming structure indeed of a hybrid mmWave beamforming structure.

Simultaneous wireless information and power transfer (SWIPT) techniques are integrated in mmWave massive MIMO-NOMA systems to maximize the energy efficiency in [8], and consequently, each user can extract both information and energy from the received RF signals by using a power splitting receiver.

In the 6G vision, low Earth orbit (LEO) satellites and diverse aerial platforms, such as UAVs, are considered to support IoT development in remote areas and in emergency situations thanks to the mobility, flexibility and good channel conditions of UAVs. UAVs have been developed for their monitoring capabilities, implemented by on board sensors, for services in agriculture or security border controls. In addition, UAVs are used as a flying base station (BS) to provide ubiquitous connectivity and effectively increase the coverage and throughput of wireless systems through the optimization of UAVs' positions and trajectories [9].

Recently, UAVs have been considered to act as relays for cooperative communications due to the high probability to establish LoS links with the ground terminals. UAV-assisted relaying systems operate according to the two classical types of transmission protocols, namely decode-and-forwarding (DF) and amplify-and-forwarding (AF) [10].

Several papers consider UAV communications combined with orthogonal multiple access (OMA) to maximize the energy efficiency or throughput by optimizing the source/relay transmit power and the UAV speed and trajectory design as, e.g., in [11]. The paper [12] focuses on a UAV full duplex (FD) relay with joint beamforming and power allocation to optimize the instantaneous data rate when the UAV flight follows a circular trajectory. FD relaying allows for a relay node to simultaneously transmit and receive in the same band, unlike half-duplex (HD) mode [10,13]. Therefore, a natural choice is to combine FD relays with NOMA to enhance spectral efficiency.

A UAV-supported clustered-NOMA system for the 6G-enabled IoT is detailed in [14], where the numerous terminals are partitioned into clusters and the UAV provides services to the clusters by using wireless-powered communication (WPC) to optimize the uplink average achievable sum rate of all terminals by designing the UAVs' trajectory.

In [15], a multiple-UAV-aided NOMA scheme is proposed to improve spectral and energy efficiency for cellular uplinks. In particular, half of users are partitioned in clusters served by multiple UAV relays, and the other ones communicate with the BS directly. A location-based user pairing (UP) scheme associates the clustered users with the multiple-UAV-aided NOMA to minimize the resource allocation problem.

However, all the aforementioned works focus on UAV-relay-aided NOMA without considering mmWave communications. The severe signal power attenuation in the mmWave band impacts UAV connectivity performance, especially when very long communications distances exist between the ground users and the associated UAV which serves them.

Therefore, this paper addresses the design of a UAV-enabled FD relaying network in the mmWave band to aid an uplink NOMA cellular system.

In detail, a multiple-UAV-relaying network supports NOMA technology, and hybrid mmWave beamforming is considered at the base station (BS), which can estimate the DoA information of UAVs to improve the overall sum rate of the system. Indeed, the DoAs are unknown and the MUltiple SIgnal Classification (MUSIC) method [16] considered in this paper estimates directly the DoAs at the hybrid mmWave beamformer.

In [17], a beamspace MUSIC algorithm is used to estimate path directions for mmWave channel estimation problem showing that the hybrid precoding structure can avoid the spectrum ambiguity and maximize the number of resolvable path directions. The mmWave channel estimation problem is also considered in [18] by using MUSIC for the hybrid analog/digital beamformer with 2D co-prime arrays where the directions can be uniquely estimated by finding the common peaks of the 2 decomposed subarrays.

Deep-learning approaches are considered to evaluate the angle-of-arrival (AoA) information in the uplink of an mmWave communication system based on MUSIC to enhance classification accuracy in [19]. The above papers use the MUSIC algorithm to derive the DoA information in mmWave hybrid beamforming, but they do not deploy an UAV-relay network, as in this paper.

To the best of our knowledge, this paper is the first contribution that considers jointly (i) the UL mmWave communications, (ii) the hybrid beamforming with DoA estimation based on the MUSIC method and (iii) multiple UAVs acting as an aerial BS to relay NOMA transmissions and, consequently, achieve better data rate for users who can suffer severe channel conditions. Moreover, a novel maximization design of the overall sum-rate is proposed for the uplink mmWave transmission of the multiple-UAV-relay-aided NOMA 6G system.

The contributions of this paper can be summarized as follows:

- We consider a multiple-UAV-aided NOMA network where each UAV acts as a mobile FD relay in mmWave UL for 6G cellular systems;
- We propose the use of the MUSIC algorithm at the hybrid beamforming to detect the DoA estimations for each UAV and improve the performance of the UAV-aided NOMA cellular system;
- We propose an optimization procedure in order to maximize the average achievable sum rate of the UL mmWave UAV relay network by taking into accounts LoS obstruction, channel time-varying condition, the DoA information at the hybrid beamforming due to the different spatial directions of UAVs, as well as the requirements in terms of quality of service (QoS) for each user.

The paper is organized as follows. Section 2 presents the proposed system model, channel model, the mmWave hybrid beamforming, the MUSIC algorithm and the problem formulation to optimize the global UL sum rate of the UAV-network-aided NOMA cellular system. In Section 3, the numerical results providing a comparison with the NOMA cellular system without the use of multi-UAV relays are shown. Finally, Section 4 concludes the paper and outlines future research activities.

2. System Model and Problem Formulation

A UAV-enabled full-duplex relaying system is considered to aid an uplink mmWave NOMA cellular system consisting in a BS, with N users and K UAVs, as shown in Figure 1. The N users are randomly distributed in the cell of radius R, and the BS is located at the center of the cell.

We assume that the BS is unable to deliver the superimposed signals to the NOMA users in a far subarea of the cell because the link between users and BS is negligible due to severe blockage.

Each UAV acts as a DF relay to help data transmission between the BS and users and operates in FD mode.

We suppose that K UAVs are flying at height H_k in such a way that all links from UAVs to BS are LOS channels. The elevation angles between the BS and the UAVs are denoted as $(\theta_1, \theta_2, ..., \theta_k)$, and $\varphi_{n,k}$ is the elevation angle between the n-th user (named as user equipment UE_n) and UAV_k, as shown in Figure 1.

2.1. Path Loss Model

The links from ground users and UAV are LOS or no LOS (NLOS) channels due to the presence of buildings, vegetation etc., which can obstruct the signals propagation. In detail, the channel gain between the k-th UAV and the BS is, according to [20–22],

$$\rho_{k,BS} = \zeta_{LOS}(d_{k,BS})^{-\alpha}, \tag{1}$$

and if the UAV_k is in LOS with the $n-th$ user, the channel gain is defined as

$$\rho_{n,k} = \zeta_{LOS}(d_{n,k})^{-\alpha}, \tag{2}$$

where $d_{k,BS}$ and $d_{n,k}$ represent the distances between UAV_k and BS and between UAV_k and UE_n, respectively. ζ_{LOS} denotes the additional attenuation factor of the LOS channel at the reference distance $d_{ref} = 1$ m, and α is the path loss exponent at the air channel. UAVs are in LOS with probability $p_{LOS}(d) = \frac{1}{1+Cexp(-(B\varphi_{n,k}-C))}$, where B and C are constants related to the environment, whereas UAVs are in NLOS according to the complementary probability $p_{NLOS}(d) = 1 - p_{LOS}(d)$.

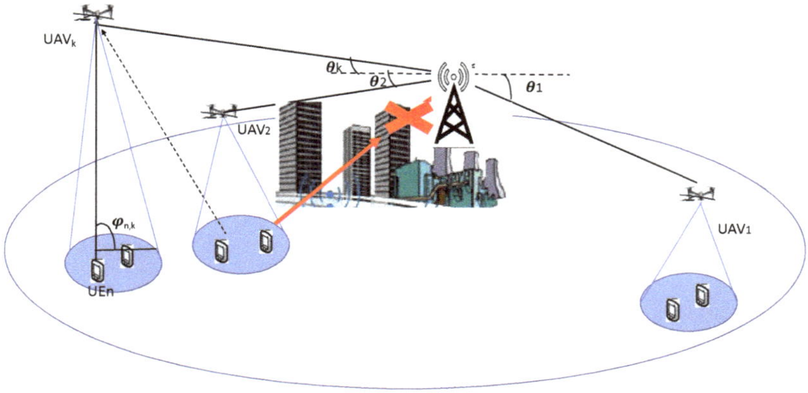

Figure 1. Multiple-UAV-relay-network-aided mmWave NOMA cellular system.

2.2. MmWave Hybrid Beamforming

In mmWave communications, large phase arrays are usually adopted to overcome the high propagation losses, and in combining with NOMA, mmWave beamforming is used to increase beam gain and serve multiple users. Usually, hybrid analog/digital beamforming is adopted in mmWave NOMA communications, where the precoding is performed in hybrid mode by combining the digital baseband precoding with an analog RF beamforming driven by a limited number of RF chains. This hybrid analog/digital beamforming is a cost-effective solution due to the use of massive antennas with limited RF chains. It can be easily implemented through the use of analog phase shifters together with the abilities of digital precoding, which allow the beams to be directed towards the desired user and remove inter-user interferences [23,24].

In particular, we consider the hybrid beamforming architecture at the BS with a number N_{RF} of RF chains exploiting NOMA in each RF chain and spatial division multiple access (SDMA) between RF chains [8], as shown in Figure 2 .

From the angle domain perspective, the knowledge of DoAs plays a fundamental role.

In this paper, the UAVs' angle information are discovered by using the MUSIC algorithm implemented at the hybrid mmWave beamforming. We assume that the number of UAVs does not exceed the number of available RF chains, i.e., $K \leq N_{RF}$.

The received signal at the BS after RF and digital beamforming combining can be expressed as [19,24]

$$\mathbf{y} = \mathbf{D}^{\mathbf{H}}\mathbf{A}^{\mathbf{H}}\mathbf{H}\mathbf{s} + \mathbf{D}^{\mathbf{H}}\mathbf{A}^{\mathbf{H}}\mathbf{w}, \tag{3}$$

where $\mathbf{H}_{N_R \times K}$ is the channel matrix (detailed in Equation (6)), N_R is the number of receiving antennas at the BS, $\mathbf{s}$ denotes the transmit signal, $\mathbf{w}$ is the zero-mean independent and identically distributed (i.i.d.) Gaussian white noise vector with power σ_w^2 and $\mathbf{A}_{N_R \times N_{RF}}$, $\mathbf{D}_{N_{RF} \times K}$ are the analog and digital beamforming matrices (see Equations (4) and (5)), respectively, with $N_{RF} < N_R$.

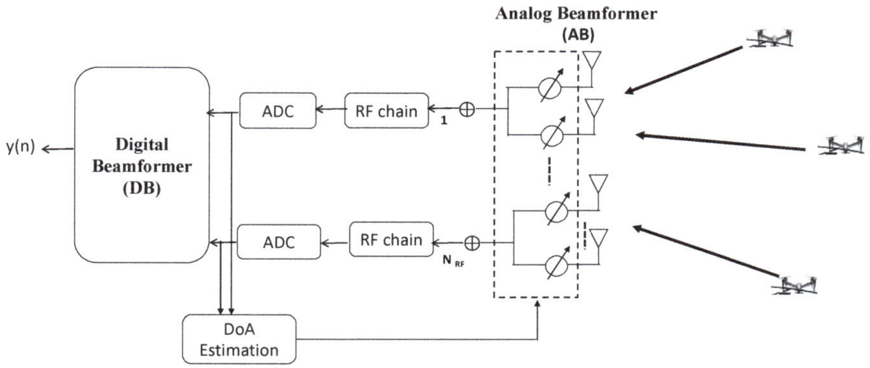

Figure 2. Hybrid mmWave beamformer with DoAs estimation.

In partially-connected hybrid mmWave beamforming, the antenna array with N_R elements can be organized into groups, called subarrays, and each subarray connected to one RF chain processes each received signal by a phase shifter. Then, all of them are added up as shown in Figure 2.

Therefore, the analog matrix $\mathbf{A}_{N_R \times N_{RF}}$ of Equation (3) is a diagonal matrix where

$$v_{A,j} = \frac{1}{\sqrt{N_R}}[e^{j\phi_{j,1}}, e^{j\phi_{j,2}} ... e^{j\phi_{j,N_R}}] \tag{4}$$

is the vector of subarray j with $j = 1, ..N_{RF}$; ϕ is chosen from a uniform distribution in the range of $[-\pi/2, \pi/2]$. The RF signal passes through N_{RF} parallel RF chains. It is down-converted and then the digital beamforming operation follows. The zero-forcing scheme can be used as a digital beamformer as

$$\mathbf{D} = (\mathbf{H_A^H H_A})^{-1} \mathbf{H_A^H} \tag{5}$$

where $\mathbf{H_A} = \mathbf{A^H H}$ considers the actual channel matrix $\mathbf{H}$ filtered by the analog beamforming matrix $\mathbf{A}$.

2.3. Channel Model

We consider a ray-tracing channel model widely used in mm-Wave communications with a limited number of L scattering paths, as, due to the spatial sparsity in the mm-wave channel, it is expected that the propagation paths are along a small number of directions [19,25,26].

Therefore, each column of the channel matrix $\mathbf{H}$ is defined as

$$\mathbf{h_k} = \sqrt{\frac{1}{L\rho_{k,BS}}} \sum_{i=1}^{L} g_{k,i} \mathbf{a}(\theta_{k,i}) \tag{6}$$

where $g_{k,i}$ is the complex gain of the i-path due to small-scale fading, $\rho_{k,BS}$ is the path loss between the BS and the UAV$_k$ and $\theta_{k,i} \in [-\pi/2, ..., \pi/2]$ is the angle of arrival of the $i-th$ path of the $k-th$ UAV.

Without loss of generality, we assume a uniform linear array (ULA) at the BS for simplicity, and the steering vector can be expressed as:

$$\mathbf{a}(\theta_{k,i}) = [1, e^{j\frac{2\pi}{\lambda} l sin(\theta_{k,i})}, ...e^{j(N_R-1)\frac{2\pi}{\lambda} l sin(\theta_{k,i})}] \tag{7}$$

where λ is the signal wavelength, and l is the distance among antenna elements, where $l \leq \lambda/2$.

In the following, we assume that in Equation (6) the variations of the channel are only caused by the path gains $g_{k,i}$, and the path angles remain unchanged according to the mmWave channel measurements in [27].

2.4. MUSIC Algorithm

The DoA information is estimated by MUSIC spectral estimation on the filtered version of the received signal, as shown in Figure 2, [19,25,26]. The MUSIC algorithm developed by Schmidt [16] is an eigenstructure-based DOA-finding method and, similar to other parametric algorithms such as ESPRIT, has demonstrated a superior resolution with respect to the non-parametric methods.

By considering a partially connected hybrid mmWave beamforming, after the analog beamforming and the analog-to-digital conversion, the baseband signal is [19,28]:

$$\mathbf{y} = \mathbf{A^H H s} + \mathbf{A^H w} \quad (8)$$

By performing eigenvalue decomposition on the covariance matrix $\mathbf{R_{yy}}$ of the output vector $\mathbf{y}$ of the virtual array, we have:

$$\mathbf{R_{yy}} = E_S \Lambda_S E_S^H + \sigma_w^2 E_w E_w^H$$

where $\mathbf{E_s}$ is the $(N_{RF} \times K)$ matrix of the signal eigenvectors corresponding to the K largest eigenvalues, $\mathbf{E_w}$ is the $(N_{RF} \times (N_{RF} - K))$ noise subspace matrix with eigenvectors corresponding to the smallest $(N_{RF} - K)$ singular values, and $\mathbf{\Lambda_s}$ is the $(K \times K)$ diagonal matrix containing the K largest eigenvalues $\lambda_1 > \lambda_2, \ldots > \lambda_K$ of $\mathbf{R}_{yy}$. It is clear that the signal subspace E_s and the noise subspace E_w are orthogonal.

The MUSIC algorithm utilizes the orthogonality between the two complementary spaces to estimate the spatial signal. Therefore, the DoA estimation $\hat{\theta_k}$ consists of finding the values of θ, whereby the filtered vector $Aa_D(\theta)$ is related to the signal subspace of $\mathbf{R}_{yy}$, where

$$\mathbf{a_D}(\theta) = \frac{1}{\sqrt{N_R}}[1, e^{j\frac{2\pi}{\lambda}lsin(\theta)}, \ldots. e^{j\frac{2\pi}{\lambda}(N_R-1)lsin(\theta)}]$$

is the array manifold vector of the virtual array.

By using the definition of a pseudo-spectrum of the MUSIC algorithm, the estimated DoA of the emitter direction can be calculated by maximizing the function

$$P_{MU}(\hat{\theta}) = \frac{1}{\mathbf{a_D^H}(\theta)\mathbf{A E_w E_w^H A^H a_D}(\theta)}, \quad (9)$$

which provides high resolution of angle separation.

2.5. Problem Formulation

Considering the NOMA method in each beam, intra-beam superposition coding at the transmitter and SIC at the receiver are performed. In the case of uplink mmWave NOMA, the users begin to transmit uplink signals x_1 and x_2 at the same time and in the same frequency band. For the proposed multiple-UAV relay network shown in Figure 1, each UAV decodes the mixed signals of the two users in its beam (called, for example, UE_1 and UE_2) and then transmits the superimposed signal $\mathbf{s}$ to the BS according to

$$\mathbf{s} = \sqrt{\mathbf{P_1}}\hat{x_1} + \sqrt{\mathbf{P_2}}\hat{x_2}, \quad (10)$$

where $\hat{x_1}$ and $\hat{x_2}$ are the decoded signals of UE_1 and UE_2, respectively, and $\mathbb{E}[|\hat{x_1}|^2] = \mathbb{E}[|\hat{x_2}|^2] = 1$ and $P_1 + P_2$ cannot exceed the maximum transmission power of the UAV.

In the conventional NOMA with a single-antenna, usually the information of the user with a lower channel gain is decoded first to maximize the sum rate. In contrast, in mmWave NOMA, the decoding order depends on both channel gain and beamforming gain. Without loss of generality, assuming that UE_1 has a better channel condition with respect to UE_2 in the area covered by UAV_k [4,29], the achievable rates are

$$R_1 = log_2(1 + \frac{|\mathbf{h_1^H U}|^2 P_1}{|\mathbf{h_2^H U}|^2 P_2 + \sigma_k^2}) \tag{11}$$

$$R_2 = log_2(1 + \frac{|\mathbf{h_2^H U}|^2 P_2}{\sigma_k^2}), \tag{12}$$

and the achievable rate from UAV_k is

$$R_{UAV_k} = log_2(1 + \frac{|\mathbf{h_k^H W}|^2 P_{UAV_k}}{\sigma_w^2}), \tag{13}$$

where P_1, P_2 and P_{UAV_k} are the transmission power of UE_1, UE_2 and UAV_k respectively, $\mathbf{h_1}$, $\mathbf{h_2}$ are the channel response vector between the UE_1, UE_2 and UAV_k, $\mathbf{h_k}$ is the channel response vector between the UAV_k and the BS, $\mathbf{U}$ and $\mathbf{W}$ represent jointly the analog and digital precoding matrices at the UAV_k and at the BS, whereas σ_k^2 and σ_w^2 are the power of the zero-mean additive Gaussian white noise at the UAV_k and BS, respectively.

Therefore, the available rate received at the BS considering the UAV_k acting as a relay is

$$R_{sum} = min(R_1 + R_2, R_{UAV_k}). \tag{14}$$

To maximize the uplink average achievable sum rate of all terminals of the multiple-UAV-aided NOMA mmWave system by dynamically tracking the DoAs of multiple UAVs, the optimization problem can be formulated as

$$\begin{aligned} &\underset{\theta_k}{\text{Maximize}} \sum_{k=1}^{K} R_{sum} \\ &\quad s.t. \\ &\quad\quad R_1 \geq \tilde{r} \\ &\quad\quad R_2 \geq \tilde{r} \\ &\quad\quad R_{sum} \geq \tilde{r} \\ &\quad\quad P_1 + P_2 \leq P_{UAV_k} \\ &\quad\quad P_{UAV_1} + P_{UAV_2} + \ ... \ P_{UAV_k} \leq P \end{aligned} \tag{15}$$

where $\tilde{r}$ denotes the minimal data rate constraint for each user and the last constraint indicates the transmitted power constraint with P being the maximum total transmitted power.

This optimization problem is very complicated to be solved directly because the problem is non-convex and may not be converted to a convex problem with simple manipulations. Consequently, in order to validate the effectiveness of the proposed optimization problem and the accuracy of the derived analytical model, we resort to numerical simulations.

3. Simulation Results

In this section, we evaluate the performance of the proposed network of multiple UAV relays supporting the uplink mmWave NOMA system. In the considered system, UAV relays are located in different positions with different spatial directions with respect to the BS, and hybrid mmWave beamforming with MUSIC technique is adopted to derive the different DoAs in order to optimize the achievable uplink sum-rate. In detail, we simulate three different scenarios, i.e., an UL mmWave NOMA without a UAV network, a single UAV and a network of multiple UAVs supporting the UL mmWave NOMA. The cellular network has an area of 300×300 m^2 with the BS located in the center. A total of 32 users are randomly and uniformly distributed in this area.

The BS is equipped with an ULA of $N_R = 64$ antennas and $N_{RF} = 4$ RF chains to simultaneously serve a number K of UAVs with $K \leq N_{RF}$. The UAVs deployed in the cell have different DoAs at the BS uniformly distributed between $[-\pi/2, \ldots. \pi/2]$, and each UAV covers a pair of users who have poor connections with the BS due to a severe blockage condition. The UAV has $N_{UR} = 16$ receiving antennas which transmits towards the BS with one antenna.

We choose to associate the considered blocked user pair not to the closest UAV in term of distance, but to the one offering the best communication performance, as especially in a very dense urban scenario, the nearest available UAV may be in NLOS visibility. An accurate DoA estimation at the BS can monitor the fluctuations of the UAV motion and consequently improve the performance of the overall uplink sum rate.

The main parameters for an urban scenario are summarized in Table 1 [20,27].

Table 1. Simulation parameters.

Number of users	N = 32
Number of UAVs	K = 4
Urban LOS probability parameters	C = 9.6117, B = 0.1581
The additional attenuation factor for LOS channel	$\zeta_{LOS} = 10^{-6.14}$
The additional attenuation factor for NLOS channel	$\zeta_{NLOS} = 10^{-7.2}$
The path loss exponent	$\alpha_{LOS} = 2$ $\alpha_{NLOS} = 2.92$
UAV deployment height H	[100–300] m
UAV transmit power	$P_{UAV} = 20$ dBm
User transmit power	$P_{UE} = 10$ dBm
Minimal rate constraint for users	$\tilde{r} = 3$ bps/Hz

In Figure 3, the achievable sum rate is shown in terms of the varying signal-to-noise ratio (SNR) for the multi-UAV network with respect to a single UAV scheme in the case of an LOS environment. The multi-UAV scheme outperforms the single UAV scheme even for low SNR values, as the channels from users to each UAV are in LOS in these simulations realizations and consequently demand low power to transmit towards the UAV.

In the same figure, the case of a multi-UAV scheme without DoA information acquisition is highlighted to validate the performance of the MUSIC method. Indeed, the DoA estimation error is an important problem due to the UAV mobility and dynamic channel variations. However, a more accurate DoA estimation occurs when the UAV is far from the BS [30], and in particular the use of the MUSIC technique allows both the DoA and received powers to be estimated more accurately with respect to other methods [19,24,31].

Figure 4 shows the achievable sum rate results with respect to SNR when multiple UAVs are considered for both the case of LOS and NLOS environments. The MUSIC method at the BS to acquire the DoA information of the UAVs is used. In these simulation results, the heights of UAVs are not considered too high, because even if the links have a higher probability to be in LOS, the impact of the increased distance between the pair of users and the serving UAV decreases the overall link budget, and the links can have a worse SNR. The results are compared with the performance of the UL NOMA mmWave system without the multi-UAV network, i.e, the users can communicate directly by NOMA mmWave link to the BS. The NOMA mmWave link to BS is considered only for evaluating the performance comparison as, practically, the user pair covered by each UAV has a severe blockage and cannot communicate with the BS. The achievable sum rate of the multiple-UAV network achieves a considerably better performance than the direct NOMA mmWave connection, as shown in Figure 4, providing a improvement of about 6 dB for high SNR values.

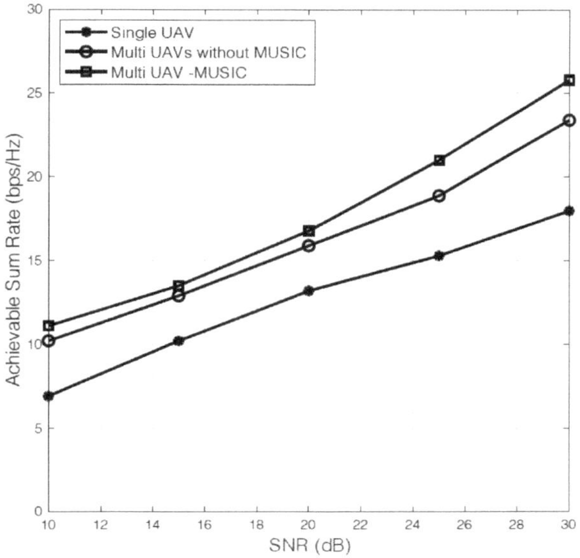

Figure 3. Achievable rate for multiple UAVs and a single UAV in LOS environment.

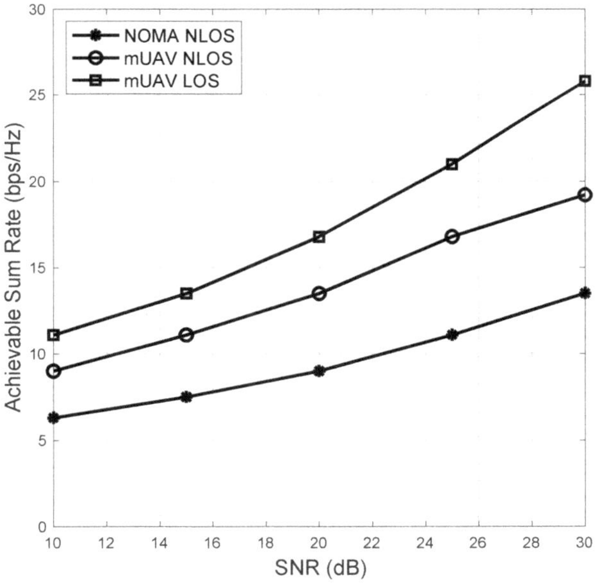

Figure 4. Comparison between multi-UAV-aided NOMA and terrestrial NOMA cellular systems.

4. Conclusions

The paper proposes a multi-UAV network where each UAV, acting as a relay, carries out the NOMA method for users experiencing severe blocking to the BS in mmWave communications. Hybrid beamforming architecture is commonly combined with mmWave transmission. We have investigated how the angle of arrival information can be estimated by the MUSIC technique performed after the analog part and the analog-to-digital conversion in the hybrid beamformer to derive the DoA of each UAV. The estimation of DoAs allows the specific antenna multi-beam targeting to be formed to each UAV. This provides an intrinsic mitigation of the UAV fluctuation effects, enhancing the UL signal reception, and leads to the substantial improvement of the system sum rate, as confirmed by simulation results. We have investigated the problem of how to maximize the sum rate of the multiple-UAV-aided NOMA mmWave system, where each pair of users is served by an UAV and where we need to find the hybrid beamforming vector to steer towards each UAV. MUSIC allows DoAs and received power to be estimated more accurately. This results in the enhanced overall sum rate (close to 50%) of the multiple-UAV-aided NOMA network with respect to the NOMA cellular system without the UAV network. In the future, research should investigate minimizing the energy consumption of a multi-UAV network by discovering efficient power associations of ground users with UAV trajectories.

Funding: This research received no external funding.

Institutional Review Board Statement: Not applicable.

Informed Consent Statement: Not applicable.

Conflicts of Interest: The author declares no conflict of interest.

References

1. Busari, S.A.; Huq, K.M.S.; Mumtaz, S.; Dai, L.; Rodriguez, J. Millimeter-Wave Massive MIMO Communication for Future Wireless Systems: A Survey. *IEEE Commun. Surv. Tutor.* **2018**, *20*, 836–869. [CrossRef]
2. Ding, Z.; Lei, X.; Karagiannidis, G.K.; Schober, R.; Yuan, J.; Bhargava, V.K. A Survey on Non-Orthogonal Multiple Access for 5G Networks: Research Challenges and Future Trends. *IEEE J. Sel. Areas Commun.* **2017**, *35*, 2181–2195. [CrossRef]
3. Zeng, M.; Hao, W.; Dobre, O.A.; Poor, H.V. Energy-Efficient Power Allocation in Uplink mmWave Massive MIMO with NOMA. *IEEE Trans. Veh. Technol.* **2019**, *68*, 3000–3004. [CrossRef]
4. Xiao, Z.; Zhu, L.; Choi, J.; Xia, P.; Xia, X. Joint Power Allocation and Beamforming for Non-Orthogonal Multiple Access (NOMA) in 5G Millimeter Wave Communications. *IEEE Trans. Wirel. Commun.* **2018**, *17*, 2961–2974. [CrossRef]
5. Wang, B.; Dai, L.; Wang, Z.; Ge, N.; Zhou, S. Spectrum and Energy-Efficient Beamspace MIMO-NOMA for Millimeter-Wave Communications Using Lens Antenna Array. *IEEE J. Sel. Areas Commun.* **2017**, *35*, 2370–2382. [CrossRef]
6. Fang, F.; Zhang, H.; Cheng, J.; Leung, V.C.M. Energy-Efficient Resource Allocation for Downlink Non-Orthogonal Multiple Access Network. *IEEE Trans. Commun.* **2016**, *64*, 3722–3732. [CrossRef]
7. Pierucci, L. The quality of experience perspective toward 5G technology. *IEEE Wirel. Commun.* **2015**, *22*, 10–16. [CrossRef]
8. Dai, L.; Wang, B.; Peng, M.; Chen, S. Hybrid Precoding-Based Millimeter-Wave Massive MIMO-NOMA with Simultaneous Wireless Information and Power Transfer. *IEEE J. Sel. Areas Commun.* **2019**, *37*, 131–141. [CrossRef]
9. Masaracchia, A.; Nguyen, L.D.; Duong, T.Q.; Yin, C.; Dobre, O.A.; Garcia-Palacios, E. Energy-Efficient and Throughput Fair Resource Allocation for TS-NOMA UAV-Assisted Communications. *IEEE Trans. Commun.* **2020**, *68*, 7156–7169. [CrossRef]
10. Simoni, R.; Jamali, V.; Zlatanov, N.; Schober, R.; Pierucci, L.; Fantacci, R. Buffer-Aided Diamond Relay Network With Block Fading and Inter-Relay Interference. *IEEE Trans. Wirel. Commun.* **2016**, *15*, 7357–7372. [CrossRef]
11. Zeng, Y.; Zhang, R.; Lim, T.J. Throughput Maximization for UAV-Enabled Mobile Relaying Systems. *IEEE Trans. Commun.* **2016**, *64*, 4983–4996. [CrossRef]
12. Song, Q.; Zheng, F.; Zeng, Y.; Zhang, J. Joint Beamforming and Power Allocation for UAV-Enabled Full-Duplex Relay. *IEEE Trans. Veh. Technol.* **2019**, *68*, 1657–1671. [CrossRef]
13. Chiti, F.; Fantacci, R.; Pierucci, L. Energy Efficient Communications for Reliable IoT Multicast 5G/Satellite Services. *Future Internet* **2019**, *11*, 164. [CrossRef]
14. Na, Z.; Liu, Y.; Shi, J.; Liu, C.; Gao, Z. UAV-supported Clustered NOMA for 6G-enabled Internet of Things: Trajectory Planning and Resource Allocation. *IEEE Internet Things J.* **2020**. [CrossRef]
15. Wang, J.; Liu, M.; Sun, J.; Gui, G.; Gacanin, H.; Sari, H.; Adachi, F. Multiple Unmanned-Aerial-Vehicles Deployment and User Pairing for Nonorthogonal Multiple Access Schemes. *IEEE Internet Things J.* **2021**, *8*, 1883–1895. [CrossRef]
16. Schmidt, R. Multiple emitter location and signal parameter estimation. *IEEE Trans. Antennas Propag.* **1986**, *34*, 276–280. [CrossRef]
17. Guo, Z.; Wang, X.; Heng, W. Millimeter-Wave Channel Estimation Based on 2-D Beamspace MUSIC Method. *IEEE Trans. Wirel. Commun.* **2017**, *16*, 5384–5394. [CrossRef]
18. Shufeng, L.; Guangjing, C.; Libiao, J.; Hongda, W. Channel estimation based on the PSS-MUSIC for millimeter-wave MIMO systems equipped with co-prime arrays. *EURASIP J. Wirel. Commun. Netw.* **2020**, *17*, 17. [CrossRef]
19. Antón-Haro, C.; Mestre, X. Learning and Data-Driven Beam Selection for mmWave Communications: An Angle of Arrival-Based Approach. *IEEE Access* **2019**, *7*, 20404–20415. [CrossRef]
20. Boschiero, M.; Giordani, M.; Polese, M.; Zorzi, M. Coverage Analysis of UAVs in Millimeter Wave Networks: A Stochastic Geometry Approach. In Proceedings of the 2020 International Wireless Communications and Mobile Computing (IWCMC), Limassol, Cyprus, 15–19 June 2020; pp. 351–357. [CrossRef]

21. Xue, Z.; Wang, J.; Ding, G.; Wu, Q.; Lin, Y.; Tsiftsis, T.A. Device-to-Device Communications Underlying UAV-Supported Social Networking. *IEEE Access* **2018**, *6*, 34488–34502. [CrossRef]
22. Andrews, J.G.; Bai, T.; Kulkarni, M.N.; Alkhateeb, A.; Gupta, A.K.; Heath, R.W. Modeling and Analyzing Millimeter Wave Cellular Systems. *IEEE Trans. Commun.* **2017**, *65*, 403–430. [CrossRef]
23. Lialios, D.I.; Ntetsikas, N.; Paschaloudis, K.D.; Zekios, C.L.; Georgakopoulos, S.V.; Kyriacou, G.A. Design of True Time Delay Millimeter Wave Beamformers for 5G Multibeam Phased Arrays. *Electronics* **2020**, *9*, 1331. [CrossRef]
24. Shu, F.; Qin, Y.; Liu, T.; Gui, L.; Zhang, Y.; Li, J.; Han, Z. Low-Complexity and High-Resolution DOA Estimation for Hybrid Analog and Digital Massive MIMO Receive Array. *IEEE Trans. Commun.* **2018**, *66*, 2487–2501. [CrossRef]
25. Khawaja, W.; Ozdemir, O.; Guvenc, I. UAV Air-to-Ground Channel Characterization for mmWave Systems. In Proceedings of the 2017 IEEE 86th Vehicular Technology Conference (VTC-Fall), Toronto, ON, Canada, 24–27 September 2017; pp. 1–5. [CrossRef]
26. Alkhateeb, A.; El Ayach, O.; Leus, G.; Heath, R.W. Channel Estimation and Hybrid Precoding for Millimeter Wave Cellular Systems. *IEEE J. Sel. Top. Signal Process.* **2014**, *8*, 831–846. [CrossRef]
27. Akdeniz, M.R.; Liu, Y.; Samimi, M.K.; Sun, S.; Rangan, S.; Rappaport, T.S.; Erkip, E. Millimeter Wave Channel Modeling and Cellular Capacity Evaluation. *IEEE J. Sel. Areas Commun.* **2014**, *32*, 1164–1179. [CrossRef]
28. Trigka, M.; Mavrokefalidis, C.; Berberidis, K. Full snapshot reconstruction in hybrid architecture antenna arrays. *J. Wirel. Commun. Netw.* **2020**, *2020*, 243. [CrossRef]
29. Souto, V.; De Souza, R.; Uchôa-Filho, B. Tx-Rx Initial Access and Power Allocation for Uplink NOMA-mmWave Communications. In Proceedings of the XXXVIII SimpÓSIO Brasileiro de TelecomunicaçõEs E Processamento de Sinais-SBrT 2020, Florianópolis, Brazil, 22–25 November 2020.
30. Miao, W.; Luo, C.; Min, G.; Wu, L.; Zhao, T.; Mi, Y. Position-Based Beamforming Design for UAV Communications in LTE Networks. In Proceedings of the ICC 2019—2019 IEEE International Conference on Communications (ICC), Shanghai, China, 15 July 2019; pp. 1–6. [CrossRef]
31. Dastgahian, M.S.; Ghomash, H.K. MUSIC-based approaches for hybrid millimeter-wave channel estimation. In Proceedings of the 2016 8th International Symposium on Telecommunications (IST), Tehran, Iran, 27–28 September 2016; pp. 266–271. [CrossRef]

applied sciences

MDPI

Review

Development and Prospect of UAV-Based Aerial Electrostatic Spray Technology in China

Yali Zhang [1,2], Xinrong Huang [1,2], Yubin Lan [2,3], Linlin Wang [2,3], Xiaoyang Lu [1,2], Kangting Yan [1,2], Jizhong Deng [1,2,*] and Wen Zeng [1,2,*]

[1] College of Engineering, South China Agricultural University, Wushan Road, Guangzhou 510642, China; ylzhang@scau.edu.cn (Y.Z.); xrhuang@stu.scau.edu.cn (X.H.); luxiaoyang@stu.scau.edu.cn (X.L.); ktyan@stu.scau.edu.cn (K.Y.)

[2] National Center for International Collaboration Research on Precision Agricultural Aviation Pesticide Spraying Technology, Wushan Road, Guangzhou 510642, China; ylan@scau.edu.cn (Y.L.); wlinlin@stu.scau.edu.cn (L.W.)

[3] College of Electronic Engineering and College of Artificial Intelligence, South China Agricultural University, Wushan Road, Guangzhou 510642, China

* Correspondence: jz-deng@scau.edu.cn (J.D.); zengwen@scau.edu.cn (W.Z.)

Abstract: Aerial electrostatic spray technology for agriculture is the integration of precision agricultural aviation and electrostatic spray technology. It is one of the research topics that have been paid close attention to by scholars in the field of agricultural aviation. This study summarizes the development of airborne electrostatic spray technology for agricultural use in China, including the early research and exploration of Chinese institutions and researchers in the aspects of nozzle structure design optimization and theoretical simulation. The research progress of UAV-based aerial electrostatic spray technology for agricultural use in China was expounded from the aspects of nozzle modification, technical feasibility study, influencing mechanism of various factors, and field efficiency tests. According to the current development of agricultural UAVs and the characteristics of the farmland environment in China, the UAV-based aerial electrostatic spray technology, which carries the airborne electrostatic spray system on the plant protection UAVs, has a wide potential in the future. At present, the application of UAV-based aerial electrostatic spray technology has yet to be further improved due to several factors, such as the optimization of the test technology for charged droplets, the impact of UAV rotor wind field, comparison study on charging modes, and the lack of technical accumulation in the research of aerial electrostatic spray technology. With the continuous improvement of the research system of agricultural aviation electrostatic spray technology, UAV-based electrostatic spray technology will give play to the advantages in increasing the droplets deposition on the target and reducing environmental pollution from the application of pesticides. This study is capable of providing a reference for the development of the UAV-based agricultural electrostatic spray technology and the spray equipment.

Keywords: agricultural aviation; UAV; plant protection; review; electrostatic spray technology; droplet deposition; aerial pesticides application; charge to mess ratio

Citation: Zhang, Y.; Huang, X.; Lan, Y.; Wang, L.; Lu, X.; Yan, K.; Deng, J.; Zeng, W. Development and Prospect of UAV-Based Aerial Electrostatic Spray Technology in China. *Appl. Sci.* **2021**, *11*, 4071. https://doi.org/10.3390/app11094071

Academic Editor: Sylvain Bertrand

Received: 12 April 2021
Accepted: 28 April 2021
Published: 29 April 2021

1. Introduction

Electrostatic spray technology for agricultural aviation is the application of traditional electrostatic spray technology in an airborne platform. It is one of the research topics that scholars in the field of agricultural aviation always pay attention to. Agricultural aerial electrostatic spray technology is mainly based on induction, corona, and contact charging methods to charge the droplets. Under the action of high voltage static electricity, charged droplets make rapid directional deposition along the electric field line in the air and settle on the target [1,2]. For this reason, aerial electrostatic spray can effectively reduce drift loss during aerial pesticide application, improve droplet deposition, and mitigate environmental pollution [3,4]. Carlton et al. [5] from the Agricultural Research Service of the United States Department of Agriculture (USDA-ARS) designed an electric rotating electrostatic nozzle in 1966, which was the first scientific research institution in the world to carry out research on aerial electrostatic spray technology for agriculture. Carlton also obtained the invention patent of the aerial electrostatic spray technology in 1999 [6]. The patent was certified by the United States Department of Agriculture as the world's first and only commercially proven airborne electrostatic spray system for agriculture [7]. SES (Spectrum Stack Sprayers, Inc., Houston, TX, USA) has been awarded an exclusive license to manufacture and market this innovative technology. Subsequently, scientific research institutions in the United States, Brazil, Canada, Switzerland, and China have conducted a large number of studies to gradually improve the research and application of electrostatic spray in agricultural aviation [8–10].

The research of electrostatic spray technology for agricultural aviation started relatively late in China but developed rapidly. Especially in recent years, the rapid development of plant protection UAVs in China has attracted scholars to make a lot of attempts on the research and application of UAV-based electrostatic spray technology. This study introduced the early exploration of aerial electrostatic spray technology in China from the aspects of nozzle structure design optimization, theoretical simulation, and field experiment with manned aircraft platforms. Then, the research progress of UAV-based electrostatic spray technology in China was emphatically expounded upon, and the existing problems were discussed as well. It is proposed that future research should be carried out in the measurement technology for charged droplets, the influence of UAV rotor wind field on charged droplets, comparative study of various charging methods, and other aspects, so as to increase the accumulation of research on aerial electrostatic spray technology based on an agricultural UAV platform. This study can provide a reference for the development of the agricultural aviation electrostatic spray technology and the spray equipment.

2. Early Exploration of Aerial Electrostatic Spray Technology in China

In the 1970s, China began to study the new technology of electrostatic spraying for agricultural application [11]. In 1977, Shenyang Spray Factory used hand-held sprayers to carry out field spraying experiments on the seedlings and young trees of cloves and found that the application of electrostatic spraying treatment could increase the amount of pesticides applied per unit area of plants while reducing the labor intensity [12]. After the application of electrostatic spray technology in hand-held and knapsack sprayers, it has also achieved success in the application of ground agricultural machinery equipment in greenhouses, orchards, and other operating scenarios [13,14]. A large number of studies on the effects of electrostatic spray equipment have been carried out on droplet size, droplet deposition distribution in each part of crop canopy, nozzle structure, and working parameters on droplet quality, etc. [15–19].

In 2005, China introduced the electrical parts and nozzles of the aerial electrostatic spray system from the SES company and carried out simulation tests and flight tests [20], marking the beginning of China's research in the field of aerial electrostatic spraying. Two years later, universities and scientific research institutions in China began to comprehensively study the aerial electrostatic spray technology based on manned aircraft platforms in aspects of hardware structure improvement, mechanism of influencing factors, and theoretical simulation.

2.1. Hardware Structure Improvement

Hardware structure improvement research refers to the adjustment or structural modification of system parameters such as electrode material, nozzle size, rotor number, and nozzle position of aerial electrostatic spray systems [21].

In 2007, Ru et al. [22] introduced a structured design on a double-nozzle for an aerial electrostatic sprayer and theoretically analyzed the space field induced by the double-nozzle and the impact on the size and charging droplets from the space field.

Zhou et al. [23] improved the design of an aerial electrostatic single-nozzle from the aspects of an electrostatic electrode, nozzle material, nozzle processing technology, connection mode of high voltage wire and electrode, rotating screw joints, overflow valve body, and so on, in order to meet the requirements for application and large-scale production.

In terms of the electrode of electrostatic nozzles, Ru et al. [24] modified the original cylindrical electrode to a cone-shaped electrode according to the features of aerial electrostatic spraying. The effect of charging voltage on the charge to mass ratio and deposition distribution of the new aerial electrostatic system was tested under simulated flight conditions. It was found that the electrostatic spray was beneficial to increase the average deposition of charged droplets on the lateral, bottom, and back sides of the neutral target significantly. Jin et al. [25] improved the ring electrode of an aerial electrostatic nozzle and conducted an experimental study on the droplet size and the charging effect of droplets after the improvement. Results showed that droplet size was influenced by nozzle diameter, spraying pressure, and charging voltage, of which the spray pressure indicated the strongest effect, and charging voltage showed the weakest effect. The charge to mass ratio increased with the increase in voltage, reaching a maximum of 2.09 mC/kg, and tends to saturation at 8 kV. The charge to mass ratio decreased with the increase in droplet size, but the change was not rapid.

2.2. Research on Mechanism of Influencing Factors

The settling process of charged droplets is restricted by environmental factors, controllable factors, and target parameters during spraying operation, thus affecting the operation quality of the electrostatic spraying system. Among these influencing factors, environmental factors include temperature, humidity, wind speed, soil, etc.; controllable parameters include charging voltage, spraying flow, spraying pressure, flight altitude, flight speed, etc.; target parameters include crop objects, target morphology, leaf inclination, and insect pest types and habits, and so on [21].

Yang et al. [26] studied the influence of different crosswinds wind speed conditions and electrostatic voltage on charging characteristics through indoor simulation of natural wind and constant wind environment, providing a basis for strengthening the anti-drift ability of charged spray droplets in the settling process. Chen et al. [27] analyzed the characteristics of the electrostatic field induced by the ring electrode with the help of Fluent software (Version 6.3). It was found that the higher charging voltage or smaller electrode spacing (when the distance between the induction electrode ring and the nozzle was less than 10 mm) could effectively improve the charge effect and spray quality.

3. Rapid Progress of UAV-Based Electrostatic Spray Technology

3.1. Research Background

In 2014, China's agricultural pesticide application was still dominated by large fixed-wing aircraft, supplemented by rotary-wing UAVs [28]. However, since 2015, with the urbanization process in China, the rural labor force population has become less and less. Meanwhile, with the improvement of living standards, automatic pesticide application tools have gradually been accepted by farmers [29]. Nowadays, plant protection UAVs have maintained a booming trend in the field of agricultural plant protection. China has taken an internationally leading position in terms of technology, quantity, and product types of agricultural plant protection UAVs [30].

Figure 1 shows the increasing number of plant protection UAVs and operating areas covered in China from 2014 to 2020. The number of plant protection UAVs has reached 106,000 units with an operation area covered of 64 million hm^2 in 2020. In this context, Chinese scholars began to put forward the idea of applying aerial electrostatic spray technology to UAVs, while there are few reports on this topic worldwide.

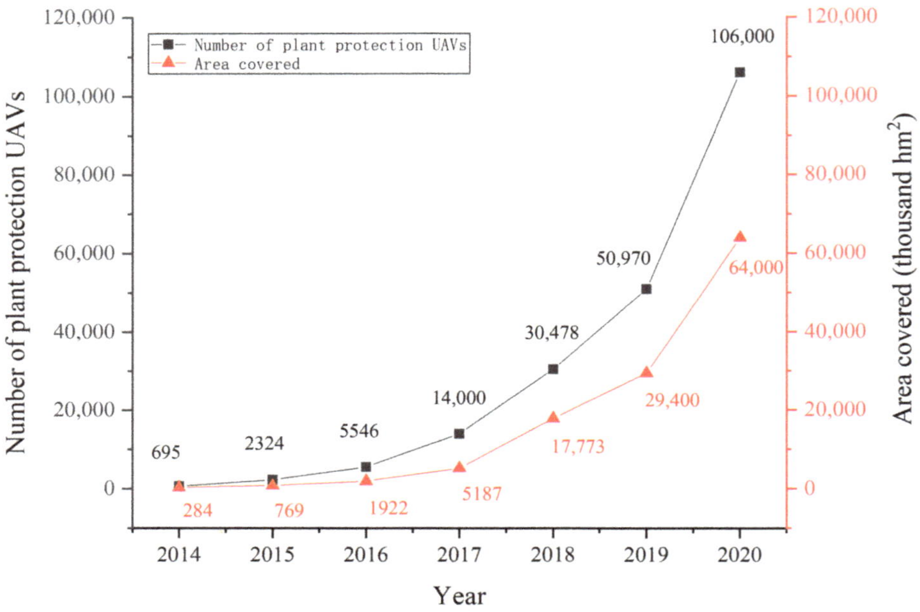

Figure 1. Increasing number of plant protection UAVs and operating area covered in China (2014–2020).

3.2. Beginning of the Research on UAV-Based Electrostatic Spray Technology

In 2015, Ru et al. [31] tried the first combination of electrostatic spray technology and plant protection UAV with an XY8D unmanned helicopter and carried out field tests in a rice field, as shown in Figure 2. A preliminary experiment on spray pressure, flow rate, and charging voltage were conducted to determine the optimal parameters for field tests. Field spraying test results showed that the droplet deposition and coverage rate on the target crop was effectively improved by the UAV-based electrostatic spraying when compared with non-electrostatic spraying. The influence of environmental factors and physical properties of the liquid was not verified in this experiment. However, the first study of carrying the electrostatic spray system on a UAV was of great value. Chinese scholars then began to make continuous innovations and breakthroughs in UAV-based electrostatic spray technology.

Figure 2. Rice field test with an XY8D UAV-based electrostatic spray system [31].

3.3. Continuous Optimization of UAV-Based Electrostatic Spray Technology

Chinese researchers have carried out a series of optimization work from the aspects of nozzle modification, test effect comparison, and mechanism of influencing factors in order to meet the precise operation requirements of UAV-based electrostatic spray system in practical application and to improve the droplet deposition on the target crops.

In terms of nozzles modification, Wang et al. [32] improved the conventional nozzles for multi-rotor plant protection UAVs. An inductive type electrostatic centrifugal nozzle was developed by combining agricultural electrostatic spray technology with a centrifugal atomizing nozzle. Besides, the spraying flow stability test and droplet deposition effect of the system were studied. When the charging voltage of the nozzle reaches 8 kV, the charge to mass ratio reaches the maximum value of 0.59 mC/kg.

In terms of comparison with the test results of non-electrostatic spray, Jin et al. [33] from Nanjing Forestry University designed an electrostatic spray system for the AF-118 helicopter. Through the effective spraying width and droplets deposition characteristics, it is found that the effective spraying width of electrostatic spray was smaller than that of non-electrostatic spray. The total droplet deposition number of electrostatic spray (9–36 drops/cm^2) was higher than that of non-electrostatic spray (6–26 drops/cm^2). Lian [34] used YG-6 six-rotor UAV to carry an electrostatic spray system. Through an indoor performance test, the optimal operating parameter combination of the system was determined (spraying height is 50 cm, charging voltage is 9 kV, spraying pressure was 0.3 MPa). This parameter combination was used to test the electrostatic spray effect of outdoor UAVs. The result showed that the average deposition density of droplets sprayed above the target by electrostatic spray was 16.1 more/cm^2, 13.6% higher than that by non-electrostatic spray. The average sediment density in the middle was 28 more/cm^2, which increased by 32.6%. Cai [35] developed an aerial electrostatic spray system based on an F-50 plant protection UAV. The spray system adopts contact charge, and the experimental research is carried out according to the factors of the rotor wind field, such as wind speed, charge to mass ratio, flight height, and flight speed. Through field spraying experiments, as shown in Figure 3, it was found that the flight height is an important factor affecting the deposition amount and the horizontal deposition uniformity but has little effect on the vertical deposition uniformity. Zhao et al. [36] proposed a method of charging the liquid in two isolated water tanks with positive and negative charges respectively by a high-voltage electrostatic generator based on the contact charge mode to solve the problem of insufficient adsorption rate of droplets on the target back when using an aerial electrostatic spray. Aerial electrostatic spraying test stand and UAV electrostatic spray system were designed, which proved that it was feasible to develop a charge transfer loop in space to improve the adsorption performance of droplets. An electrostatic physical model of aerial electrostatic spray based on charge transfer space loop is shown in Figure 4. Zhang et al. [37] developed a fan-shaped induction electrostatic spray system based on a six-rotor UAV and defined the corresponding operating parameters (spray height 50 cm, spray pressure 0.3 MPa, and charging voltage 9 kV). Compared with non-electrostatic spray, the electrostatic spray had more concentrated droplet deposition and smaller drift. The average droplet deposition density at the top of the electrostatic spray sampling device was 16.1 drops/cm^2 higher than that of non-electrostatic spray, and the deposition density in the middle was 28 drops/cm^2 higher than that of non-electrostatic spray. Therefore, aerial electrostatic spray could significantly improve droplet deposition and prevent drift.

Figure 3. UAV paddy field spraying operations [35].

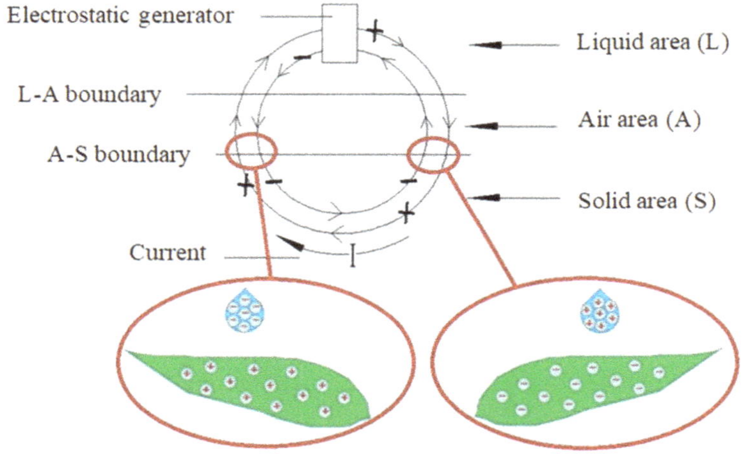

Figure 4. Electrostatics physical model of aerial electrostatic spraying based on charge transfer space loop [36].

In terms of the mechanism of influencing factors, Bu [38] designed an electrostatic spray system based on the FR-200 large-load unmanned helicopter with a maximum load of 80 kg. The charge and spray characteristics of the electrostatic spray system were studied, and the prediction models of the charge to mass ratio and droplet size were established. The characteristics of deposition and drift of electrostatic spray of heavy load unmanned helicopter were analyzed by field experiments. It was concluded that the charge voltage, flight height, and crosswind wind speed were the main factors affecting the drift and deposition of an electrostatic spray of FR-200. The results showed that the charge voltage had the greatest influence, followed by the crosswind wind speed, and the flight height had the least influence. Based on the research results, the mathematical model was established between the center distance of droplet mass and droplet drift rate and the charged voltage, flight height, and crosswind wind speed. Wu et al. [39] introduced response surface analysis (RSM) into the optimization design of spray parameters of Electrostatic spray system of UAVs. A response surface model with injection pressure and nozzle diameter as design variables and droplet charge to mass ratio as optimization objectives was constructed. The performance tests of the new electrostatic spray system under different nozzle diameters, spray pressures, and electrostatic voltages were carried out. It showed that the nozzle met the theoretical requirements of optimal biological particle size and the requirements of hydraulic spray control for most crop diseases and insect pests. The validity of the model was also proved. In addition, multivariate analysis of variance showed that nozzle diameter, spray pressure, and electrostatic voltage have important effects on performance. The influence of the two factors on the deposition density and uniformity is the spray pressure, the nozzle diameter, and the optimal combination of the factor levels to obtain the best results. Chinese scholars also carried out research on multiple operating parameters and their interaction effects on the performance of aerial electrostatic spray systems. Lu et al. [40] simulated the spray performance of UAV-based electrostatic spray systems at different flight heights by measuring the droplet diameter under different nozzle apertures and system pressures and obtained the optimal flight parameter combination. Zhao et al. [41] studied the spray deposition with three factors: spraying duration, charging voltage, and flight height. The experimental results showed that the back-front ratio of droplets on the back and front of leaf targets could reach 158.8% under indoor conditions, and the droplet size on the back was smaller than that on the front. The number of droplets increased with the accumulation of spraying time without affecting the back-front ratio. Higher charging voltage and lower spraying height for the aerial electrostatic spray system can achieve a better deposition effect and higher back-front ratio. Wang et al. [42] established theoretical equations of droplet group charge based on the water inductive charging theory and then studied the effects of electrode figuration (electrode ring diameter, electrode spacing, spray pressure, and charge voltage) on droplet charging and spray performance through experiments. Under the action of an electrostatic field, the droplet size decreased obviously with the increase in charged voltage. The charge performance improved with the decrease in electrode ring diameter. Lan et al. [43] studied the impact of electrode materials on the deposition characteristics of an aerial electrostatic system with different orifice sizes, system pressures, and charging voltages. The results showed that red copper was the best electrode material. Li et al. [44] simulated five factors, such as temperature, humidity, electrode ring diameter, electrostatic voltage, and nozzle flow, using a BP model based on Neuroshell software and studied the effects on the charge to mass ratio of airborne electrostatic droplets. The final linear regression model indicated that the charging voltage and flow rate were the two main influencing factors on the droplet charge to mass ratio.

3.4. Spray Efficiency Experiments of UAV-Based Electrostatic Spray Technology

Wang et al. [45] designed a bipolar contact oil-powered single rotor aerial electrostatic spray system for plant protection UAVs. The static electricity system was used to spray the static electricity oil agent and the conventional water-based chemical agent. Besides, the spray droplet deposition distribution and the control effect of wheat aphid and rust were tested. The result showed that the deposition per unit area was 0.0486 $\mu g/cm^2$, the standard deviation of deposition amounts was 0.015 $\mu g/cm^2$, and the coefficient of variation was 30.43%. The distribution uniformity of droplet deposition is obviously better than the other two treatments, and the prevention and control of diseases and insect pests in the wheat field showed a good control effect. The aphid control effect was 87.92% on the seventh day after spraying, which was significantly higher than that of the conventional spray treatment (76.43% with the electrostatic oil agent and 66.47% with the conventional water-based agent). Liu [46] designed a high-voltage electrostatic generator with smaller volume and mass based on the optimized structure parameters of the cone-shaped electrode and the determined dosage form of the special electrostatic liquid agent for aviation. It was more suitable for plant protection UAVs, aimed at the current problems such as the poor charging effect of droplets, poor applicability of aerial spraying agents, complex high-voltage electrostatic generator system, and heavy high-voltage electrostatic generator. In addition, a set of electrostatic spray systems that can be applied to six-rotor and single-rotor plant protection UAVs was developed, as shown in Figure 5. The system was mounted on the plant protection UAV and carried out the experiment in the cotton and rice-growing areas of Changji, Xinjing, Shihezi, Xinjing, and Ledong, Hainan. The results of spraying operation of hybrid breeding rice growth regulator and defoliating cotton agent by plant protection UAV showed that when spraying rice growth regulator by plant protection UAV, the effect of electrostatic spraying could be up to 20% higher than that of non-static spraying, and the effects of spraying different chemicals were different. When sprayed the cotton defoliant, the coverage rate of droplet increased by 140%, the defoliation efficacy increased by 12.22%, and the cotton boll opening rate increased by 18.55%.

Figure 5. HY-B-15L single-rotor plant protection unmanned helicopter for cotton defoliant spray (**a**) and TXA616 plant protection UAV for rice growth regulator spray (**b**) [46].

3.5. Summary

To sum up, a large number of field experiments have been conducted in China to test the actual performance of the electrostatic spray system based on UAVs. For the convenience of comparative analysis, Table 1 summarizes the researches on UAV-based electrostatic spray systems. From the perspective of the test scheme, the system pressure, flight height and speed, charging voltage, rotor wind speed, and their effects on each other were studied. Chinese scholars have also completed a lot of work using theoretical simulations of the spray system, the improvement of the electrostatic nozzle, and the optimization of the aerial chemicals. In terms of the mode of charge, most studies adopt the induction mode of charge with the highest safety. The inductive charging voltage is usually between 2 kV and 15 kV, and the electrode making and insulation methods are easy to be realized, which is the most developed method of charging droplets at present [47]. In addition, contact charging has also been tried in China, which has also achieved good results [34–36,41,45]. However, it needs to keep the absolute insulation of the spray system in the charging process, which puts forward high requirements for the design method and safety.

Figure 6 illustrates the key achievements of the aerial electrostatic spray technology and the research groups over a timeline. Since 2015, China's UAV-based aerial electrostatic spray technology has developed rapidly. In general, although the UAV-based electrostatic spray technology is a new technology, it is gradually being improved. Compared with the electrostatic spray system on manned aircraft, plant protection UAVs have more development prospects in China at present. With the operation area covered exceeding 67 million hm^2, the plant protection UAV has developed from an early experimental product into a common agricultural production machine in China [48]. It is of great practical significance to study the fusion and influence mechanism of UAV and aerial electrostatic spray systems, which contributes to the field of agricultural aviation plant protection in the future.

Table 1. Study on UAV-based Electrostatic Spray System.

Test Device	Charged Mode	Test Target	Test Method	Evaluating Indicator	Test Environment	Researchers and Affiliation
XY8D UAV	Induction	Rice	Different flight height	Droplet deposition characteristics	Field test	Ru, et al., Nanjing Forestry University
AF-811 unmanned helicopter	Induction	Copy paper	Comparison of electrostatic spray and non-electrostatic spray	Effective spray amplitude and droplet deposition characteristics	Outdoor	Jin, et al., Nanjing Forestry University
YG20-6 type six-rotor UAV	Induction	Water sensitive paper	Comparison of electrostatic spray and non-electrostatic spray	Measurement of effective spray amplitude, droplet penetration, and droplet deposition	Outdoor	Lian, Heilongjiang Bayi Agricultural University
F-50 plant-based unmanned helicopter	Contact	Sample tape, rice	Comparison of different charge modes (non-electrostatic, induction electrostatic, contact electrostatic)	Droplet deposition measurement and coefficient of variation of deposition amount	Field test	Cai, Jiangsu University
HY-B-15L single rotor plant protection unmanned helicopter, TXA micro six-rotor UAV	Induction	Crossbreeding rice, Cotton picked by machine	Comparison of different water agents, number of rotors, and crops	Droplet deposition characteristics and pharmacodynamics	Field test	Liu, South China Agricultural University
Integrated spray test stand	Induction	Water sensitive paper	Comparison of electrostatic spray and non-electrostatic spray	Density of droplet deposition	Indoor	Zhang et al., Heilongjiang Bayi Agricultural University
3WQF120-12 oil-operated single-rotor plant protection UAV	Contact	wheat	Comparison between electrostatic oil agent and conventional water-based agent, electrostatic spray and conventional spray system	Distribution uniformity of droplet deposition and its control effect on wheat aphid and rust	Field test	Wang et al., China Agriculture University
FR-200 type large-load unmanned helicopter	Induction	Water sensitive paper	Charged voltage, flight speed, flight height, and crosswind wind speed	Droplet deposition characteristics and droplet drift rate	Field test	Bu, Jiangsu University

Table 1. *Cont.*

Test Device	Charged Mode	Test Target	Test Method	Evaluating Indicator	Test Environment	Researchers and Affiliation
response surface methodology (RSM)	Induction	-	Nozzle diameter, nozzle pressure, and electrostatic voltage	Droplet deposition characteristics and charge to mass ratio	Simulation	Wu et al., Guizhou University
Gaoke M45 plant protection UAV	Contact	Mock target	Comparison of electrostatic spray and non-electrostatic spray	Droplet deposition characteristics	Outdoor	Zhao et al., Shandong University of Technology
W730S four-rotor semi-automatic plant protection UAV	Induction	Effusion cylinder	-	Droplet deposition measurement and spray flow stability	Indoor	Wang et al., Hangzhou Dianzi University
An indoor spray bar electrostatic spray system test platform	Induction	Water sensitive paper	Nozzle diameter, nozzle pressure, and spraying height	Droplet deposition characteristics	Indoor	Lu et al., Guizhou University
Aerial electrostatic spraying system suspension test bench	Contact	Art Paper	Spraying height, spray time, and charging voltage	Droplet deposition characteristics and back-front ratio	Indoor	Zhao et al., Shandong University of Technology
Charge aerosol spray test bench	Induction	White plastic board	Charging voltage, spray pressure, electrode ring diameter, and electrode spacing	Spray charge-to-mass ratio and droplet size	Indoor	Wang et al., Jiangsu University
Fixed spray simulator	Induction	Dracaena	Electrode material, charging voltage, spray pressure, and nozzle diameter	Droplet deposition	Indoor	Lan et al., South China Agricultural University
Neuroshell	Induction	-	Temperature, humidity, electrode ring diameter, charging voltage, and nozzle flow	Spray charge-to-mass ratio	Simulation	Li et al., Heilongjiang Bayi Agricultural University

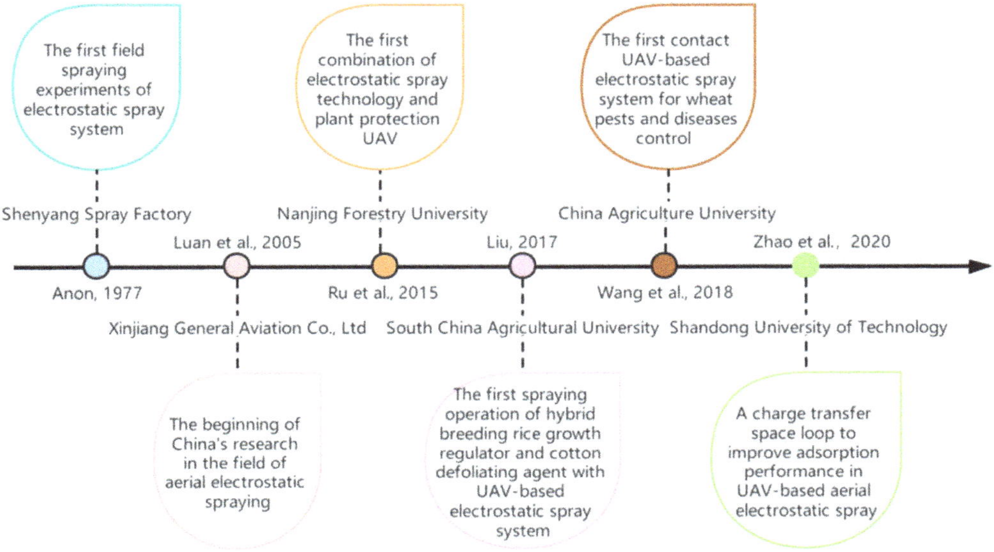

Figure 6. Key milestones and research groups working on aerial electrostatic spray technology.

4. Analysis and Prospects

The exploration of UAV-based electrostatic spray technology expands a new research perspective for the research of agricultural aviation electrostatic spray technology. The continuous improvement of the technology brings new opportunities for the application of aerial electrostatic spray systems on commercial plant protection UAVs in China. However, according to the current research progress, still, the following key technical elements need further exploration.

4.1. Measurement Technology of Charged Droplets

Droplet charge to mass ratio is a term associated with aerial electrostatic spray technology, and the measurement results of charge to mass ratio provide a reference for the evaluation of electrostatic spray system performance. In the absence of interference (e.g., indoors), the droplet charge to mass ratio is positively correlated with the deposition effect [49]. However, when working in an outdoor environment, there is usually a big difference between the droplet charge to mass ratios at the nozzle end and the terminal target, which requires the terminal measuring device to evaluate the charge amount. However, due to the absence of electrical grounding, the device that collects charged droplets often results in an inconsistent result with the actual effect of settling on the target [50]. For the electrostatic spray technology, the electric field intensity varies with the plant form, liquid conductivity, and environmental factors. For example, the electric field near the tip or terminal part of the leaf is the strongest [51]. Law [52] reported that gaseous discharges between sharp leaf tips and incoming charged spray clouds had been shown to limit deposition.

In many studies, aerial electrostatic spray had a better deposition effect compared with conventional electrostatic spray and non-electrostatic spray, but there are not a few reports that the effect of electrostatic spray is not satisfactory [1,49,53]. However, it remains to be further confirmed whether the difference in operation effect is caused by a sharp reduction in the charge to mass ratio of charge droplets, faulty experimental design scheme, or environmental influence. In recent years, more detection methods such as laser particle size analyzer have been continuously applied to the measurement of droplets effect in

China [54,55]. However, there are many problems such as high manufacturing cost, large measurement error, and complex measurement process. Therefore, it is necessary to develop a droplet characteristic detection technology suitable for aerial electrostatic spray systems to monitor the real state of charged droplets.

4.2. Impact of UAV Rotor Wind Field on Charged Droplets

The wind field on a manned fixed-wing aircraft causes charged droplets to settle in the direction of flight towards the area over which the aircraft passes. However, the wind field of a multi-rotor plant protection UAV is chaotic and changeable. It is not a regular wind field in a single direction. Moreover, due to the dual interference of ambient wind and rotor wind field, the charged droplet deposition process is more complicated in the actual operation scenario. Existing studies have basically ignored the working state of bipolar electrostatic spray system under the action of multiple rotor wind fields, the attracting process of positive and negative charged droplets, and the influence of droplets on the humidity of inductive charging electrodes. However, the impact is huge in that the characteristic of charged droplet property undoubtedly loses if the attraction of the charged droplets with positive and negative polarity is affected. In addition, wet electrodes will also make the electrically charged performance worse. Therefore, the influence of UAV rotor wind fields on charged droplets should be paid more attention in future research.

PIV (Particle Image Velocimetry) and other techniques can be used to simulate the droplet settling state under the influence of the rotor wind field, natural environmental wind, multiple gradient crosswinds, and other factors, so as to establish a theoretical system for reducing the influence of the rotor UAV downwash wind field and even utilizing the wind field action.

4.3. Comparative Study of Various Charging Methods

The induction charging mode was determined from the early stage of aerial electrostatic spray technology research in the United States, and it has been adopted in commercial electrostatic spray systems. Induction charging has the advantages of low charging voltage and low application threshold. It is a safe and effective method for charging droplets. Most Chinese scholars have also applied the induction charging method in their research. However, in addition to induction, there are also contact and corona ways to charge droplets. Corona charging voltage is very high, up to 30 kV. It can be used for conductive and non-conductive liquids with low insulation requirements. The contact charging voltage is required to be between corona type and induction type, but the insulation requirements are higher. Although the corona type with high charging voltage and the contact type with high insulation requirements still need a lot of basic research to clarify the charging mechanism and eliminate application risks, the disadvantages of the induction charging mode with poor charging effect cannot be ignored. Previous studies have shown that the contact charging method is able to generate a larger target current when compared with the induction charging method [47]. In recent years, there have been reports on contact charging methods in the research of UAV-based electrostatic spray technology with gratifying test results. In the future, it is necessary to carry out comparative studies of various charging methods in order to evaluate the operation effect of aerial electrostatic spray technology under different charging methods.

4.4. Accumulation of Aerial Electrostatic Spray Technology Research Based on Agricultural UAVs

During the early stage of China's agricultural aviation electrostatic spray technology research, the design and test were carried out on manned fixed-wing aircraft and helicopter platforms. In recent years, plant protection UAVs have provided Chinese growers with a lower barrier to entry and a higher level of applicability. Chinese scholars have carried on beneficial exploration for UAV-based electrostatic spray technology according to different UAV models, field crops, operating parameters, and electrostatic nozzle parameters, but the main research emphasis is still on the contrast test of electrostatic spray and non-electrostatic spray to verify the operation effect of charged droplets. The research accumulation of airborne electrostatic spray technology based on agricultural UAV platforms is still less. It is not mature at the application level because the morphological characteristics of various crops and the farmland environment need more experimental data support. For example, (1) Relevant studies on the system construction, composition, and weight control of the electrostatic spray system lack continuity; (2) whether the operation with aerial electrostatic spray system is affected by surrounding facilities such as high-voltage lines, or whether it is incompatible with the flight control system of high-precision and fully autonomous plant protection UAV, is still unknown; (3) at present, the practical application of aerial electrostatic spray system in the field is limited to spraying water and water-based pesticides, and there is a lack of in-depth discussion on the research of the special electrostatic liquid pesticides and the electrical conductivity of pesticides for aerial spraying application; (4) working parameters of the spray system, such as charging voltage, system pressure and spraying speed, selection of flight speed and altitude, characteristics of aviation agents, and environmental factors such as temperature, humidity, and wind speed, all affect the settling process of charged droplets. Therefore, it is necessary to observe the droplet characteristics by studying their interaction effects; (5) the development trend of most commercial agricultural UAVs in China is integrating pesticide spraying, seed sowing, and fertilizer spreading together with fully autonomous and high-precision operation. Therefore, it is necessary to consider whether the integration of aerial electrostatic spray system, UAV working systems, and flight control system will cause mutual interference.

5. Conclusions

The development of aerial electrostatic spray technology in China, especially UAV-based aerial electrostatic spray technology, was analyzed in this review from nozzle modification, technical feasibility tests, mechanisms influencing each factor, and field spray efficiency tests. According to the literature retrieved, the research of China's aerial electrostatic spray technology in the past five years has focused on the innovative exploration of UAVs as a platform. Combined with the current development of agricultural plant protection and industrial application practice, UAV-based aerial electrostatic spray technology has wider developmental potential in China. In the future, the development plans should be made around the basic research, field test, commercialization, demonstration, and service guidance. In addition, researchers are recommended to pay attention to the integrated design of UAV and aerial electrostatic spray systems and to formulate the application standard of aerial electrostatic spray technology in China.

Author Contributions: Conceptualization, Y.Z., Y.L., and X.H.; methodology, X.H., Y.Z., and W.Z.; software, X.L. and W.Z.; validation, X.H. and W.Z.; formal analysis, X.H. and L.W.; investigation, Y.Z., X.H., and L.W.; resources, X.H.; data curation, X.H.; writing—original draft preparation, X.H., K.Y., and Y.Z.; writing—review and editing, Y.Z. and K.Y.; visualization, X.L.; supervision, W.Z. and J.D.; project administration, Y.Z., Y.L., and J.D.; funding acquisition, Y.Z. All authors have read and agreed to the published version of the manuscript.

Funding: This research was funded by Key Field Research and Development Plan of Guangdong Province, China, grant number 2019B020221001, Science and Technology Plan Project of Guangdong Province, China, grant number 2018A050506073, Guangdong Modern Agricultural Industry Generic Key Technology Research and Development Innovation Team Project, grant number 2020KJ133, National Key Research and Development Program, grant number 2018YFD0200304, and the 111 Project, grant number D18019.

Institutional Review Board Statement: Not applicable.

Informed Consent Statement: Not applicable.

Data Availability Statement: Not applicable.

Conflicts of Interest: The authors declare no conflict of interest.

References

1. Zhang, Y.L.; Lan, Y.B.; Fritz, B.K.; Xue, X.Y. Development of aerial electrostatic spraying systems in the United States and applications in China. *Trans. Chin. Soc. Agric. Eng.* **2016**, *32*, 1–7, (In Chinese with English abstract). [CrossRef]
2. Martin, D.E.; Latheef, M.A. Efficacy of electrostatically charged glyphosate on ryegrass. *J. Electrost.* **2017**, *90*, 45–53. [CrossRef]
3. Tavares, R.M.; Cunha, J.P.; Alves, T.C.; Bueno, M.R.; Silva, S.M.; Zandonadi, C.H. Electrostatic spraying in the chemical control of Triozoida limbata (Enderlein) (Hemiptera: *Triozidae*) in guava trees (*Psidium guajava* L.). *Pest Manag. Sci.* **2016**, *73*, 1148–1153. [CrossRef] [PubMed]
4. Martin, D.E.; Latheef, M.A.; López, J.D. Electrostatically charged aerial application improved spinosad deposition on early season cotton. *J. Electrost.* **2019**, *97*, 121–125. [CrossRef]
5. Carlton, J.B.; Isler, D.A. Development of a device to charge aerial sprays electrostatically. *Agric. Aviat.* **1966**, *8*, 44–51.
6. Carlton, J.B. Technique to Reduce Chemical Usage and Concomitant Drift from Aerial Sprays. United States Department of. Agriculture Patents No. 5975425, 2 November 1999.
7. Spectrum Electrostatic Sprayers, Inc. Available online: http://spectrumsprayer.com/company.html (accessed on 11 May 2020).
8. Inculet, I.I.; Fischer, J.K. Electrostatic aerial spraying. *IEEE Trans. Ind. Appl.* **1989**, *25*, 558–562. [CrossRef]
9. Da Cunha, J.P.A.R.; Barizon, R.R.M.; Ferracini, V.L. Assalin, Spray drift and caterpillar and stink bug control from aerial applications with electrostatic charge and atomizer on soybean crop. *Eng. Agrícola* **2017**, *37*, 1163–1170. [CrossRef]
10. Martin, D.E.; Latheef, M.A.; Lopez, J.D.; Duke, S.E. Aerial Application Methods for Control of Weed Species in Fallow Farmlands in Texas. *Agronomy* **2020**, *10*, 1764. [CrossRef]
11. Anon. Introduce a new technology of pesticide application-electrostatic spray. *For. Sci. Technol.* **1974**, *1*, 18.
12. Anon. Preliminary test of pesticide electrostatic spray. *For. Sci. Technol.* **1978**, *2*, 16–17.
13. He, X.K.; Yan, K.R.; Chu, J.Y.; Wang, J.; Zeng, A.J.; Liu, Y.J. Design and testing of the automatic target detecting, electrostatic, air assisted, orchard sprayer. *Nongye Gongcheng Xuebao Trans. Chin. Soc. Agric. Eng.* **2003**, *19*, 78–80, (In Chinese with English abstract).
14. Yu, Y.C.; Wang, B.H.; Shi, J.Z.; Li, X.F. Design and experimental study of combined-charging hydraulic electrostatic spraying box. *Trans. Chin. Soc. Agric. Eng.* **2005**, *21*, 85–88, (In Chinese with English abstract).
15. Zhang, J.; Zheng, J.Q. Experiment on the variations of droplet diameter distribution and local flow rate over radial direction for the electrostatic spray process. *Trans. Chin. Soc. Agric. Eng.* **2009**, *25*, 104–109, (In Chinese with English abstract). [CrossRef]
16. Jia, W.D.; Hu, H.C.; Chen, L.; Chen, Z.G.; Wei, X.H. Performance experiment on spray atomization and droplets deposition of wind-curtain electrostatic boom spray. *Trans. Chin. Soc. Agric. Eng.* **2015**, *31*, 53–59, (In Chinese with English abstract). [CrossRef]
17. Zhou, L.F.; Zhang, L.; Xue, X.Y.; Qin, W.C.; Sun, Z.; Zhou, Q.Q.; Chen, C. Design and test of double air channel auxiliary electrostatic sprayer. *Jiangsu Agric. Sci.* **2017**, *45*, 192–196, (In Chinese with English abstract). [CrossRef]
18. Ma, X.; Guo, L.J.; Wen, Z.C.; Wei, Y.H.; Xiao, R.H.; Zeng, H. Atomization characteristics of multi-nozzle electrostatic spray and field experiment. *Trans. Chin. Soc. Agric. Eng.* **2020**, *36*, 73–82, (In Chinese with English abstract). [CrossRef]
19. Dai, Q.F.; Hong, T.S.; Song, S.R.; Li, Z.; Chen, J.Z. Influence of pressure and pore diameter on droplet parameters of hollow cone nozzle in pipeline spray. *Trans. Chin. Soc. Agric. Eng.* **2016**, *32*, 97–103, (In Chinese with English abstract). [CrossRef]
20. Luan, H.; Zhang, Q.; Wang, W.X. Installation and modification of Z03K000B electrostatic spray system and flight tests. *Xin-jiang Reclam. Technol.* **2006**, *5*, 46–47, (In Chinese with English abstract).
21. Zhang, Y.L.; Huang, X.R.; Wang, L.L.; Deng, J.Z.; Zeng, W.; Lan, Y.B.; Muhammad, N.T. Progress in foreign agricultural aviation electrostatic spray technologies and references for China. *Trans. Chin. Soc. Agric. Eng.* **2021**, *37*, 50–59, (In Chinese with English abstract). [CrossRef]
22. Ru, Y.; Zheng, J.Q.; Zhou, H.P.; Shu, C.R. Design and experiment of double nozzle of aerial electrostatic sprayer. *Trans. Chin. Soc. Agric. Mach.* **2007**, *38*, 58–61.
23. Zhou, H.P.; Ru, Y.; Shu, C.R.; Jia, Z.C. Improvement and experiment of aerial electrostatic spray device. *Trans. Chin. Soc. Agric. Eng.* **2012**, *28*, 7–12, (In Chinese with English abstract). [CrossRef]
24. Ru, Y.; Jin, L.; Zhou, H.P.; Shu, C.R. Effect of high-voltage electrostatic field on droplet charging based on cone-shaped electrode. *High Volt. Eng.* **2014**, *40*, 2721–2727, (In Chinese with English abstract). [CrossRef]
25. Jin, L.; Ru, Y.; Sun, M.L.; Jia, Z.C.; Wang, B.X. Performance experiments of aerial electrostatic nozzle with cone shaped electrode. *J. Nanjing For. Univ. (Nat. Sci. Ed.)* **2015**, *39*, 155–160, (In Chinese with English abstract). [CrossRef]
26. Yang, Z.; Niu, M.M.; Li, J.; Xu, X.; Sun, Z.Q.; Xue, K.P. Influence of lateral wind and electrostatic voltage on spray drift of electrostatic sprayer. *Trans. Chin. Soc. Agric. Eng.* **2015**, *31*, 39–45, (In Chinese with English abstract). [CrossRef]
27. Chen, H.L.; Zhao, Y.C. Electrostatic induction field and charging property of droplet in electrostatic spraying process. *High Volt. Eng.* **2010**, *36*, 2519–2524, (In Chinese with English abstract). [CrossRef]
28. Zhang, D.Y.; Lan, Y.B.; Chen, L.P.; Wang, X.; Liang, D. Current status and future trends of agricultural aerial spraying technology in China. *Trans. Chin. Soc. Agric. Mach.* **2014**, *45*, 53–59, (In Chinese with English abstract). [CrossRef]
29. Tian, Z.W.; Xue, X.Y.; Li, L.; Cui, L.F.; Wang, G.; Li, Z.J. Research status and prospects of spraying techology of plant-protection unmanned aerial vehicle. *J. Chin. Agric. Mech.* **2019**, *40*, 37–45, (In Chinese with English abstract). [CrossRef]
30. Li, J.Y.; Lan, Y.B.; Shi, Y.Y. Research progress on airflow characteristics and field pesticide application system of rotary-wing UAV. *Trans. Chin. Soc. Agric. Eng.* **2018**, *34*, 104–118, (In Chinese with English abstract). [CrossRef]

31. Ru, Y.; Jin, L.; Jia, Z.C.; Bao, R.; Qian, X.D. Design and experiment on electrostatic spraying system for unmanned aerial ve-hicle. *Trans. Chin. Soc. Agric. Eng.* **2015**, *32*, 42–47, (In Chinese with English abstract). [CrossRef]
32. Wang, Y.T.; Wu, K.H. Research on an electrostatic spraying system for multi-rotor plant protection UAV. *Jiangsu Agric. Sci.* **2020**, *48*, 225–230, (In Chinese with English abstract). [CrossRef]
33. Jin, L.; Ru, Y. Research on UAV-based aerial electrostatic spraying system. *J. Agric. Mech. Res.* **2016**, *38*, 227–230, (in Chinese with English abstract). [CrossRef]
34. Lian, Q. Experimental Research on Electrostatic Spray System of Multi-Rotor UAV. Master's Thesis, Heilongjiang Bayi Agricultural University, Daqing, China, 2016.
35. Cai, Y.L. Design and Test of Contact Electrostatic Spray System for Low-Altitude UAV. Master's Thesis, Jiangsu University, Zhenjiang, China, 2017.
36. Zhao, D.N.; Lan, Y.B.; Shen, W.G.; Wang, S.Z.; Abhishek, D. Development of a charge transfer space loop to improve ad-sorption performance in aerial electrostatic spray. *Int. J. Agric. Biol. Eng.* **2020**, *13*, 50–55. [CrossRef]
37. Yanliang, Z.; Qi, L.; Wei, Z. Design and test of a six-rotor unmanned aerial vehicle (UAV) electrostatic spraying system for crop protection. *Int. J. Agric. Biol. Eng.* **2017**, *10*, 68–76. [CrossRef]
38. Bu, J.Z. Experimental Study on Deposition and Drift Characteristics of Electrostatic Spray for Large-Load Unmanned Helicopter. Master's Thesis, Jiangsu University, Zhenjiang, China, 2019.
39. Wu, Y.; Lu, J.; Wang, Y. Research on Multiresponse Robustness Optimization for Unmanned Aerial Vehicle Electrostatic Spray System. *Fundam. Mach. Theory Mech.* **2020**, *77*, 719–728. [CrossRef]
40. Lu, J.J.; Chen, J.D.; Wu, Y.D.; Wang, B. Agricultural aviation research aeronautical electrostatic spray system performance. *J. Agric. Mech. Res.* **2019**, *41*, 174–179, (In Chinese with English abstract). [CrossRef]
41. Zhao, D.N.; Lan, Y.B.; Shen, W.G. Building of aerial electrostatic spraying system and exploration on the influencing factors of droplet deposition effect. *J. Agric. Mech. Res.* **2021**, *43*, 204–207, (In Chinese with English abstract). [CrossRef]
42. Wang, J.F.; Gu, W.K.; Wang, X.Y.; Chu, J.J. Experimental study on charging performance and spray characteristics of induction charged spray. *High Volt. Eng.* **2015**, *41*, 504–509, (In Chinese with English abstract). [CrossRef]
43. Lan, Y.B.; Zhang, H.Y.; Wen, S.; Li, S.H. Analysis and experiment on atomization characteristics and spray deposition of electrostatic nozzle. *Trans. Chin. Soc. Agric. Mach.* **2018**, *49*, 130–139, (In Chinese with English abstract). [CrossRef]
44. Li, Y.F.; Hu, J.; Li, Q.D.; Zhao, M.M. Study on influencing factors of airborne electrostatic fog trapping. *Trans. Chin. Soc. Agric. Mach.* **2019**, *41*, 171–176, (In Chinese with English abstract). [CrossRef]
45. Wang, S.L.; He, X.K.; Song, J.L.; Zhong, C.S.; Wang, Z.C.; Qi, P.; Ling, Y. Charging and spraying performance test of bipolar contact electrostatic spraying system for unmanned aerial vehicle. *Trans. Chin. Soc. Agric. Eng.* **2018**, *34*, 82–89, (In Chinese with English abstract). [CrossRef]
46. Liu, W.L. Design and Experiment of Electrostatic Spraying System for Plant Protection Unmanned Aerial Vehicle. Master's Thesis, South China Agricultural University, Guangzhou, China, 2017.
47. Zhao, S.; Castle, G.P.; Adamiak, K. Comparison of conduction and induction charging in liquid spraying. *J. Electrost.* **2005**, *63*, 871–876. [CrossRef]
48. Agropages. China's Crop Protection UAV Development Report 2020. Available online: http://news.agropages.com/News/NewsDetail---38284.htm (accessed on 17 March 2021).
49. Kirk, I.W.; Hoffmann, W.C.; Carlton, J.B. Aerial electrostatic spray system performance. *Trans. ASAE* **2001**, *44*, 1089–1092. [CrossRef]
50. Martin, D.E.; Latheef, M.A. Aerial electrostatic spray deposition and canopy penetration in cotton. *J. Electrost.* **2017**, *90*, 38–44. [CrossRef]
51. Dai, Y.; Law, S. Modeling the transient electric field produced by a charged pollen cloud entering a flower. In Proceedings of the IAS '95. Conference Record of the 1995 IEEE Industry Applications Conference Thirtieth IAS Annual Meeting, Orlando, FL, USA, 8–12 October 1995; Institute of Electrical and Electronics Engineers (IEEE): New York, NY, USA, 2002; Volume 2, pp. 1395–1402.
52. Law, S.E. Electrostatic Pesticide Spraying: Concepts and Practice. *IEEE Trans. Ind. Appl.* **1983**, *2*, 160–168. [CrossRef]
53. Martin, D.E.; Lopez, J.D.; Hoffmann, W.C.; Fritz, B.K.; Lan, Y. Field Evaluation of Spinosad Aerial Applications for Thrips Control on Cotton1. *Southwest. Èntomol.* **2007**, *32*, 221–228. [CrossRef]
54. Zhang, R.R.; Chen, L.P.; Lan, Y.B.; Xu, G.; Kan, J.; Zhang, D.Y. Development of a deposit sensing system for aerial spraying application. *Trans. Chin. Soc. Agric. Mach.* **2014**, *45*, 123–127, (In Chinese with English abstract). [CrossRef]
55. He, Y.; Wu, J.J.; Fang, H.; Zheng, Q.S.; Xiao, S.P.; Cen, H.Y. Research on deposition effect of droplets based on plant protection unmanned aerial vehicle: A review. *J. Zhejiang Univ. (Agric. Life Sci.)* **2018**, *44*, 392–398, (In Chinese with English abstract). [CrossRef]

MDPI
St. Alban-Anlage 66
4052 Basel
Switzerland
www.mdpi.com

Applied Sciences Editorial Office
E-mail: applsci@mdpi.com
www.mdpi.com/journal/applsci

www.ingramcontent.com/pod-product-compliance
Lightning Source LLC
LaVergne TN
LVHW070052120526
838202LV00102B/2217
* 9 7 8 3 0 3 6 5 7 8 2 4 8 *